Constants Involving π and e

$\pi = 3.141\ 593\ (+00)$	$\sqrt{\pi} = 1.772\ 454\ (+00)$	$\sqrt[3]{\pi} = 1.464\ 592\ (+00)$
$e = 2.718\ 282\ (+00)$	$\sqrt{e} = 1.648\ 721\ (+00)$	$\sqrt[3]{e} = 1.395\ 612\ (+00)$
$\log \pi = 4.971\ 499\ (-01)$	$\log e = 4.342\ 945\ (-01)$	$\ln \pi = 1.144\ 730\ (+00)$

Selected Physical Constants (Appendix A)

Angular velocity of earth	$7.27200\ (-05)$ rad-s^{-1}	Gravitation constant (G)	$6.67200\ (-11)$ N-m^2-kg^{-2}
Avogadro's number (N_A)	$6.02205\ (+26)$ kmol^{-1}	Gravitational acceleration (g)	$9.80665\ (+00)$ m-s^{-2}
Boltzmann constant (k)	$1.38066\ (-23)$ J-K^{-1}	Kilomole volume (standard) (V_0)	$2.24138\ (+01)$ m^3-atm-kmol^{-1}
Density of dry air, 0°C	$1.29290\ (+00)$ kg-m^{-3}	Linear velocity of earth	$2.97700\ (+04)$ m-s^{-1}
Density of earth (mean)	$5.52200\ (+03)$ kg-m^{-3}	Magnetic field constant (μ_0)	$1.25664\ (-06)$ N-A^{-2}
Density of mercury, 0°C	$1.35955\ (+04)$ kg-m^{-3}	Speed of sound, 0°C	$3.31360\ (+02)$ m-s^{-1}
Density of water, 4°C	$1.00000\ (+03)$ kg-m^{-3}	Speed of light in vacuum	$2.99792\ (+08)$ m-s^{-1}
Electric field constant (ϵ_0)	$8.85420\ (-12)$ C^2-N^{-1}-m^{-2}	Universal gas constant (R)	$8.31441\ (+03)$ J-K^{-1}-kmol^{-1}
Electronvolt (eV)	$1.60219\ (-19)$ J	Volume of earth	$1.08300\ (+21)$ m^3

Selected Conversion Factors* (Appendix B)

Acre	$4.04686\ (+03)$ m^2	Inch (in.)	$2.54000\ (-02)$ m
Angstrom (Å)	$1.00000\ (-10)$ m	Kilogram-force (kgf)	$9.80665\ (+00)$ N
Atmosphere (atm, phs.)	$1.01325\ (+05)$ Pa	Knot (int.)	$5.14444\ (-01)$ m-s^{-1}
Atmosphere (at, tec.)	$9.80665\ (+04)$ Pa	Light-year	$9.46055\ (+15)$ m
Astronomical unit	$1.49598\ (+11)$ m	Liter (l)	$1.00000\ (-03)$ m^3
Bar (b)	$1.00000\ (+05)$ Pa	Mile (U.S., nat)	$1.85200\ (+03)$ m
British thermal unit (Btu, IST)	$1.05506\ (+03)$ J	Mile (U.S., stu)	$1.60934\ (+03)$ m
British thermal unit (Btu. tec)	$1.05500\ (+03)$ J	Ounce (U.S., avd)	$2.83495\ (-02)$ kg
British thermal unit (Btu, thm)	$1.05435\ (+03)$ J	Ounce (U.S., liq)	$2.957353\ (-05)$ m^3
Bushel (U.S.)	$3.52391\ (-02)$ m^3	Pint (U.S., liq)	$4.73176\ (-04)$ m^3
Calorie (cal, IST)	$4.18680\ (+00)$ J	Poise	$1.00000\ (-01)$ N-s-m^{-2}
Calorie (cal, tec)	$4.18600\ (+00)$ J	Pound (avd)	$4.53592\ (-01)$ kg
Calorie (cal, thm)	$4.18400\ (+00)$ J	Pound-force (lbf)	$4.44822\ (+00)$ N
Circular mil	$5.06707\ (-10)$ m^2	Poundal (pd)	$1.38255\ (-01)$ N
Dyne	$1.00000\ (-05)$ N	Psi (l	
Erg	$1.00000\ (-07)$ J	Quar	
Foot (ft)	$3.04800\ (-01)$ m	Quar	
Gallon (U.S., liq)	$3.78541\ (-03)$ m^3	Slug	
Grain	$6.47989\ (-05)$ kg	Stok	-s^{-1}
Horsepower (hp, FPS)	$7.45699\ (+02)$ W	Torr	$1.33322\ (+02)$ Pa
Horsepower (hp, MKS)	$7.35499\ (+02)$ W	Yard (yd)	$9.14400\ (-01)$ m

*Abbreviations:

avd = avoirdupois	nat = nautical	tec = technical	MKS = metric system
liq = liquid	phs = physical	thm = thermochemical	FPS = English system
	stu = statute	IST = international steam tables	SI = international system

J. WELLS

Handbook of Physical Calculations

DEFINITIONS · FORMULAS · TECHNICAL
APPLICATIONS · PHYSICAL TABLES
CONVERSION TABLES · GRAPHS
DICTIONARY OF PHYSICAL TERMS

Jan J. Tuma, Ph.D.

Professor of Engineering
Arizona State University

Second Enlarged and
Revised Edition

McGraw-Hill Book Company

New York St. Louis San Francisco Auckland Bogotá
Hamburg Johannesburg London Madrid Mexico Montreal
New Delhi Panama Paris São Paulo Singapore
Sydney Tokyo Toronto

Library of Congress Cataloging in Publication Data
Tuma, Jan J.
   Handbook of physical calculations.

   Bibliography: p.
   Includes index.
   1. Physics—Handbooks, manuals, etc.
2. Physical measurements—Handbooks, manuals, etc.
I. Title
QC61.T85   1983   530'.02'02   82-9002
ISBN 0-07-065439-5

Copyright © 1983, 1976 by McGraw-Hill, Inc. All
rights reserved. Printed in the United States of
America. Except as permitted under the United
States Copyright Act of 1976, no part of this publication may be reproduced or distributed in any
form or by any means, or stored in a data base or
retrieval system, without the prior written permission of the publisher.
1234567890   VHVH   898765432

ISBN 0-07-065439-5

*The editors for this book were Harold B. Crawford and Ruth L. Weine
and its production was supervised by Teresa F. Leaden.
It was printed and bound by Von Hoffman Press.*

CONTENTS

7.   ELECTROSTATICS AND ELECTRIC CURRENT

8.   MAGNETISM AND ELECTRODYNAMICS

9.   VIBRATION AND ACOUSTICS

10.   GEOMETRICAL AND WAVE OPTICS

APPENDIX A.   PHYSICAL TABLES

APPENDIX B.   CONVERSION TABLES

PREFACE TO
THE SECOND EDITION

Shortly after publication of the first edition, it became apparent that the scope of use of this handbook is much broader than originally anticipated. The book, which initially was designed as a reference book for practicing engineers architects, and technologists to use in their professional work, became also a solution manual used by engineers preparing for and taking the Professional Registration Examination.

This wide use plus the constant demand for additional reprintings resulted in the publisher's requesting the preparation of a revised and enlarged second edition. In the preparation of this new edition an effort was made to preserve the telescopic form of presentation, which was found so useful by many users, to improve the material in the first edition, and to include new material requested by users.

The major revisions and additions appear in Chaps. 2, 3, 4, 5, and 7, dealing with the dynamics of mechanical systems, fluid and soil mechanics, the analysis of elastic and non-elastic systems, and transients in electric currents. Also, significant changes and additions appear in Appendix A, which presents revised and new tables of physical constants of metals, plastics, rocks, soils, and other engineering materials.

In closing, the author expresses his gratitude to the many users who helped to improve the handbook by suggesting corrections and additions, and earnestly solicits comments and recommendations for improvements and future additions.

Tempe, Arizona   *Jan J. Tuma*

PREFACE TO
THE FIRST EDITION

This handbook presents in one volume a concise summary of the major definitions, formulas, tables, and examples of elementary and intermediate technology physics. It was prepared to serve as a desk-top reference book for practicing engineers, architects, and technologists. Both primer and refresher, it provides simple, easy-to-grasp fundamentals with emphasis on practical applications. No mathematics beyond the elementary calculus is used and calculus as such is used only where absolutely necessary.

The content of the book is divided into four distinct parts, each fulfilling a specific function and all four together serving a single purpose: *solution of physical problems.*

The first part covers applications of technical physics. It is divided into ten chapters arranged in a standard sequence and corresponding to the material required in the physics courses of the undergraduate engineering, architectural, and technology curricula. The numerous solved problems illustrate the principles and provide the user with a wealth of ready-made solutions to many important problems.

The second part consists of physical tables, presented in Appendix A. This part gives a tabular summary of major physical constants; densities of solids, liquids, and gases; mechanical properties of structural materials; viscosity and friction coefficients; thermal, electrical, acoustical, and optical properties of technical materials; and static and inertia functions of sections and of homogeneous solids. In the presentation of this appendix an effort was made to lay out tables in facing spreads that constitute complete conceptual as well as visual correspondence.

The third part offers systems of units and their relationships, and consists of the second appendix. Appendix B contains a glossary of symbols of units, a summary of basic and derived units of the SI, MKS and FPS systems, and matrix tables of conversion factors stating their relationships. All conversion factors are expressed in seven digits supplemented by modified scientific notation.

The fourth part, the Index, is an extensive alphabetic dictionary of key words, **units of measure, and physical and material constants referred to** the respective page or table **number and is** supplemented by additional information such as chemical composition, system designation, symbolic representation, origin, or use as appropriate.

The form of presentation has many special characteristics allowing easy and rapid location of the desired information and permitting the indexing of this information.

1. Each statement in the book is a coded sentence designated by a position number and a key word.
2. Sentences form logical sequences and their lengths allow speed reading.
3. All formulas are presented in general form and their applications are illustrated by numerical examples where this is desirable.
4. Special cases are presented in catalog form to facilitate their location and comparison.
5. Each chapter includes applied sections such as useful approximations, simple machines, thermal engines, electric circuits, lenses, mirrors, etc.

In the selection of *units of measure* an effort was made to build a bridge of transition from the FPS system to the MKS system and finally to the SI (international system). To accomplish this end, the kilogram-force and the thermochemical calorie have been introduced in many problems and tables; their conversion to their SI counterparts can always be accomplished by means of the exact relations:

1 kilogram-force = 9.806 650 newton
1 thermochemical calorie = 4.184 000 joule

which, with the general availability of small pocket calculators, is a routine process.

The contents of this book are based on extensive reference material forming a body of technical knowledge known as technology physics. Space limitations prevent the inclusion of a complete list of references, yet a great effort was made to credit those sources which were directly used. Permissions to quote from these sources have been freely granted by copyright owners and are acknowledged with thanks.

Boulder, Colorado   *Jan J. Tuma*

HANDBOOK OF
PHYSICAL
CALCULATIONS

1
FOUNDATIONS
OF PHYSICAL
CALCULATIONS

1.01 PHYSICAL QUANTITIES AND UNITS

(1) Quantitative Measurements

(a) **Physical quantities** such as length, weight, temperature, etc., are measured by comparison with quantities of the same kind called *units of measure*.

(b) **Magnitude of a physical quantity** Q is specified by the numerical factor N (dimensionless number) defining the number of units and the symbol of unit u designating the dimensions. Analytically,

$$Q = N \cdot u$$

examples:

The mass of 15 kg is specified by $N = 15$ and $u = \text{kg} = \text{kilogram-mass}$.

The velocity of 100 m/s is specified by $N = 100$ and $u = \text{m/s} = \text{meter per second}$.

(2) Basic and Derived Units

(a) **Origin of units.** The various units of measure (such as meter, kilogram, degree Celsius, etc.) are by no means prescribed by nature but are products of human selection (national or international conventions).

(b) **Classification.** As physical quantities are of the basic type (length, mass, time, temperature, etc.) and of the derived type (volume, velocity, work, etc.), their units are also designated as basic units and derived units.

(c) **Systems of units.** Out of several systems of units, three systems are introduced in Appendix B.06. They are the English system of units (FPS), the metric system of units (MKS), and the International System of Units (SI).*

(d) **Conversion factors.** The relationship of two units of the same kind is given by the conversion factor (dimensionless number). Specific conversion factors are given in the text where the respective unit appears first, and they are all assembled in Appendix B.†

example:

1 foot = 0.3048 meter

and inversely,

$$1 \text{ meter} = \frac{1}{0.3048} \text{ foot} = 3.2808 \text{ feet}$$

*FPS = foot-pound-second; MKS = meter-kilogram-second; SI = meter-kilogram-second.
†Conversion relations presented in this book are based on standards defined by the International Bureau of Weights and Measures (IBWM), Sévres, France; the International Organization for Standardization (IOS), Geneva, Switzerland; the National Bureau of Standards (NBS), Washington, D.C.; and the National Aeronautics and Space Administration (NASA), Washington, D.C. For more detailed information refer to:

IOS Report 31, Part I, "SI Units," 1956, Part II, "Units of Periodic and Related Phenomena," 1958, Part III, "Units of Mechanics," 1960, Part IV, "Units of Heat," 1960, Part V, "Units of Electricity and Magnetism," 1963.

NBS Misc. Publication 286, "Units of Weight and Measure, Definitions and Tables of Equivalents," 1967 (out-of-print, available in libraries only).

NBS LC-1035, "Units and Systems of Weights and Measures—Their Origin, Development and Present Status," 1976 (available free from the National Bureau of Standards).

NASA | Publication SP-7012, "The International System of Units, Physical Constants and Conversion Factors," revised, E.A. Mechtly, 1973.

1.02 INTERNATIONAL SYSTEM OF UNITS

(1) Establishment of the System

(a) **Convention.** At the Eleventh General Conference on Weights and Measures of 1960, the metric system (with some modifications) was given the name "International System of Units" and the abbreviation "SI" (for Système International) to be used in all languages.

(b) **Recommendation.** The members of this conference recommended the adoption of the SI system for all scientific, technical, practical, and teaching purposes.

(2) Units of the SI System

(a) **Types of units.** The SI units are classified as basic, derived, supplementary, and associated.

(b) **Basic units** of the SI system are

meter	= m	= unit of length (Sec. 2.01)
kilogram	= kg	= unit of mass (Sec. 3.02)
second	= s	= unit of time (Sec. 3.01)
ampere	= A	= unit of electric current (Sec. 7.03)
kelvin	= K	= unit of temperature difference (Sec. 6.01)
candela	= cd	= unit of luminous intensity (Sec. 10.01)
mole	= mol	= unit of amount of substance (Sec. 6.02)

(c) **Derived units** of the SI system are the multiples, products, or quotients of the basic units. The most commonly used derived units are

newton	= N	= unit of force (Sec. 3.02)	= $kg\text{-}m/s^2$
joule	= J	= unit of energy (work) (Sec. 3.05)	= N-m
watt	= W	= unit of power (Sec. 3.05)	= J/s
pascal	= Pa	= unit of pressure (Sec. 5.01)	= N/m^2
coulomb	= C	= unit of electrical quantity (Sec. 7.01)	= A-s
volt	= V	= unit of electric potential (Sec. 7.02)	= W/A
ohm	= Ω	= unit of electric resistance (Sec. 7.03)	= V/A
farad	= F	= unit of electric capacitance (Sec. 7.02)	= C/V
henry	= H	= unit of electric inductance (Sec. 8.03)	= Ω-s
weber	= Wb	= unit of magnetic flux (Sec. 8.01)	= V-s
tesla	= T	= unit of magnetic flux density (Sec. 8.01)	= Wb/m^2
lumen	= lm	= unit of luminous flux (Sec. 10.01)	= cd-sr
lux	= lx	= unit of illumination (Sec. 10.01)	= lm/m^2

(d) **Supplementary units** of the SI system are

radian	= rad	= unit of plane angle (Sec. 2.01)
steradian	= sr	= unit of solid angle (Sec. 10.01)

(e) **Decimal multiples and fractions** of the SI units, defined by their factor, prefix, and symbol, are given in Appendix B.07.

(f) Associated units of the SI system are

liter	= l	= unit of volume (Sec. 5.01)	$= 10^{-3}\,\text{m}^3$
kilogram-force	= kgf	= unit of force (Sec. 3.02)	$= 9.807\,\text{N}$
dyne	= dyne	= unit of force (Sec. 3.02)	$= 10^{-5}\,\text{N}$
bar	= b	= unit of pressure (Sec. 5.01)	$= 10^5\,\text{N/m}^2$
tech. atmosphere	= at	= unit of pressure (Sec. 5.01)	$= 10^4\,\text{kgf/m}^2$
torr	= torr	= unit of pressure (Sec. 5.01)	$= 13.595\,\text{kgf/m}^2$
phys. atmosphere	= atm	= unit of pressure (Sec. 5.01)	$= 760\,\text{torr}$
erg	= erg	= unit of energy (Sec. 3.05)	$= 10^{-7}\,\text{N-m}$
horsepower (MKS)	= hp	= unit of power (Sec. 3.05)	$= 735.498\,\text{W}$

1.03 SCALARS AND VECTORS

(1) Definitions

(a) **Physical quantities** are basically of two types: scalars and vectors.

(b) **Scalar** is a quantity defined by a magnitude only. Examples of scalars are mass, length, time, and temperature.

(c) **Vector** is a quantity defined by a magnitude and a direction (line of action and sense). Examples of vectors are force, moment, displacement, velocity, and acceleration.

(2) Representation of Scalars

(a) **Symbolically**, a scalar is designated by italic letters $a, b, c, \ldots, A, B, C, \ldots, \alpha, \beta, \gamma, \ldots$.

(b) **Geometrically**, a scalar is represented by a straight-line segment drawn to a certain scale in units of the scalar.

(c) **Numerically**, a scalar is defined as a product of a numerical factor and a symbol of units.

(3) Representation of Vectors

(a) **Symbolically**, a vector is designated by boldface letters $\mathbf{a}, \mathbf{b}, \mathbf{c}, \ldots, \mathbf{A}, \mathbf{B}, \mathbf{C}, \ldots$.

(b) **Geometrically**, a vector is represented by a directed segment (Fig. 1.03–1) such as

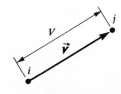

$$\mathbf{V} = \vec{ij}$$

where the segment ij is the vector length (magnitude), the straight line given by i and j is called the *line of action*, and the arrow indicates the *sense* of the vector.

Fig. 1.03–1

(c) **Numerically**, a vector is defined as a product of the vector length (which is a scalar) and the symbol of direction called the *unit vector*, i.e.,

$$\mathbf{V} = V\mathbf{v}$$

where V = product of a numerical factor and the symbol of units and $\mathbf{v}$ = dimensionless symbol of direction.

(d) **In technical calculations** a vector is frequently given by its components related to a selected system of coordinate axes and the vector notation is abandoned. In this book no vector notation is used.

1.04 NUMERICAL CALCULATIONS*

(1) Operations with Physical Quantities

(a) **Addition and subtraction.** Only quantities of the same kind and in the same units can be added and subtracted.

examples:

$$100 \text{ m} + 5 \text{ m} = 105 \text{ m} \qquad 100 \text{ m} + 5 \text{ cm} = 100 \text{ m} + 0.05 \text{ m} = 100.05 \text{ m}$$

(b) **Multiplication and division** of quantities of different kinds are admissible, but each term must be stated in or converted to the same system of units.

examples:

$$15 \text{ kgf} \times 25 \text{ m} = 375 \text{ kgf-m, but}$$

$$15 \text{ kgf} \times 80 \text{ ft} = 15 \text{ kgf} \times (80 \times 0.3048) \text{ m} = 365.76 \text{ kgf-m}$$

(c) **In equations** defining physical relationships, both sides must be expressed in the same units.

(2) Significant Figures

(a) **Measured data** inherently are not exact, and if recorded in decimal notation they consist of a finite set of decimal digits called *significant figures*, the last of which is called the *doubtful figure.*

(b) **Zero** is a significant figure if surrounded by other digits, or if specifically designated as such; otherwise the zeros are used to fix the decimal point and hence are not significant.

examples:

$123.4 \ldots$ four significant figures (1, 2, 3, 4)

$0.056 \ldots$ two significant figures (5, 6)

(c) **Scientific notation.** Any number N can be expressed in the form

$$N = q \times 10^n$$

where q = number between 1 and 10 and n = integer.

examples:

$$5,130 = 5.13 \times 10^3 \qquad 0.083 = 8.3 \times 10^{-2}$$

(d) **Conversion factors** are frequently expressed in a modified form of scientific notation, where the power of 10 is represented by its signed exponent only (Appendix B).

example:

$$5,130 = 5.13 \times 10^3 = 5.13 \quad (+03) \qquad 0.083 = 8.3 \times 10^{-2} = 8.3 \quad (-02)$$

*J. J. Tuma, "Technology Mathematics Handbook," McGraw-Hill, New York, 1975, pp. 254–257.

(3) Retention of Significant Figures

(a) Rounding off a number

(a) Rounding off a number is the process of rejecting (dropping) one or several of its last digits. In rejecting figures, the last figure retained should be increased by 1 if the figure rejected (dropped) is 5 or greater. If, however, several numbers with final 5s are involved in a given operation, only half of these 5s should be rounded off.

example:

Given number	Rounded off to		
	Four figures	Three figures	Two figures
73.589	73.59	73.6	74
10,232	10,230	10,200	10,000
329,350	329,400	329,000	330,000

(b) Addition and subtraction. The number of significant figures of the sum or difference should be rounded off by eliminating any digit resulting from operations on broken column on the right as shown below.

examples:

```
 201.3              201.3
   1.05            - 1.05
  21.76           - 21.76
   0.0013         - 0.0013
 ─────────        ─────────
 224.1113 ≐ 224.1  178.4887 ≐ 178.5
```

(c) Multiplication and division. The number of significant figures of the product or of the quotient should be rounded off to a number of significant figures equal to that of the least accurate term involved in the calculation.

examples:

$$3.14159 \times 21.13 = 66.38179 \doteq 66.38$$

$$3.14159 : 21.13 = 0.14868 \doteq 0.1487$$

where 21.13 is the least accurate number (four significant figures) and therefore both results are rounded off to four significant figures.

(d) Squares and cubes. The number of significant figures of a square or of a cube of a number N should be rounded off to the number of significant figures of N.

examples:

$$2.13^2 = 4.5369 \doteq 4.54 \qquad 2.13^3 = 9.6636 \doteq 9.66$$

(e) Square roots and cube roots. The number of significant figures of a square root or of a cube root of a number N should be rounded off to the number of significant figures of N.

examples:

$$\sqrt{2.13} = 1.4595 \doteq 1.46 \qquad \sqrt[3]{2.13} = 1.2866 \doteq 1.29$$

(4) Numerical Errors

(a) **Absolute error** ϵ is the difference between the true (correct) value N (assumed to be known) and the approximate value $\bar{N}$ (obtained by measurements or calculations).

$$\epsilon = N - \bar{N}$$

which is positive if $\bar{N}$ is less than N; otherwise it is negative.

(b) **Relative error** $\bar{\epsilon}$ is the ratio of the absolute error ϵ to the true value N.

$$\bar{\epsilon} = \frac{\epsilon}{N} = 1 - \frac{\bar{N}}{N}$$

(c) **Percent error** $\bar{\epsilon}\%$ is defined as

$$\bar{\epsilon}\% = (100\bar{\epsilon})\% = 100\left(1 - \frac{\bar{N}}{N}\right)\%$$

(d) **Classification.** Five types of basic errors may occur in numerical calculations: inherent error ϵ_i, truncation error ϵ_t, round-off error ϵ_r, interpolation error ϵ_p, and approximation error ϵ_a.

(e) **Inherent error** is the error in the initial data based on inaccurate measurements, observations, or recordings.

(f) **Truncation error** ϵ_t is the error caused by representing a function by a series of few terms.

example:

The correct value of $N = \sin\dfrac{\pi}{2} = 1.00000$

The approximate value of N computed by series expansion is

$$\bar{N} = \frac{\pi}{2} - \frac{(\pi/2)^3}{3!} + \frac{(\pi/2)^5}{5!} - \frac{(\pi/2)^7}{7!} + \cdots$$

If only the first term of the series is used, $\epsilon_t = N - \bar{N} = 1.00000 - \dfrac{\pi}{2} = -0.57080 \qquad (-57\%)$

If the first two terms are used, $\epsilon_t = N - \bar{N} = 1.00000 - \dfrac{\pi}{2} + \dfrac{(\pi/2)^3}{3!} = +0.07516 \qquad (+7.5\%)$

If the first three terms are used, $\epsilon_t = N - \bar{N} = 1.00000 - \dfrac{\pi}{2} + \dfrac{(\pi/2)^3}{3!} - \dfrac{(\pi/2)^5}{5!} = -0.00453 \qquad (-0.5\%)$

Finally, if the first four terms of the series are used, $\epsilon_t = 0.00015$ and the truncation error becomes insignificant.

(g) **Round-off error** is the error introduced by rounding off a decimal number.

example:

If $\pi = 3.14159$ is rounded off to $\pi = 3.14$, then $\epsilon_r = 3.14159 - 3.14 = 0.00159$ and $\bar{\epsilon}_r\% = \dfrac{0.00159}{3.14159} \times 100 = 0.05\%$.

(h) **Interpolation error** is the error introduced by the approximation of a value by its interpolative equivalent.

example:

The circumference of a circle of diameter $D = 10\,\mathrm{m}$ is $C_{10} = D\pi = 31.42\,\mathrm{m}$ and of a circle of $D = 11\,\mathrm{m}$ is $C_{11} = D\pi = 34.56\,\mathrm{m}$. By linear interpolation, the circumference of a circle of $D = 10.6\,\mathrm{m}$ is $C_{10.6} \cong C_{10} + (C_{11} - C_{10}) \times 0.6 = 33.30\,\mathrm{m}$, but the exact value is $C_{10.6} = 10.6\pi = 33.31\,\mathrm{m}$. Hence $\epsilon_p = 33.31 - 33.30 = 0.01\,\mathrm{m}$ or $\bar{\epsilon}_p\% = 0.03\%$.

(i) Approximation error is the error introduced by the approximation of a constant or a function by a selected value.

example:

If $\pi = 3.14159$ is approximated as $\pi \cong \frac{22}{7} = 3.14286$, then

$$\bar{\epsilon}_a \% = \frac{3.14159 - 3.14286}{3.14159} \times 100 = 0.04\%$$

If π is approximated by $\pi \cong \frac{355}{113} = 3.14159$, then $\bar{\epsilon}_a \% = 0.00\%$ (no error up to the fifth decimal place).

(j) Numerical accuracy of technical calculations is governed by all these errors, and the result of calculations cannot be more accurate than the given data. Since technical data are seldom known with an accuracy greater than $\pm 0.2\%$, the computation performed should remain within the same range of error.

1.05   USEFUL APPROXIMATIONS

(1) Algebraic Approximations

(a) Square root of large numbers.   If $N = a^2 + \Delta$ and Δ is very small compared to a^2, then

$$\sqrt{N} = \sqrt{a^2 + \Delta} \cong a + \frac{\Delta}{2a}$$

where a^2 = number to be selected by inspection or from standard tables of squares.

example:

$$\sqrt{364} = \sqrt{19^2 + 3} \cong 19 + \frac{3}{2 \times 19} \doteq 19.08$$

which should satisfy (approximately) $(19.08)^2 \doteq 364$

where $\doteq$ signifies a rounded-off result.

(b) Cube root of large numbers.   If $N = a^3 + \Delta$ and Δ is very small compared to a^3, then

$$\sqrt[3]{N} = \sqrt[3]{a^3 + \Delta} \cong a + \frac{\Delta}{3a^2}$$

where a^3 = number to be selected by inspection or from standard tables of cubes.

example:

$$\sqrt[3]{364} = \sqrt[3]{7^3 + 21} \cong 7 + \frac{21}{3 \times 7^2} \doteq 7.14$$

which should satisfy (approximately) $(7.14)^3 \doteq 364$

where $\doteq$ again signifies a rounded-off result.

(2) Transcendental Approximations*

(a) Trigonometric functions and their inverses.

Approximation*	Interval	Conversion
$\sin x \cong x$ $\sin^{-1} x \cong x$	$[-0.11, 0.11]$	$0.11 \text{ rad} \equiv 6.30°$
$\cos x \cong 1$ $\cos^{-1} 1 \cong x$	$[-0.06, 0.06]$	$0.06 \text{ rad} \equiv 3.44°$
$\tan x \cong x$ $\tan^{-1} x \cong x$	$[-0.08, 0.08]$	$0.08 \text{ rad} \equiv 4.58°$

*Error of less than $\pm 0.2\%$.

(b) Hyperbolic functions and their inverses

Approximation*	Interval	Conversion
$\sinh x \cong x$ $\sinh^{-1} x \cong x$	$[-0.22, 0.22]$	$0.22 = 0.070\pi$
$\cosh x \cong 1$ $\cosh^{-1} 1 \cong x$	$[-0.06, 0.06]$	$0.06 = 0.019\pi$
$\tanh x \cong x$ $\tanh^{-1} x \cong x$	$[-0.18, 0.18]$	$0.18 = 0.057\pi$

*Error of less than $\pm 0.2\%$.

(3) Area Approximations*

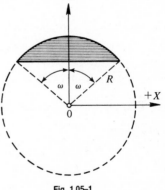

Fig. 1.05–1

(a) Area of circular segment of Fig. 1.05–1 is

$$A \approx \frac{2R^2\omega^3}{3}(1 - 0.2\omega^2 + 0.02\omega^4)$$

where 2ω is the central angle in radians. For $2\omega = \pi$ (half circle) the error of this approximation is 3.3 percent, and it decreases rapidly as 2ω decreases. For $2\omega < \pi/2$ the error is less than 0.1 percent.

(b) Area of elliptical segment of Fig. 1.05–2 for $x < a/2$ is

$$A \approx ab\left(\frac{\pi}{2} - \frac{x}{a} - \frac{x^3}{6a^3}\right) - xy$$

where x, y are the coordinates of P. For $x > a/2$ use the exact formula,

$$A = ab \cos^{-1}\frac{x}{a} - xy$$

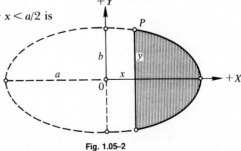

Fig. 1.05–2

*Ibid., pp. 276–277.

(c) Areas of hyperbolic segment of Fig. 1.05–3 for $x > 2a$ is

$$A \approx xy - ab\left(\ln \frac{2x}{a} - \frac{a^2}{4x^2}\right)$$

where x, y are the coordinates of P. For $x < 2a$ use the exact formula,

$$A = xy - ab \cosh^{-1} \frac{x}{a}$$

(d) Area of parabolic segment of Fig. 1.05–4 is exactly

$$A = \frac{(x_1 - x_2)(y_1 - y_2)^2}{6(y_1 + y_2)}$$

where x_1, y_1 and x_2, y_2 are the coordinates of P_1 and P_2, respectively, $(x_1 \neq x_2)$.

If $x_1 = x_2$, then

$$A = \frac{4x_1 y_1}{3}$$

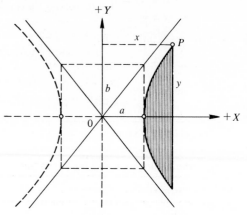

Fig. 1.05–3

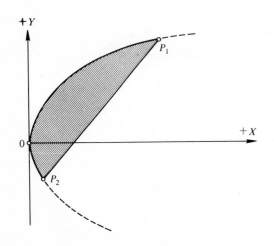

Fig. 1.05–4

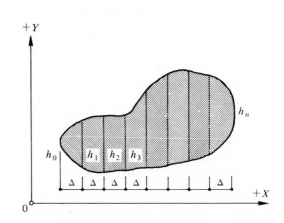

Fig. 1.05–5

(e) Irregular area of Fig. 1.05–5 is first divided into n vertical strips by equidistant parallel chords of lengths $h_0, h_1, h_2, \ldots, h_n$ and then evaluated by one of the approximate formulas given below.

(α) *Trapezoidal rule* (n = even or odd): $\qquad A \approx \Delta\left(\dfrac{h_0}{2} + h_1 + h_2 + \cdots + \dfrac{h_n}{2}\right)$

(β) *Simpson's rule* (n = even): $\qquad A \approx \dfrac{\Delta}{3}(h_0 + 4h_1 + 2h_2 + 4h_3 + 2h_4 + \cdots + h_n)$

where h_0 and/or h_n may be zero.

(4) Linear Interpolation of Tabulated Values*

(a) **Definition.** The process of finding a value of a function between two tabulated values by a procedure other than the evaluation of the function generating these values is called interpolation.

(b) **Linear interpolation.** If N_a and N_b are two tabulated values representing $N(x)$ at $x = a$ and $x = b$, then the approximate value of N_p at $x = p$ is

$$N_p = N_a + \frac{N_b - N_a}{b - a}(p - a) = N_a + f\Delta \qquad a < p < b$$

where $N_b - N_a = \Delta$ is the tabular difference, $b - a = s$ is the interval of interpolation, and $(p - a)/s = f$ is the interpolation factor (decimal fraction).

example:

The circumference of a circle of diameter $D = 4$ is $C_4 = 4\pi = 12.5664$ and of a circle of diameter $D = 5$ is $C_5 = 5\pi = 15.7080$. Then by linear interpolation the circumference of a circle of diameter $D = 4.25$ is

$$D = 12.5664 + \frac{15.7080 - 12.5664}{5 - 4}(4.25 - 4) = 13.3518$$

which is in a good agreement with the exact value $4.25\pi = 13.3518$.

(c) **Limitations.** The application of linear interpolation is limited to densely spaced values which can be represented graphically by a smooth curve.

(5) Nonlinear Interpolation of Tabulated Values

(a) **Lagrange interpolation formulas.** When the tabulated values are located at equally spaced intervals, so that

$$x_{k+1} - x_k = s$$

is a constant value, Lagrange n-point interpolation formulas offer a convenient and accurate solution. The two most commonly used formulas of this type are given below.

(b) **Three-point formula.** If N_1, N_2, and N_3 are three tabulated values representing $N(x)$ at x_1, x_2, and x_3, respectively, the approximate value of N_p which lies between N_2 and N_3 at $x_p = p$ is

$$N_p = \alpha \left(-\frac{N_1}{1 + u} + \frac{2N_2}{u} + \frac{N_3}{1 - u} \right) \qquad x_2 < p < x_3$$

where $u = (x_p - x_2)/s$ and $\alpha = \frac{1}{2}(1 + u)u(1 - u)$.

(c) **Five-point formula.** Similarly, if N_1, N_2, ..., N_5 are five tabulated values representing $N(x)$ at $x_1, x_2, \ldots, x_5$, respectively, the approximate value of N_p which lies between N_3 and N_4 at $x_p = p$ is

$$N_p = \beta \left(\frac{N_1}{2 + u} - \frac{4N_2}{1 + u} + \frac{6N_3}{u} + \frac{4N_4}{1 - u} - \frac{N_5}{2 - u} \right) \qquad x_3 < p < x_4$$

where $u = (x_p - x_3)/s$ and $\beta = (2 + u)(1 + u)u(1 - u)(2 - u)/24$.

*Ibid., pp. 255–256.

example:

In terms of the tabulated values,

$$\begin{aligned}
\sin 63° &= 0.891\ 006\ 524 \dots\dots\dots\dots\dots\dots\dots\dots\dots\dots\dots\dots N_1 \\
\sin 64° &= 0.898\ 794\ 046 \dots\dots\dots\dots N_1 \dots\dots\dots\dots N_2 \\
\sin 65° &= 0.906\ 307\ 787 \dots\dots\dots\dots N_2 \dots\dots\dots\dots N_3 \\
\sin 66° &= 0.913\ 545\ 458 \dots\dots\dots\dots N_3 \dots\dots\dots\dots N_4 \\
\sin 67° &= 0.920\ 504\ 854 \dots\dots\dots\dots\dots\dots\dots\dots\dots\dots N_5
\end{aligned}$$

the approximate value of $\sin 65°24'13''$ is $0.909\ 262\ 218$ (by the three-point formula) and $0.909\ 262\ 344$ (by the five-point formula), where $x_p = 65°24'13'' = 65.403\ 611\ 111°$, $u = 0.403\ 611\ 111$, $\alpha = 0.168\ 931\ 041$, and $\beta = 0.054\ 017\ 081$.

The comparison of these results with the display on an electronic calculator (which is $0.909\ 262\ 343\ 7$) shows the great accuracy of these formulas.

2
STATICS OF
RIGID BODIES

2.01 BASIC CONCEPTS OF STATICS

(1) Definitions

(a) **Material particle** is defined as a small amount of matter assumed to occupy a single point in space (it has no dimensions).

(b) **Mechanics** is the branch of physics concerned with the investigation of the effects of forces on systems of material particles (bodies). According to the state of these particles (of these bodies), the designation of mechanics of solids, liquids, or gases is used.

(c) **Rigid body** is a system of particles which are in a given and fixed position relative to each other (distance between two particles in a rigid body remains unchanged).

(d) **Statics of rigid bodies** investigates the equilibrium of rigid bodies under the action of stationary forces (forces which are in a fixed position on or in the body).

(e) **Position of a particle** (of a body) in space is defined in terms of lengths of lines and of angles between these lines.

(2) Measurement of Length

(a) **SI unit of length** is 1 meter, designated by the symbol m, and defined as exactly $1\,650\,763.73$ times the wavelength of the orange light emitted in a vacuum when a gas consisting of the pure krypton nuclide of mass number 86 is excited by an electric discharge.*

(b) **Multiples and fractions of 1 meter** frequently used in technology are

$$1 \text{ kilometer} = 1\,\text{km} = 10^{3}\,\text{m} \qquad 1 \text{ millimeter} = 1\,\text{mm} = 10^{-3}\,\text{m}$$
$$1 \text{ decimeter} = 1\,\text{dm} = 10^{-1}\,\text{m} \qquad 1 \text{ micrometer} = 1\,\mu\text{m} = 10^{-6}\,\text{m}$$
$$1 \text{ centimeter} = 1\,\text{cm} = 10^{-2}\,\text{m} \qquad 1 \text{ angstrom} = 1\,\text{Å} = 10^{-10}\,\text{m}$$

(c) **FPS unit of length** is 1 foot, designated by ft, and defined as exactly 0.3048 meter.

(d) **Multiples and fractions of 1 foot** frequently used in technology are

$$1 \text{ league (U.S. land)} = 3 \text{ miles (U.S. statute)} = 15{,}840 \text{ ft}$$
$$1 \text{ mile (U.S. statute)} = 8 \text{ furlongs} = 320 \text{ rods} = 5{,}280 \text{ ft}$$
$$1 \text{ furlong} = 40 \text{ rods} = 660 \text{ ft} \qquad 1 \text{ rod (perch or pole)} = 16.5 \text{ ft}$$
$$1 \text{ yard} = 3 \text{ ft} = 36 \text{ in} \qquad 1 \text{ ft} = 12 \text{ in}$$
$$1 \text{ pace} = 2.5 \text{ ft} \qquad 1 \text{ span} = 0.75 \text{ ft}$$
$$1 \text{ hand} = 4 \text{ in} \qquad 1 \text{ palm} = 3 \text{ in}$$

(e) **Three basic conversion relations** are

$$1 \text{ mile (U.S. statute)} = 1.609 \text{ km} \qquad 1 \text{ km} = 0.6214 \text{ mile (U.S. statute)}$$
$$1 \text{ ft} = 0.3048 \text{ m} \qquad 1 \text{ m} = 3.281 \text{ ft}$$
$$1 \text{ in} = 2.540 \text{ cm} \qquad 1 \text{ cm} = 0.3937 \text{ in}$$

(f) **Surveying measures** (also Gunter's chain measures) are

$$1 \text{ mile (U.S. statute)} = 80 \text{ chains} = 320 \text{ rods} = 5{,}280 \text{ ft}$$
$$1 \text{ chain} = 100 \text{ links} = 66 \text{ ft} = 4 \text{ rods} = 20.117 \text{ m}$$
$$1 \text{ rod} = 1 \text{ pole} = 1 \text{ perch} = 33/2 \text{ ft} = 5.029 \text{ m}$$
$$1 \text{ vara (California)} = 33 \text{ in} \qquad 1 \text{ vara (Texas)} = 33\tfrac{1}{3} \text{ in}$$

*The older and obviously less accurate but physically more descriptive definition of 1 meter is 1/40,000,000 of the equatorial circumference of the earth.

(g) Nautical measures (also called marine measures) are

      1 league (U.S. naut.) = 3 miles (U.S. naut.) = 18,228.3 ft
      1 mile (U.S. naut.) = 6,076.1 ft = 1,852 m
      1 cable = 120 fathoms = 720 ft = 219.5 m
      1 fathom = 6 ft = 1.83 m

(h) Light-year equals the distance traveled by light in a vacuum in one calendar year.

      1 L.Y. = 9.46055×10^{15} m = 3.10386×10^{16} ft

(i) Astronomical unit equals the length of the major semiaxis of the earth's orbit around the sun.

      1 A.U. = 1.49598×10^{11} m = 4.90807×10^{11} ft

(j) Parsec is the length at which 1 astronomical unit subtends an angle of 1 second.

      1 parsec = 3.08374×10^{16} m = 1.01173×10^{17} ft

(3) Measurement of Angles

(a) SI unit of plane angle is 1 radian, designated by the symbol rad, and defined as exactly the plane angle subtended at the center of a circle by an arc whose length is equal to the radius (Fig. 2.01–1).

(b) Alternative units of plane angle are 1 degree, 1 grad, and 1 revolution, designated as 1°, 1 grad, and 1 rev, respectively.

(c) One degree (1°) is the plane angle subtended at the center of a circle by 1/360 of the circumference. One minute (1′) is 1/60 of a degree, and 1 second (1″) is 1/60 of a minute.

(d) One grad (1 grad) is the plane angle subtended at the center of a circle by 1/400 of the circumference.

(e) One revolution (1 rev) is the plane angle subtended by the full circumference of the circle (also called the *perigon*).

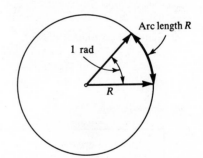

Fig. 2.01–1

(f) Basic conversion relations of the angular units are

$$1 \text{ radian} = \frac{180}{\pi} \text{ degrees} = \frac{200}{\pi} \text{ grads} = \frac{1}{2\pi} \text{ revolution}$$

$$1 \text{ degree} = \frac{\pi}{180} \text{ radian} = \frac{100}{90} \text{ grads} = \frac{1}{360} \text{ revolution}$$

$$1 \text{ grad} = \frac{90}{100} \text{ degree} = \frac{\pi}{200} \text{ radian} = \frac{1}{400} \text{ revolution}$$

$$1 \text{ revolution} = 360 \text{ degrees} = 400 \text{ grads} = 2\pi \text{ radians}$$

(4) Coordinate Systems

(a) Three coordinate systems are used in this book to specify the position of a particle (or a body) in space. They are the rectangular (cartesian) system, the cylindrical system, and the spherical system.

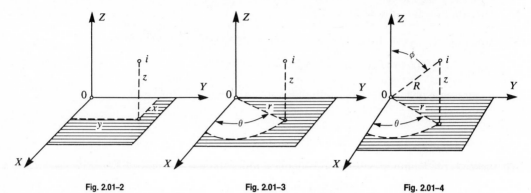

| Fig. 2.01–2 | Fig. 2.01–3 | Fig. 2.01–4 |

(b) Rectangular coordinate system. The position of a particle i is given by three mutually perpendicular distances x, y, z (coordinates) measured from three mutually perpendicular YZ, ZX, XY planes, respectively. The lines of intersection of these planes are the coordinate axes X, Y, Z, and the point of their intersection is called the *origin* 0. The right-handed system is shown in Fig. 2.01–2.

(c) Cylindrical coordinate system. The position of a particle i is given by two polar coordinates r and θ in the polar plane (usually the XY plane) and by the vertical coordinate z normal to the polar plane. The X axis from which the polar angle θ is measured in the direction shown in Fig. 2.01–3 is called the *polar axis*. The origin 0 from which the radius is measured in the polar plane is called the *pole*.

(d) Spherical coordinate system. The position of a particle i is given by the spherical radius R measured from the origin 0 and two position angles θ and ϕ measured from the X and Z axis as shown in Fig. 2.01–4. The position of the X axis is arbitrary but the Z axis is always normal to the XY plane.

(e) Relationships

Cartesian	Cylindrical	Spherical
$x = x$ $y = y$ $z = z$	$x = r \cos \theta$ $y = r \sin \theta$ $z = z$	$x = R \cos \theta \sin \phi$ $y = R \sin \theta \sin \phi$ $z = R \cos \phi$
$r = \sqrt{x^2 + y^2}$ $\theta = \tan^{-1} y/x$ $z = z$	$r = r$ $\theta = \theta$ $z = z$	$r = R \sin \phi$ $\theta = \theta$ $z = R \cos \phi$
$R = \sqrt{x^2 + y^2 + z^2}$ $\theta = \tan^{-1} y/x$ $\phi = \cos^{-1} \dfrac{z}{\sqrt{x^2 + y^2 + z^2}}$	$R = \sqrt{r^2 + z^2}$ $\theta = \theta$ $\phi = \cos^{-1} \dfrac{z}{\sqrt{x^2 + y^2}}$	$R = R$ $\theta = \theta$ $\phi = \phi$

2.02 STATIC FORCE

(1) Definitions and Units

(a) **Static force** acting on a particle i is interpreted as a push (acting toward the particle) or a pull (acting from the particle) and is designated by F.

(b) **MKS unit of static force** is 1 kilogram-force (derived unit), designated by the symbol kgf, and defined exactly as the weight of a particular cylinder of platinum-iridium alloy (called the *international prototype*, which is preserved in a vault in Sèvres, France, by the International Bureau of Weights and Measures) at latitude 45° and sea level.

(c) **Relation of 1 kgf to 1 newton** (derived unit of force) is given in Sec. 3.02.

(d) **Multiples and fractions of 1 kgf** frequently used in technology are

$$1 \text{ ton-force} = 1 \text{ tf} = 1{,}000 \text{ kgf} \qquad 1 \text{ gram-force} = 1 \text{ gf} = 10^{-3} \text{ kgf}$$

(e) **FPS unit of static force** is 1 pound-force, designated by the symbol lbf, and defined as exactly 0.453 592 37 kgf.

(f) **Multiples and fractions of 1 lbf** frequently used in technology are

$$1 \text{ ton-force} = 2{,}000 \text{ lbf} \qquad 1 \text{ ounce-force} = \tfrac{1}{16} \text{ lbf}$$

$$1 \text{ kip-force} = 1{,}000 \text{ lbf}$$

(g) **Two basic conversion relations** are

$$1 \text{ lbf} = 0.4536 \text{ kgf} \qquad 1 \text{ kgf} = 2.2046 \text{ lbf}$$

$$1 \text{ ouncef} = 28.35 \text{ gf} \qquad 1 \text{ gf} = 0.0353 \text{ ouncef}$$

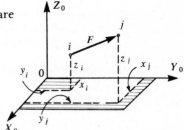

Fig. 2.02–1

(2) Characteristics

(a) **Force** F is a single-headed vector given by its magnitude, direction (line of action and sense), and point of application.

(b) **Graphical representation** of a force is a directed segment, whose initial point is called the *point of application i* and whose end point is called the *terminus j* (Fig. 2.02–1). The magnitude F of the force is the length of the segment ij.

(c) **Direction of a force** is given by the angles $\alpha_F, \beta_F, \gamma_F$, which are measured from lines parallel to the coordinate axes through i to the line of action of F (Fig. 2.02–2).

$$\cos \alpha_F = \frac{x_j - x_i}{F} \qquad \cos \beta_F = \frac{y_j - y_i}{F} \qquad \cos \gamma_F = \frac{z_j - z_i}{F}$$

where $F = $ length ij.

(d) **Relation between the direction cosines** of a force is given by

$$\cos^2 \alpha_F + \cos^2 \beta_F + \cos^2 \gamma_F = 1$$

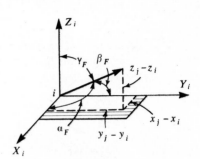

Fig. 2.02–2

(e) Uniqueness. Consequently the direction of a force is given completely and uniquely by two direction angles (any two). The third angle is then computed from (*c*).

(f) Sign convention. A force acting from the particle (pull) is positive. A force acting on the particle (push) is negative (Fig. 2.02–3).

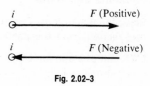

Fig. 2.02–3

(3) Resolution of Force

(a) Definition. The resolution of a given force is the replacement of this force by two or several forces called *components*, which are equivalent in static action to the given force.

(b) Force polygon. A single force can always be resolved into *n* components (any number of components), forming with the given force a closed polygon in which *n* − 1 components can be selected arbitrarily (are independent) and one (any one) is given as the closure of the polygon (is dependent) (Fig. 2.02–4).

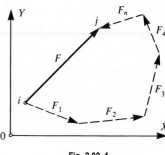

Fig. 2.02–4

(c) Types of force polygon. The force polygon can be a planar polygon or a space polygon.

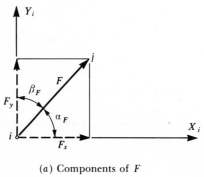
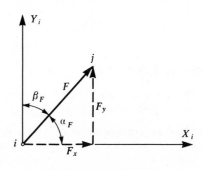

(*a*) Components of *F*   (*b*) Force triangle of *F*

Fig. 2.02–5

(d) Orthogonal resolution of a force *F* in a plane (Fig. 2.02–5*a*) is represented by two components F_x, F_y parallel to the *X*, *Y* axes, respectively, and forming with the given force a force triangle (Fig. 2.02–5*b*).

$$F_x = F \cos \alpha_F \qquad F_y = F \sin \alpha_F \qquad \text{or} \qquad F_x = F \sin \beta_F \qquad F_y = F \cos \beta_F$$

and

$$F = \sqrt{F_x^2 + F_y^2} \qquad \tan \alpha_F = \frac{F_y}{F_x} \qquad \tan \beta_F = \frac{F_x}{F_y}$$

where α_F, β_F = direction angles.

(e) Orthogonal resolution of a force F in space is represented by three components F_x, F_y, F_z parallel to the X, Y, Z axes, respectively, forming with the given force F a closed spatial polygon (in which F_x, F_y, F_z are the edges of a parallelepiped of which F is the diagonal) (Fig. 2.02–6).

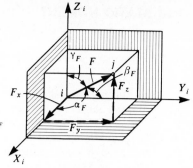

Fig. 2.02–6

$$F_x = F \cos \alpha_F \qquad F_y = F \cos \beta_F \qquad F_z = F \cos \alpha_F$$

and $\qquad F = \sqrt{F_x^2 + F_y^2 + F_z^2}$

where $\alpha_F, \beta_F, \gamma_F$ = direction angles discussed in Sec. 2.02–2c.

example:

If the force F of Fig. 2.02–6 is given by its magnitude $F = 20$ kgf and by two direction angles $\alpha_F = 60°$ and $\beta_F = 45°$, the third direction angle γ_F, according to Sec. 2.02–2d, is computed from

$$\cos \gamma_F = \sqrt{1 - \cos^2 \alpha_F - \cos^2 \beta_F} = \sqrt{1 - 0.500^2 - 0.707^2} = 0.500$$

and thus $\gamma_F = 60°$. Then $F_x = F \cos \alpha_F = 20 \times 0.500 = 10$ kgf,

$F_y = F \cos \beta_F = 20 \times 0.707 = 14.14$ kgf, $F_z = F \cos \gamma_F = 20 \times 0.500 = 10$ kgf,

and the result must satisfy

$$F = \sqrt{F_x^2 + F_y^2 + F_z^2} = 20 \text{ kgf}$$

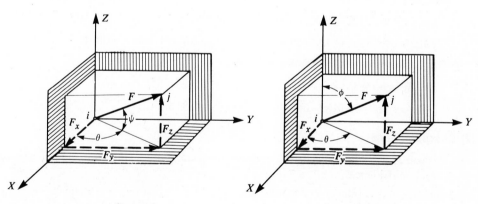

Fig. 2.02–7   **Fig. 2.02–8**

(f) Cylindrical system. If a force is given by its magnitude F, vertical component F_z, and polar angle θ (Fig. 2.02–7), then

$$\cos \alpha_F = \cos \theta \cos \psi \qquad \cos \beta_F = \sin \theta \sin \psi \qquad \cos \gamma_F = \sin \psi$$

where $\psi = \sin^{-1}(F_z/F)$.

(g) Spherical system. If a force is given by its magnitude F and two spherical angles θ, ϕ (Fig. 2.02–8), then

$$\cos \alpha_F = \cos \theta \sin \phi \qquad \cos \beta_F = \sin \theta \sin \phi \qquad \cos \gamma_F = \cos \phi$$

2.03 STATIC MOMENT

(1) Definitions and Units

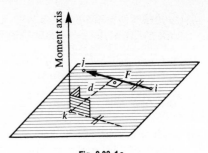

Fig. 2.03–1a

(a) **Static moment** M acting on a particle k is defined as the product of a static force F and the normal distance d (moment arm) from the particle k (moment center) to that force (Fig. 2.03–1a), and is interpreted as the tendency to rotate the particle about an axis through k normal to the plane of the particle and of the force.

$$M = F \cdot d$$

(b) **MKS unit of static moment** is 1 kilogram-force × meter, designated by the symbol kgf-m, and defined exactly as the product of 1 kilogram-force and 1 meter.

(c) **Relation of 1 kgf-m to 1 newton-meter** is given in Sec. 3.02–4a.

(d) **Multiples and fractions of 1 kgf-m** frequently used in technology are

1 ton-force × meter = 1 tf-m × 1,000 kgf-m

1 gram-force × centimeter = 1 gf-cm = 10^{-5} kgf-m

(e) **FPS unit of static moment** is 1 pound-force × foot, designated by the symbol 1 lbf-ft, and defined exactly as 0.138 254 954 kgf-m.

(f) **Multiples and fractions of 1 lbf-ft** frequently used in technology are

1 ton-force × foot = 2,000 lbf-ft   1 pound-force × inch = $\frac{1}{12}$ lbf-ft

1 kip-force × foot = 1,000 lbf-ft   1 ounce-force × inch = $\frac{1}{192}$ lbf-ft

(g) **Two basic conversion relations** are

1 lbf-ft = 0.1383 kgf-m   1 kgf-m = 7.2330 lbf-ft

1 ouncef-in. = 72.01 gf-cm   1 gf-cm = 0.0139 ouncef-in.

(2) Characteristics

(a) **Moment** M is a double-headed vector given by its magnitude, direction (line of action and sense), and point of application.

(b) **Graphical representation of a moment** $M = F \cdot d$ is a semicircle in the plane of the particle k and the force F, with center at k and arrow indicating the sense (Fig. 2.03–1b), or is a directed segment, whose initial point is called the *point of application* k and whose end point is called the *terminus* l (Fig. 2.03–1b).*

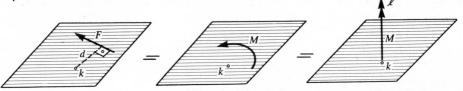

Fig. 2.03–1b

*The script letter in the figure is used where l is used in the text; this avoids possible confusion with the numeral one (1).

(c) Magnitude of a moment (in a general position) is the length of the segment kl (Fig. 2.03–2).

$$M = \sqrt{(x_k - x_l)^2 + (y_k - y_l)^2 + (z_k + z_l)^2}$$

where x_k, y_k, z_k and x_l, y_l, z_l = coordinates of the end points k and l. These coordinates must be expressed in units of moment.

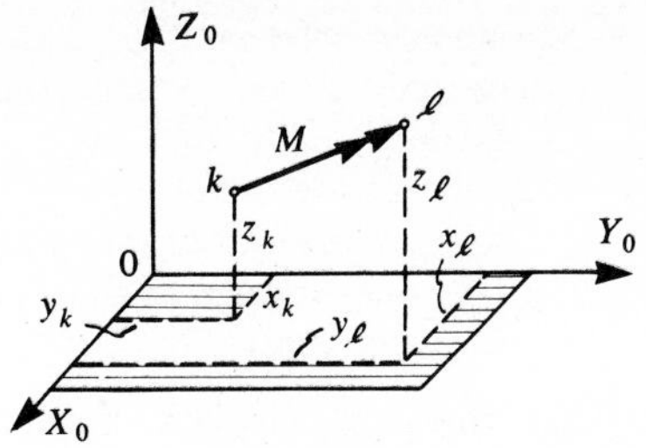

(d) Direction of a moment is given by the direction cosines of the angles $\alpha_M, \beta_M, \gamma_M$, which are measured from lines parallel to the coordinate axes through k to the line of action of M (Fig. 2.03–3).

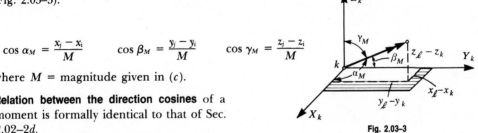

Fig. 2.03–2

$$\cos \alpha_M = \frac{x_j - x_i}{M} \qquad \cos \beta_M = \frac{y_j - y_i}{M} \qquad \cos \gamma_M = \frac{z_j - z_i}{M}$$

where M = magnitude given in (c).

(e) Relation between the direction cosines of a moment is formally identical to that of Sec. 2.02–2d.

(f) Uniqueness. As in the case of force (Sec. 2.02–2e) the direction of a moment is given completely and uniquely by two direction angles (any two). The third angle is then computed from (e).

Fig. 2.03–3

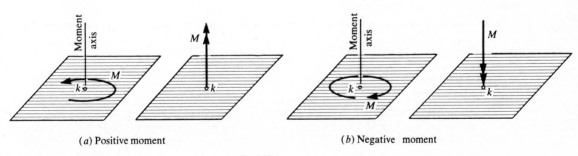

(a) Positive moment   (b) Negative moment

Fig. 2.03–4

(g) Sign convention. A moment acting counterclockwise is positive; a moment acting clockwise is negative. The right-hand rule as shown in Fig. 2.03–4 can be also used.

(3) Resolution of Moment

(a) Definition. The resolution of a given moment is the replacement of this moment by two or several moments called *components*, which are equivalent in the static action to the given moment.

(b) Moment polygon. A single moment can always be resolved into n (any number) components, forming with the given moment a closed polygon, in which $n-1$ components can be selected arbitrarily (are independent) and one (any one) is given as the closure of the polygon (is dependent) (Fig. 2.03–5).

(c) Types of moment polygons. The moment polygon can be a planar polygon or a space polygon.

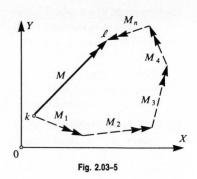

Fig. 2.03–5

(d) Orthogonal resolution of moment in a plane is represented by two components M_x, M_y parallel to the X, Y axes, respectively, forming with the given moment a right triangle (Fig. 2.03–6).

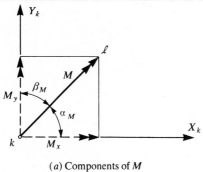
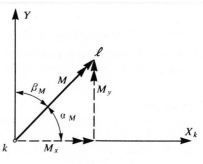

(*a*) Components of *M*   (*b*) Moment triangle of *M*

Fig. 2.03–6

$$M_x = M \cos \alpha_M \qquad M_y = M \sin \alpha_M \qquad \text{or} \qquad M_x = M \sin \beta_M \qquad M_y = M \cos \beta_M$$

and

$$M = \sqrt{M_x^2 + M_y^2}$$

where α_M, β_M = direction angles of M.

(e) Orthogonal resolution of moment in space is represented by three components M_x, M_y, M_z parallel to the X, Y, Z axes, respectively, forming with the given moment a closed spatial polygon (in which M_x, M_y, M_z are the edges of a parallelepiped of which M is the diagonal) (Fig. 2.03–7).

$$M_x = M \cos \alpha_M \qquad M_y = M \cos \beta_M$$

$$M_z = M \cos \gamma_M$$

and

$$M = \sqrt{M_x^2 + M_y^2 + M_z^2}$$

where α_M, β_M, γ_M = direction angles discussed in Sec. 2.03–2d.

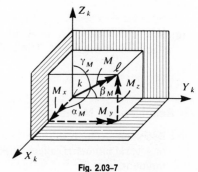

Fig. 2.03–7

(4) Analytical Relations

(a) Case 1: Force given by F, α_F, and i. The moment of a force (given by its magnitude F, direction angle α, and its point of application i) about another point k (Fig. 2.03–8) in the same plane is

$$M = F[(x_i - x_k)\sin\alpha_F - (y_i - y_k)\cos\alpha_F] = F \cdot d$$

in which the moment arm is

$$d = (x_i - x_k)\sin\alpha_F - (y_i - y_k)\cos\alpha_F$$

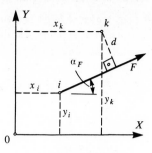

Fig. 2.03–8

(b) Case 2: Force given by F_x, F_y, and i. The moment of a force (given by its components F_x, F_y and its point of application i) about another point k (Fig. 2.03–9) in the same plane is

$$M = F_y(x_i - x_k) - F_x(y_i - y_k) = F \cdot d$$

from which the moment arm is

$$d = \frac{F_y(x_i - x_k) - F_x(y_i - y_k)}{\sqrt{F_x^2 + F_y^2}}$$

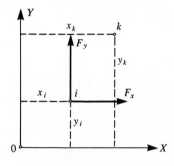

Fig. 2.03–9

example:

If the force of Fig. 2.03–9 is given by $F_x = 100$ kgf and $F_y = 200$ kgf, the coordinates of the point of application i are $x_i = 6$ m, $y_i = 9$ m, and if the coordinates of the moment center k are $x_k = 8$ m, $y_k = 32$ m, then the moment of F about k is

$$M = 200(6 - 8) - 100(9 - 32) = 1{,}900 \text{ kgf-m}$$

The moment arm is then

$$d = \frac{M}{F} = \frac{1{,}900}{\sqrt{100^2 + 200^2}} = \frac{1{,}900}{223.6} = 8.5 \text{ m}$$

2.04 SYSTEMS OF FORCES

(1) Definitions

(a) Systems of forces (sets of forces) are classified as collinear, coplanar, and spatial systems.

(b) Collinear system of forces consists of forces of identical lines of action but not necessarily of the same sense.

(c) Coplanar system of forces consists of forces whose lines of action are in one plane.

(d) Spatial system of forces consists of forces whose lines of action are not in one plane.

(e) Composition (reduction) of a system of forces is the replacement of a given system by the simplest static equivalent called the *resultant*, which can be a force or a couple, or a force and a couple, or finally it can also be zero (state of static equilibrium).

(2) Classification of Coplanar Force Systems

(a) Four types of force systems may occur in the plane. They are the collinear systems, concurrent systems, parallel systems, and general systems.

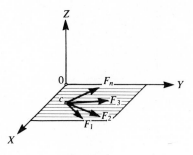

Fig. 2.04–1

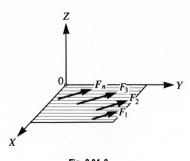

Fig. 2.04–2

(b) Coplanar concurrent force system consists of forces whose lines of action are in one plane and intersect in one point called the *concurrence* (Fig. 2.04–1).

(c) Coplanar parallel force system consists of forces whose lines of action are parallel in one plane (Fig. 2.04–2).

(d) Coplanar general force system consists of forces whose lines of action are in one plane but are not all parallel and are not all concurrent (Fig. 2.04–3).

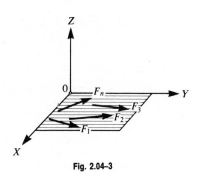

Fig. 2.04–3

(3) Classification of Spatial Force Systems

(a) Four types of force systems may be recognized in space; they are collinear systems, concurrent systems, parallel systems, and general systems.

(b) Spatial concurrent force system consists of forces which intersect in one point, called the *concurrence*, and whose lines of action are not all in one plane (Fig. 2.04–4).

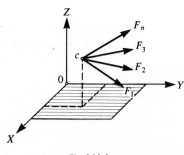

Fig. 2.04–4

(c) Spatial parallel force system consists of forces whose lines of action are parallel and are not all in one plane (Fig. 2.04–5).

(d) Spatial general force system consists of forces whose lines of action are not all in one plane, are not all parallel, and are not all concurrent (Fig. 2.04–6).

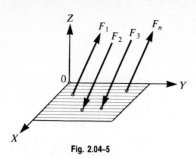

Fig. 2.04-5

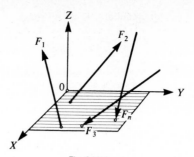

Fig. 2.04-6

2.05 SYSTEMS OF COPLANAR FORCES

(1) Coplanar Concurrent Forces

(a) Resultant F of two concurrent forces F_1, F_2 is the diagonal of the parallelogram of these forces (Fig. 2.05-1).

$$F = \sqrt{F_1^2 + F_2^2 + 2F_1F_2 \cos \omega}$$

where ω = angle between F_1 and F_2.

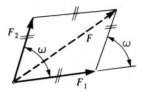

Fig. 2.05-1

(b) Triangle law. The resultant of two concurrent forces forms with these forces a closed triangle which can be converted into a right triangle by means of the orthogonal components of F_1 and F_2, as shown in Fig. 2.05-2.

$$F = \sqrt{(F_{1x} + F_{2x})^2 + (F_{1y} + F_{2y})^2}$$
$$= \sqrt{(F_1 \cos \alpha_1 + F_2 \cos \alpha_2)^2 + (F_1 \sin \alpha_1 + F_2 \sin \alpha_2)^2}$$

where α_1, α_2 = direction angles of F_1, F_2, respectively.

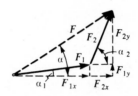

Fig. 2.05-2

example:

If the forces of Fig. 2.05-1 are $F_1 = 100$ kgf, $F_2 = 200$ kgf and their direction angles are $\alpha_1 = 30°$ and $\alpha_2 = 60°$, then $\omega = \alpha_2 - \alpha_1 = 30°$, and according to (a) the resultant is

$$F = \sqrt{100^2 + 200^2 + 2 \times 100 \times 200 \times \cos 30°} = 291 \text{ kgf}$$

or according to (b) in terms of

$$F_x = F_1 \cos \alpha_1 + F_2 \cos \alpha_2 = 100 \times 0.866 + 200 \times 0.500 = 186.6 \text{ kgf}$$
$$F_y = F_1 \sin \alpha_1 + F_2 \sin \alpha_2 = 100 \times 0.500 + 200 \times 0.866 = 223.2 \text{ kgf}$$

the resultant is

$$F = \sqrt{F_x^2 + F_y^2} = \sqrt{186.6^2 + 223.2^2} = 291 \text{ kgf}$$

(c) Direction of the resultant of two forces in Fig. 2.05-2 is given by

$$\tan \alpha = \frac{F_{1y} + F_{2y}}{F_{1x} + F_{2x}} = \frac{F_1 \sin \alpha_1 + F_2 \sin \alpha_2}{F_1 \cos \alpha_1 + F_2 \cos \alpha_2}$$

where α = direction angle of F.

(d) Resultant F of n coplanar concurrent forces
$F_1, F_2, \ldots, F_n$ (Fig. 2.05–3) is the force closing their force polygon (Fig. 2.05–4). Algebraically,

$$F = \sqrt{F_x^2 + F_y^2}$$

where

$$F_x = F_{1x} + F_{2x} + \cdots + F_{nx} = F_1 \cos \alpha_1 + F_2 \cos \alpha_2 + \cdots + F_n \cos \alpha_n$$

$$F_y = F_{1y} + F_{2y} + \cdots + F_{ny} = F_1 \sin \alpha_1 + F_2 \sin \alpha_2 + \cdots + F_n \sin \alpha_n$$

and $\alpha_1, \alpha_2, \ldots, \alpha_n =$ direction angles of the given forces.

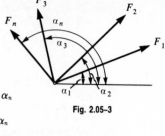

Fig. 2.05–3

(e) Direction of the resultant of n forces in Fig. 2.05–3 is given by

$$\tan \alpha = \frac{F_y}{F_x}$$

where $\alpha =$ direction angle of F (Fig. 2.05–4).

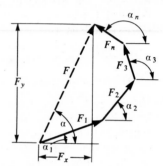

Fig. 2.05–4

(f) Varignon's theorem. The algebraic sum of moments of two concurrent forces F_1 and F_2 about a chosen moment center 0 in their plane of action equals the moment of their resultant F about the same moment center (Fig. 2.05–5). This statement is valid for any number of concurrent forces.

$$M = F_1 \cdot d_1 + F_2 \cdot d_2 + \cdots + F_n \cdot d_n = F \cdot d$$

from which the moment arm is

$$d = \frac{F_1 \cdot d_1 + F_2 \cdot d_2 + \cdots + F_n \cdot d_n}{F}$$

(g) Equilibrium. A coplanar concurrent force system $F_1, F_2, \ldots, F_n$ is in a state of static equilibrium when the resultant of all forces is equal to zero (the closed-force polygon consists only of $F_1, F_2, \ldots, F_n$). Algebraically,

$$F_{1x} + F_{2x} + \cdots + F_{nx} = 0 \qquad \left(\sum_{m=1}^{n} F_{mx} = 0 \right)$$

$$F_{1y} + F_{2y} + \cdots + F_{ny} = 0 \qquad \left(\sum_{m=1}^{n} F_{my} = 0 \right)$$

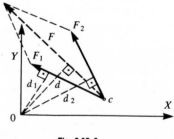

Fig. 2.05–5

where $F_{1x}, F_{2x}, \ldots, F_{nx} =$ components of all the forces along the X axis and $F_{1y}, F_{2y}, \ldots, F_{ny} =$ components of the same forces along the Y axis.

(2) Coplanar Parallel Forces

(a) **Couple** C is the moment of two parallel forces of equal magnitude and opposite sense (Fig. 2.05–6); that is, if $F_1 = -F_2 = F$, then the couple is

$$C = F \cdot d$$

where d = normal distance of their lines of action.

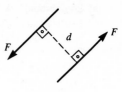

Fig. 2.05–6

(b) **Independence.** The couple C of Fig. 2.05–6 is a vector normal to the plane of F_1, F_2 and independent of its moment center; i.e., the moment of F_1 and F_2 is the same value about any point in their plane.

(c) **Resultant of a coplanar parallel force system** is a force or a couple, or is zero (Fig. 2.05–7a).

(d) **Force resultant** F of a coplanar parallel force system (Fig. 2.05–7a) is the algebraic sum of all forces of the system and is parallel to these forces.

$$F = F_1 + F_2 + \cdots + F_n$$

(e) **Position** of the resultant F with respect to an arbitrary point 0 is

$$d = \frac{F_1 \cdot d_1 + F_2 \cdot d_2 + \cdots + F_n \cdot d_n}{F}$$

where $d_1, d_2, \ldots, d_n$ = moment arms measured from 0.

example:

In Fig. 2.05–7a if $F_1 = 400\,\text{kgf}$, $F_2 = 600\,\text{kgf}$, $F_3 = -500\,\text{kgf}$ (acting in the opposite direction) and $d_1 = 30\,\text{m}$, $d_2 = 20\,\text{m}$, $d_3 = 10\,\text{m}$, then the resultant is

$$F = 400 + 600 - 500 = 500\,\text{kgf}$$

and its position is

$$d = \frac{400 \times 30 + 600 \times 20 - 500 \times 10}{500} = 38\,\text{m}$$

as shown in Fig. 2.05–7b.

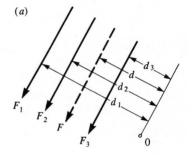

(a)

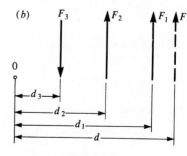

(b)

Fig. 2.05–7

(f) **Resultant couple** C. If the force resultant of a coplanar parallel force system is zero and the moment of the given forces about an arbitrary point 0 is not zero, then the resultant of the system is a couple and the magnitude of this couple is the same for any point of the system plane.

(g) Equilibrium. A coplanar parallel force system $F_1, F_2, \ldots, F_n$ is in a state of static equilibrium when the resultant force and the resultant couple are equal to zero. Algebraically,

$$F_1 + F_2 + \cdots + F_n = 0 \qquad \left(\sum_{m=1}^{n} F_m = 0 \right)$$

$$F_1 \cdot d_1 + F_2 \cdot d_2 + \cdots + F_n \cdot d_n = 0 \qquad \left(\sum_{m=1}^{n} M_m = 0 \right)$$

where $d_1, d_2, \ldots, d_n$ = moment arms of the respective forces with respect to an arbitrarily selected point in the system plane.

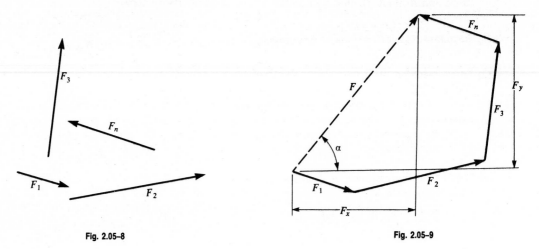

Fig. 2.05–8   Fig. 2.05–9

(3) Coplanar General Forces

(a) Resultant of a coplanar general force system is a force F or a couple C, or is zero (Sec. 2.05–3g).

(b) Magnitude of F (Fig. 2.05–8) is the force closing the polygon of the given forces (Fig. 2.05–9).

$$F = \sqrt{F_x^2 + F_y^2}$$

where

$$F_x = F_{1x} + F_{2x} + \cdots + F_{nx} = F_1 \cos \alpha_1 + F_2 \cos \alpha_2 + \cdots + F_n \cos \alpha_n$$

$$F_y = F_{1y} + F_{2y} + \cdots + F_{ny} = F_1 \sin \alpha_1 + F_2 \sin \alpha_2 + \cdots + F_n \sin \alpha_n$$

and $\alpha_1, \alpha_2, \ldots, \alpha_n$ = direction angles of the respective forces.

(c) Direction of F in Fig. 2.05–9 is given by

$$\tan \alpha = \frac{F_y}{F_x}$$

where α = direction angle of F.

(d) Position of F is given by the coordinates of point i,

$$x_i = \frac{F_{1y} \cdot x_1 + F_{2y} \cdot x_2 + \cdots + F_{ny} \cdot x_n}{F_{1y} + F_{2y} + \cdots + F_{ny}}$$

$$y_i = -\frac{F_{1x} \cdot y_1 + F_{2x} \cdot y_2 + \cdots + F_{nx} \cdot y_n}{F_{1x} + F_{2x} + \cdots + F_{nx}}$$

where $F_{1x}, F_{2x}, \ldots, F_{nx}$ and $F_{1y}, F_{2y}, \ldots, F_{ny}$ = components of the given forces and $x_1, x_2, \ldots, x_n$ and $y_1, y_2, \ldots, y_n$ = coordinates of their points of applications.

(e) Moment M of coplanar general forces about a chosen moment center O in their plane of action equals the algebraic sum of the moments of all the given forces.

(f) Moment resultant C. When the resultant force of a coplanar force system is zero and the moment of the given forces about an arbitrary point O is not zero, then the resultant of the system is a couple C and the magnitude of this couple is the same for any point in the system plane.

(g) Equilibrium. A coplanar general force system $F_1, F_2, \ldots, F_n$ is in a state of static equilibrium when the resultant force and the resultant couple are equal to zero. Algebraically,

$$F_{1x} + F_{2x} + \cdots + F_{nx} = 0 \qquad \left(\sum_{m=1}^{n} F_{mx} = 0\right)$$

$$F_{1y} + F_{2y} + \cdots + F_{ny} = 0 \qquad \left(\sum_{m=1}^{n} F_{my} = 0\right)$$

$$F_1 \cdot d_1 + F_2 \cdot d_2 + \cdots + F_n \cdot d_n = 0 \qquad \left(\sum_{m=1}^{n} M_m = 0\right)$$

where $d_1, d_2, \ldots, d_n$ = moment arms of the respective forces with respect to an arbitrarily selected point in the system plane.

2.06 SYSTEMS OF SPATIAL FORCES

(1) Spatial Concurrent Forces

(a) Resultant F of several spatial concurrent forces $F_1, F_2, \ldots, F_n$ of Fig. 2.04–4 is the force closing their force polygon. Algebraically,

$$F = \sqrt{F_x^2 + F_y^2 + F_z^2}$$

where

$$F_x = F_{1x} + F_{2x} + \cdots + F_{nx} = F_1 \cos \alpha_1 + F_2 \cos \alpha_2 + \cdots + F_n \cos \alpha_n$$

$$F_y = F_{1y} + F_{2y} + \cdots + F_{ny} = F_1 \cos \beta_1 + F_2 \cos \beta_2 + \cdots + F_n \cos \beta_n$$

$$F_z = F_{1z} + F_{2z} + \cdots + F_{nz} = F_1 \cos \gamma_1 + F_2 \cos \gamma_2 + \cdots + F_n \cos \gamma_n$$

and $\alpha_1, \beta_1, \gamma_1, \alpha_2, \beta_2, \gamma_2, \ldots, \alpha_n, \beta_n, \gamma_n$ = direction angles of the respective forces.

(b) Direction of the resultant F is given by

$$\cos \alpha_F = \frac{F_x}{F} \qquad \cos \beta_F = \frac{F_y}{F} \qquad \cos \gamma_F = \frac{F_z}{F}$$

where F_x, F_y, F_z = magnitudes of the respective components of F and α_F, β_F, γ_F = direction angles of F (Sec. 2.02–3e).

(c) Moment about an axis. Moments of the resultant F about the coordinate axes X, Y, Z are, respectively,

$$M_x = F_z \cdot y_c - F_y \cdot z_c \qquad \text{(about } X)$$

$$M_y = F_x \cdot z_c - F_z \cdot x_c \qquad \text{(about } Y)$$

$$M_z = F_y \cdot x_c - F_x \cdot y_c \qquad \text{(about } Z)$$

where x_c, y_c, z_c = coordinates of the concurrence (Fig. 2.06–1).

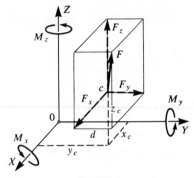

(d) Moment about a point. Moment of the resultant F about the moment center 0 is a vector of magnitude

$$M = \sqrt{M_x^2 + M_y^2 + M_z^2}$$

whose direction is given by

$$\cos \alpha_M = \frac{M_x}{M} \qquad \cos \beta_M = \frac{M_y}{M} \qquad \cos \gamma_M = \frac{M_z}{M}$$

Fig. 2.06–1

where M_x, M_y, M_z = moments of F about the X, Y, Z axes, respectively, defined in (c), and α_M, β_M, γ_M = direction angles of M (Sec. 2.03–3e).

(e) Moment arm. Since this moment must equal the product of F and its moment arm, this moment arm is therefore

$$d = \frac{\sqrt{M_x^2 + M_y^2 + M_z^2}}{\sqrt{F_x^2 + F_y^2 + F_z^2}} = \frac{M}{F}$$

and is normal to the line of action of F.

example:

If F in Fig. 2.06–1 is given by its components $F_x = 100$ kgf, $F_y = 200$ kgf, $F_z = 300$ kgf and by the coordinates of its point of application $x_c = 60$ m, $y_c = 40$ m, $z_c = 20$ m, then the moments of F about the coordinate axes are, respectively,

$$M_x = 300 \times 40 - 200 \times 20 = 8{,}000 \text{ kgf-m}$$

$$M_y = 100 \times 20 - 300 \times 60 = -16{,}000 \text{ kgf-m}$$

$$M_z = 200 \times 60 - 100 \times 40 = 8{,}000 \text{ kgf-m}$$

the magnitude of the moment resultant is

$$M = \sqrt{8{,}000^2 + 16{,}000^2 + 8{,}000^2} = 19{,}596 \text{ kgf-m}$$

and the moment arm is

$$d = \frac{M}{F} = \frac{19{,}596}{\sqrt{100^2 + 200^2 + 300^2}} = 52.37 \text{ m}$$

(f) Equilibrium. A spatial concurrent force system $F_1, F_2, \ldots, F_n$ is in a state of static equilibrium when the resultant of all forces is equal to zero (the closed-force polygon consists only of $F_1, F_2, \ldots, F_n$). Algebraically,

$$F_{1x} + F_{2x} + \cdots + F_{nx} = 0 \qquad \left(\sum_{m=1}^{n} F_{mx} = 0 \right)$$

$$F_{1y} + F_{2y} + \cdots + F_{ny} = 0 \qquad \left(\sum_{m=1}^{n} F_{my} = 0 \right)$$

$$F_{1z} + F_{2z} + \cdots + F_{nz} = 0 \qquad \left(\sum_{m=1}^{n} F_{mz} = 0 \right)$$

where $F_{1x}, F_{2x}, \ldots, F_{nx}, F_{1y}, F_{2y}, \ldots, F_{ny}$, and $F_{1z}, F_{2z}, \ldots, F_{nz}$ = components of all the forces with respect to the X, Y, Z axes.

(g) Two forces. Two concurrent forces in space are in a state of static equilibrium if and only if they are collinear, equal in magnitude, and opposite in sense, that is, $F_1 = -F_2$.

(h) Three forces. Three concurrent forces in space are in a state of static equilibrium if and only if they are coplanar and their force polygon is a closed triangle.

(2) Spatial Parallel Forces

(a) Resultant of a spatial parallel force system is a force F or a couple C, or is zero (Sec. 2.06–2f).

(b) Resultant force F is the algebraic sum of all forces of the system and is parallel to the given forces (Fig. 2.06–2).

$$F = F_1 + F_2 + \cdots + F_n$$

(c) Position of the resultant force of this system with respect to two orthogonal axes U, V in the plane normal to the system (Fig. 2.06–2) is given by

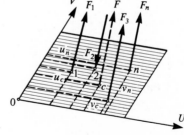

Fig. 2.06-2

$$u_c = -\frac{F_1 \cdot u_1 + F_2 \cdot u_2 + \cdots + F_n \cdot u_n}{F}$$

$$v_c = \frac{F_1 \cdot v_1 + F_2 \cdot v_2 + \cdots + F_n \cdot v_n}{F}$$

where $u_1, v_1, u_2, v_2, \ldots, u_n, v_n$ = coordinates of the points $1, 2, \ldots, n$ (points of intersection of $F_1, F_2, \ldots, F_n$ with the normal plane UV).

(d) Resultant couple C. If the resultant force of the spatial parallel force system is zero and the moment of the given system about an axis in the plane normal to the system is not zero, then the resultant of this system is a couple of magnitude

$$C = \sqrt{C_u^2 + C_v^2}$$

where $\qquad C_u = F_1 \cdot v_1 + F_2 \cdot v_2 + \cdots + F_n \cdot v_n \qquad C_v = -F_1 \cdot u_1 - F_2 \cdot u_2 - \cdots - F_n \cdot u_n$

and $u_1, v_1, u_2, v_2, \ldots, u_n, v_n$ have the same meaning as in (c). The direction of C is given by

$$\tan \delta = \frac{C_v}{C_u}$$

where δ = direction angle of C with respect to U.

(e) Position of the resultant couple C is immaterial since the couple is the same for any axis of direction δ in any plane normal to the system.

(f) Equilibrium. A spatial parallel force system $F_1, F_2, \ldots, F_n$ is in a state of static equilibrium when the resultant force and the resultant couple are equal to zero. Algebraically,

$$F_1 + F_2 + \cdots + F_n = 0 \qquad \left(\sum_{m=1}^{n} F_m = 0 \right)$$

$$F_1 \cdot v_1 + F_2 \cdot v_2 + F_n \cdot v_n = 0 \qquad \left(\sum_{m=1}^{n} M_{mu} = 0 \right)$$

$$F_1 \cdot u_1 + F_2 \cdot u_2 + F_n \cdot u_n = 0 \qquad \left(\sum_{m=1}^{n} M_{mv} = 0 \right)$$

where $u_1, v_1, u_2, v_2, \ldots, u_n, v_n$ = same meaning as in (c).

(3) Spatial General Forces

(a) Resultant of a spatial general force system is a force F and a couple C, or a force F, or a couple C, or is zero (Sec. 2.04–3).

(b) Magnitude of F is the force closing the polygon of given forces.

$$F = \sqrt{F_x^2 + F_y^2 + F_z^2}$$

where

$$F_x = F_{1x} + F_{2x} + \cdots + F_{nx} = F_1 \cos \alpha_1 + F_2 \cos \alpha_2 + \cdots + F_n \cos \alpha_n$$

$$F_y = F_{1y} + F_{2y} + \cdots + F_{ny} = F_1 \cos \beta_1 + F_2 \cos \beta_2 + \cdots + F_n \cos \beta_n$$

$$F_z = F_{1z} + F_{2z} + \cdots + F_{nz} = F_1 \cos \gamma_1 + F_2 \cos \gamma_2 + \cdots + F_n \cos \gamma_n$$

and $\alpha_1, \beta_1, \gamma_1, \alpha_2, \beta_2, \gamma_2, \ldots, \alpha_n, \beta_n, \gamma_n$ = direction angles of the respective forces.

(c) Direction of F is given by

$$\cos \alpha_F = \frac{F_x}{F} \qquad \cos \beta_F = \frac{F_y}{F} \qquad \cos \gamma_F = \frac{F_z}{F}$$

where $\alpha_F, \beta_F, \gamma_F$ = direction angles of F.

(d) Position of F is given by the coordinates of point i, which can be selected arbitrarily; i.e., the direction of F is uniquely defined by (c) but the position of F is arbitrary.

(e) Magnitude of C of these forces is

$$C = \sqrt{C_x^2 + C_y^2 + C_z^2}$$

where

$$C_x = - F_{1y} \cdot z_1 - F_{2y} \cdot z_2 - \cdots - F_{ny} \cdot z_n + F_{1z} \cdot y_1 + F_{2z} \cdot y_2 + \cdots + F_{nz} \cdot y_n$$

$$C_y = F_{1x} \cdot z_1 + F_{2x} \cdot z_2 + \cdots + F_{nx} \cdot z_n - F_{1z} \cdot x_1 - F_{2z} \cdot x_2 - \cdots - F_{nz} \cdot x_n$$

$$C_z = - F_{1x} \cdot y_1 - F_{2x} \cdot y_2 - \cdots - F_{nx} \cdot y_n + F_{1y} \cdot x_1 + F_{2y} \cdot x_2 + \cdots + F_{ny} \cdot y_n$$

are the moments of the components of the given forces about axes parallel to the X, Y, Z axes, and $x_1, y_1, z_1, x_2, y_2, z_2, \ldots, x_n, y_n, z_n$ = coordinates of the points of application of the given forces with respect to the point i.

(f) Direction of C is given by

$$\cos \alpha_C = \frac{C_x}{C} \qquad \cos \beta_C = \frac{C_y}{C} \qquad \cos \gamma_C = \frac{C_z}{C}$$

where α_C, β_C, γ_C = direction angles of C.

(g) Position of C is the same point i arbitrarily selected in (d).

(h) Wrench. For any spatial general force system there exists one and only one line along which F and C are collinear, i.e.,

$$\alpha_F = \alpha_C \qquad \beta_F = \beta_C \qquad \gamma_F = \gamma_C$$

(i) Equilibrium. A spatial general force system $F_1, F_2, \ldots, F_n$ is in a state of static equilibrium when the resultant force and the resultant couple are equal to zero. Algebraically,

$$\sum_{m=1}^{n} F_{mx} = 0 \qquad \sum_{m=1}^{n} F_{my} = 0 \qquad \sum_{m=1}^{n} F_{mz} = 0$$

$$\sum_{m=1}^{n} M_{mx} = 0 \qquad \sum_{m=1}^{n} M_{my} = 0 \qquad \sum_{m=1}^{n} M_{mz} = 0$$

where $\displaystyle\sum_{m=1}^{n} M_{mx}, \sum_{m=1}^{n} M_{my}, \sum_{m=1}^{n} M_{mz}$ = algebraic sums of the moments of the forces of the system about the X, Y, Z axes, respectively.

(4) Special Relations

(a) Angle ϕ between two moments M_1, M_2 (Fig. 2.06–4) which are given by their orthogonal components (Sec. 2.02–3e) is

$$\phi = \cos^{-1} \frac{F_{1x}F_{2x} + F_{1y}F_{2y} + F_{1z}F_{2z}}{F_1 F_2}$$

They are *parallel* if

$$\frac{F_{1x}}{F_1} = \frac{F_{2x}}{F_2} \qquad \frac{F_{1y}}{F_1} = \frac{F_{2y}}{F_2} \qquad \frac{F_{1z}}{F_1} = \frac{F_{2z}}{F_2}$$

They are *normal* if

$$0 = F_{1x}F_{2x} + F_{1y}F_{2y} + F_{1z}F_{2z}$$

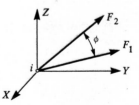

Fig. 2.06–3

(b) Angle ϕ between two moments M_1, M_2 (Fig. 2.06–4) which are given by their orthogonal components (Sec. 2.03–3e) is

$$\phi = \cos^{-1} \frac{M_{1x}M_{2x} + M_{1y}M_{2y} + M_{1z}M_{2z}}{M_1 M_2}$$

They are *parallel* if

$$\frac{M_{1x}}{M_1} = \frac{M_{2x}}{M_2} \qquad \frac{M_{1y}}{M_1} = \frac{M_{2y}}{M_2} \qquad \frac{M_{1z}}{M_1} = \frac{M_{2z}}{M_2}$$

They are *normal* if

$$0 = M_{1x}M_{2x} + M_{1y}M_{2y} + M_{1z}M_{2z}$$

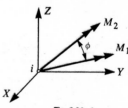

Fig. 2.06–4

(c) Normal distance d from point i to force F_j (Fig. 2.06–5) is

$$d = \frac{1}{F_j} \sqrt{D_1{}^2 + D_2{}^2 + D_3{}^2}$$

where $a_1 = x_j - x_i,\ a_2 = y_j - y_i,\ a_3 = z_j - z_i$ are constants in determinants,

$$D_1 = \begin{vmatrix} a_1 & F_{jx} \\ a_2 & F_{jy} \end{vmatrix} \qquad D_2 = \begin{vmatrix} a_2 & F_{jy} \\ a_3 & F_{jz} \end{vmatrix} \qquad D_3 = \begin{vmatrix} a_3 & F_{jz} \\ a_1 & F_{jx} \end{vmatrix}$$

and j is any point on the line of action of the force.

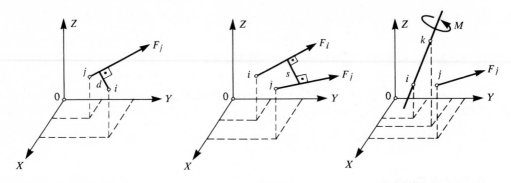

Fig. 2.06–5   Fig. 2.06–6   Fig. 2.06–7

(d) Shortest distance s between two skew forces F_i, F_j (Fig. 2.06–6) is

$$s = \frac{a_1 E_1 + a_2 E_2 + a_3 E_3}{\sqrt{E_1{}^2 + E_2{}^2 + E_3{}^2}}$$

where a_1, a_2, a_3 are the same as in (c) above,

$$E_1 = \begin{vmatrix} F_{iy} & F_{jy} \\ F_{iz} & F_{jz} \end{vmatrix} \qquad E_2 = \begin{vmatrix} F_{iz} & F_{jz} \\ F_{ix} & F_{jx} \end{vmatrix} \qquad E_3 = \begin{vmatrix} F_{ix} & F_{jx} \\ F_{iy} & F_{jy} \end{vmatrix}$$

and i, j are any points on the lines of action of F_i, F_j, respectively.

(e) Moment M of force F_j about an axis given by two points i, k (Fig. 2.06–7) is

$$M = \frac{b_1 M_1 + b_2 M_2 + b_3 M_3}{\sqrt{b_1{}^2 + b_2{}^2 + b_3{}^2}}$$

where a_1, a_2, a_3 are the same as in (c) above, $b_1 = x_k - x_i,\ b_2 = y_k - y_i,\ b_3 = z_k - z_i,$

$$M_1 = \begin{vmatrix} a_2 & F_{jy} \\ a_3 & F_{jz} \end{vmatrix} \qquad M_2 = \begin{vmatrix} a_3 & F_{jz} \\ a_1 & F_{jx} \end{vmatrix} \qquad M_3 = \begin{vmatrix} a_1 & F_{jx} \\ a_2 & F_{jy} \end{vmatrix}$$

and j is any point on the line of action of the force.

2.07 DISTRIBUTED FORCES

(1) Area Forces

(a) Definition. Area force is a force vector distributed over a plane area. Since a vector can always be resolved into a normal component (perpendicular to the area) and a tangential component (in the plane of the area), area forces are classified as normal and tangential forces.

(b) Intensity of a uniformly distributed area force is defined as the force per unit area; i.e.,

$$f_n = \frac{F_n}{A} \qquad f_t = \frac{F_t}{A} \qquad \left\{ \begin{array}{l} F_n, F_t \text{ in kgf} \\ A \text{ in m}^2 \\ f_n, f_t \text{ in kgf/m}^2 \end{array} \right\}$$

where F_n, F_t = normal and tangential components of F, A = area, and f_n, f_t = normal and tangential intensities, respectively.

(c) Variation. The intensities f_n and f_t may be constants (uniformly distributed area force) given by the formulas of (b) or they may vary according to a certain rule. For example, the snow load on a sloped roof may be of constant intensity, but the wind load on the same roof may vary according to a certain function (obtained by field measurements).

(d) Pressure and uplift. The intensity of the normal area force is called the *pressure*, and if acting from the area it is called the *uplift* (suction).

(e) MKS unit of pressure (uplift) is 1 kilogram-force per square meter, designated by the symbol kgf/m² (see also Sec. 4.03).

(f) FPS unit of pressure (uplift) is 1 pound-force per square foot, designated by the symbol lbf/ft² (see also Sec. 4.03), related to 1 kgf/m² as

$$1 \text{ lbf/ft}^2 = 4.8824 \text{ kgf/m}^2 \qquad 1 \text{ kgf/m}^2 = 0.2048 \text{ lbf/ft}^2$$

More extensive relations (to other units of pressure) are given in Appendix B.

(2) Static Functions of Area

(a) Static moments of plane area A about the X, Y axes (Fig. 2.07–1), respectively, are

$$Q_x = \int_A y\,dA \qquad Q_y = \int_A x\,dA \qquad \left\{ \begin{array}{l} Q_x, Q_y \text{ in m}^3 \\ x, y \text{ in m} \\ dA \text{ in m}^2 \end{array} \right\}$$

where $dA = dx\,dy$ = element of the area, x, y = coordinates of dA, and each integral is taken over the entire area A.

(b) Centroid c of a plane area is defined as the point of intersection of lines about which the static moments of this area are zero; i.e., the static moment of A about an axis through c is zero.

(c) Coordinates of the centroid of the area are

$$x_c = \frac{Q_y}{A} \qquad y_c = \frac{Q_x}{A} \qquad \left\{ \begin{array}{l} Q_x, Q_y \text{ in m}^3 \\ A \text{ in m}^2 \\ x, y \text{ in m} \end{array} \right\}$$

where A = area and Q_x, Q_y = static moments defined in (a).

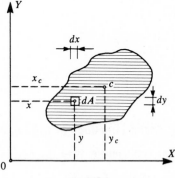

Fig. 2.07–1

(d) Particular cases. If the area has one axis of symmetry, then its centroid lies on this axis. If the area has two axes of symmetry, then its centroid is the point of intersection of these axes of symmetry. Centroidal coordinates of the most frequently used plane areas are given in Appendix A.

(e) Composite areas. The area of any geometrical shape can be divided exactly or approximately into a finite number of basic areas, and the static moments and coordinates of the centroid of the total area can be computed in terms of the static functions of these parts. Analytically,

$$A = A_1 + A_2 + \cdots + A_n = \sum_{m=1}^{n} A_m$$

$$Q_x = A_1 \cdot y_1 + A_2 \cdot y_2 + \cdots + A_n \cdot y_n = \sum_{m=1}^{n} A_m \cdot y_m$$

$$Q_y = A_1 \cdot x_1 + A_2 \cdot x_2 + \cdots + A_n \cdot x_n = \sum_{m=1}^{n} A_m \cdot x_m$$

$$x_c = \frac{\sum_{m=1}^{n} A_m \cdot x_m}{\sum_{m=1}^{n} A_m} \qquad y_c = \frac{\sum_{m=1}^{n} A_m \cdot y_m}{\sum_{m=1}^{n} A_m}$$

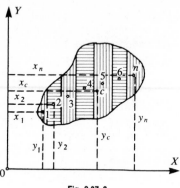

Fig. 2.07-2

where $A_1, A_2, \ldots, A_n$ = areas of subdivisions and $x_1, y_1, x_2, y_2, \ldots, x_n, y_n$ = coordinates of the centroids of these subdivisions (Fig. 2.07–2).

example:

The area of Fig. 2.07–3 can be subdivided into three rectangles of areas $A_1 = 10 \times 4 = 40 \, \text{cm}^2$, $A_2 = 2 \times 16 = 32 \, \text{cm}^2$, and $A_3 = 10 \times 2 = 20 \, \text{cm}^2$, and the total area is then $A = 40 + 32 + 20 = 92 \, \text{cm}^2$.

The static moments of A are

$$Q_x = 40 \times 2 + 32 \times 8 + 20 \times 15 = 636 \, \text{cm}^3$$

$$Q_y = 40 \times 7 + 32 \times 1 + 20 \times 7 = 452 \, \text{cm}^3$$

and the coordinates of the centroid are

$$x_c = \frac{452}{92} = 4.9 \, \text{cm} \qquad y_c = \frac{636}{92} = 6.9 \, \text{cm}$$

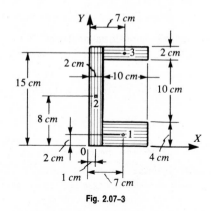

Fig. 2.07-3

(3) Static Functions of Normal Area Forces

(a) Static moments of the normal area forces about the X, Y axes, respectively, are

$$Q_x = \int_A y f_n \, dA \qquad Q_y = \int_A x f_n \, dA \qquad \begin{cases} Q_x, \, Q_y \text{ in kgf-m} \\ f_n \text{ in kgf/m}^2 \end{cases}$$

where dA, x, y have the same meaning as in Sec. 2.07–2a and f_n = intensity of the normal force defined in Sec. 2.07–1b.

(b) Center of pressure (uplift) is the point of intersection of the lines about which the static moments of the normal pressure (uplift) are zero.

(c) Coordinates of the center of pressure i (uplift) are

$$x_i = \frac{Q_y}{F_n} \qquad y_i = \frac{Q_x}{F_n}$$

where Q_x, Q_y = static moments defined in (a) and

$$F_n = \int_A f_n \, dA \qquad \{F_n \text{ in kgf}\}$$

is the resultant of pressure (uplift).

(4) Body Forces

(a) Definition. Body force is a force vector distributed over the entire volume of a body. Weight of the body is the most typical body force. Since a vector can always be resolved into three rectangular components, the body forces are frequently represented by such components.

(b) Intensity of a uniformly distributed body force is defined as the force per unit volume; i.e.,

$$f_x = \frac{F_x}{V} \qquad f_y = \frac{F_y}{V} \qquad f_z = \frac{F_z}{V} \qquad \left\{ \begin{array}{l} F_x, F_y, F_z \text{ in kgf} \\ V \text{ in m}^3 \\ f_x, f_y, f_z \text{ in kgf/m}^3 \end{array} \right\}$$

where F_x, F_y, F_z = components of F, V = volume of the body, and f_x, f_y, f_z = intensities.

(c) Variation. The intensities f_x, f_y, f_z may be constants (uniformly distributed body forces), given by the formulas of (b), or they may vary according to a certain rule.

(d) MKS unit of body force is 1 kilogram-force per cubic meter, designated by the symbol kgf/m^3 (see also Sec. 5.01).

(e) FPS unit of body force is 1 pound-force per cubic foot, designated by the symbol lbf/ft^3 (see also Sec. 5.01), related to 1 kgf/m^3 as

$$1 \text{ lbf/ft}^3 = 16.018 \text{ kgf/m}^3 \qquad 1 \text{ kgf/m}^3 = 0.06243 \text{ lbf/ft}^3$$

More extensive relations to other units of body force are given in Appendix B.

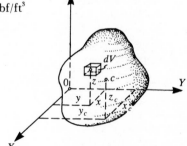

Fig. 2.07–4

(5) Static Functions of Volume

(a) Static moments of a body of volume V about the coordinate planes XY, YZ, ZX (Fig. 2.07–4), respectively, are

$$Q_{xy} = \int_V z \, dV \qquad Q_{yz} = \int_V x \, dV \qquad Q_{zx} = \int_V y \, dV \qquad \left\{ \begin{array}{l} Q_{xy}, Q_{yz}, Q_{zx} \text{ in m}^4 \\ x, y, z \text{ in m} \\ dV \text{ in m}^3 \end{array} \right\}$$

where $dV = dx\,dy\,dz$ = element of the volume, x, y, z = coordinates of dV, and each integral is taken over the entire volume V.

(b) Centroid c of the volume is defined as the point of intersection of planes about which the static moments of the volume are zero.

(c) Coordinates of centroid c of the volume are

$$x_c = \frac{Q_{yz}}{V} \qquad y_c = \frac{Q_{zx}}{V} \qquad z_c = \frac{Q_{xy}}{V} \qquad \left\{ \begin{array}{l} Q_{xy},\ Q_{yz},\ Q_{zx}\ \text{in m}^4 \\ x,\ y,\ z\ \text{in m} \\ V\ \text{in m}^3 \end{array} \right\}$$

where V = volume and Q_{xy}, Q_{yz}, Q_{zx} = static moments defined in (a).

(d) Composite volumes. The volume of any geometrical shape can be subdivided exactly or approximately into a finite number of basic volumes, and the static moments and coordinates of the centroid of the total volume can be computed in terms of these parts.   Analytically (Fig. 2.07–5),

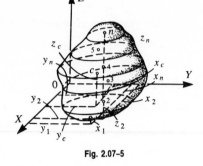

$$V = V_1 + V_2 + \cdots + V_n = \sum_{m=1}^{n} V_m$$

$$Q_{xy} = V_1 \cdot z_1 + V_2 \cdot z_2 + \cdots + V_n \cdot z_n = \sum_{m=1}^{n} V_m \cdot z_m$$

$$Q_{yz} = V_1 \cdot x_1 + V_2 \cdot x_2 + \cdots + V_n \cdot x_n = \sum_{m=1}^{n} V_m \cdot x_m$$

$$Q_{zx} = V_1 \cdot y_1 + V_2 \cdot y_2 + \cdots + V_n \cdot y_n = \sum_{m=1}^{n} V_m \cdot y_m$$

$$x_c = \frac{\sum_{m=1}^{n} V_m \cdot x_m}{\sum_{m=1}^{n} V_m} \qquad y_c = \frac{\sum_{m=1}^{n} V_m \cdot y_m}{\sum_{m=1}^{n} V_m} \qquad z_c = \frac{\sum_{m=1}^{n} V_m \cdot z_m}{\sum_{m=1}^{n} V_m}$$

Fig. 2.07–5

where V_1, V_2,..., V_n = volumes of subdivision and x_1, y_1, z_1, x_2, y_2, z_2, ..., x_n, y_n, z_n = centroidal coordinates of these subdivisions.

(e) Particular cases. If a body has one plane of symmetry, then its centroid is on this plane.   If the body has two planes of symmetry, then its centroid is on the line of intersection of these two planes.   If the body has three planes of symmetry, then its centroid is the point of intersection of these three planes.   Centroidal coordinates of the most frequently used volumes are given in Appendix A.

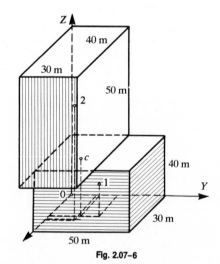

example:

The volume of the body of Fig. 2.07–6 can be subdivided into two parallelepipeds of volumes $V_1 = 30 \times 50 \times 40 = 60{,}000$ m³   and $V_2 = 40 \times 30 \times 50 = 60{,}000$ m³, and the total volume is then $V = 60{,}000 + 60{,}000 = 120{,}000$ m³.

The static moments of V are

$$Q_{xy} = 60{,}000 \times 20 + 60{,}000 \times 65 = 5{,}100{,}000 \text{ m}^4$$

$$Q_{yz} = 60{,}000 \times 15 + 60{,}000 \times 20 = 2{,}100{,}000 \text{ m}^4$$

$$Q_{zx} = 60{,}000 \times 25 + 60{,}000 \times 15 = 2{,}400{,}000 \text{ m}^4$$

and the coordinates of the centroid are

$$x_c = \frac{2{,}100{,}000}{120{,}000} = 17.5 \text{ m} \qquad y_c = \frac{2{,}400{,}000}{120{,}000} = 20 \text{ m} \qquad z_c = \frac{5{,}100{,}000}{120{,}000} = 42.5 \text{ m}$$

Fig. 2.07–6

(6) Inertia Functions of Plane Area

(a) Centroidal inertia functions of plane area A in X_c, Y_c axes (Fig. 2.07–7) are

$$I_{cxx} = \int_A y^2 \, dA \qquad I_{cyy} = \int_A x^2 \, dA$$

$$I_{cxy} = I_{cyx} = \int_A xy \, dA$$

$$\left\{ \begin{array}{l} x, y \text{ in m} \\ I_{cxx}, I_{cyy} \text{ in m}^4 \\ I_{cxy}, I_{cyx} \text{ in m}^4 \\ J_c \text{ in m}^4 \\ dA \text{ in m}^2 \end{array} \right\}$$

$$J_c = \int_A (x^2 + y^2) \, dA = I_{cxx} + I_{cyy}$$

where X_c, Y_c = centroidal axes; I_{cxx} = moment of inertia of A about X_c; I_{cyy} = moment of inertia of A about Y_c; I_{cxy}, I_{cyx} = products of inertia of A in X_c, Y_c; J_c = polar moment of inertia in X_c, Y_c; $dA = dx \, dy$ = element of A; x, y = coordinates of dA in X_c, Y_c; and each integral is taken over the entire area A.

(b) Centroidal radii of gyration of the same area in X_c, Y_c are

$$k_{cxx} = \sqrt{\frac{I_{cxx}}{A}} \qquad k_{cyy} = \sqrt{\frac{I_{cyy}}{A}}$$

$$k_{cxy} = k_{cyx} = \sqrt{\frac{I_{cxy}}{A}}$$

$$k_c = \sqrt{\frac{J_c}{A}} = \sqrt{k_{cxx}^2 + k_{cyy}^2}$$

where A = area.

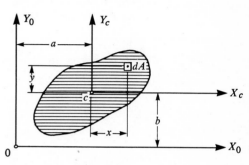

Fig. 2.07–7

(c) Particular cases.
If the area has one axis of symmetry, then the moment of inertia of the entire area about this axis equals twice the moment of inertia of the area on one side of this axis about this axis and the product of inertia of the entire area in axes, of which one is the axis of symmetry, is zero. If the area has two mutually perpendicular axes of symmetry, the moment of inertia of the entire area about one of these axes of symmetry equals 4 times the moment of inertia of the area in one quadrant about the respective axis, and the product of inertia of the entire area in these axes is zero. Inertia functions of the most frequently used areas are given in Appendix A.

(d) Parallel axes theorem.
If the centroidal inertia functions of a given area A are known (Fig. 2.07–7), the inertia functions of the same area in axes X_0, Y_0 parallel to the centroidal axes X_c, Y_c are

$$I_{0xx} = b^2 A + I_{cxx} \qquad I_{0yy} = a^2 A + I_{cyy}$$

$$I_{0xy} = abA + I_{cxy} \qquad J_0 = (a^2 + b^2)A + J_0$$

and analogically, the *radii of gyration* in X_0, Y_0 of the same area A in terms of their counterparts in X_c, Y_c are

$$k_{0xx} = \sqrt{b^2 + k_{cxx}^2} \qquad k_{0yy} = \sqrt{a^2 + k_{cyy}^2}$$

$$k_{0xy} = \sqrt{ab + k_{cxy}^2} \qquad k_0 = \sqrt{k_{cxx}^2 + k_{cyy}^2}$$

where a, b = coordinates of the centroid of A in X_0, Y_0.

(e) Composite areas.
The area of any geometric figure can be subdivided exactly or approximately into a finite number of basic areas (Fig. 2.07–2) and the inertia functions of the

entire area can be expressed in terms of the inertia functions of these basic areas. Analytically,

$$I_{cxx} = \sum_{m=1}^{n} y_{cm}{}^2 A_m + \sum_{m=1}^{n} I_{mxx} \qquad I_{cyy} = \sum_{m=1}^{n} x_{cm}{}^2 A + \sum_{m=1}^{n} I_{myy}$$

$$I_{cxy} = \sum_{m=1}^{n} x_{cm} y_{cm} + \sum_{m=1}^{n} I_{mxy} \qquad J_c = \sum_{m=1}^{n} (x_{cm}{}^2 + y_{cm}{}^2) A_m + \sum_{m=1}^{n} J_m$$

where x_{cm}, y_{cm} = coordinates of the centroid of basic area A_m; $m = 1, 2, \ldots, n$; and I_{mxx}, I_{myy}, I_{mxy}, J_m = inertia functions of A_m in the centroidal axes of that area.

(7) Variations of Inertia Functions

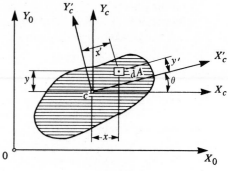

(a) Inertia functions in rotated axes. If the reference axes X_c, Y_c are rotated about c by the right-hand angle θ and designated as X'_c, Y'_c (Fig. 2.07–8), then the inertia functions in these rotated axes may be expressed in terms of the inertia functions in the initial axes as

$$I'_{cxx} = I_{cxx} \cos^2\theta - I_{cxy} \sin 2\theta + I_{cyy} \sin^2\theta$$
$$I'_{cxy} = \tfrac{1}{2}(I_{cxx} - I_{cyy}) \sin 2\theta + I_{cxy} \cos 2\theta$$
$$I'_{cyy} = I_{cxx} \sin^2\theta + I_{cxy} \sin 2\theta + I_{cyy} \cos^2\theta$$
$$J'_c = J_c = I_{cxx} + I_{cyy} = I'_{cxx} + I'_{cyy}$$

Fig. 2.07–8

which shows that the polar moment of inertia is *invariant* and equals the sum of the moments of inertia about two mutually perpendicular axes (any two perpendicular axes).

(b) Principal inertia functions. For any point in the plane of A there are two mutually perpendicular axes in which the products of inertia equal zero. At c, these axes, given by

$$\theta_1 = \tfrac{1}{2} \tan^{-1} \frac{2I_{cxy}}{I_{cyy} - I_{cxx}} \qquad \theta_2 = \frac{\pi}{2} + \theta_1$$

are called the *principal axes* U_c, V_c, and the moments of inertia of A about these axes,

$$I_{cuu} = I'_{max} = \tfrac{1}{2}(I_{cxx} + I_{cyy}) + \tfrac{1}{2}\sqrt{(I_{cxx} - I_{cyy})^2 + 4I_{cxy}}$$
$$I_{cvv} = I'_{min} = \tfrac{1}{2}(I_{cxx} + I_{cyy}) - \tfrac{1}{2}\sqrt{(I_{cxx} - I_{cyy})^2 + 4I_{cxy}}$$

are called the *principal moments of inertia.*

(c) Particular cases. For a symmetrical area the axis of symmetry is the principal axis. If the principal moments of inertia of a given area are known with respect to their principal axes U_c, V_c, then the inertia functions in any set of orthogonal axes A_c, B_c are

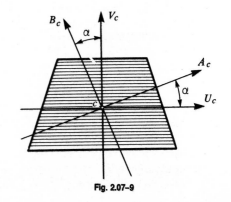

$$I_{caa} = I_{cuu} \cos^2\alpha + I_{cvv} \sin^2\alpha$$
$$I_{cab} = \tfrac{1}{2}(I_{cuu} - I_{cvv}) \sin 2\alpha$$
$$I_{cbb} = I_{cuu} \sin^2\alpha + I_{cvv} \cos^2\alpha$$

where α = position angle of A_c, B_c with respect to U_c, V_c (Fig. 2.07–9).

Fig. 2.07–9

2.08 FRICTION

(1) Concept and Classification

(a) **Concept.** Whenever two solid bodies which are in contact slide or tend to slide on each other, a force is developed in their tangential plane which resists or tends to resist their motion. This force is called *dry friction*, and in the case of area contact it is a distributed force.

(b) **Classification.** Dry friction is classified as *static friction* (bodies in contact are at rest), *kinetic friction* (bodies slide relative to each other), or *rolling friction* (bodies roll relative to each other).

(c) **Limiting friction.** The force or moment required to overcome static friction and start motion is called *limiting friction*.

(2) Laws of Dry Friction

(a) **Contact area.** The total amount of friction which can be developed is independent of the magnitude of the contact area.

(b) **Contact pressure.** The total amount of friction which can be developed on the contact area is directly proportional to the resultant of contact pressure.

(c) **Static and kinetic friction.** At low velocities of sliding, kinetic friction is independent of velocity; however, the force necessary to start sliding (to overcome friction) is greater than the force necessary to maintain sliding.

(3) Parameters of Friction

(a) **Analytically**, the laws of dry friction, known as *Coulomb's laws*, can be expressed as

$$F_s = \mu_s N \qquad F_k = \mu_k N \qquad \left\{ \begin{array}{l} F_s,\ F_k \text{ in kgf} \\ \mu_s,\ \mu_k \text{ dimensionless} \\ N \text{ in kgf} \end{array} \right\}$$

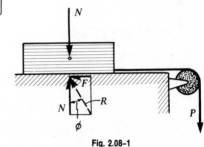

Fig. 2.08–1

where F_s = static friction, μ_s = coefficient of static friction, F_k = kinetic friction, μ_k = coefficient of kinetic friction, and N = resultant of contact pressure.

(b) **Coefficients of friction** μ_s and μ_k are experimental constants depending largely on the roughness of the mating surfaces. The numerical values of these coefficients for the most frequently encountered engineering materials are given in Appendix A.

(c) **Angle of friction** is the angle between the resultant R (of F and N) and the normal to the plane of friction (Fig. 2.08–1). Analytically,

$$R_s = \sqrt{N^2 + F_s^2} = N\sqrt{1 + \mu_s^2} \qquad \tan \phi_s = \frac{F_s}{N} = \mu_s$$

$$R_k = \sqrt{N^2 + F_k^2} = N\sqrt{1 + \mu_k^2} \qquad \tan \phi_k = \frac{F_k}{N} = \mu_k$$

where ϕ = angle of friction and the subscripts s and k identify static and kinetic frictions, respectively.

(4) Friction on a Horizontal Plane

(a) Tensile force.
If the block of weight W (Fig. 2.08–2) is acted upon by a tensile force P, then the static friction is

$$F_s = \mu_s(W - P\cos\alpha)$$

and if $F_s = P\sin\alpha$, motion of the block is impending, which requires that

$$P = \frac{\mu_s W}{\sin\alpha + \mu_s\cos\alpha}$$

or

$$\mu_s = \frac{P\sin\alpha}{W - P\cos\alpha}$$

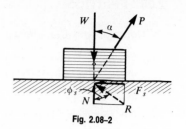

Fig. 2.08–2

(b) Compressive force.
If the same block is acted upon by a compressive force P (Fig. 2.08–3), then

$$F_s = \mu_s(W + P\cos\alpha)$$

and if $F_s = P\sin\alpha$, motion is impending, which requires that

$$P = \frac{\mu_s W}{\sin\alpha - \mu_s\cos\alpha}$$

or

$$\mu_s = \frac{P\sin\alpha}{W + P\cos\alpha}$$

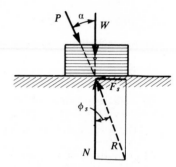

Fig. 2.08–3

(5) Friction on an Inclined Plane

(a) Tensile force, $\alpha < 90°$.
If the block of weight W (Fig. 2.08–4) is acted upon by a tensile force P, then the static friction is

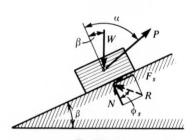

Fig. 2.08–4

$$F_s = \mu_s(W\cos\beta - P\cos\alpha)$$

and if $F_s = P\sin\alpha - W\sin\beta$, motion of the block is impending, which requires that

$$P = \frac{W\sin\beta + \mu_s W\cos\beta}{\sin\alpha + \mu_s\cos\alpha} \qquad \text{or} \qquad \mu_s = \frac{P\sin\alpha - W\sin\beta}{W\cos\beta - P\cos\alpha}$$

(b) Tensile force, $\alpha = 90°$.
If the force P is parallel to the plane, then the limiting case requires that

$$P = W(\sin\beta + \mu_s\cos\beta) \qquad \text{or} \qquad \mu_s = \frac{P - W\sin\beta}{W\cos\beta}$$

(c) Tensile force, $\alpha = 90° + \beta$.
If the force P is parallel to the base, then the limiting case requires that

$$P = \frac{W(\sin\beta + \mu_s\cos\beta)}{\cos\beta - \mu_s\sin\beta} \qquad \text{or} \qquad \mu_s = \frac{P\cos\beta - W\sin\beta}{W\cos\beta + P\sin\alpha}$$

(d) Compressive force, $\alpha < 90°$. If the block of weight W (Fig. 2.08–5) is acted upon by a compressive force, then the static friction is

$$F_s = \mu_s(W \cos \beta + P \cos \alpha)$$

and if $F_s = P \sin \alpha - W \sin \beta$, motion of the block is impending, which requires that

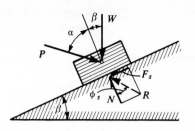

Fig. 2.08–5

$$P = \frac{W(\sin \beta + \mu_s \cos \beta)}{\sin \alpha - \mu_s \cos \alpha} \qquad \text{or} \qquad \mu_s = \frac{P \sin \alpha - W \sin \beta}{W \cos \beta + P \cos \alpha}$$

(e) Compressive force, $\alpha = 90°$. If the force P in Fig. 2.08–5 is parallel to the plane, then the limiting case is governed by formulas (b).

(f) Compressive force, $\alpha = 90° - \beta$. If the force P in Fig. 2.08–5 is parallel to the base, then the limiting case is governed by formulas (c).

(g) Zero force. If the force of Fig. 2.08–4 is absent, then the static friction is

$$F_s = \mu_s W \cos \beta$$

and if $F_s = W \sin \beta$, the motion of the block is impending, which requires that

$$\tan \beta = \mu_s \qquad \text{or} \qquad \beta = \tan^{-1} \mu_s$$

and β is said to be the *angle of repose*.

(h) Angle of repose is the maximum angle to which the plane can be raised before the block resting on it begins to slip under the action of its own weight.

(6) Symmetrical Wedge Friction

(a) Driving in. The force required to drive in the wedge of Fig. 2.08–6 is

$$P = 2Q \frac{\tan (\alpha + \phi_s)}{1 - \tan \phi_s \tan (\alpha + \phi_s)}$$

where P = force applied on the top, Q = lateral constraints, and ϕ_s = angle of friction.

Fig. 2.08–6

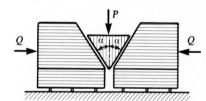

(b) Self-locked wedge. The wedge is self-locked if $P = 0$ and

$$\alpha = \phi_s$$

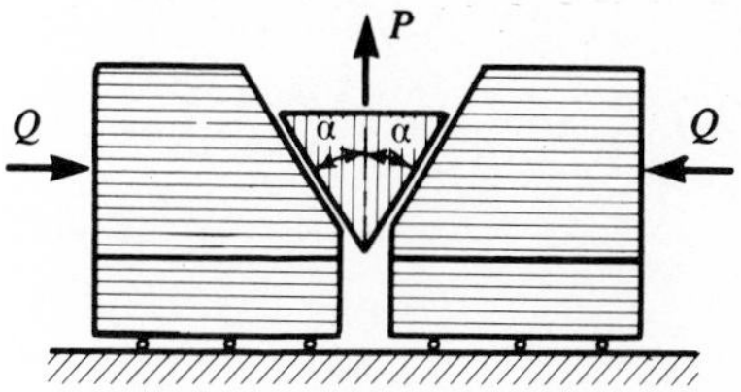

(c) Releasing. The force required to release the locked wedge of Fig. 2.08–7 is

Fig. 2.08–7

$$P = 2Q \frac{\tan (\alpha - \phi_s)}{1 + \tan \phi_s \tan (\alpha - \phi_s)}$$

(7) Unsymmetrical Wedge Friction

(a) Lifting. The force required to lift the block of weight W in Fig. 2.08–8 is

$$P = W[\tan (\alpha + \phi_{1s}) + \tan (\beta + \phi_{2s})]$$

where ϕ_{1s} = angle of friction between the block and the wedge and ϕ_{2s} = angle of friction between the wedge and the base.

Fig. 2.08–8

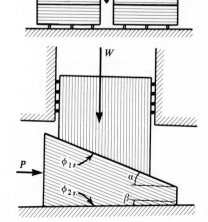

(b) Self-locking and slipping. If P is removed, the wedge of Fig. 2.08–8 remains self-locked if

$$\alpha + \beta = \phi_{1s} + \phi_{2s}$$

and slips if

$$\alpha + \beta > \phi_{1s} + \phi_{2s}$$

(c) Releasing. The force required to release the locked wedge in Fig. 2.08–9 is

$$P = W[\tan(\phi_{1s} - \alpha) + \tan(\phi_{2s} - \beta)]$$

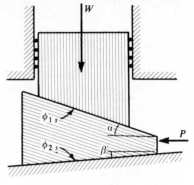

Fig. 2.08–9

(8) Disk, Pivot, and Journal Bearing Friction

(a) Notation. Equations of torque transmission by contact friction for six particular cases are given below where

N = normal force (kgf)	W = vertical force (kgf)
R = external radius (m)	r = internal radius (m)
α = pitch cone angle (rad)	ω = angular speed (rad/s)
t = thickness of lubricant (m)	μ_v = absolute viscosity (kgf-s/m^2)

μ_s = coefficient of static friction (dimensionless)
T_s = friction torque on new surfaces (kgf-m)
T'_s = friction torque on worn-in surfaces (kgf-m)

(b) Disk friction. The moment transmitted by the circular disks of Fig. 2.08–10 under the constant compressive force is

$$T_s = \frac{2\mu_s NR}{3} \qquad \text{or} \qquad T'_s = \frac{\mu_s NR}{2}$$

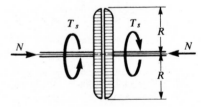

Fig. 2.08–10

(c) Pivot friction. If the contact area is the annulus of Fig. 2.08–11 the moment transmitted under the constant compressive force is

$$T_s = \frac{2\mu_s N(R^3 - r^3)}{3(R^2 - r^2)} \qquad \text{or} \qquad T'_s = \frac{\mu_s N(R + r)}{2}$$

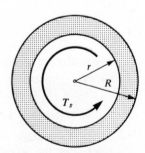

Fig. 2.08–11

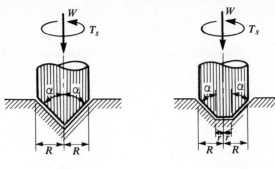

Fig. 2.08-12   Fig. 2.08-13

(d) Conical pivot friction. The moment transmitted by the conical pivot of Fig. 2.08–12 under the vertical force is

$$T_s = \frac{2\mu_s WR}{3 \sin \alpha} \qquad \text{or} \qquad T'_s = \frac{\mu_s WR}{2 \sin \alpha}$$

(e) Truncated conical pivot friction. If the cone is truncated as shown in Fig. 2.08–13,

$$T_s = \frac{2\mu_s W(R^3 - r^3)}{3(R^2 - r^2) \sin \alpha} \qquad \text{or} \qquad T'_s = \frac{\mu_s W(R + r)}{2 \sin \alpha}$$

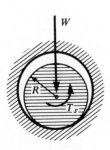
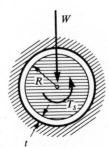

Fig. 2.08-14   Fig. 2.08-15

(f) Dry journal bearing friction. The friction moment of a dry journal bearing under the action of vertical force (Fig. 2.08–14) is

$$T_s = \frac{\pi \mu_s WR}{2} \qquad \text{or} \qquad T'_s = \frac{4\mu_s WR}{\pi}$$

(g) Lubricated journal bearing friction. If the lubricant forms a complete layer between the surfaces (Fig. 2.08–15),

$$T_s = \frac{2\pi \mu_v R^3 L \omega}{t}$$

where L = length of shaft in contact.

Statics of Rigid Bodies **45**

(9) Screw Friction

(a) Geometry. A screw thread (Fig. 2.08–16a) is actually an inclined plane wrapped around a cylinder of radius r. The base of this plane is $2\pi R$, h is the pitch (rise), R is the mean radius of the thread, and μ_s is the coefficient of static friction (Fig. 2.08–16b). Screws are classified as *square-threaded* (Fig. 2.08–16c) and *sharp-threaded* (Fig. 2.08–16d). In square-threaded screws,

$$\phi_s = \tan^{-1}\mu_s$$

and in sharp-threaded screws,

$$\phi_s = \tan^{-1}\left(\mu_s\cos\frac{\beta}{2}\right) \qquad \{\beta \text{ in rad}\}$$

where μ_s = coefficient of static friction and β = pitch angle.

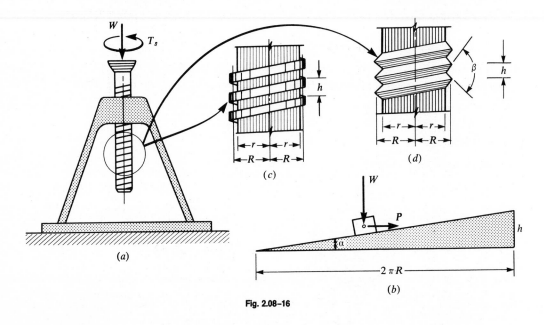

Fig. 2.08–16

(b) Raising of load. The frictional moment (torque T_s) which must be overcome to raise the load W is

$$T_s = WR\tan(\alpha + \phi_s)$$

where α = helix angle (Fig. 2.08–15) and ϕ_s = angle of static friction.

(c) Self-locking and slipping. If $\phi_s > \alpha$, the screw is self-locked. If $\phi_s = \alpha$, the screw is at the verge of unwinding. If $\phi_s < \alpha$, the screw is slipping, and an active moment

$$M = WR\tan(\alpha - \phi_s)$$

must be applied to stop the slipping.

(d) Lowering of load. The friction moment (torque T_s) which is required to lower the load W in a self-locked screw is

$$T_s = WR\tan(\phi_s - \alpha)$$

(10) Belt Friction

(a) **Belt and pulley.** When a belt (rope, line) and a pulley are used to raise or lower a load, the total friction developed between them along their area of contact is

$$F_s = W(e^{\alpha\mu_s} - 1) \qquad \left\{ \begin{array}{l} F_s,\, W \text{ in kgf} \\ \alpha \text{ in rad} \end{array} \right\}$$

where W = load, e = base of natural logarithms (2.718 281. . .), α = angle of contact (angle of wrap), and μ_s = coefficient of static friction (Fig. 2.08–17).

(b) **Raising and lowering of load.** The force required to raise the load is

$$F_R = We^{\alpha\mu_s}$$

and the force required to lower the load is

$$F_L = We^{-\alpha\mu_s}$$

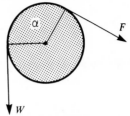

Fig. 2.08–17

(c) **Other applications.** The relationships given above hold equally well for noncircular sections, and if a rope is wrapped n times around a shaft or drum the angle of wrap is always

$$\alpha = 2\pi n$$

(11) Belt-Drive Friction

(a) **Force equations.** When flat belts and V belts are used to transmit power from one shaft to another, the tension T_1 in the tight side (driving side) and the tension T_2 in the loose side (driven side) (Fig. 2.08–18) are, respectively,

$$T_1 = \frac{ME}{a(E-1)} + \rho v^2 \qquad \left\{ \begin{array}{l} T_1,\, T_2 \text{ in kgf} \\ M \text{ in kgf-m} \\ a \text{ in m} \\ \rho \text{ in kg/m} \\ v \text{ in m/s} \\ \alpha,\, \theta \text{ in rad} \end{array} \right\}$$

$$T_2 = \frac{M}{a(E-1)} + \rho v^2$$

where $E = e^{\alpha\mu_s/[\sin(\theta/2)]}$, M = acting moment, a = radius of driving shaft, α = angle of wrap on driving shaft, θ = groove angle for V belt, ρ = mass of belt per unit length, and v = speed of belt.

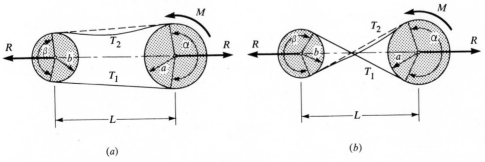

(a)   (b)

Fig. 2.08–18

(b) Special conditions. For flat belts $\theta = 2\pi$ and E reduces to

$$E = e^{\alpha\mu_s}$$

For low speed,

$$T_1 = \frac{ME}{a(E-1)} \qquad T_2 = \frac{M}{a(E-1)} \qquad \frac{T_1}{T_2} = E$$

(c) Reactive force of the belt drive (Fig. 2.08–18) is

$$R = \frac{M(E+1)}{a(E-1)} \cos\left(\frac{\alpha}{2} - \frac{\pi}{2}\right)$$

(d) Angles of wrap for the open belt (Fig. 2.08–18a) and for the crossed belt (Fig. 2.08–18b) are, respectively,

$$\alpha = \pi + 2\sin^{-1}\frac{a-b}{L} \qquad\qquad \alpha + \beta = \pi$$

$$\alpha = \pi + 2\sin^{-1}\frac{a+b}{L} \qquad\qquad \alpha = \beta$$

where L = distance between centers of shafts and a, b = radii of shafts.

(e) Power transmitted by the belt drive at low speed is

$$P = (T_1 - T_2)v = T_1\left(1 - \frac{1}{E}\right)v \qquad\qquad \{P \text{ in kfg-m/s}\}$$

The definition, units, and conversion factors of P are given in Sec. 3.05–6.

3
DYNAMICS OF
RIGID BODIES

3.01 KINEMATICS OF A PARTICLE

(1) Classification and Definitions

(a) Motion of a rigid body is described as a continuous change in the position of all or of some particles of the body.

(b) Dynamics is concerned with the study of this motion and is traditionally divided into two parts called *kinematics* and *kinetics*.

(c) Kinematics is the branch of dynamics concerned with the space-time characteristics of motion without regard to causes producing the motion.

(d) Kinetics is the branch of dynamics concerned with the force-space-time characteristics of motion with regard to causes producing, sustaining, or opposing the motion.

(e) Kinematic quantities describing the motion of a particle are the displacement, velocity, and acceleration, all of which are functions of time or are constants.

(f) Path of motion is the line traced by the moving particle in a given interval of time. This path can be a straight line (rectilinear motion), or a plane curve (plane curvilinear motion), or a space curve (spatial curvilinear motion).

(2) Concept and Units of Time

(a) Time is a scalar, defined as the measure of a succession of events related to an arbitrarily selected datum (beginning, start). The instruments used for measuring time are called *chronometers* (clocks).

(b) SI unit of time is 1 second, designated by the symbol s, and defined as exactly 9 192 631 770 vibrations of radiation of the cesium atom in the microwave radar region.*

(c) Multiples of 1 second are

$$1 \text{ minute} = 60 \text{ seconds}$$

$$1 \text{ hour} = 60 \text{ minutes} = 3{,}600 \text{ seconds}$$

$$1 \text{ day} = 24 \text{ hours} = 1{,}440 \text{ minutes} = 86{,}400 \text{ seconds}$$

where the abbreviations for hour and minute are h and min, respectively.

(d) Mean solar day is the time interval between two successive noons averaged over a long period of years. In terms of the mean solar day,

$$1 \text{ year} = 365.242\,198\,79 \text{ days}$$

where the abbreviations for year and day are a (for *annus*) and d (for *dies*), respectively.

(e) FPS unit of time is also 1 second, designated as sec, and is exactly equal to 1 second of the SI system.

(3) Rectilinear Motion, General Case

(a) Instantaneous position i of a particle moving along the X axis in Fig. 3.01–1 is the distance $s = x$ (also called the *linear displacement*) traveled by this particle in the time interval t.

(b) Time-displacement diagram of Fig. 3.01–2 gives the graphical relation between the time of travel t and the displacement x.

Fig. 3.01–1

---

*The older and less exact definition of 1 second (ephemeris second) is 1/31 556 925.974 7 of the tropical year of 1900, January, 0 day, and 12 hours ephemeris time.

(c) Average velocity $\bar{v}$ of this particle during the time interval Δt is the ratio of the distance Δx traveled during this time interval to Δt.

$$\bar{v} = \frac{\text{distance traveled}}{\text{time interval}} = \frac{\Delta x}{\Delta t} = \tan \bar{\alpha} \qquad \{\bar{v} \text{ in m/s}\}$$

where $\bar{\alpha}$ = angle shown in Fig. 3.01–2.

example:

If a particle travels a straight-line path of length 100 m in 10 s, then its average velocity $\bar{v}$ = 100/10 = 10 m/s.

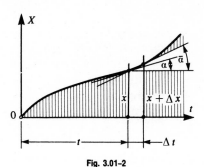

Fig. 3.01–2

(d) Instantaneous velocity v of this particle is the time rate of change in its position at a given instant t.

$$v = \frac{dx}{dt} = \dot{x} = \tan \alpha \qquad \{v \text{ in m/s}\}$$

where dx/dt = first derivative of x with respect to time t and α = angle between the tangent to the time-displacement curve and a line parallel to the t axis in Fig. 3.01–2. If Δt is very small, $v \cong \bar{v}$ and $\alpha \cong \bar{\alpha}$.

(e) Velocity diagram of Fig. 3.01–3 gives the graphical relation between velocity v and time of travel t. In this diagram each vertical coordinate is the slope of the time-displacement curve in Fig. 3.01–2.

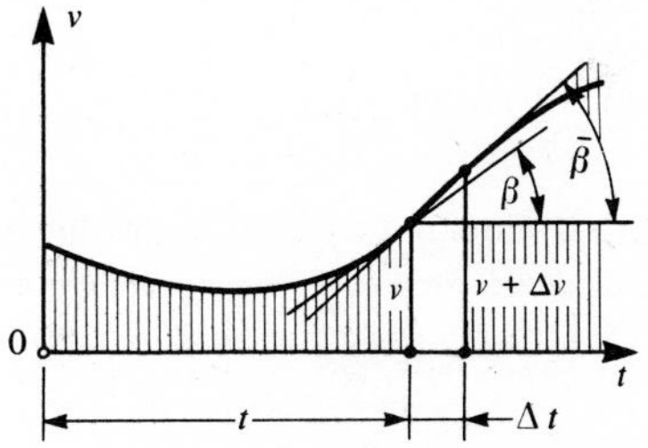

(f) Unit of velocity in the SI system is 1 meter per second, designated as m/s; in the FPS system it is 1 foot per second, designated as ft/sec.

(g) Average acceleration $\bar{a}$ of the particle during the time interval Δt is the change in velocity Δv divided by Δt.

$$\bar{a} = \frac{\text{change in velocity}}{\text{time interval}} = \frac{\Delta v}{\Delta t} = \tan \bar{\beta} \qquad \{\bar{a} \text{ in m/s}^2\}$$

where $\bar{\beta}$ = angle shown in Fig. 3.01–3.

Fig. 3.01–3

example:

If a particle moves on a straight-line path and its velocity at t_1 = 10 s is 50 m/s and at t_2 = 20 s is 100 m/s, then its average acceleration $\bar{a}$ = $\frac{100 - 50}{20 - 10}$ = 5 m/s^2.

(h) **Instantaneous acceleration** a of this particle is the time rate of change in its velocity at a given instant t.

$$a = \frac{dv}{dt} = \frac{d^2x}{dt^2} = \ddot{x} = \tan\beta \qquad \{a \text{ in m/s}^2\}$$

where dv/dt = first derivative of v with respect to time t, d^2x/dt^2 = second derivative of x with respect to time t, and β = angle between the tangent to the time-velocity curve and a line parallel to the t axis in Fig. 3.01–3. If Δt is very small, $a \cong \bar{a}$ and $\beta \cong \bar{\beta}$.

(i) **Unit of acceleration** in the SI system is 1 meter per second per second, designated as m/s², and in the FPS system is 1 foot per second per second, designated as ft/sec².

(4) Rectilinear Motion, Particular Cases

(a) **Classification.** Rectilinear motions of a particle are classified as

(α) *Uniform motion* (v = constant, $a = 0$)
(β) *Uniformly accelerated motion* (v = variable, a = constant)
(γ) *Nonuniformly accelerated motion* (v = variable, a = variable)

(b) **Notation.** Equations of motion for four types of rectilinear motion are given in (c) to (f), where s = linear position coordinate, v = linear velocity, and a = linear acceleration, all at the instant t, all in the XY plane; s_0, v_0, and a_0 are their counterparts at $t = 0$, called the *initial conditions*, some or all of which may be equal to zero.

(c) **Uniform motion** ($v = v_0$, $a = 0$).

$s = s_0 + v_0 t$	$v = v_0 = \dfrac{s - s_0}{t}$	$a = 0$
$t = \dfrac{s - s_0}{v_0}$	$s_0 = s - v_0 t$	$v_0 = v$

The time-velocity diagram of this motion is shown in Fig. 3.01–4.

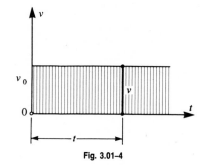

Fig. 3.01–4

(d) **Uniformly accelerated motion** ($v_0 = 0$, $a = a_0$).

$s = s_0 + \dfrac{a_0 t^2}{2}$	$v = a_0 t$	$a = a_0 = \dfrac{v}{t}$
$\quad = s_0 + \dfrac{vt}{2}$	$\quad = \dfrac{2(s - s_0)}{t}$	$\quad = \dfrac{2(s - s_0)}{t^2}$
$\quad = s_0 + \dfrac{v^2}{2a_0}$	$\quad = \sqrt{2a_0(s - s_0)}$	$\quad = \dfrac{v^2}{2(s - s_0)}$
$t = \dfrac{v}{a_0}$	$s_0 = s - \dfrac{a_0 t^2}{2}$	$v_0 = 0$
$\quad = \dfrac{2(s - s_0)}{v}$	$\quad = s - \dfrac{vt}{2}$	
$\quad = \sqrt{\dfrac{2(s - s_0)}{a_0}}$	$\quad = s - \dfrac{v^2}{2a_0}$	

The time-velocity diagram of this motion is shown in Fig. 3.01–5.

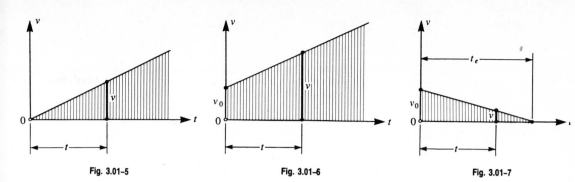

| Fig. 3.01–5 | Fig. 3.01–6 | Fig. 3.01–7 |

(e) Uniformly accelerated motion ($v_0 \neq 0$, $a = a_0$).

$s = s_0 + v_0 t + \dfrac{a_0 t^2}{2}$	$v = v_0 + a_0 t$	$a = a_0 = \dfrac{v - v_0}{t}$
$= s_0 + \dfrac{(v_0 + v)t}{2}$	$= \dfrac{2(s - s_0)}{t} - v_0$	$= \dfrac{2(s - s_0 - v_0 t)}{t^2}$
$= s_0 + \dfrac{v^2 - v_0^2}{2a_0}$	$= \sqrt{v_0^2 + 2a_0(s - s_0)}$	$= \dfrac{v^2 - v_0^2}{2(s - s_0)}$
$t = \dfrac{v - v_0}{a_0}$	$s_0 = s - \dfrac{(v_0 + v)t}{2}$	$v_0 = v - a_0 t$
$= \dfrac{2(s - s_0)}{v + v_0}$	$= s - \dfrac{(2v_0 + a_0 t)t}{2}$	$= \dfrac{2(s - s_0)}{t} - v$
$= \dfrac{-v_0 + \sqrt{v_0^2 + 2a_0(s - s_0)}}{a_0}$	$= s - v_0 t - \dfrac{a_0 t^2}{2}$	$= \sqrt{v^2 - 2a_0(s - s_0)}$

The time-velocity diagram is shown in Fig. 3.01–6.

(f) Uniformly decelerated motion ($v_0 \neq 0$, $a = a_0$).

$s = s_0 + v_0 t - \dfrac{a_0 t^2}{2}$	$v = v_0 - a_0 t$	$a = a_0 = \dfrac{v_0 - v}{t}$
$= s_0 + \dfrac{(v_0 + v)t}{2}$	$= \dfrac{2(s - s_0)}{t} - v_0$	$= \dfrac{2(s_0 + v_0 t - s)}{t^2}$
$= s_0 + \dfrac{v_0^2 - v^2}{2a_0}$	$= \sqrt{v_0^2 - 2a_0(s - s_0)}$	$= \dfrac{v_0^2 - v^2}{2(s - s_0)}$
$t = \dfrac{v_0 - v}{a_0}$	$s_0 = s - \dfrac{(v_0 + v)t}{2}$	$v_0 = v + a_0 t$
$= \dfrac{2(s - s_0)}{v + v_0}$	$= s - \dfrac{(2v_0 - a_0 t)t}{2}$	$= \dfrac{2(s - s_0)}{t} - v$
$= \dfrac{v_0 - \sqrt{v_0^2 - 2a_0(s - s_0)}}{a_0}$	$= s - v_0 t + \dfrac{a_0 t^2}{2}$	$= \sqrt{v^2 + 2a_0(s - s_0)}$

The particle stops at $t_e = v_0/a_0$, where $v_e = 0$, $s_e = s_0 + v_0^2/2a_0$. In these expressions, $t_e =$ terminal time and $s_e =$ terminal displacements. The time-displacement diagram of this motion is shown in Fig. 3.01–7.

example:

If a particle travels on a straight path with $v_0 = 30$ m/s and with constant deceleration $a_0 = 2$ m/s², then at $t = 10$ s,

$$s = s_0 + v_0 t - \frac{a_0 t^2}{2} = 0 + 30 \times 10 - \frac{2 \times 10^2}{2} = 200 \text{ m}$$

$$v = v_0 - a_0 t = 30 - 2 \times 10 = 10 \text{ m/s}$$

and it stops at

$$t_e = \frac{v_0}{a_0} = \frac{30}{2} = 15 \text{ s} \qquad s_e = s_0 + \frac{v_0^2}{2a_0} = 0 + \frac{30^2}{2 \times 2} = 225 \text{ m}$$

(5) Circular Motion, General Case

(a) Instantaneous position i of a particle moving along a circle of radius R in the XY plane (Fig. 3.01–8) is given by the position coordinates as

$$x = R \cos \theta \qquad y = R \sin \theta \qquad \begin{Bmatrix} x, y, R \text{ in m} \\ \theta \text{ in rad} \end{Bmatrix}$$

where the position angle $\theta = \theta(t) =$ function of time t.

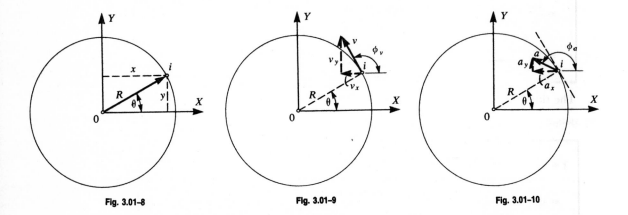

Fig. 3.01–8   Fig. 3.01–9   Fig. 3.01–10

(b) Instantaneous velocity v of this particle is a vector tangential to the circular path at i (Fig. 3.01–9) acting in the direction of motion. The components of v are

$$v_x = \dot{x} = -R\dot{\theta} \sin \theta = -y\omega \qquad \begin{Bmatrix} v_x, v_y \text{ in m/s} \\ \dot{\theta} = \omega \text{ in rad/s} \end{Bmatrix}$$

$$v_y = \dot{y} = R\dot{\theta} \cos \theta = x\omega$$

where $\dot{x} = dx/dt$ and $\dot{y} = dy/dt =$ linear velocities along the X and Y axes, respectively, and $\dot{\theta} = d\theta/dt = \omega =$ angular velocity. The SI and FPS unit of the angular velocity is 1 radian/second.

(c) Magnitude and direction of v are given by

$$v = \sqrt{v_x^2 + v_y^2} = R\omega \qquad \tan \phi_v = \frac{v_y}{v_x} = \frac{\dot{y}}{\dot{x}} = -\frac{x}{y}$$

which shows that the velocity vector is always normal to the radius of the circle.

(d) Instantaneous acceleration a of this particle is a vector deviating from the tangent line at i (Fig. 3.01–10). The components of a are

$$a_x = \ddot{x} = -R\ddot{\theta}\sin\theta - R\dot{\theta}^2\cos\theta = -y\alpha - x\omega^3$$

$$a_y = \ddot{y} = R\ddot{\theta}\cos\theta - R\dot{\theta}^2\sin\theta = x\alpha - y\omega^2$$

$$\left\{\begin{array}{l} a_x, a_y \text{ in m/s}^2 \\ \dot{\theta} = \omega \text{ in rad/s} \\ \ddot{\theta} = \alpha \text{ in rad/s}^2 \end{array}\right\}$$

where $\ddot{x} = d^2x/dt^2$ and $\ddot{y} = d^2y/dt^2$ = linear accelerations along the X and Y axes, respectively, and $\ddot{\theta} = d^2\theta/dt^2 = d\omega/dt = \alpha$ = angular acceleration. The SI and FPS unit of angular acceleration is 1 radian/second^2.

(e) Magnitude and direction of a are given by

$$a = \sqrt{a_x^2 + a_y^2} = R\sqrt{\alpha^2 + \omega^4} \qquad \tan\phi_a = \frac{a_y}{a_x} = \frac{\ddot{y}}{\ddot{x}} = \frac{y\omega^2 - x\alpha}{x\omega^2 + y\alpha}$$

which shows that the velocity vector and the acceleration vector are not collinear.

(6) Circular Motion, Particular Cases

(a) Classification. Circular motions of a particle are classified as

(α) *Uniform motion* (ω = constant, α = 0).
(β) *Uniformly accelerated motion* (ω = variable, α = constant)
(γ) *Nonuniformly accelerated motion* (ω = variable, α = variable)

(b) Notation. Analytical relations for two types of circular motion are given in (c) and (d), where

x, y = linear position coordinates
v_x, v_y = components of linear velocity
a_x, a_y = components of linear acceleration
R = radius of circular path

θ = position angle
ω = angular velocity
α = angular acceleration
$s = R\theta$ = arc of circular path

all at the instant t in the XY plane, and $s_0, x_0, y_0, v_{0x}, v_{0y}, a_{0x}, a_{0y}, \theta_0, \omega_0, \alpha_0$ are their counterparts at $t = 0$, called the *initial conditions*.

(c) Uniform motion ($\theta_0 \neq 0$).

$x = R\cos(\theta_0 + \omega_0 t)$	$y = R\sin(\theta_0 + \omega_0 t)$
$v_x = -R\omega_0\sin(\theta_0 + \omega_0 t) = -y\omega_0$	$v_y = R\omega_0\cos(\theta_0 + \omega_0 t) = x\omega_0$
$a_x = -R\omega_0^2\cos(\theta_0 + \omega_0 t) = -x\omega_0^2$	$a_y = -R\omega_0^2\sin(\theta_0 + \omega_0 t) = -y\omega_0^2$

from which

$$\theta = \theta_0 + \omega_0 t \qquad \omega = \omega_0 \qquad \alpha = 0$$

$$s = R(\theta_0 + \omega_0 t) \qquad v = R\omega_0 \qquad a = \frac{v_0^2}{R}$$

and

$$\tan\phi_v = -\cot\theta \qquad \tan\phi_a = \tan\theta$$

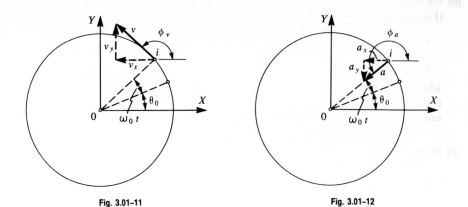

| Fig. 3.01–11 | Fig. 3.01–12 |

These relations show that the velocity vector v is always tangential to the circle (Fig. 3.01–11) and is of constant magnitude $R\omega_0$, and that the acceleration vector a is always directed toward the center of the circle (centripetal acceleration) and is of constant magnitude v^2/R (Fig. 3.01–12).

(d) Uniformly accelerated motion ($\theta_0 \neq 0$, $\omega_0 \neq 0$).

$$x = R \cos \left(\theta_0 + \omega_0 t + \frac{\alpha_0 t^2}{2} \right) \qquad y = R \sin \left(\theta_0 + \omega_0 t + \frac{\alpha_0 t^2}{2} \right)$$

$$v_x = -R(\omega_0 + \alpha_0 t) \sin \left(\theta_0 + \omega_0 t + \frac{\alpha_0 t^2}{2} \right)$$

$$v_y = R(\omega_0 + \alpha_0 t) \cos \left(\theta_0 + \omega_0 t + \frac{\alpha_0 t^2}{2} \right)$$

$$a_x = -R\alpha_0 \sin \left(\theta_0 + \omega_0 t + \frac{\alpha_0 t^2}{2} \right) - R(\omega_0 + \alpha_0 t)^2 \cos \left(\theta_0 + \omega_0 t + \frac{\alpha_0 t^2}{2} \right)$$

$$a_y = R\alpha_0 \cos \left(\theta_0 + \omega_0 t + \frac{\alpha_0 t^2}{2} \right) - R(\omega_0 + \alpha_0 t)^2 \sin \left(\theta_0 + \omega_0 t + \frac{\alpha_0 t^2}{2} \right)$$

from which

$$\theta = \theta_0 + \omega_0 t + \frac{\alpha_0 t^2}{2} \qquad \omega = \omega_0 + \alpha_0 t \qquad \alpha = \alpha_0$$

$$s = R\theta \qquad v = R\omega \qquad a = R\sqrt{\alpha_0^2 + \omega^4}$$

and

$$\tan \phi_v = -\cot \theta \qquad \tan \phi_a = \frac{\alpha_0 \cos \theta - \omega^2 \sin \theta}{-\alpha_0 \sin \theta - \omega^2 \cos \theta} \qquad \tan \psi_a = \frac{\omega^2}{\alpha_0}$$

where ϕ_v, ϕ_a, ψ_a = angles shown in Figs. 3.01–13 to 3.01–15, respectively.

These relations show that the velocity vector v is again tangential to the circle (Fig. 3.01–13) and its magnitude increases linearly with time, and that the acceleration vector is directed inward (Fig. 3.01–14) and its magnitude is increasing also with time. As shown in Fig. 3.01–15 the acceleration vector can also be resolved into a constant tangential component $a_t = R\alpha_0$ and a normal component $a_n = R\omega^2 = R(\omega_0 + \alpha_0 t)^2$, which is a quadratic function of time.

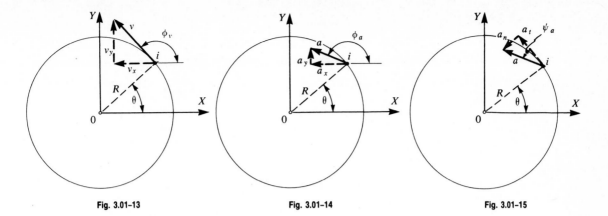

Fig. 3.01–13   Fig. 3.01–14   Fig. 3.01–15

(7) General Plane Motion

(a) Cartesian coordinates. If the position of a moving particle i in the XY plane is given by the cartesian coordinates $x = x(t)$, $y = y(t)$, then

$$v_x = \dot{x} \qquad v_y = \dot{y} \qquad v = \sqrt{v_x^2 + v_y^2} \qquad \tan \phi_v = \frac{v_y}{v_x}$$

$$a_x = \ddot{x} \qquad a_y = \ddot{y} \qquad a = \sqrt{a_x^2 + a_y^2} \qquad \tan \phi_a = \frac{a_y}{a_x}$$

where v_x, v_y = components of velocity v (m/s); a_x, a_y = components of acceleration a (m/s^2); and ϕ_v, ϕ_a = position angles of the velocity and acceleration vectors, respectively. As shown in Fig. 3.01–16, the velocity vector is again tangential to the path, whereas the acceleration vector deviates from the path as indicated in Fig. 3.01–17. The direction functions of the velocity and acceleration are, respectively,

$$\cos \phi_v = \frac{v_x}{v} \qquad \sin \phi_v = \frac{v_y}{v}$$

$$\cos \phi_a = \frac{a_x}{a} \qquad \sin \phi_a = \frac{a_y}{a}$$

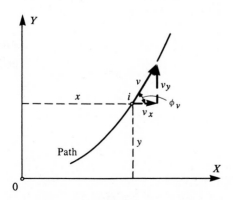

Fig. 3.01–16

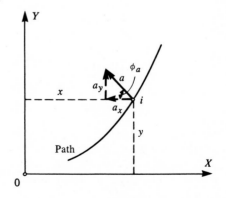

Fig. 3.01–17

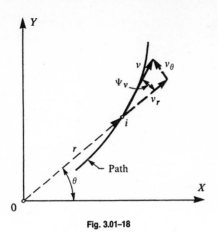

Fig. 3.01–18

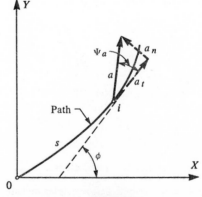

Fig. 3.01–19

(b) Polar coordinates. If the position of the same particle is given by the polar coordinates $r = r(t)$, $\theta = \theta(t)$, as shown in Fig. 3.01–18, then

$$v_r = \dot{r} \qquad\qquad v_\theta = r\dot{\theta} \qquad\qquad v = \sqrt{v_r^2 + v_\theta^2} \qquad\qquad \tan \psi_v = \frac{v_\theta}{v_r}$$

$$a_r = \ddot{r} - r\dot{\theta}^2 \qquad a_\theta = r\ddot{\theta} + 2\dot{r}\dot{\theta} \qquad a = \sqrt{a_r^2 + a_\theta^2} \qquad \tan \psi_a = \frac{a_\theta}{a_r}$$

where $\dot{r} = dr/dt$, $\ddot{r} = d^2r/d^2t$, $\dot{\theta} = d\theta/dt$, and $\ddot{\theta} = d^2\theta/d^2t$ are the radial and angular velocity and acceleration, respectively; v_r, v_θ = radial and transverse components of velocity;
a_r, a_θ = radial and transverse components of acceleration; and ψ_r, ψ_θ = position angles
shown in Figs. 3.01–18 and 3.01–19.

example:

If a particle travels on a circular path of $r = R$ and $\theta = \omega t$, where ω is constant; then $v_r = 0$, $v_\theta = R\omega$,
$a_r = -R\omega^2$, $a_\theta = 0$, $\psi_v = 90°$, and $\psi_a = -180°$. This example shows the advantage of the polar
coordinates (compare the solution of Sec. 3.01–6c).

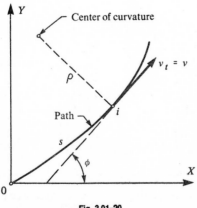

Fig. 3.01–20

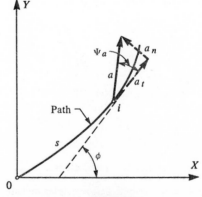

Fig. 3.01–21

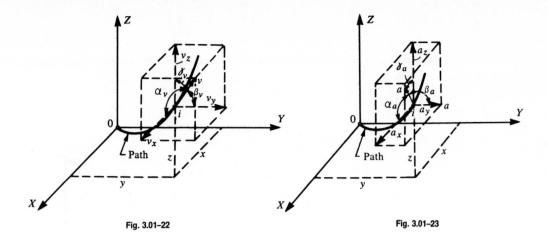

Fig. 3.01–22   Fig. 3.01–23

(c) Curvilinear coordinates. Finally, the same motion in terms of the curvilinear coordinate $s = s(t)$, which is measured along the path (Fig. 3.01–20), is defined by

$$v_t = \dot{s} \qquad v_n = 0 \qquad v = v_t \qquad\qquad \tan \psi_v = 0$$

$$a_t = \ddot{s} \qquad a_n = \frac{\dot{s}^2}{\rho} \qquad a = \sqrt{a_t^2 + a_n^2} \qquad \tan \psi_a = \frac{a_n}{a_t}$$

where $\dot{s} = ds/dt$, $\ddot{s} = d^2s/d^2t$ are the curvilinear velocity and acceleration; v_t, v_n = tangential and normal components of velocity; a_t, a_n = tangential and normal components of acceleration; ρ = radius of curvature of the path; and ψ_v, ψ_a = position angles defined in Figs. 3.01–20 and 3.01–21.

(8) General Space Motion

(a) Cartesian coordinates. The space motion of a particle i in the cartesian coordinates $x = x(t)$, $y = y(t)$, $z = z(t)$ is defined (Fig. 3.01–22) by

$$v_x = \dot{x} \qquad v_y = \dot{y} \qquad v_z = \dot{z} \qquad v = \sqrt{v_x^2 + v_y^2 + v_z^2}$$

$$a_x = \ddot{x} \qquad a_y = \ddot{y} \qquad a_z = \ddot{z} \qquad a = \sqrt{a_x^2 + a_y^2 + a_z^2}$$

where $\dot{x} = dx/dt$, $\ddot{x} = d^2x/d^2t$, . . . , are the derivatives of the position coordinates, which in turn are the velocity components and the acceleration components v_x, v_y, v_z and a_x, a_y, a_z, parallel to the respective coordinate axis. The direction cosines of the velocity vector and of the acceleration vector are

$$\cos \alpha_v = v_x/v \qquad \cos \beta_v = v_y/v \qquad \cos \gamma_v = v_z/v$$

$$\cos \alpha_a = a_x/a \qquad \cos \beta_a = a_y/a \qquad \cos \gamma_a = a_z/a$$

where α, β, γ = direction angles measured from the respective axis to the vector as shown in Figs. 3.01–22 and 3.01–23.

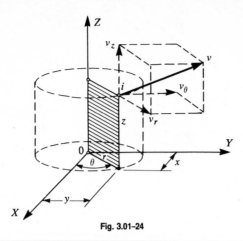

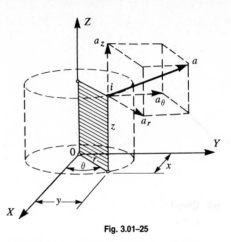

Fig. 3.01–24   Fig. 3.01–25

(b) Cylindrical coordinates. If the position of the same particle i is given by the cylindrical coordinates of Sec. 2.01–4c as $r = r(t)$, $\theta = \theta(t)$, $z = z(t)$, then the velocity and the acceleration (Figs. 3.01–24 and 3.01–25) become

$$v_r = \dot{r} \qquad v = r\dot{\theta} \qquad v_z = \dot{z} \qquad v = \sqrt{v_r{}^2 + v_\theta{}^2 + v_z{}^2}$$

$$a_r = \ddot{r} - r\dot{\theta}^2 \qquad a_\theta = r\ddot{\theta} + 2\dot{r}\dot{\theta} \qquad a_z = \ddot{z} \qquad a = \sqrt{a_r{}^2 + a_\theta{}^2 + a_z{}^2}$$

where $\dot{r} = dr/dt$, $\ddot{r} = d^2r/d^2t, \ldots$, are the derivatives of the position coordinates, which in turn form the components of the velocity v_r, v_θ, v_z and of the acceleration a_r, a_θ, a_z. The relationships of the cylindrical components of v to their cartesian counterparts and vice versa are

$$v_r = v_x \cos\theta + v_y \sin\theta \qquad\qquad v_x = v_r \cos\theta - v_\theta \sin\theta$$

$$v_\theta = -v_x \sin\theta + v_y \cos\theta \qquad\qquad v_y = v_r \sin\theta + v_\theta \cos\theta$$

$$v_z = v_z \qquad\qquad\qquad\qquad\qquad v_z = v_z$$

and identical relationships hold for the components of acceleration (Sec. 2.01–4e).

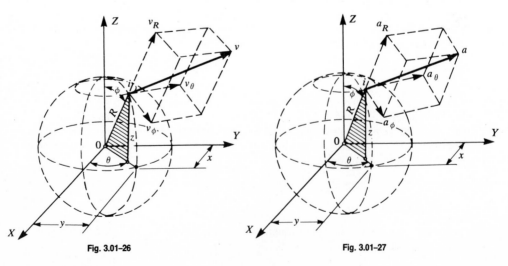

Fig. 3.01–26   Fig. 3.01–27

(c) Spherical coordinates. In the spherical coordinates $R = R(t)$, $\theta = \theta(t)$, $\phi = \phi(t)$ of Sec. 2.01–4d the motion of the same particle i (Figs. 3.01–26 and 3.01–27) is defined by

$$v_R = \dot{R} \qquad\qquad a_R = \ddot{R} - \dot{R}^2 \sin^2 \phi - R\dot{\phi}^2$$

$$v_\theta = R\dot{\theta} \sin \phi \qquad\qquad a_\theta = 2\dot{R}\dot{\theta} \sin \phi + R\ddot{\theta} \sin \phi + 2R\dot{\theta}\dot{\phi} \cos \phi$$

$$v_\phi = R\dot{\phi} \qquad\qquad a_\phi = 2\dot{R}\dot{\phi} + R\ddot{\phi} - R\dot{\theta}^2 \sin \phi \cos \phi$$

where the single and double overdots have the same meaning as in (b) above, v_R, v_θ, v_ϕ = spherical components of the velocity v, and a_R, a_θ, a_ϕ = spherical components of the acceleration a. The relationships of the spherical components of v to their counterparts in the cartesian system and vice versa are

$$v_R = v_x \cos \theta \sin \phi - v_y \sin \theta \sin \phi + v_z \cos \phi$$

$$v_\theta = -v_x \sin \theta + v_y \cos \theta$$

$$v_\phi = v_x \cos \theta \cos \phi + v_y \sin \theta \cos \phi - v_z \sin \phi$$

$$v_x = v_R \cos \theta \sin \phi - v_\theta \sin \theta + v_\phi \cos \theta \cos \phi$$

$$v_y = v_R \sin \theta \sin \phi + v_\theta \cos \theta + v_\phi \sin \theta \cos \phi$$

$$v_z = v_R \cos \phi - v_\phi \sin \phi$$

and identical relationships hold for the components of a.

example:

If a particle travels along a circular helix (Fig. 3.01–28) of a constant radius R and of a constant pitch $2\pi h$ with a constant angular velocity $\dot{\theta} = \omega$, then, in the cartesian coordinates,

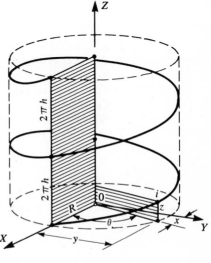

Fig. 3.01–28

$x = R \cos \omega t$	$y = R \sin \omega t$	$z = h\omega t$
$v_x = -R\omega \sin \omega t$	$v_y = R\omega \cos \omega t$	$v_z = h\omega$
$a_x = -R\omega^2 \cos \omega t$	$a_y = -R\omega^2 \sin \omega t$	$a_z = 0$

and in the cylindrical coordinates,

$r = R$	$\theta = \omega t$	$z = h\omega t$
$v_r = 0$	$v_\theta = c\omega$	$v_z = h\omega$
$a_r = -R\omega^2$	$a_\theta = 0$	$a_z = 0$

where $c = \sqrt{R^2 + h^2}$. This comparison shows again the advantage of the introduction of special coordinates.

If in a particular case $2\pi h = 10$ m, $R = 50$ m, $\omega = \pi/100$ rad/s, then $h = 1.592$ m, $c = 50.025$ m, and at a given time $t = 100$ s,

$r = 50$ m	$\theta = 3.142$ rad	$z = 5.000$ m
$v_r = 0$	$v_\theta = 1.572$ m/s	$v_z = 0.050$ m/s
$a_r = -0.049$ m/s^2	$a_\theta = 0$	$a_z = 0$

3.02 KINETICS OF A PARTICLE

(1) Newton's Fundamental Laws

(a) **Kinetics of a particle**, defined as the systematic study of the causes producing, maintaining, and resisting motion of a particle, is based on three laws called *Newton's fundamental laws of mechanics*.

(b) **Newton's first law.** A particle (body) remains at rest or in a uniform rectilinear (straight-line) motion (Sec. 3.01–4) unless compelled by unbalanced causes (forces and/or moments) to change this state.

(c) **Newton's second law.** The acceleration (Sec. 3.01–3) of a particle (of a body) acted upon by an unbalanced cause (force or moment) is proportional to the cause and has its direction.

$$F_x = \frac{d(mv_x)}{dt} = ma_x \qquad F_y = \frac{d(mv_y)}{dt} = ma_y \qquad F_z = \frac{d(mv_z)}{dt} = ma_z$$

where the mass m is assumed to be constant, v_x, v_y, v_z = components of the velocity, a_x, a_y, a_z = components of the acceleration, and F_x, F_y, F_z = components of the resultant of all forces acting on the particle.

(d) **Newton's third law.** The forces of action and of reaction at the point of contact of two particles (two bodies) are equal in magnitude and opposite in direction.

(e) **Condition.** The validity of these laws is based on the condition that all quantities are related to a fixed frame of reference, which is at rest or in uniform motion.

(f) **Principles of static equilibrium** are the consequence of Newton's second law, since when

$$a_x = 0 \qquad a_y = 0 \qquad a_z = 0$$

then

$$F_x = 0 \qquad F_y = 0 \qquad F_z = 0$$

$$M_x = 0 \qquad M_y = 0 \qquad M_z = 0$$

and the particle (body) is in a state of static equilibrium (at rest or in uniform motion).

(2) Newton's Law of Gravitation

(a) **Attraction of two particles.** Every particle of mass m_1 in the universe attracts every other particle of mass m_2 with a gravitational force F, which is directly proportional to the product of their masses, is inversely proportional to the square of the distance between them, and acts along the straight line joining them.

(b) **Attraction of two bodies.** Since a body is a system of particles and the forces of attraction between these particles within the body are in equilibrium, the gravitational force F of two bodies of masses m_1 and m_2 is also directly proportional to the product of their masses, is inversely proportional to the square of the distance between their mass centers, and acts along the straight line joining these centers (Fig. 3.02–1).

(c) **Gravitational force** F is, analytically,

$$F = G \frac{m_1 \cdot m_2}{r^2} \qquad \left\{ \begin{array}{l} m_1, m_2 \text{ in kg} \\ r \text{ in m} \\ G \text{ in } (N-m^2)/kg^2 \end{array} \right\}$$

where r = distance between the particles (bodies) and G = universal constant of gravitation (Appendix A.01).

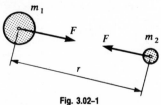

Fig. 3.02–1

(d) Weight of a body W is the attraction of the earth on a body at its surface, defined analytically as

$$W = G\frac{Mm}{R^2} = mg \qquad \left\{\begin{array}{l} m \text{ in kg} \\ g \text{ in m/s}^2 \end{array}\right\}$$

where M = mass of the earth, m = mass of the body, R = radius of earth, and g = acceleration due to gravity.

(e) Acceleration g is an experimental quantity which varies with geographical location. In technology g is assumed to be a constant,

$$g = 9.80665 \text{ m/s}^2 = 32.17405 \text{ ft/sec}^2$$

which is the measured value of g at latitude 45° and sea level (Appendix A.05).

(f) Approximate values of g used in technology are

$$g \cong \frac{51}{52} \times 10 \doteq 9.81 \text{ m/s}^2 \qquad \text{or} \qquad g \cong \frac{222}{69} \times 10 \doteq 32.17 \text{ ft/sec}^2$$

(g) Mass of a body is then determined experimentally as

$$m = \frac{W}{g} \qquad \{m \text{ in kg}\}$$

where W = apparent weight of the body weighed on a calibrated spring balance attached to the surface of the earth.

(3) Units of Mass

(a) SI unit of mass is 1 kilogram, designated by kg, and defined as the mass of a particular cylinder of platinum-iridium alloy, called the *international prototype kilogram*, which is preserved in a vault in Sèvres, France, by the International Bureau of Weights and Measures.

(b) Multiples and fractions of 1 kg used in technology are

$$1 \text{ ton} = 1\,000 \text{ kilograms} \qquad 1 \text{ gram} = 10^{-3} \text{ kilogram} \qquad 1 \text{ carat} = 2 \times 10^{-4} \text{ kilogram}$$

The metric ton is also sometimes spelled tonne.

(c) FPS unit of mass is 1 pound, designated by lb, and defined as the mass of a particular cylinder of platinum alloy kept in the Tower of London and called the *Imperial Standard Pound.*

(d) Multiples and fractions of 1 lb used in technology are

1 ton (short) = 2,000 pounds	1 dram = 1/256 pound
1 ounce = 1/16 pound	1 slug = 32.17 pounds
1 ton (long) = 2,240 pounds	1 grain = 1/7,000 pound

(e) Relations of the SI and FPS units of mass are

1 ton (SI) = 2,205 pounds	1 ton (short) = 907.2 kilograms
1 kilogram = 2.205 pounds	1 ton (long) = 1 016 kilograms
1 gram = 0.0353 ounce	1 pound = 0.453 6 kilogram
1 carat = 0.00706 ounce	1 ounce = 28.35 grams

All FPS units of mass introduced in (c) and (d) above are called the *avoirdupois units.* The *troy (apothecary) units* of mass are listed and defined in Appendix B.

(4) Units of Force

(a) **SI unit of force** is 1 newton (after Isaac Newton), designated by N, and defined as that force which gives to a mass of 1 kilogram an acceleration of 1 meter per second per second.

$$1 \text{ newton} = 1 \text{ N} = 1 \text{ kg-m/s}^2$$

(b) **MKS unit of force** is 1 kilogram-force, designated by kgf, and defined as that force which gives to the mass of one kilogram (Sec. 3.02–3a) an acceleration of 9.806 65 meters per second per second.

$$1 \text{ kilogram-force} = 1 \text{ kgf} = 9.806\,65 \text{ kg-m/s}^2$$

(c) **Multiples and fractions of 1 N and 1 kgf** used in technology are

$$1 \text{ ton-force} = 1\,000 \text{ kgf} = 9\,807 \text{ newtons}$$

$$1 \text{ kilogram-force} = 1 \text{ kgf} = 9.807 \text{ newtons}$$

$$1 \text{ newton} = 1 \text{ N} = 0.102\,0 \text{ kilogram-force}$$

$$1 \text{ dyne} = 1 \text{ dyn} = 10^{-5} \text{N} = 0.001\,0 \text{ gram-force}$$

(d) **FPS unit of force** is 1 poundal, designated by pd, and defined as that force which gives to a mass of 1 pound an acceleration of 1 foot per second per second.

$$1 \text{ poundal} = 1 \text{ pd} = 1 \text{ lb-ft/sec}^2$$

(e) **U.S. customary unit of force** is 1 pound-force, designated by lbf, and defined as that force which gives to the mass of one pound (Sec. 3.02–3c) an acceleration of 32.174 05 feet per second per second.

$$1 \text{ pound-force} = 1 \text{ lbf} = 32.174\,05 \text{ lb-ft/sec}^2$$

(f) **Multiples and fractions of 1 pd and 1 lbf** used in technology are

$$1 \text{ ton-force} = 2,000 \text{ lbf} = 64,348 \text{ poundals}$$

$$1 \text{ pound-force} = 1 \text{ lbf} = 32.17 \text{ poundals}$$

$$1 \text{ poundal} = 1 \text{ pd} = 0.031\,1 \text{ pound-force}$$

$$1 \text{ ounce-force} = 1 \text{ ozf} = 2.011 \text{ poundals}$$

(g) **Relations of the units of force** are

$$1 \text{ newton} = 0.224\,8 \text{ pound-force} \qquad 1 \text{ pound-force} = 4.448 \text{ newtons}$$

$$1 \text{ kilogram-force} = 2.205 \text{ pound-force} \qquad 1 \text{ pound-force} = 0.453\,6 \text{ kilogram-force}$$

(h) **Computation of mass.** If the weight of a body W and the acceleration due to gravity g are given, then the mass of the same body is computed as $m = W/g$.

examples:

If $W = 2\,000$ kgf and $g = 9.81$ m/s^2, then

$$m = \frac{2\,000 \text{ kgf}}{9.81 \text{ m/s}^2} = \frac{2\,000 \text{ kg} \times 9.81 \text{ m/s}^2}{9.81 \text{ m/s}^2} = 2\,000 \text{ kg}$$

If $W = 2,000$ lbf and $g = 32.17$ ft/sec^2, then

$$m = \frac{2,000 \text{ lbf}}{32.17 \text{ ft/sec}^2} = \frac{2,000 \text{ lb} \times 32.17 \text{ ft/sec}^2}{32.17 \text{ ft/sec}^2} = 2,000 \text{ lb}$$

(i) **Computation of weight.** If the mass of a body m and the acceleration due to gravity g are given, then the weight of the same body is computed as $W = mg$.

examples:

If $m = 2\,000$ kg and $g = 9.81$ m/s^2, then

$$W = 2\,000 \text{ kg} \times 9.81 \text{ m/s}^2 = 19\,620 \text{ N} = 2\,000 \text{ kgf}$$

If $m = 2,000$ lb and $g = 32.17$ ft/sec, then

$$W = 2,000 \text{ lb} \times 32.17 \text{ ft/sec}^2 = 64,340 \text{ pd} = 2,000 \text{ lbf}$$

Thus f following kg and lb is equivalent to $g = 9.81$ m/s^2 and $g = 32.17$ ft/sec^2, respectively.

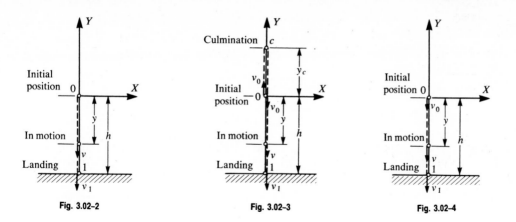

Fig. 3.02–2   Fig. 3.02–3   Fig. 3.02–4

(5) Gravitational Force, Free Vertical Motion

(a) Free fall (Fig. 3.02–2).   Near the surface of the earth the magnitude of the acceleration due to gravity varies only a little with altitude (Appendix A.05) and in ordinary technical calculations is assumed to be $g = 9.81$ m/s^2 (directed vertically downward).   Hence a body dropped from rest at a height h above ground level falls (if the air resistance is neglected) in a rectilinear uniformly accelerated motion defined by

$$y = -\frac{gt^2}{2} \qquad v = -gt \qquad a = -g$$

related to the origin 0.   The motion terminates when the body hits the ground at $t_1 = \sqrt{2h/g}$, with the terminal velocity $v_1 = \sqrt{2hg}$.   As is apparent from these relations, free fall is independent of the volume, shape, and mass of the body (a conclusion which is correct in vacuum only).   In reality, the air resistance is approximately proportional to the projected area of the body on the plane normal to the path of motion and increases with speed.

(b) Projection upward (Fig. 3.02–3).   If the same body is projected upward at the same height h with linear velocity v_0, then it moves first in a rectilinear uniformly decelerated (retarded) motion defined by

$$y = v_0 t - \frac{gt^2}{2} \qquad v = v_0 - gt \qquad a = -g$$

At $t_c = v_0/g$, it reaches the culmination point c at $y_c = v_0^2/2g$, where its velocity $v_c = 0$ and it immediately begins its free fall as described in (a).   As it passes through the point of ejection 0, its velocity equals $-v_0$ (negative of the initial velocity), and the time interval required to fall from c to 0 equals the time required to rise from 0 to c.   Finally, at $t_1 = (v_0 + \sqrt{v_0^2 + 2hg})/g$, the body reaches the ground with the terminal velocity $v_1 = -\sqrt{v_0^2 + 2hg}$.   The time t_1 includes the time required to rise from 0 to c and the time required to fall from c to 1 (total time of motion).

example:

If $h = 100$ m and $v_0 = 20$ m/s, then

$$t_e = \frac{20}{9.81} = 2 \text{ s} \qquad v_e = 0 \qquad y_e = \frac{20^2}{2 \times 9.81} = 20.38 \text{ m}$$

and

$$t_1 = \frac{20 + \sqrt{20^2 + 2 \times 100 \times 9.81}}{9.81} = 7 \text{ s} \qquad v_1 = -\sqrt{20^2 + 2 \times 100 \times 9.81} = -48.6 \text{ m/s}$$

where the minus sign indicates that the body falls in the negative direction of the Y axis.

(c) Projection downward (Fig. 3.02–4). If the same body is projected downward from the same height with the linear velocity v_0, then it moves in a rectilinear uniformly accelerated motion defined by

$$y = v_0 t - g\frac{t^2}{2} \qquad v = v_0 - gt \qquad a = -g$$

and reaches the ground at $t_1 = (v_0 + \sqrt{v_0^2 + 2hg})/g$, with the terminal velocity $v_1 = -\sqrt{v_0^2 + 2hg}$, which is the same as in case (*b*) above. In these equations the numerical value of v_0 is negative.

example:

If $h = 100$ m and $v_0 = -20$ m/s, then

$$t_1 = \frac{-20 + \sqrt{(-20)^2 + 2 \times 100 \times 9.81}}{9.81} = 3 \text{ s} \qquad v_1 = -\sqrt{(-20)^2 + 2 \times 100 \times 9.81} = -48.6 \text{ m/s}$$

(6) Gravitational Force, Motion of Projectile

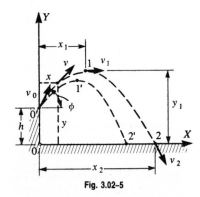

Fig. 3.02–5

(a) Basic relations (Fig. 3.02–5). If a projectile of weight W is fired from the height h above the ground with the initial velocity v_0 at an angle ϕ measured from the horizon, then the kinetic equations of motion (Newton's second law) are

$$F_x = ma_x \qquad F_y = ma_y$$

and in terms of given values,

$$0 = \frac{W}{g}a_x \qquad -W = \frac{W}{g}a_y$$

(b) Equations of motion derived from these relations are

$$a_x = 0 \qquad a_y = -g$$

$$v_x = v_0 \cos \phi \qquad v_y = v_0 \sin \phi - gt$$

$$x = (v_0 \cos \phi)t \qquad y = (v_0 \sin \phi)t - \frac{gt^2}{2} + h$$

and the path of motion (trajectory) is a 2° parabola defined by

$$y = h - \frac{g}{2}\left(\frac{x}{v_0 \cos \phi}\right)^2 + x \tan \phi$$

(c) Vertex. At the instant $t_1 = v_0 \sin \phi / g$ the projectile reaches the vertex of its path (maximum-height–point 1) given by

$$x_1 = \frac{v_0^2 \sin 2\phi}{2g} \qquad y_1 = h + \frac{v_0^2 \sin^2 \phi}{2g}$$

where $\qquad v_{1x} = v_0 \cos \phi \qquad v_{1y} = 0$

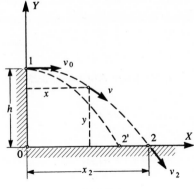

(d) Terminus. At the instant

$$t_2 = \frac{v_0 \sin \phi}{g} + \sqrt{\frac{2h}{g} + \left(\frac{v_0 \sin \phi}{g}\right)^2}$$

the projectile reaches the terminus of its path (landing-point 2) given by

$$x_2 = \frac{v_0^2 \sin 2\phi}{2g} + v_0 \cos \phi \sqrt{\frac{2h}{g} + \left(\frac{v_0 \sin \phi}{g}\right)^2} \qquad y_2 = -h$$

where $\qquad v_{2x} = v_0 \cos \phi \qquad v_{2y} = -\sqrt{2hg + v_0^2 \sin^2 \phi}$

Fig. 3.02-6

(e) Zero angle (Fig. 3.02-6). If $\phi = 0$, then

$$x = v_0 t \qquad y = -\frac{gt^2}{2} + h \qquad v_x = v_0 \qquad v_y = -gt$$

and the trajectory is a parabola with vertex at 1,

$$y = h - \frac{g}{2}\left(\frac{x}{v_0}\right)^2$$

At $t_2 = \sqrt{2h/g}$,

$$x_2 = v_0 \sqrt{\frac{2h}{g}} \qquad y_2 = 0 \qquad v_{2x} = v_0 \qquad v_{2y} = \sqrt{2hg}$$

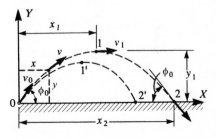

(f) Zero height (Fig. 3.02-7). If $h = 0$, then the equations of motion given in (b) of this section are also valid but the equation of the path reduces to

$$y = x \tan \phi - \frac{g}{2}\left(\frac{x}{v_0 \cos \phi}\right)^2$$

At $t_1 = \dfrac{v_0 \sin \phi}{g}$, $\qquad\qquad$ At $t_2 = \dfrac{2v_0 \sin \phi}{g}$,

$$x_1 = \frac{v_0^2 \sin 2\phi}{2g} \qquad y_1 = \frac{v_0^2 \sin^2 \phi}{2g} \qquad x_2 = \frac{v_0^2 \sin 2\phi}{g} \qquad y_2 = 0$$

$$v_{1x} = v_0 \cos \phi \qquad v_{1y} = 0 \qquad\qquad v_{2x} = v_0 \cos \phi \qquad v_{2y} = -v_0 \sin \phi$$

which shows the symmetry of the motion about the axis of the parabola.

Fig. 3.02-7

example:

If $h = 0$, $v_0 = 100$ m/s, and $\phi = 45°$, then

At $t_1 = \dfrac{100 \times 0.7071}{9.81} = 7.2$ s,

At $t_2 = 2t_1 = 14.4$ s,

$x_1 = 509.7$ m $y_1 = 204.8$ m

$x_2 = 2x_1 = 1{,}019.4$ m $y_1 = 0$

(g) Maximum range (Fig. 3.02–7). If $\phi = 45°$, the length of the shot

$$x_2 = \frac{v_0{}^2}{g}$$

is called the *maximum range*.

(h) Ballistic curves. The heights and ranges of the trajectories shown in Figs. 3.02–5 to 3.02–7 are those for motions in vacuum. In reality they are always reduced by air resistance acting against the motion. The computation of this resistance is very complicated and must be based on experimental measurements. The reduced trajectories are called *ballistic curves*, shown by dashed lines in the respective figures.

(7) Constant Force, Guided Rectilinear Motion

(a) Elevator problem. The force exerted by the body of weight W on the floor of an elevator in Fig. 3.02–8 may be one of the following types.

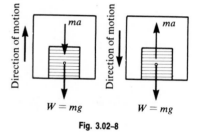

Fig. 3.02–8

CASE I:

If the elevator is at rest or in a uniform motion ($a = 0$, upward or downward), the compressive force $F = W$ (weight of the body).

CASE II:

If the elevator moves with a constant acceleration a, then

$$F = W \pm \frac{W}{g} a = m(g \pm a)$$

where W and a are taken as positive values, and the signs are $(+)$ for upward motion and $(-)$ for downward motion.

example:

If $W = 100$ kgf and $a = 2$ m/s²,

For upward motion: $F = 100 + (100/9.81) \times 2 = 120.4$ kgf

For downward motion: $F = 100 - (100/9.81) \times 2 = 79.6$ kgf

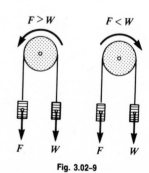

Fig. 3.02–9

(b) Pulley problem. If a rope passing over a pulley is acted upon by a load W on one side, a weight F on the other side, as shown in Fig. 3.02–9, and the friction between the rope and the pulley is disregarded, the acceleration of the rope a and the tension S in the rope may be of the following types.

CASE I:

If $F = W$, the rope is at rest or in uniform motion, $a = 0$, and $S = W$, where W is taken as a positive value.

CASE II:

If $F \neq W$, the rope moves with a constant acceleration

$$a = \frac{(F-W)g}{F+W}$$

in the direction of F if $a > 0$ and in the direction of W if $a < 0$. The tension in the rope in both cases is

$$S = \frac{2FW}{F+W}$$

example:

If $F = 100\,\text{kgf}$ and $W = 200\,\text{kgf}$, then

$$a = \frac{(100-200)\times 9.81}{300} = -3.27\ \text{m/s}^2$$

in the direction of W and $S = 2 \times 100 \times 200/300 = 133.3\,\text{kgf}$.

(8) Constant Force, Guided Circular Motion

(a) **Horizontal plane motion (Fig. 3.02–10).** If the body of weight W tied to the end of a cord whirls in a circle of radius R on a smooth horizontal plane (friction is negligible) with a constant velocity v_0, then the centripetal force acting toward the center of the circle is

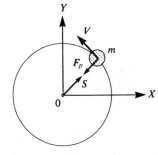

$$F_p = ma_n = \frac{W}{g}\frac{v_0^2}{R} = S \qquad \{S \text{ in kgf}\}$$

and according to Newton's third law equals the centrifugal force acting in the opposite direction. The tension S in the cord also equals F_p.

Fig. 3.02–10

(b) **Period** τ of the circular motion is the time required to complete 1 revolution.

$$\tau = \frac{2\pi R}{v_0} = \frac{2\pi}{\omega_0} \qquad \left\{ \begin{array}{l} \pi \text{ in rad} \\ \omega_0 \text{ in rad/s} \\ \tau \text{ in s} \end{array} \right\}$$

where ω_0 = constant angular speed ($\omega_0 = v_0/R$).

(c) **Frequency** f of the circular motion is the number of revolutions per unit of time.

$$f = \frac{v_0}{2\pi R} = \frac{\omega_0}{2\pi} = \frac{1}{\tau} \qquad \{f \text{ in rev/s}\}$$

and is the reciprocal value of the period τ.

(d) **Centripetal and centrifugal forces** defined in (a) in terms of τ or f are

$$F_p = F_f = \frac{4\pi^2 WR}{g\tau^2} = \frac{4\pi^2 Wf^2 R}{g}$$

example:

If $W = 100\,\text{kgf}$, $R = 10\,\text{m}$, and $v_0 = 20\,\text{m/s}$, then

$$\omega_0 = \frac{20}{10} = 2\,\text{rad/s} \qquad \tau = \frac{2 \times 3.14 \times 10}{20} = 3.14\,\text{s} \qquad f = \frac{1}{3.14} = 0.32\,\text{rev/s}$$

and

$$F_p = \frac{100 \times 20^2}{9.81 \times 10} = 407.7\,\text{kgf} \quad \text{which is also the tension } S \text{ in the cord.}$$

(e) Banking of curves of highways, race tracks, and railbeds is done to reduce the danger that a vehicle may slide off as it goes around the curve. Two forces acting on such a vehicle are its weight W and the reaction of the roadway N (Fig. 3.02–11). The resultant of these two forces must be the centripetal force

$$F_p = ma_n = \frac{Wv_0^2}{gR}$$

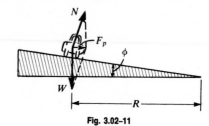

where R = radius of the curve and v_0 = linear velocity. The slope (banking) of the roadbed is then

$$\tan \phi = \frac{F_p}{W} = \frac{v_0^2}{gR}$$

Fig. 3.02–11

example:

If $R = 100\,\text{m}$, $v_0 = 20\,\text{m/s}$, and $g = 9.81\,\text{m/s}^2$, then $\tan \phi = \dfrac{20^2}{9.81 \times 100} = 0.408$ and $\phi = 22°10'$

(f) Simple pendulum (Fig. 3.02–12) consists of a ball of weight W attached to a weightless, inextensible string of length l attached to a fixed hinge 0. If the ball is displaced so that the string makes a certain angle θ_A with the vertical $\overline{C0}$ and is then suddenly released, it will swing (oscillate) indefinitely between the extreme points A and B along the arc of radius l (provided that there is no air resistance). From Fig. 3.02–12,

$$F_n = \frac{W}{g} a_n = -S + W \cos \theta \qquad F_t = \frac{W}{g} a_t = -W \sin \theta$$

where S = tension in the string. From the condition of rigidity of the string, $a_n \cong 0$, and from the geometry of circular motion, $a_t = l(d^2\theta/dt)$. For small θ ($\sin \theta \cong \theta$, $\cos \theta \cong 1$), the angular position θ, the angular velocity ω, and the angular acceleration α are

Fig. 3.02–12

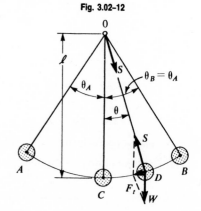

	At A	At C	At D
θ	θ_A	0	$\theta_A \cos\left(\sqrt{\frac{g}{l}}\,t\right)$
ω	0	$\pm\theta_A \sqrt{\frac{g}{l}}$	$\pm\theta_A \sqrt{\frac{g}{l}} \sin\left(\sqrt{\frac{g}{l}}\,t\right)$
α	$\pm\theta_A \frac{g}{l}$	0	$\pm\theta_A \frac{g}{l} \cos\left(\sqrt{\frac{g}{l}}\,t\right)$

The time required for the ball to swing from A to B and return to A is the period

$$\tau = 2\pi \sqrt{\frac{l}{g}}$$

which is independent of W.

example:

If $l = 39.24$ m, $g = 9.81$ m/s², and $\theta_A = 0.05$ rad $\cong 2.86°$, then in terms of $\sqrt{g/l} = 0.5$,

at D: $v_D = l\omega_D = \pm 39.24 \times 0.05 \times 0.5 \times \sin\frac{t}{2} = \pm\left(0.981 \sin\frac{t}{2}\right)$ m/s

$a_D = l\alpha_D = \pm 39.24 \times 0.05 \times 0.25 \times \cos\frac{t}{2} = \pm\left(0.490 \cos\frac{t}{2}\right)$ m/s²

at C: $v_C = l\omega_C = \pm 39.24 \times 0.05 \times 0.5 = \pm 0.981$ m/s

$a_C = l\alpha_C = 0$

and the time required to swing from A to C is

$$\frac{\text{Period}}{4} = \frac{\tau}{4} = \frac{2\pi \times 0.5}{4} = 0.785 \text{ s}$$

(g) Conical pendulum (Fig. 3.02–13) consists of a ball of weight W attached to a weightless, inextensible string of length l attached at a fixed hinge 0 while moving with a constant velocity $v_0 = R\omega_0$ along a circular path, so that the string generates the surface of a cone. Two forces acting on the ball are its weight W and the tension S in the cord. The resultant of these two forces must be the centripetal force

$$F_p = \frac{W}{g}\frac{v_0^2}{R}$$

where R = radius of the circle. The parameters of this motion are

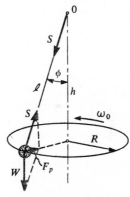

Fig. 3.02–13

Angular velocity $\omega_0 = \dfrac{v_0}{R} = \sqrt{\dfrac{g}{h}} = \sqrt{\dfrac{g}{l\cos\phi}} = \sqrt{\dfrac{g\tan\phi}{R}}$	
Period $\tau = \dfrac{1}{\text{frequency}} = \dfrac{2\pi}{\omega_0} = \dfrac{2\pi R}{v_0} = 2\pi\sqrt{\dfrac{h}{g}} = 2\pi\sqrt{\dfrac{l\cos\phi}{g}}$	
Height $h = \sqrt{l^2 - R^2} = \dfrac{g}{\omega_0^2} = g\left(\dfrac{R}{v_0}\right)^2 = g\left(\dfrac{\tau}{2\pi}\right)^2 = g\left(\dfrac{1}{2\pi f}\right)^2$	
Slope $= \tan\phi = \dfrac{R}{h} = \dfrac{R\omega_0^2}{g} = \dfrac{v_0^2}{gR} = \dfrac{R}{g}\left(\dfrac{2\pi}{\tau}\right)^2 = \dfrac{R}{g}(2\pi f)^2$	

(9) Friction, Horizontal Rectilinear Motion

(a) Acting tensile force (Fig. 3.02–14).

If the block of weight W moves on a horizontal plane under the action of a tensile force P and the coefficient of kinetic friction between the block and the plane is μ_k, then its acceleration, velocity, and displacement are, respectively,

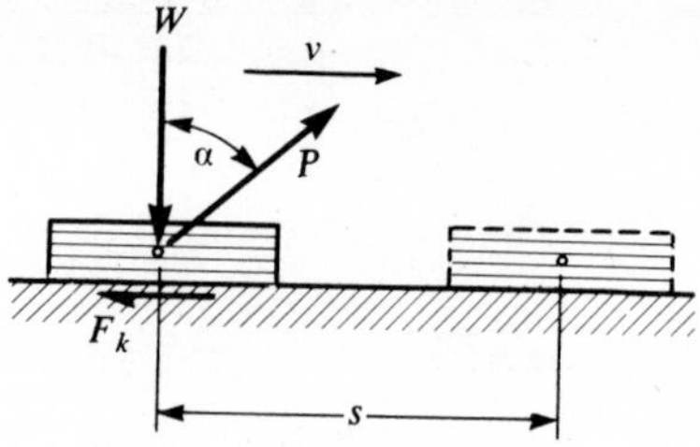

$$a = Cg \qquad v = Cgt \qquad s = \frac{Cgt^2}{2}$$

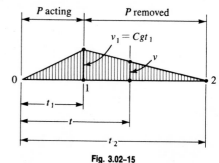

Fig. 3.02–14

where

$$C = \frac{P \sin \alpha - F_k}{W} = \frac{P \sin \alpha - \mu_k(W - P \cos \alpha)}{W}$$

For $\alpha = 90°$, P is parallel to the plane and C reduces to

$$C = \frac{P - F_k}{W} = \frac{P}{W} - \mu_k$$

(b) Removed tensile force.

If the acting tensile force P in case (a) is removed ($P = 0$) at t_1, then from that instant on the motion is decelerated.

$$a = -\mu_k g \qquad v = Cgt_1 - \mu_k g(t - t_1) \qquad s = \frac{Cgt_1^2}{2} + \frac{\mu_k g(t - t_1)^2}{2}$$

and the block stops at

$$t_2 = \frac{(C + \mu_k)t_1}{\mu_k} \qquad s_2 = \frac{(C + \mu_k)Cgt_1^2}{2\mu_k}$$

as interpreted graphically by the time-velocity diagram of Fig. 3.02–15.

example:

If $P = 1{,}000$ kgf, $W = 2{,}000$ kgf, $\mu_k = 0.1$, and $\alpha = 60°$, then

$$a = Cg = \frac{[1{,}000 \times 0.866 - 0.1(2{,}000 - 1{,}000 \times 0.5)]9.81}{2{,}000} = 3.5 \text{ m/s}^2$$

If P is removed at $t_1 = 10$ s and $s_1 = 175$ m, the block stops at

$$t_2 = \frac{(0.358 + 0.1)10}{0.1} = 45.8 \text{ s} \qquad s_2 = \frac{(0.358 + 0.1)0.358 \times 9.81 \times 10^2}{2 \times 0.1} = 804 \text{ m}$$

Fig. 3.02–15

(c) Compressive force (Fig. 3.02–14).

If the block of case (a) is acted upon by a compressive force, then its equations of motion remain formally the same as in (a) but

$$C = \frac{P \sin \alpha - F_k}{W} = \frac{P \sin \alpha - \mu_k(W + P \cos \alpha)}{W}$$

For $\alpha = 90°$, P is parallel to the plane and C is the same as in case (a).

(d) Removed compressive force. If the acting compressive force P in case (c) is removed $(P = 0)$ at t_1, then from that instant on the motion is decelerated and is governed by the equations of case (b).

example:

If $P = 1{,}000\,\text{kgf}$, $W = 2{,}000\,\text{kgf}$, $\mu_k = 0.1$, and $\alpha = 60°$, then

$$a = Cg = \frac{[1{,}000 \times 0.866 - 0.1(2{,}000 + 1{,}000 \times 0.5)]9.81}{2{,}000} = 3\,\text{m/s}^2$$

If P is removed at $t_1 = 10\,\text{s}$ and $s_1 = 150\,\text{m}$, the block stops at

$$t_2 = \frac{(0.308 + 0.1)10}{0.1} = 40.8\,\text{s} \qquad s_2 = \frac{(0.308 + 0.1)0.308 \times 9.81 \times 10^2}{2 \times 0.1} = 616\,\text{m}$$

(10) Friction, Inclined Rectilinear Motion

(a) Acting tensile force (Fig. 3.02–16). When the block of weight W moves on an inclined plane under the action of a tensile force P and the coefficient of kinetic friction between the block and the plane is μ_k, then its acceleration a, velocity v, and displacement s are, respectively,

$$a = Dg \qquad v = Dgt \qquad s = \frac{Dgt^2}{2}$$

where

$$D = \frac{P\sin\alpha - W\sin\beta - F_k}{W} = \frac{P(\sin\alpha + \mu_k \cos\alpha) - W(\sin\beta + \mu_k \cos\beta)}{W}$$

If $\alpha = 90°$, P is parallel to the inclined plane and D reduces to

$$D = \frac{P}{W} - (\sin\beta + \mu_k \cos\beta)$$

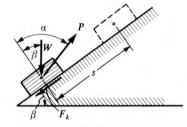

Fig. 3.02–16

(b) Removed tensile force. If the acting tensile force P in case (a) is removed $(P = 0)$ at t_1, then from that instant on the motion is decelerated.

$$a = -Eg \qquad v = Dgt_1 - Eg(t - t_1) \qquad s = \frac{Dgt_1^2}{2} + \frac{\mu_k g(t - t_1)^2}{2}$$

where

$$E = \sin\beta + \mu_k \cos\beta$$

and the block stops at

$$t_2 = \frac{(D + E)t_1}{E} \qquad s_2 = \frac{(D + E)Dgt_1^2}{2E}$$

which can be interpreted graphically by the time-velocity diagram of Fig. 3.02–15.

(c) Gravity motion (Fig. 3.02–17). Once the block stops, one of two possibilities may arise:

    (*i*) If the coefficient of static friction $\mu_s > \tan \beta$, the block remains at rest.

    (*ii*) If $\mu_s < \tan \beta$, the block begins to move under its own weight.

$$a = -Gg \qquad v = -Ggt$$

$$s = -\frac{Ggt^2}{2}$$

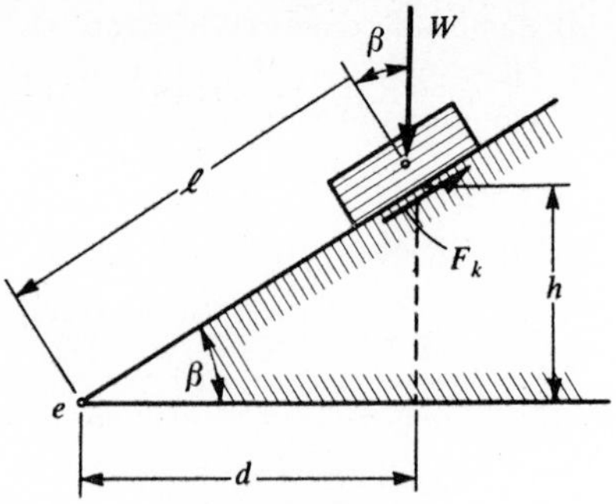

where

$$G = \sin \beta - \mu_k \cos \beta$$

The terminal values of this motion are

$$v_e = -\sqrt{2gh(1 - \mu_k \cot \beta)}$$

$$t_e = \sqrt{\frac{2h}{g \sin^2 \beta \, (1 - \mu_k \cot \beta)}}$$

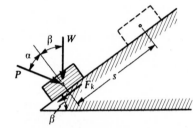

Fig. 3.02–17

(d) Acting compressive force (Fig. 3.02–18). If the block of case (*a*) is acted upon by a compressive force P, then its equations of motion remain formally the same as in (*a*) but

$$D = \frac{P \sin \alpha - W \sin \beta - F_k}{W} = \frac{P(\sin \alpha - \mu_k \cos \alpha) - W(\sin \beta + \mu_k \cos \beta)}{W}$$

Fig. 3.02–18

If $\alpha = 90°$, P is parallel to the inclined plane and D reduces to

$$D = \frac{P}{W} - (\sin \beta + \mu_k \cos \beta)$$

(e) Removed compressive force. If the acting compressive force P in case (*d*) is removed ($P = 0$) at t_1, then from that instant on the motion is decelerated and is governed by the equations of case (*b*).

(f) Projected block (Fig. 3.02–19). If the block of case (*a*) is projected up the inclined plane with an initial velocity v_0 and $P = 0$, then

$$a = -g(\mu \cos \beta - \sin \beta)$$

$$v = v_0 + at$$

and the block stops at

$$t_1 = -\frac{v_0}{a} \qquad s_1 = \frac{v_0^2}{2a}$$

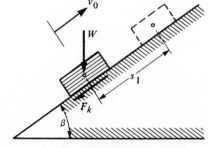

Fig. 3.02–19

(11) Variable Force

(a) **Function of time.** If the particle of mass m is acted upon by a force which is a function of time t, so that $F = f(t)$, then at $t = t_1$ (Fig. 3.02–20)

$$a_1 = \frac{f(t_1)}{m}$$

$$v_1 = v_0 + \frac{1}{m} \int_0^{t_1} f(t)\, dt$$

$$s_1 = s_0 + v_0 t_1 + \frac{1}{m} \int_0^{t_1} \int_0^{t} f(t)\, dt\, dt$$

where $s_0 =$ position coordinate at $t = 0$, $v_0 =$ velocity at $t = 0$, $\int_0^{t_1} f(t)\, dt =$ shaded area in Fig. 3.02–20, $\int_0^{t_1} \int_0^{t} f(t)\, dt\, dt =$ static moment of the same area about the vertical through 1, and the shaded area may or may not be continuous.

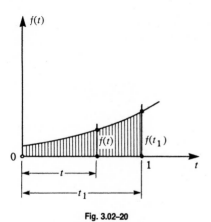

Fig. 3.02–20   Fig. 3.02–21

(b) **Function of linear coordinates.** If the same particle is acted upon by a force which is a function of linear coordinate s, so that $F = f(s)$, then at $s = s_1$ (Fig. 3.02–21)

$$a_1 = \frac{f(s_1)}{m}$$

$$v_1 = \sqrt{v_0^2 + \frac{2}{m} \int_{s_0}^{s_1} f(s)\, ds}$$

$$t_1 = \int_{s_0}^{s_1} \frac{ds}{\sqrt{v_0^2 + \dfrac{2}{m} \displaystyle\int_{s_0}^{s} f(s)\, ds}}$$

where s_0 and v_0 have the same meaning as in (a) above, t_1 is the time required to reach velocity v_1 and distance s_1, and the integrals of $f(s)$ are the areas corresponding to the respective limits of integration.

(c) Function of velocity. If in turn the force acting upon the particle is a function of velocity v, so that $F = f(v)$ (Fig. 3.02–22),

$$a_1 = \frac{f(v_1)}{m}$$

$$t_1 = m \int_{v_0}^{v_1} \frac{dv}{f(v)}$$

$$s_1 = s_0 + m \int_{v_0}^{v_1} \frac{v\, dv}{f(v)}$$

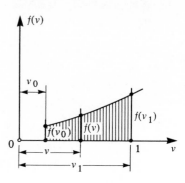

Fig. 3.02–22

where again s_0, v_0 are the initial conditions of the motion and t_1, s_1, v_1 have the same meaning as before. The integrals of these equations of motion may be complicated expressions requiring evaluation by numerical methods (Sec. 1.05–3).

example:

If a body of mass m moves along a straight line with initial velocity v_0 from the initial position given by s_0 in a medium offering resistance linearly proportional to its velocity, so that the retarding force $F = -cv$, where c is a given constant known from experiments (Fig. 3.02–23), then

$$a = -\frac{cv}{m}$$

$$s = s_0 + \frac{mv_0}{c} - \frac{mv}{c}$$

$$t = -\frac{m}{c} \ln \frac{v}{v_0}$$

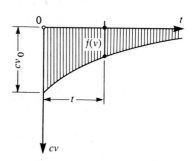

From the last relation

$$v = v_0 E \qquad s = s_0 + \frac{m}{c} v_0 (1 - E)$$

Fig. 3.02–23

where $E = e^{-ct/m}$.

In particular, if $m = 10$ kg, $c = 0.5$ kg/s, $s_0 = 0$, $v_0 = 10$ m/s,

$$a = -0.05 \text{ m/s}^2 \qquad v = 10e^{-0.05t}$$

$$s = 200\,(1 - e^{-0.05t})$$

then at $t_1 = 10$ s, $v_1 = 6.07$ m/s and $s_1 = 78.7$ m. Theoretically, this body will never stop.

(d) Function of angular coordinate. Finally, if the force is a function of the angular coordinate of the path of motion θ and this path is a circle of radius R, so that $F = f(R, \theta)$, then at $\theta = \theta_1$ (Fig. 3.02–24),

$$a_{\theta,1} = R\ddot{\theta}_1 = R\alpha_1 \qquad a_{r,1} = R\dot{\theta}_1^{\,2} = R\omega_1^{\,2}$$

$$\omega_1 = \dot{\theta}_1 = \sqrt{\left(\frac{2g}{R}\right)(1 - \cos\theta_1) + \omega_0^{\,2}}$$

$$v_1 = R\dot{\theta}_1 = \sqrt{\left(\frac{2g}{R}\right)(1 - \cos\theta_1) + v_0^{\,2}}$$

$$t_1 = \int_0^{\theta_1} \frac{d\theta}{\sqrt{\left(\frac{2g}{R}\right)(1 - \cos\theta) + \omega_0^{\,2}}}$$

where a_θ = linear tangential acceleration, a_r = linear radial acceleration, ω = angular velocity, and α = angular acceleration, as related by Fig. 3.02–25. The subscripts 0 and 1 identify the initial and arbitrary quantity at $\theta = 0$ and $\theta = \theta_1$, respectively.

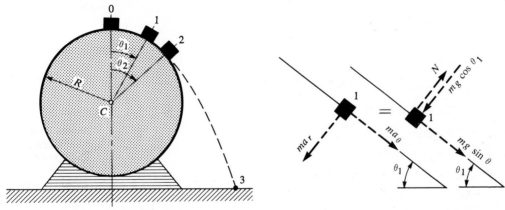

Fig. 3.02–24   **Fig. 3.02–25**

example:

A block of mass $m = 10$ kg slides from the top 0 of a smooth sphere of radius $R = 20$ m with initial velocity $v_0 = 4$ m/s. As it reaches point 1 (any point on its path), its linear velocity becomes

$$v_1 = \sqrt{2 \times 9.81 \times 20(1 - \cos\theta_1) + 4^2} = 19.81\sqrt{1.04 - \cos\theta_1}$$

The block leaves the surface of the sphere when the normal force N between the block and the sphere becomes zero (Fig. 3.02–25), so that

$$ma_r = mg\cos\theta_2$$

from which the position angle θ_2 and the departure velocity v_2 are, respectively,

$$\theta_2 = \cos^{-1}\left(\frac{2}{3} + \frac{v_0^{\,2}}{3\,Rg}\right) = 46.1°$$

$$v_2 = 11.7 \text{ m/s}$$

The motion of the block after leaving the surface of the sphere becomes the motion of the projectile described in Sec. 3.02–6d.

3.03 KINEMATICS OF RIGID BODIES

(1) Definitions and Classification

(a) Motion of a rigid body, defined as a continuous change in the position of all or some particles of the body (Sec. 3.01–1a), is classified as a *plane motion* (particles move along plane curves in parallel planes) or as a *space motion* (particles move along space curves).

(b) Pure rotation is the motion during which the particles of the body on the axis of rotation are stationary and the remaining particles move on circles in planes normal to this axis.

(c) Pure translation is the motion during which all particles move along parallel lines which, however, may be plane curves or space curves.

(d) Euler's theorem. The rotation of a rigid body about a fixed point is equivalent to the rotation of that body about an axis through this point.

(e) Chasles' theorem. The general motion of a rigid body may be represented as a superposition of translation and rotation about an arbitrary point (which is often the centroid of the body).

(f) Degrees of freedom. The number of independent coordinates required to specify the position of a system of particles or bodies is called the *number of degrees of freedom* of the system. In general, a rigid body has six degrees of freedom, and six independent coordinates are required to specify its position in space (three linear coordinates and three angular coordinates). If the motion of the body is subjected to r restrictions, called *constraints*, the number of degrees of freedom n reduces to

$$n = 6 - r$$

and only n independent coordinates are required. Thus rectilinear translation requires $n = 1$, rotation about an axis also requires $n = 1$, and general plane motion requires $n = 3$, but general space motion requires $n = 6$.

(2) Pure Rotation, General Case

(a) Instantaneous position. If a rigid body rotates about a fixed axis (Fig. 3.03–1), its position at any instant t can be defined by the angle θ through which the radius r of an arbitrary point i turns,

$$\theta = \theta(t) \qquad \{\theta \text{ in rad}\}$$

or by the arc corresponding to this angle,

$$s = r\theta(t) \qquad \left\{ \begin{matrix} r \text{ in m} \\ s \text{ in m} \end{matrix} \right\}$$

where $r = $ constant and $\theta(t)$ changes with time.

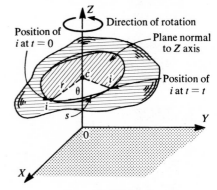

Fig. 3.03–1

(b) Average velocity. If t_1 and t_2 are two successive instants of time and θ_1 and θ_2 are the angular positions of the point i at these instants, then the *average angular velocity* of the body is

$$\bar{\omega} = \frac{\text{change in angular position}}{\text{time interval}} = \frac{\theta_2 - \theta_1}{t_2 - t_1} = \frac{\Delta\theta}{\Delta t} \qquad \{\bar{\omega} \text{ in rad/s}\}$$

and the *average linear velocity* is

$$\bar{v} = \frac{\text{change in linear position}}{\text{time interval}} = \frac{r(\theta_2 - \theta_1)}{t_2 - t_1} = \frac{r\,\Delta\theta}{\Delta t} \qquad \{\bar{v} \text{ in m/s}\}$$

(c) Instantaneous velocity of the body in pure rotation is defined as the *instantaneous angular velocity*

$$\omega = \frac{d\theta}{dt} = \dot{\theta} \qquad \{\omega \text{ in rad/s}\}$$

or as the *instantaneous linear velocity*

$$v = r\frac{d\theta}{dt} = r\dot{\theta} \qquad \{v \text{ in m/s}\}$$

(d) Average acceleration. If t_1 and t_2 are two successive instants of time and ω_1 and ω_2 are the angular velocities at these instants, then the *average angular acceleration* of the body is

$$\bar{\alpha} = \frac{\text{change in angular velocity}}{\text{time interval}} = \frac{\omega_2 - \omega_1}{t_2 - t_1} = \frac{\Delta\omega}{\Delta t} \qquad \{\bar{\alpha} \text{ in rad/s}^2\}$$

and the *average linear acceleration* is

$$\bar{a} = \sqrt{\bar{a}_t^2 + \bar{a}_n^2} = r\sqrt{\bar{\alpha}^2 + \left(\frac{\omega_1 + \omega_2}{2}\right)^4} \qquad \{\bar{a} \text{ in m/s}^2\}$$

where $\bar{a}_t$ = average tangential acceleration and $\bar{a}_n$ = average normal acceleration (pointing toward the axis of rotation).

example:

If $t_1 = 6\,\text{s}$, $t_2 = 8\,\text{s}$, $\omega_1 = 2\,\text{rad/s}$, $\omega_2 = 2.5\,\text{rad/s}$, and $r = 20\,\text{m}$, then $\Delta t = 2\,\text{s}$, $\Delta\omega = 0.5\,\text{rad/s}$, and

$$\bar{\alpha} = \frac{0.5}{2} = 0.25\,\text{rad/s}^2$$

$$= \frac{0.5 \times (1/2\pi)}{2} = 0.040\,\text{rev/s}^2$$

$$= \frac{0.5 \times (1/2\pi)}{2/3,600} = 143.20\,\text{rev/min}^2$$

$$\bar{a}_t = 20 \times 0.25 = 5\,\text{m/s}^2 \qquad \bar{a}_n = 20 \times \left(\frac{2 + 2.5}{2}\right)^2 = 101.25\,\text{m/s}^2$$

where the conversion factors used are introduced in Sec. 2.01–3f.

(e) Instantaneous acceleration of the body in pure rotation is defined as the *instantaneous angular acceleration*

$$\alpha = \frac{d^2\theta}{dt^2} = \ddot{\theta} \qquad \{\alpha \text{ in rad/s}^2\}$$

or as the *instantaneous linear acceleration*

$$a = \sqrt{a_t^2 + a_n^2} = r\sqrt{\alpha^2 + \omega^4}$$

where a_t = instantaneous tangential acceleration and a_n = instantaneous normal acceleration.

(f) Sign convention. Counterclockwise rotation is positive; clockwise rotation is negative.

(3) Pure Rotation, Particular Cases

(a) Classification. Pure rotations of a rigid body are classified as

(α) *Uniform rotation* (ω = constant, $\alpha = 0$)

(β) *Uniformly accelerated rotation* (ω = variable, α = constant)

(γ) *Nonuniformly accelerated rotation* (ω = variable, α = variable)

(b) Notation. Equations of motion for four types of pure rotation are given in (c) to (f), where

$$\theta = \text{position angle} \qquad \omega = \text{angular velocity} \qquad \alpha = \text{angular acceleration}$$

all at the instant t in the plane of the circle of rotation, and θ_0, ω_0, α_0 are their counterparts at $t = 0$, called the *initial conditions*, some or all of which may be equal to zero.

(c) Uniform rotation ($\omega = \omega_0$, $\alpha = 0$)

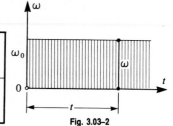

Fig. 3.03–2

$\theta = \theta_0 + \omega_0 t$	$\omega = \omega_0 = \dfrac{\theta - \theta_0}{t}$	$\alpha = 0$
$t = \dfrac{\theta - \theta_0}{\omega_0}$	$\theta_0 = \theta - \omega_0 t$	$\omega_0 = \omega$

The time-velocity diagram of this rotation is shown in Fig. 3.03–2.

(d) Uniformly accelerated rotation ($\omega_0 = 0$, $\alpha = \alpha_0$).

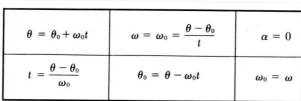

$\theta = \theta_0 + \dfrac{\alpha_0 t^2}{2}$	$\omega = \alpha_0 t$	$\alpha = \alpha_0 = \dfrac{\omega}{t}$	$t = \dfrac{\omega}{\alpha_0}$	$\theta_0 = \theta - \dfrac{\alpha_0 t^2}{2}$
$= \theta_0 + \dfrac{\omega t}{2}$	$= \dfrac{2(\theta - \theta_0)}{t}$	$= \dfrac{2(\theta - \theta_0)}{t^2}$	$= \dfrac{2(\theta - \theta_0)}{\omega}$	$= \theta - \dfrac{\omega t}{2}$
$= \theta_0 + \dfrac{\omega^2}{2\alpha_0}$	$= \sqrt{2\alpha_0(\theta - \theta_0)}$	$= \dfrac{\omega^2}{2(\theta - \theta_0)}$	$= \sqrt{\dfrac{2(\theta - \theta_0)}{\alpha_0}}$	$= \theta - \dfrac{\omega^2}{2\alpha_0}$

The time-velocity diagram of this rotation is shown in Fig. 3.03–3.

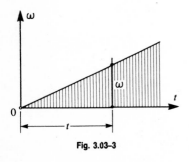

Fig. 3.03–3

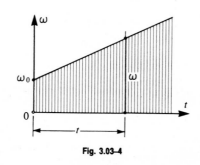

Fig. 3.03–4

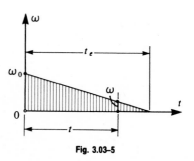

Fig. 3.03–5

(e) Uniformly accelerated rotation ($\omega_0 \neq 0$, $\alpha = \alpha_0$).

$\theta = \theta_0 + \omega_0 t + \dfrac{\alpha_0 t^2}{2}$	$\omega = \omega_0 + \alpha_0 t$	$\alpha = \alpha_0 = \dfrac{\omega - \omega_0}{t}$
$\quad = \theta_0 + \dfrac{(\omega + \omega_0)t}{2}$	$\quad = \dfrac{2(\theta - \theta_0)}{t} - \omega_0$	$\quad = \dfrac{2(\theta - \theta_0 - \omega_0 t)}{t^2}$
$\quad = \theta_0 + \dfrac{\omega^2 - \omega_0^2}{2\alpha_0}$	$\quad = \sqrt{\omega_0^2 + 2\alpha_0(\theta - \theta_0)}$	$\quad = \dfrac{\omega^2 - \omega_0^2}{2(\theta - \theta_0)}$
$t = \dfrac{\omega - \omega_0}{\alpha_0}$	$\theta_0 = \theta - \dfrac{(\omega_0 + \omega)t}{2}$	$\omega_0 = \omega - \alpha_0 t$
$\quad = \dfrac{2(\theta - \theta_0)}{\omega + \omega_0}$	$\quad = \theta - \dfrac{(2\omega_0 + \alpha_0 t)t}{2}$	$\quad = \dfrac{2(\theta - \theta_0)}{t} - \omega$
$\quad = \dfrac{-\omega_0 + \sqrt{\omega_0^2 + 2\alpha_0(\theta - \theta_0)}}{\alpha_0}$	$\quad = \theta - \omega_0 t - \dfrac{\alpha_0 t^2}{2}$	$\quad = \sqrt{\omega^2 - 2\alpha_0(\theta - \theta_0)}$

The time-velocity diagram of this rotation is shown in Fig. 3.03–4.

(f) Uniformly decelerated rotation ($\omega_0 \neq 0$, $\alpha = \alpha_0$).

$\theta = \theta_0 + \omega_0 t - \dfrac{\alpha_0 t^2}{2}$	$\omega = \omega_0 - \alpha_0 t$	$\alpha = \alpha_0 = \dfrac{\omega_0 - \omega}{t}$
$\quad = \theta_0 + \dfrac{(\omega_0 + \omega)t}{2}$	$\quad = \dfrac{2(\omega - \omega_0)}{t} - \omega_0$	$\quad = \dfrac{2(\theta_0 + \omega_0 t - \theta)}{t^2}$
$\quad = \theta_0 + \dfrac{\omega_0^2 - \omega^2}{2\alpha_0}$	$\quad = \sqrt{\omega_0^2 - 2\alpha_0(\theta - \theta_0)}$	$\quad = \dfrac{\omega^2 - \omega_0^2}{2(\theta - \theta_0)}$
$t = \dfrac{\omega_0 - \omega}{\alpha_0}$	$\theta_0 = \theta - \dfrac{(\omega_0 + \omega)t}{2}$	$\omega_0 = \omega + \alpha_0 t$
$\quad = \dfrac{2(\theta - \theta_0)}{\omega + \omega_0}$	$\quad = \theta - \dfrac{(2\omega_0 - \alpha_0 t)t}{2}$	$\quad = \dfrac{2(\theta - \theta_0)}{t} - \omega$
$\quad = \dfrac{\omega_0 - \sqrt{\omega_0^2 - 2\alpha_0(\theta - \theta_0)}}{\alpha_0}$	$\quad = \theta - \omega_0 t + \dfrac{\alpha_0 t^2}{2}$	$\quad = \sqrt{\omega^2 + 2\alpha_0(\theta - \theta_0)}$

The body stops ($\omega_e = 0$) at $t_e = \omega_0/\alpha_0$ and $\theta_e = \theta_0 + \omega_0^2/2\alpha_0$, where t_e = terminal time and θ_e = terminal position angle. The time-displacement diagram of this rotation is shown in Fig. 3.03–5.

(g) Correspondence. There is a formal correspondence between the equations of rectilinear motion of a particle (Secs. 3.01–4c to f) and those of pure rotation of a rigid body (Secs. 3.03–3c to f) based on

$$s \leftrightarrow \theta \qquad v \leftrightarrow \omega \qquad a \leftrightarrow \alpha$$

$$s_0 \leftrightarrow \theta_0 \qquad v_0 \leftrightarrow \omega_0 \qquad a_0 \leftrightarrow \alpha_0$$

(4) Plane Motion, General Case

(a) **Definition.** The plane motion of a rigid body can be defined by the motion of one of its cross sections parallel to the reference plane of motion (Fig. 3.03–6) and represented as superposition of translation and rotation.

(b) **Translation** of the body is defined by the linear motion of an arbitrary point i of the cross section (the centroid of the cross section is frequently selected for this purpose).

(c) **Rotation** of the body is defined by the angular motion of a segment $\overline{ij}$ related to the point i (for bodies with axial symmetry a particular radius perpendicular to the axis of rotation is selected for this purpose).

(d) **Position coordinates** of the rigid body referred to these two points and the fixed reference axes X, Y are

$$x_j = x_i + x_{ij} = x_i + r \cos \theta$$

$$y_j = y_i + y_{ij} = y_i + r \sin \theta$$

where x_i, y_i = coordinates of i, x_j, y_j = coordinates of j, $r = \overline{ij}$, and θ = position angle measured from the horizontal to $\overline{ij}$.

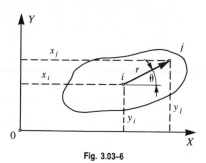

Fig. 3.03–6

(e) **Instantaneous velocity** v_j of the body is given by the components

$$v_{jx} = \frac{dx_j}{dt} = \dot{x}_j = \dot{x}_i + \dot{x}_{ij} = \dot{x}_i - r\dot{\theta} \sin \theta$$

$$v_{jy} = \frac{dy_j}{dt} = \dot{y}_j = \dot{y}_i + \dot{y}_{ij} = \dot{y}_i + r\dot{\theta} \cos \theta$$

where $\dot{x}_i = dx_i/dt$ and $\dot{y}_i = dy_i/dt$ = absolute linear velocities of i, $\dot{x}_{ij} = d(r \cos \theta)/dt = -r\dot{\theta} \sin \theta$ and $\dot{y}_{ij} = d(r \sin \theta)/dt = r\dot{\theta} \cos \theta$ = relative linear velocities of $\overline{ij}$ with respect to i, and $\dot{\theta} = \omega$ = angular velocity of $\overline{ij}$.

(f) **Instantaneous acceleration** a_j of the body is given by the components

$$a_{jx} = \frac{d^2x_j}{dt^2} = \ddot{x}_j = \ddot{x}_i + \ddot{x}_{ij} = \ddot{x}_i - r\ddot{\theta} \sin \theta - r\dot{\theta}^2 \cos \theta$$

$$a_{jy} = \frac{d^2y_j}{dt^2} = \ddot{y}_j = \ddot{y}_i + \ddot{y}_{ij} = \ddot{y}_i + r\ddot{\theta} \cos \theta - r\dot{\theta}^2 \sin \theta$$

where $\ddot{x}_i = d^2x_i/dt^2$ and $\ddot{y}_i = d^2y_i/dt^2$ = absolute linear accelerations of i, $\ddot{x}_{ij} = d^2(r \cos \theta)/dt^2 = -r\ddot{\theta} \sin \theta - r\dot{\theta}^2 \cos \theta$ and $\ddot{y}_{ij} = d^2(r \sin \theta)/dt^2 = r\ddot{\theta} \cos \theta - r\dot{\theta}^2 \sin \theta$ = relative linear accelerations of $\overline{ij}$ with respect to i, and $\ddot{\theta} = \alpha$ = angular acceleration of $\overline{ij}$.

(g) **Absolute motion** is a motion measured with respect to a fixed reference system.

(h) **Relative motion** is a motion measured with respect to a moving reference system.

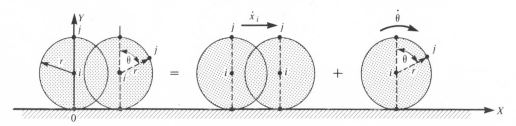

Fig. 3.03–7

(5) Plane Motion, Particular Cases

(a) Rolling. If a wheel of radius r rolls without slipping on a horizontal plane in the plane XY (Fig. 3.03–7), then the motion is described by the following equations.

$x_j = r\theta + r\cos\left(\dfrac{\pi}{2} - \theta\right)$ $= r\theta + r\sin\theta$	$v_{jx} = r\dot\theta + r\dot\theta\cos\theta$ $= r\omega + r\omega\cos\theta$	$a_{jx} = r\ddot\theta + r\ddot\theta\cos\theta - r\dot\theta^2\sin\theta$ $= r\alpha + r\alpha\cos\theta - r\omega^2\sin\theta$
$y_j = r + r\sin\left(\dfrac{\pi}{2} - \theta\right)$ $= r + r\cos\theta$	$v_{jy} = -r\dot\theta\sin\theta$ $= -r\omega\sin\theta$	$a_{jy} = -r\ddot\theta\sin\theta - r\dot\theta^2\cos\theta$ $= -r\alpha\sin\theta - r\omega^2\cos\theta$

where ω and α are angular velocity and acceleration, respectively.

example:

If $\omega = \dfrac{\pi}{2}$ rad/s and $r = 10$ m, then at $t = 2$ s, $\theta = \dfrac{\pi}{2} \times 2 = \pi$, $\alpha = 0$, and

$$x_j = 10 \times \pi + 10 \times \sin\pi = 10\pi \text{ m/s} \qquad y_j = 10 + 10 \times \cos\pi = 0$$

$$v_{jx} = 10 \times \frac{\pi}{2} + 10 \times \frac{\pi}{2} \times \cos\pi = 0 \qquad v_{jy} = -10 \times \frac{\pi}{2} \times \sin\pi = 0$$

$$a_{jx} = -10 \times \left(\frac{\pi}{2}\right)^2 \sin\pi = 0 \qquad a_{jy} = -10 \times \left(\frac{\pi}{2}\right)^2 \cos\pi = 2.5\pi^2 \text{ m/s}^2$$

which shows that at $t = 2$ s the point j has rotated $180°$ and is on the X axis. Since slipping is prevented, $v = 0$, $a_{jx} = 0$, and only the centripetal acceleration occurs.

(b) Slider-crank mechanism. If the crank $\overline{0m} = r$ of Fig. 3.03–8 rotates clockwise with a constant angular velocity ω and moves the connecting rod $\overline{mn} = l$, then the motion of the piston hinged to $\overline{mn}$ at n is described by the following equations.

$x = r(1 - \cos\omega t) \pm \dfrac{\lambda}{2} r \sin^2\omega t$	$v_{max} = r\omega\sqrt{1 + \lambda^2}$	$\theta = \omega t$
$v = r\omega\sin\omega t(1 \pm \lambda\cos\omega t)$	$a_1 = r\omega^2(1 + \lambda)$	$\theta_1 = 0, 2\pi, \ldots$
$a = r\omega^2(\cos\omega t \pm \lambda\cos 2\omega t)$	$a_2 = -r\omega^2(1 - \lambda)$	$\theta_2 = \pi, 3\pi, \ldots$

where $\lambda = r/l$, v_{max} = maximum velocity when the rod is tangent to the circle, a_1 = acceleration when m is at θ_1, and a_2 = acceleration when m is at θ_2.

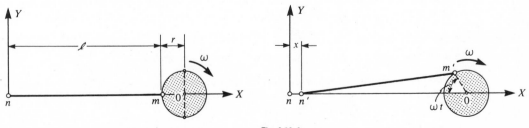

Fig. 3.03–8

example:

If $r = 10$ cm, $l = 30$ cm, and $\omega = 100$ rad/s, then at $t = \pi/400$ s, $x = 10(1-0.707) + \dfrac{1/3}{2} \times 10 \times 0.707^2 =$

3.76 cm, $v = 10 \times 100 \times 0.707(1 + 1/3 \times 0.707) = 873$ cm/s, $a = 10 \times 100^2 \times 0.707 = 70,700$ cm/s^2, and the extreme values are

$$v_{\max} = 10 \times 100\sqrt{1 + (1/3)^2} = 1,054 \text{ cm/s} \qquad a_1 = 10 \times 100^2(1 + \tfrac{1}{3}) = 133,333 \text{ cm/s}^2$$

$$a_2 = -10 \times 100^2(1 - 1/3) = -66,666 \text{ cm/s}^2$$

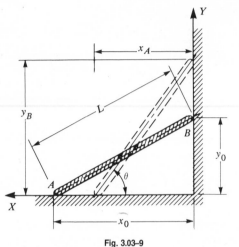

Fig. 3.03–9   Fig. 3.03–10

(c) **Sliding bar.**   If the end A of a straight bar AB of length L (Fig. 3.03–9) slides along the X axis in the XY plane with constant velocity v_0 and its end B slides along the Y axis, its angular position θ, angular velocity ω, and angular acceleration α are, respectively

$$\theta = \cos^{-1} \frac{x_0 - v_0 t}{L} \qquad \omega = \frac{v_0}{L \sin \theta} \qquad \alpha = \frac{v_0^2 \cos \theta}{L^2 \sin^3 \theta}$$

and their linear counterparts at A and B, respectively, are

$$x_A = x_0 - v_0 t \qquad v_A = -v_0 \qquad a_A = 0$$

$$y_B = (x_0 - v_0 t) \tan \theta \qquad v_B = v_0 \cot \theta \qquad a_B = \frac{v_0^2}{L \sin^3 \theta}$$

where x_0, y_0, v_0 are the initial conditions and the minus sign in front of v_0 indicates the direction of the velocity relative to the direction of the X axis.

(d) **Two gears.**   If two gears of respective radii R and r connected by a rigid arm $AB = L = R + r$ (Fig. 3.03–10) form a rotating system about the hinge A, their angular velocities ω_A, ω_B, ω_L and their angular accelerations α_A, α_B, α_L are related by

$$R\omega_A + r\omega_B = L\omega_L \qquad R\alpha_A + r\alpha_B + r\omega_A^2 + r\omega_B^2 = L\alpha_L + L\omega_L^2$$

and the linear velocity and acceleration of the moving center B are

$$v_B = L\omega_L \qquad a_B = L\alpha_L + L\omega_L^2$$

where the signs of ω and α are those defined in Sec. 3.03–2f.

(6) Instantaneous Center of Plane Motion

(a) Definition. During the plane motion of a rigid body (Sec. 3.03–4a) there is at any time an axis normal to the plane of motion which is instantaneously at rest (axis of zero linear velocity). This axis, called the *instantaneous axis*, intersects the plane of motion in a point called the *instantaneous center* (point of zero linear velocity), which may be on the body or outside the body.

(b) Position of this center at any given instant may be located by one of two possible ways. If the linear velocities v_A and v_B of points A and B, respectively, are nonparallel, the instantaneous center E is the point of intersection of the normal through A to v_A and the normal through B to v_B (Fig. 3.03–11a). If the linear velocities v_A and v_B are parallel and normal to AB, the instantaneous center E is the point of intersection of the line through A and B and the line passing through the tips of the velocity vectors (Fig. 3.03–11b).

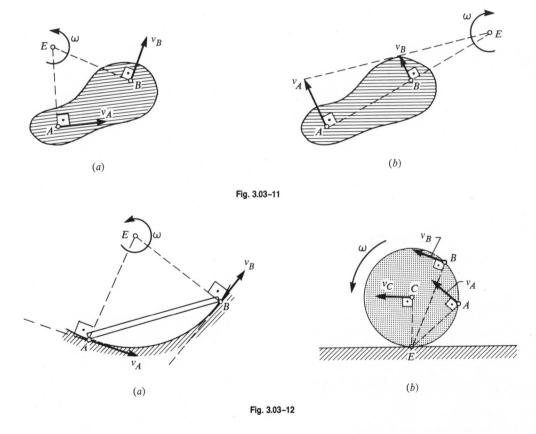

(a)   (b)

Fig. 3.03–11

(a)   (b)

Fig. 3.03–12

(c) Sliding. If the body slides on two straight or curvilinear surfaces, the instantaneous center is the intersection of the normals to these surfaces at the points of contact (Fig. 3.03–12a).

(d) Rolling. If the body rolls without sliding on a straight or curvilinear surface, the instantaneous center is the point of contact (Fig. 3.03–12b).

3.04 KINETICS OF RIGID BODIES

(1) Weight and Mass Densities

(a) **Unit weight** γ of a body, also called the *weight density*, is defined as

$$\gamma = \frac{\text{total weight}}{\text{total volume}} = \frac{W}{V} \qquad \left\{\begin{matrix} W \text{ in kgf} \\ V \text{ in m}^3 \\ \gamma \text{ in kgf/m}^3 \end{matrix}\right\}$$

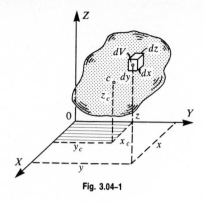

Fig. 3.04–1

(b) **Elemental weight** dW of a rigid body is the weight of its differential element, defined as

$$dW = \gamma \, dV \qquad \left\{\begin{matrix} dW \text{ in kgf} \\ dV \text{ in m}^3 \end{matrix}\right\}$$

where dV = volume of the elemental parallelepiped $dx \cdot dy \cdot dz$ shown in Fig. 3.04–1.

(c) **Unit mass** ρ of a body, also called the *mass density*, is defined as

$$\rho = \frac{\text{total mass}}{\text{total volume}} = \frac{W/g}{V} \qquad \left\{\begin{matrix} g \text{ in m/s}^2 \\ \rho \text{ in kg} \end{matrix}\right\}$$

where $g = 9.80665 \text{ m/s}^2$ = acceleration due to gravity (Sec. 3.02–2e).

(d) **Elemental mass** dm of a body is the mass of its differential element, defined as

$$dm = \rho \, dV \qquad \left\{\begin{matrix} dm \text{ in kg} \\ dV \text{ in m}^3 \end{matrix}\right\}$$

(2) Center of Mass and Gravity

(a) **Total mass** m of a body is then

$$m = \int_V dm = \int_V \rho \, dV = \rho V = \frac{W}{g}$$

where the integral is taken for the entire volume V.

(b) **Static moments of mass** of a body about the YZ, ZX, and XY planes, respectively, are

$$q_{yz} = \int_V x \, dm \qquad q_{zx} = \int_V y \, dm \qquad q_{xy} = \int_V z \, dm$$

where x, y, z = coordinates of the elemental mass dm.

(c) **Center of mass** c of a body is the point of intersection of all axes about which the static moments of mass are zero (Fig. 3.04–1). Analytically, the coordinates of the center of mass c are

$$\bar{x}_c = \frac{q_{yz}}{m} = \frac{\int_V x \, dm}{\int_V dm} \qquad \bar{y}_c = \frac{q_{zx}}{m} = \frac{\int_V y \, dm}{\int_V dm} \qquad \bar{z}_c = \frac{q_{xy}}{m} = \frac{\int_V z \, dm}{\int_V dm}$$

(d) Center of gravity of a body in a uniform gravitational field (g = constant) is identical in position to its center of mass. In this book the terms *center of mass* and *center of gravity* are interchangeable.

(e) Centroid of a body is identical in position to the center of mass if and only if the body consists of a homogeneous mass (all parts of the body have the same density).

(f) Composite body. If a body consists of parts and/or shapes, some or all of which have different densities, the body can be subdivided into parts of particular densities or shapes and the mass functions of the body can be computed in terms of these parts. Analytically (referring to Fig. 2.07–5),

$$m = m_1 + m_2 + \cdots + m_n = \sum_{p=1}^{n} m_p$$

$$q_{xy} = m_1 \cdot z_1 + m_2 \cdot z_2 + \cdots + m_n \cdot z_n = \sum_{p=1}^{n} m_p \cdot z_p$$

$$q_{yz} = m_1 \cdot x_1 + m_2 \cdot x_2 + \cdots + m_n \cdot x_n = \sum_{p=1}^{n} m_p \cdot x_p$$

$$q_{zx} = m_1 \cdot y_1 + m_2 \cdot y_2 + \cdots + m_n \cdot y_n = \sum_{p=1}^{n} m_p \cdot y_p$$

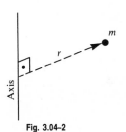

Fig. 3.04–2

$$\bar{x}_c = \frac{\sum_{p=1}^{n} m_p \cdot x_p}{\sum_{p=1}^{n} m_p} \qquad \bar{y}_c = \frac{\sum_{p=1}^{n} m_p \cdot y_p}{\sum_{p=1}^{n} m_p} \qquad \bar{z}_p = \frac{\sum_{p=1}^{n} m_p \cdot z_p}{\sum_{p=1}^{n} m_p}$$

where $m_1, m_2, \ldots, m_n$ = masses of the respective parts and $x_1, y_1, z_1, x_2, y_2, z_2, \ldots, x_n, y_n, z_n$ = coordinates of the mass centers of these parts.

example:

The mass densities of the parallelepipeds in Fig. 2.07–6 (Sec. 2.07–5e) are $\rho_1 = 120 \text{ kg/m}^3$, $\rho_2 = 60 \text{ kg/m}^3$, respectively, their masses are $m_1 = V_1 \cdot \rho_1 = 60{,}000 \times 120 = 7.2 \times 10^6 \text{ kg}$, $m_2 = V_2 \cdot \rho_2 = 60{,}000 \times 60 = 3.6 \times 10^6 \text{ kg}$, and the total mass is $m = m_1 + m_2 = 1.08 \times 10^7 \text{ kg}$. The static moments of m are

$$q_{xy} = 7{,}200{,}000 \times 20 + 3{,}600{,}000 \times 65 = 3.78 \times 10^8 \text{ kg-m}$$

$$q_{yz} = 7{,}200{,}000 \times 15 + 3{,}600{,}000 \times 20 = 1.80 \times 10^8 \text{ kg-m}$$

$$q_{zx} = 7{,}200{,}000 \times 25 + 3{,}600{,}000 \times 15 = 2.34 \times 10^8 \text{ kg-m}$$

and the coordinates of the mass center and also of the center of gravity are

$$\bar{x}_c = \frac{1.80 \times 10^8}{1.08 \times 10^7} = 16.7 \text{ m} \qquad \bar{y}_c = \frac{2.34 \times 10^8}{1.08 \times 10^7} = 21.7 \text{ m} \qquad \bar{z}_c = \frac{3.78 \times 10^8}{1.08 \times 10^7} = 35 \text{ m}$$

whereas the coordinates of the centroid computed in Sec. 2.07–5e are $x_c = 17.5 \text{ m}$, $y_c = 20 \text{ m}$, and $z_c = 42.5 \text{ m}$.

(g) Particular cases. If a body has one plane of symmetry, its mass center is on this plane. If the body has two planes of symmetry, its mass center is on the line of intersection of these planes. If the body has three planes of symmetry, its mass center is the point of intersection of these three planes. Coordinates of mass centers of the most frequently used homogeneous bodies are given in Appendix A.

(3) Inertia Functions

(a) **Moment of inertia of a particle of mass** m (Fig. 3.04–2) with respect to a given axis is

$$I = r^2 m$$

where r = normal distance of the particle to this axis.

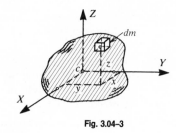

(b) **Moment of inertia of a body of mass** m (Fig. 3.04–3) with respect to the mass center axes X_c, Y_c, Z_c are

$$I_{cx} = \int_V (y^2 + z^2)\, dm \qquad \text{(about } X_c \text{ axis)}$$

$$I_{cy} = \int_V (z^2 + x^2)\, dm \qquad \text{(about } Y_c \text{ axis)}$$

$$\left\{ \begin{array}{l} I_{cx},\ I_{cy},\ I_{cz} \text{ in kg-m}^2 \\ x,\ y,\ z \text{ in m} \\ dm \text{ in kg} \end{array} \right\}$$

$$I_{cz} = \int_V (x^2 + y^2)\, dm \qquad \text{(about } Z_c \text{ axis)}$$

Fig. 3.04–3

where x, y, z = coordinates of the elemental mass $dm = \rho\, dV = \rho\, dx\, dy\, dz$ in the X_c, Y_c, Z_c axes and the integrals are taken for the entire volume of the body.

(c) **Polar moment of inertia of a body of mass** m (Fig. 3.04–3) with respect to the same axes is

$$J_c = \int_V (x^2 + y^2 + z^2)\, dm = \tfrac{1}{2}(I_{cx} + I_{cy} + I_{cz})$$

where the symbols and units are the same as in (b) above.

(d) **Mass radii of gyration of a body** are

$$k_{cx} = \sqrt{\frac{I_{cx}}{m}} \qquad k_{cy} = \sqrt{\frac{I_{cy}}{m}} \qquad k_{cz} = \sqrt{\frac{I_{cz}}{m}} \qquad \{k_{cx}, k_{cy}, k_{cz} \text{ in m}\}$$

where m = total mass of the body.

(e) **Parallel axis theorem.** If the inertial moment I_c of a body of mass m with respect to a mass center axis is known (Fig. 3.04–4), the inertial moment I_0 of this body with respect to any parallel axis is

$$I_0 = r^2 m + I_c$$

where r = normal distance of these two axes.

(f) **Particular cases.** If a body has one plane of symmetry, the moment of inertia of the mass of the body about a given axis in this plane of symmetry equals twice the moment of inertia of the mass located on one side of this plane about the given axis. If the body has two mutually perpendicular planes of symmetry, the moment of inertia of the mass of the body about the line of intersection of these planes equals 4 times the moment of inertia of the mass located in one quadrant of these planes about the line of intersection. Moments of inertia and radii of gyration of the most frequently used homogeneous bodies are given in Appendix A.

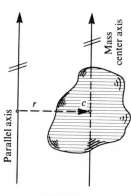

Fig. 3.04–4

(g) Composite body. If a body consists of parts some or all of which have different mass densities and/or shapes, the moments of inertia and the radii of gyration of the whole body can be computed in terms of the inertia functions of these parts. Analytically (referring to Fig. 2.07–5),

$$I_{0x} \sum_{p=1}^{n} (y_p^2 + z_p^2)m_p + \sum_{p=1}^{n} I_{px} \qquad I_{0y} = \sum_{p=1}^{n} (z_p^2 + x_p^2)m_p + \sum_{p=1}^{n} I_{py} \qquad I_{0z} = \sum_{p=1}^{n} (x_p^2 + y_p^2)m_p + \sum_{p=1}^{n} I_{pz}$$

where the summation symbols have the same meaning as in Sec. 3.04–2f, I_{0x}, I_{0y}, I_{0z} = total moments of inertia of the mass of the body with respect to the X_0, Y_0, Z_0 axes, respectively, $I_{1x}, I_{1y}, I_{1z}, I_{2x}, I_{2y}, I_{2z}, \ldots, I_{nx}, I_{ny}, I_{nz}$ = moments of inertia of the respective parts related in each case to the mass center axes of that part, and $x_1, y_1, z_1, x_2, y_2, z_2, \ldots, x_n, y_n, z_n$ = coordinates of these mass centers in the X_0, Y_0, Z_0 axes.

example:

For the body of Fig. 2.07–6 (Sec. 2.07–5e) in terms of $\rho_1 = 120 \text{ kg/m}^3$ and $\rho_2 = 60 \text{ kg/m}^3$, $I_{0x} = (y_1^2 + z_1^2)m_1 + I_{1x} + (y_2^2 + z_2^2)m_2 + I_{2x} = 2.688 \times 10^{10} \text{ kg-m}^2$, where

$$y_1 = 25 \text{ m} \qquad z_1 = 20 \text{ m} \qquad m_1 = 7.2 \times 10^6 \text{ kg}$$
$$y_2 = 15 \text{ m} \qquad z_2 = 65 \text{ m} \qquad m_2 = 3.6 \times 10^6 \text{ kg}$$

and from Appendix A,

$$I_{1x} = \frac{m_1}{12}(b_1^2 + c_1^2) = \frac{7.2 \times 10^6}{12}(50^2 + 40^2) = 2.46 \times 10^9 \text{ kg-m}^2$$

$$I_{2x} = \frac{m_2}{12}(b_2^2 + c_2^2) = \frac{3.6 \times 10^6}{12}(30^2 + 50^2) = 1.02 \times 10^9 \text{ kg-m}^2$$

The radius of gyration is

$$k_{0x} = \sqrt{\frac{I_{0x}}{m}} = \sqrt{\frac{2.688 \times 10^{10}}{1.08 \times 10^7}} = 49.9 \text{ m}$$

The calculation of I_{0y}, I_{0z} and of k_{0y}, k_{0z} is similar.

(4) Linear Momentum and Impulse

(a) Linear momentum G of a particle of mass m moving with linear velocity v with respect to a fixed point (Fig. 3.04–5) is

$$G = \text{mass} \times \text{velocity} = m \cdot v \qquad \begin{cases} m \text{ in kg} \\ v \text{ in m/s} \\ G \text{ in kg-m/s} \end{cases}$$

where G = vector acting in the direction of v.

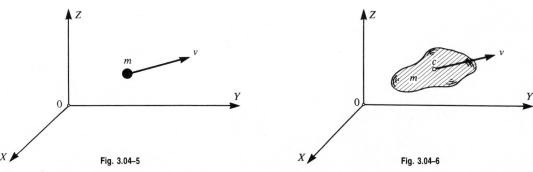

Fig. 3.04–5   Fig. 3.04–6

(b) Linear momentum G of a rigid body in motion is the sum of the linear momenta of all the particles of the body (Fig. 3.04–6). Symbolically,

$$G = \sum_{p=1}^{n} m_p \cdot v = m \cdot v \qquad \{v \text{ in m/s}\}$$

where m_p = mass of a particle of the body, v = linear velocity of this particle and also of the whole body, and m = total mass of the body.

(c) Units of linear momentum G are

$$1 \text{ newton} \times \text{second} = 1 \text{ N-s} = 1 \text{ kilogram} \times \text{meter/second} = 1 \text{ kg-m/s}$$
$$= 0.102 \text{ kilogram-force} \times \text{second} = 0.102 \text{ kgf-s}$$
$$1 \text{ poundal} \times \text{second} = 1 \text{ pd-sec} = 1 \text{ pound} \times \text{foot/second} = 1 \text{ lb-ft/sec}$$
$$= 0.031 \text{ pound-force} \times \text{second} = 0.031 \text{ lbf-sec}$$

(d) Average force $\bar{F}$ equals the ratio of the change in G to the time interval Δt during which the change takes place.

$$\bar{F} = \frac{\Delta G}{\Delta t} = \frac{G_2 - G_1}{t_2 - t_1} = \frac{m(v_2 - v_1)}{t_2 - t_1} \qquad \{F \text{ in kgf}\}$$

where G_1, G_2 = linear momenta and v_1, v_2 = linear velocities at the instants t_1, t_2, respectively.

(e) Linear impulse $\bar{F} \Delta t$ equals the product of the force (acting at the mass center of the body) by the time of action.

$$\bar{F} \Delta t = \bar{F}(t_2 - t_1) = m(v_2 - v_1) = G_2 - G_1$$

where $\bar{F} \Delta t$ = a vector acting in the direction of the force and the force is assumed to be constant.

example:

If a force $\bar{F} = 1,000$ kgf is applied at $t_1 = 0$ on a body of mass $m = 4,000$ kg which is at rest, then at $t_2 = 10$ s the linear impulse produced by this force in this time interval is

$$\bar{F} \times (t_2 - \cancel{t_1})^0 = m(v_2 - \cancel{v_1})^0$$

from which the linear velocity of the body at t_2 is

$$v_2 = \frac{\bar{F} \times t_2}{m} = \frac{1,000 \times 9.81 \times 10}{4,000} = 24.5 \text{ m/s}$$

where $\bar{F} = 1,000$ kgf $= 1,000 \times 9.81$ kg-m/s².

(5) Central Collision

(a) Impact is the phenomenon of collision of two bodies during a very short time interval. The magnitude of the force involved and the duration of the impact depends on the shape of the bodies, their velocities, and their elastic properties.

(b) Straight central impact of two or several bodies which are axially symmetrical with respect to a straight line is their impact along this line.

(c) Conservation of linear momentum. If two or more perfectly elastic bodies experience the straight-line impact described above (Fig. 3.04–7), then the sum of their linear momenta before the impact equals the sum of their linear momenta after the impact; i.e., their total linear momentum is conserved (remains the same).

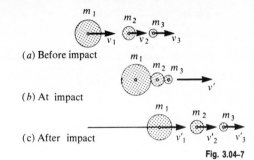

(a) Before impact

(b) At impact

(c) After impact

Fig. 3.04–7

$$m_1v_1 + m_2v_2 + \cdots + m_nv_n = m_1v_1' + m_2v_2' + \cdots + m_nv_n'$$

where $m_1, m_2, \ldots, m_n$ = respective masses, $v_1, v_2, \ldots, v_n$ = their linear velocities before the impact, and $v_1', v_2', \ldots, v_n'$ = their velocities after the impact.

(d) Coefficient of restitution e is a material constant (pure number), determined experimentally by dropping a small sphere on a smooth rigid plane from a height h and measuring the height of the rebound h' (Fig. 3.04–8).

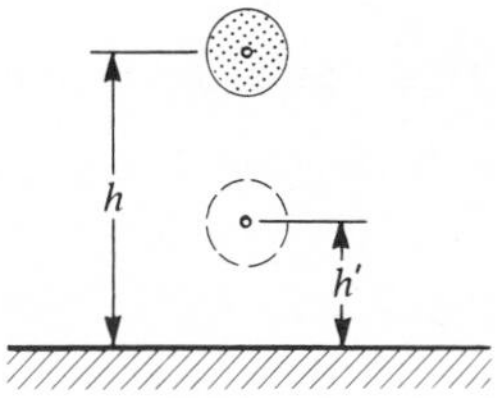

Fig. 3.04–8

$$e = \sqrt{\frac{h'}{h}}$$

For perfectly plastic bodies, $e = 0$ (no rebound); for semielastic bodies, e = number between 0 and 1 (partial rebound); and for perfectly elastic bodies, $e = 1$ (complete rebound).

(e) Case I: Plastic central impact. The common velocity v' of two perfectly plastic bodies of masses m_1, m_2, which collided in central impact with preimpact velocities v_1, v_2 and did not separate after the impact, is

$$v' = a_1v_1 + a_2v_2$$

where $a_1 = \dfrac{m_1}{m_1 + m_2}$ and $a_2 = \dfrac{m_2}{m_1 + m_2}$.

(f) Case II: Semielastic central impact. The separate velocities v_1', v_2' of two semielastic bodies of masses m_1, m_2 and of a known coefficient of restitution e, which collided in central impact with preimpact velocities v_1, v_2, are, after separation,

$$v_1' = v_1 - b_1(v_1 - v_2) \qquad v_2' = v_2 + b_2(v_1 - v_2)$$

where $b_1 = \dfrac{1 + e}{1 + m_1/m_2}$ and $b_2 = \dfrac{1 + e}{1 + m_2/m_1}$.

(g) Case III: Elastic central impact. The separate velocities v_1', v_2' of two perfectly elastic bodies of masses m_1, m_2, which collided in central impact with preimpact velocities v_1, v_2, are

$$v_1' = (a_1 - a_2)v_1 + 2a_2v_2 \qquad v_2' = 2a_1v_1 - (a_1 - a_2)v_2$$

where $a_1 = m_1/(m_1 + m_2)$ and $a_2 = m_2/(m_1 + m_2)$.

example:

If two bodies of $W_1 = 1,000\,\text{kgf}$ and $W_2 = 2,000\,\text{kgf}$ experience straight central impact with velocities $v_1 = 30\,\text{m/s}\ (\rightarrow)$ and $v_2 = -40\,\text{m/s}\ (\leftarrow)$, respectively (Fig. 3.04–9), then if $e = 0$ (plastic impact),

$$v' = \frac{1,000 \times 30}{3,000} + \frac{2,000 \times (-40)}{3,000} = -16.7\,\text{m/s} \qquad (\leftarrow)$$

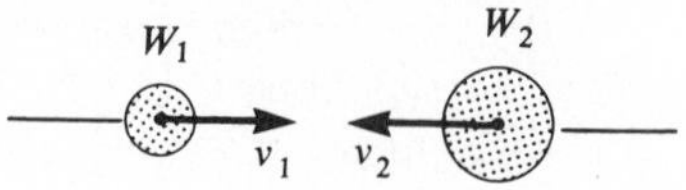

If $e = 0.3$ (semielastic impact),

$$v_1' = 30 - \frac{(1+0.3) \times 70}{1+0.5} = -30.7\,\text{m/s} \qquad (\leftarrow)$$

$$v_2' = -40 + \frac{(1+0.3) \times 70}{1+2} = -9.7\,\text{m/s} \qquad (\leftarrow)$$

If $e = 1$ (elastic impact),

$$v_1' = (0.33 - 0.66) \times 30 + 2 \times 0.66 \times (-40) = -62.7\,\text{m/s} \qquad (\leftarrow)$$

$$v_2' = 2 \times 0.33 \times 30 - (0.33 - 0.66) \times (-40) = 6.6\,\text{m/s} \qquad (\rightarrow)$$

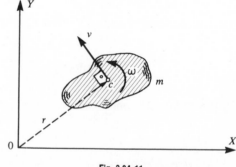

Fig. 3.04–9

(6) Angular Momentum and Impulse

(a) Angular momentum H of a particle of mass m moving with linear velocity v normal to the position radius r measured from a fixed point 0 (Fig. 3.04–10) is

$$H = \text{radius} \times \text{mass} \times \text{velocity} = r \cdot m \cdot v \qquad \left\{ \begin{array}{l} r \text{ in m} \\ m \text{ in kg} \\ v \text{ in m/s} \\ H \text{ in kg-m}^2/\text{s} \end{array} \right\}$$

where H = angular momentum vector normal to the plane of r and v.

(b) Angular momentum H of a body of mass m moving with linear velocity v normal to the position radius r of its mass center measured from a fixed point 0 and rotating about its mass center axis normal to the plane of r and v with angular velocity ω (Fig. 3.04–11) is

$$H = \underbrace{\text{radius} \times \text{mass} \times \text{velocity}}_{r \cdot m \cdot v} + \underbrace{\text{mass moment of inertia} \times \text{angular velocity}}_{I \cdot \omega} \qquad \left[\begin{array}{l} I \text{ in kg-m}^2 \\ \omega \text{ in rad/s} \\ H \text{ in kg-m}^2/\text{s} \end{array} \right]$$

where H = angular momentum vector normal to the plane of r and v and I = mass moment of inertia of the body about the mass center axis normal to the same plane (plane of motion of the body). For pure rotation, $H = I \cdot \omega$; for pure translation, $H = r \cdot m \cdot v$.

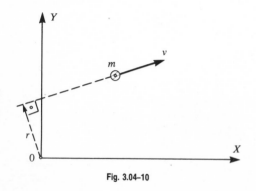

Fig. 3.04–10

Fig. 3.04–11

(c) Units of angular momentum H are

$$1 \text{ newton} \times \text{meter} \times \text{second} = 1 \text{ N-m-s} = 1 \text{ kilogram} \times \text{meter}^2/\text{second} = 1 \text{ kg-m}^2/\text{s}$$
$$= 0.102 \text{ kilogram-force} \times \text{meter} \times \text{second}$$
$$= 0.102 \text{ kgf-m-s}$$

$$1 \text{ poundal} \times \text{foot} \times \text{second} = 1 \text{ pd-ft-sec} = 1 \text{ pound} \times \text{foot}^2/\text{second} = 1 \text{ lb-ft}^2/\text{sec}$$
$$= 0.031 \text{ pound-force} \times \text{foot} \times \text{second}$$
$$= 0.031 \text{ lbf-ft-sec}$$

(d) Average moment $\bar{M}$ equals the ratio of the change in H to the time interval Δt during which the change takes place.

$$\bar{M} = \frac{\Delta H}{\Delta t} = \frac{H_2 - H_1}{t_2 - t_1} = \frac{rm(v_2 - v_1)}{t_2 - t_1} + \frac{I(\omega_2 - \omega_1)}{t_2 - t_1}$$

where H_1, H_2 = angular momenta, v_1, v_2 = linear velocities of the mass centers, and ω_1, ω_2 = angular velocities at the instants t_1, t_2 respectively.

(e) Angular impulse $\bar{M} \, \Delta t$ equals the product of the average moment (acting on the body) by the time of action.

$$\bar{M} \, \Delta t = \bar{M}(t_2 - t_1) = rm(v_2 - v_1) + I(\omega_2 - \omega_1) = H_2 - H_1$$

where $\bar{M} \, \Delta t$ = a vector normal to the plane of motion and $\bar{M}$ is assumed to be constant.

example:

If a moment $\bar{M} = 80,000 \text{ kgf-m}$ is applied at $t_1 = 0$ on a body of mass moment of inertia $I = 100,000 \text{ kg-m}^2$ (which is at rest), then at $t_2 = 10 \text{ s}$ the angular impulse produced by this moment in this time interval is

$$\bar{M} \times (t_2 - \cancel{t_1})^0 = I(\omega_2 - \cancel{\omega_1})^0$$

from which the angular velocity of the body at t_2 is

$$\omega_2 = \frac{\bar{M} \times t_2}{I} = \frac{80,000 \times 9.81 \times 10}{100,000} = 78.5 \text{ rad/s}$$

(f) Conservation of angular momentum. The angular momentum H of a body in motion is conserved (is constant, does not change with time) if the resultant of moments acting on the particle is zero.

$$H = \text{constant} \qquad \bar{M} = 0$$

(7) Equations of Motion

(a) Time derivatives of G (Sec. 3.04–4b) and of H (Sec. 3.04–6b) are

$$\frac{dG}{dt} = \dot{G} = m \cdot \frac{dv}{dt} = m \cdot a = F$$

$$\frac{dH}{dt} = \dot{H} = r \cdot m \cdot \frac{dv}{dt} + I \cdot \frac{d\omega}{dt} = \underbrace{r \cdot m \cdot a}_{r \cdot F} + \underbrace{I \cdot \alpha}_{M_c} = M$$

where F = resultant force acting at the mass center c, $M = r \cdot F + M_c$ = resultant moment at the reference point 0, consisting of the moment of F about 0 (r = arm) and of the moment M_c acting at the mass center c.

(b) Component forms of these time derivatives, called the *equations of plane motion of a rigid body* (Fig. 3.04–12), are

$$m \cdot a_x = F_x \qquad m \cdot a_y = F_y$$

$$-y \cdot m \cdot a_x + x \cdot m \cdot a_y + I \cdot \alpha = M$$

where a_x, a_y = linear accelerations of the body, α = its angular acceleration, m = its mass, I = its mass moment of inertia about the mass center axis normal to the XY plane, x, y = coordinates of mass center c, F_x, F_y = components of the resultant force at c, and M = resultant moment at 0.

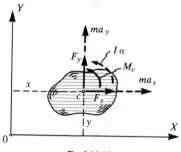

Fig. 3.04–12

(c) Reduced equations. If the third equation of motion in (b) is related to the mass center,

$$m \cdot a_x = F_x \qquad m \cdot a_y = F_y \qquad I \cdot \alpha = M$$

which is a more convenient form of the equations of motion.

(d) Uniformly accelerated motion. If F_x, F_y, and M in (c) are constants, then

$a_x = \ddot{x} = \dfrac{F_x}{m}$	$a_y = \ddot{y} = \dfrac{F_y}{m}$	$\alpha = \ddot{\theta} = \dfrac{M}{I}$
$v_x = \dot{x} = \dfrac{F_x t}{m} + v_{0x}$	$v_y = \dot{y} = \dfrac{F_y t}{m} + v_{0y}$	$\omega = \dot{\theta} = \dfrac{Mt}{I} + \omega_0$
$s_x = x = \dfrac{F_x t^2}{2m} + v_{0x} t + s_{0x}$	$s_y = y = \dfrac{F_y t^2}{2m} + v_{0y} t + s_{0y}$	$\theta = \dfrac{Mt^2}{2I} + \omega_0 t + \theta_0$

where s_{0x}, s_{0y}, θ_0, v_{0x}, v_{0y}, ω_0 = initial conditions of the motion and $M = M_c$.

(8) Rolling of Circular Disks, Cylinders, and Spheres

(a) Horizontal plane. If a body of revolution of weight W and mass moment of inertia I is projected along a straight line on a horizontal plane with initial linear velocity v_0 and initial angular velocity ω_0, then three distinct cases arise depending on the relation of v_0 to ω_0 and the magnitude of the coefficient μ_k of kinetic friction (for I, refer to Appendix A).

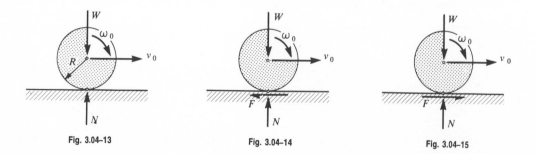

Fig. 3.04–13   Fig. 3.04–14   Fig. 3.04–15

(b) Case I: Condition $v_0 = R\omega_0$. The motion is a uniform motion without slipping at the point of contact (Fig. 3.04–13), described by

$$a = 0 \qquad v = v_0 \qquad s = v_0 t$$

$$\alpha = 0 \qquad \omega = \omega_0 \qquad \theta = \omega_0 t$$

(c) Case II: Condition $v_0 > R\omega_0$. The motion is first a uniformly decelerated motion, caused by the friction acting against the motion at the point of contact (Fig. 3.04–14), described by

$$a = -\mu_k g \qquad v = v_0 - \mu_k g t \qquad s = v_0 t - \frac{\mu_k g t^2}{2}$$

$$\alpha = \frac{\mu_k W R}{I} \qquad \omega = \omega_0 + \frac{\mu_k W R t}{I} \qquad \theta = \omega_0 t + \frac{\mu_k W R t^2}{2I}$$

At $t_1 = \dfrac{v_0 - R\omega_0}{\lambda g \mu_k}$ the slipping stops and the motion is uniform with

$$v_1 = v_0 - \frac{v_0 - R\omega_0}{\lambda} \qquad \omega_1 = \frac{v_1}{R}$$

where $\lambda = 1 + R^2 W / Ig$.

(d) Case III: Condition $v_0 < R\omega_0$. In this case the direction of friction is reversed and the object is first in a uniformly accelerated motion (Fig. 3.04–15), described by

$$a = \mu_k g \qquad v = v_0 + \mu_k g t \qquad s = v_0 t + \frac{\mu_k g t^2}{2}$$

$$\alpha = \frac{-\mu_k W R}{I} \qquad \omega = \omega_0 - \frac{\mu_k W R t}{I} \qquad \theta = \omega_0 t - \frac{\mu_k W R t^2}{2I}$$

At $t_1 = \dfrac{R\omega_0 - v_0}{\lambda g \mu_k}$ the slipping stops and the motion is uniform with

$$v_1 = v_0 + \frac{R\omega_0 - v_0}{\lambda} \qquad \omega_1 = \frac{v_1}{R}$$

where $\lambda = 1 + R^2 W / Ig$.

(e) Inclined plane. If a body of revolution described in (a) rolls down under the action of gravity without slipping on an inclined plane of angle β, then its equations of motion along this plane (Fig. 3.04–16) are

$$a = \frac{gR^2 W \sin\beta}{gI + R^2 W} \qquad v = \frac{gR^2 W \sin\beta}{gI + R^2 W} t \qquad s = \frac{gR^2 W \sin\beta}{2(gI + R^2 W)} t^2$$

and the motion terminates at $s_e = l$, where

$$t_e = \sqrt{\frac{2l(gI + R^2 W)}{gR^2 W \sin\beta}} \qquad v_e = \sqrt{\frac{2ghR^2 W}{gI + R^2 W}}$$

(f) Condition of rolling without slipping in case (e) is

$$\tan\beta_{max} = \frac{\mu_k(gI + R^2 W)}{gI} = \mu_k\left(1 + \frac{R^2 m}{I}\right)$$

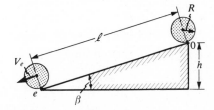

Fig. 3.04–16

example:

If the body in Fig. 3.04–16 is a circular disk of $R = 10$ m, $W = 1,000$ kgf, and $I = \dfrac{WR^2}{2g} = 50,000$ kg-m² which rolls on a plane of $l = 100$ m and $\mu_k = 0.2$,

$$\tan \beta_{max} = 0.2\left(1 + \frac{10^2 \times 1,000}{50,000}\right) = 0.6 \qquad \beta_{max} \cong 31°$$

and the terminal values are then

$$t_e = \sqrt{\frac{2 \times 100(9.81 \times 50,000 + 10^2 \times 1,000 \times 9.81)}{9.81 \times 10^2 \times 1,000 \times 9.81 \times 0.515}} = 7.7 \text{ s}$$

$$v_e = \sqrt{\frac{2 \times 9.81 \times 51.5 \times 10^2 \times 1,000 \times 9.81}{9.81 \times 50,000 + 10^2 \times 1,000 \times 9.81}} = 26 \text{ m/s}$$

3.05 WORK, ENERGY, POWER

(1) Work of a Force

(a) Definition. The work U_F of a force F is a scalar, defined as the product of the linear displacement (path) s and the component F_s of the force along this path (Fig. 3.05–1).

$$U_F = s \cdot F_s = s \cdot (F \cos \alpha)$$

where α = angle of F with s. For $\alpha = 0°$, $U_F = s \cdot F$; and for $\alpha = 90°$, $U_F = s \cdot 0 = 0$.

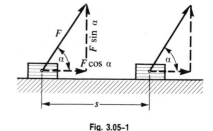

Fig. 3.05–1

(b) SI unit of U_F is 1 joule (after James Prescott Joule), designated by the symbol J, and defined exactly as the work done by the force of 1 newton moving a particle (body) a distance of 1 meter in the direction of the force.

$$1 \text{ joule} = 1 \text{ J} = 1 \text{ meter} \times \text{newton} = 1 \text{ m-N}$$
$$= 0.102 \text{ meter} \times \text{kilogram-force} = 0.102 \text{ m-kgf}$$

(c) Multiples and fractions of 1 joule frequently used in technology are

$$1 \text{ erg} = 1 \text{ centimeter} \times \text{dyne} = 10^{-7} \text{ joule}$$
$$1 \text{ meter} \times \text{kilogram-force} = 1 \text{ m-kgf} = 9.807 \text{ joules}$$

(d) FPS unit of U_F is 1 foot × pound-force, designated by the symbol ft-lbf, and defined exactly as the work done by a force of 1 pound-force moving a particle (body) a distance of 1 foot in the direction of the force.

$$1 \text{ foot} \times \text{pound-force} = 1 \text{ ft-lbf} = 32.17 \text{ foot-poundals} = 32.17 \text{ ft-pd}$$

(e) Two basic conversion relations are

$$1 \text{ ft-lbf} = 1.356 \text{ J} \qquad\qquad 1 \text{ J} = 0.7376 \text{ ft-lbf}$$
$$1 \text{ ft-lbf} = 0.1383 \text{ m-kgf} \qquad\qquad 1 \text{ m-kgf} = 7.2330 \text{ ft-lbf}$$

(f) Work of a force resultant. If several forces $F_1, F_2, \ldots, F_n$ are acting on a body in motion, the resultant work is the product of path s and the component F_s (along the path) of their resultant

$$U_F = s \cdot (F_1 \cos \alpha_1 + F_2 \cos \alpha_2 + \cdots + F_n \cos \alpha_n) = s \cdot F_s$$

where $\alpha_1, \alpha_2, \ldots, \alpha_n$ = direction angles of the particular forces measured from s.

(g) Circular path. If the force F moves along a circular path of radius R (Fig. 3.05–2), then

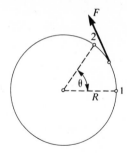

$$U_F = s \cdot F_s = R\theta(F \cos \alpha) = R\theta F$$

where θ = angle of the radii of the end points 1 and 2 of the path and $\alpha = 0$, $\cos \alpha = 1$.

(h) Curvilinear path. If the force F of a constant direction angle ϕ moves along a curvilinear plane path (Fig. 3.05–3), then

Fig. 3.05–2

$$U_F = \int_{s_1}^{s_2} F_s \cdot ds = \int_{x_1}^{x_2} F_x \, dx + \int_{y_1}^{y_2} F_y \, dy$$

where ds = element of path, dx and dy = projections of ds in the X and Y axes, respectively,

$$F_x = F \cos \phi \qquad F_y = F \sin \phi$$

and x_1, y_1, x_2, y_2 = coordinates of the end points of the path.

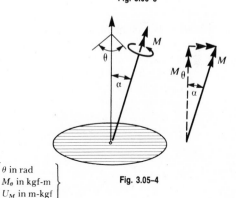

Fig. 3.05–3

(2) Work of a Moment

(a) Definition. The work U_M of a moment M is a scalar, defined as the product of the angular displacement (rotation) θ and the component M_θ of the moment along the axis of θ (Fig. 3.05–4).

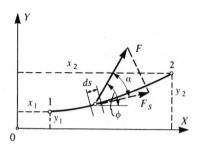

$$U_M = \theta \cdot M_\theta = \theta \cdot (M \cos \alpha) \qquad \begin{cases} \theta \text{ in rad} \\ M_\theta \text{ in kgf-m} \\ U_M \text{ in m-kgf} \end{cases}$$

Fig. 3.05–4

where α = direction angle of M measured from the axis of θ. For $\alpha = 0°$, $U_M = \theta \cdot M$; for $\alpha = 90°$, $U_M = \theta \cdot 0 = 0$.

(b) SI and FPS units of U_M and their conversion factors are the same as in Secs. 3.05–1b to e.

(c) Work of a moment resultant. If several moments $M_1, M_2, \ldots, M_n$ are acting on a body in motion, the resultant work is the product of the rotation θ and the component M_θ (along the axis of θ) of the resultant moment.

$$U_M = \theta \cdot (M_1 \cos \alpha_1 + M_2 \cos \alpha_2 + \cdots + M_n \cos \alpha_n) = \theta \cdot M_\theta$$

where $\alpha_1, \alpha_2, \ldots, \alpha_n$ = direction angles of the respective moments measured from the axis of θ.

(3) Kinetic Energy

(a) **Definition.** The energy of a body is its capacity to perform work. It is a scalar quantity measured in the same units as work (Secs. 3.05–1b to e).

(b) **Classification.** The two most important types of mechanical energy are kinetic energy (energy of motion) and potential energy (energy of position).

(c) **Translatory kinetic energy** T_v of a body in linear motion is

$$T_v = \tfrac{1}{2} mv^2 \qquad \left\{ \begin{array}{l} T_v \text{ in m-kgf} \\ m \text{ in kg} \\ v \text{ in m/s} \end{array} \right\}$$

where m = mass and v = linear velocity of the body.

(d) **Rotary kinetic energy** T_ω of a body in rotation about its mass center axis is

$$T_\omega = \tfrac{1}{2} I\omega^2 \qquad \left\{ \begin{array}{l} T_\omega \text{ in m-kgf} \\ I \text{ in kg-m}^2 \\ \omega \text{ in rad/s} \end{array} \right\}$$

where I = mass moment of inertia about the axis of rotation and ω = angular velocity of the body.

(e) **Work of a force** F required to translate a body of mass m along a straight path from position 1 to position 2 equals the change in the kinetic energy of this body between these two positions (Fig. 3.05–5).

$$\underbrace{(s_2 - s_1) \cdot (F \cos \alpha)}_{U_F} = \underbrace{\frac{m}{2}(v_2^2 - v_1^2)}_{\Delta T_v}$$

where U_F = work (Sec. 3.05–1a), ΔT_v = change in translatory kinetic energy (Sec. 3.05–3c), s_1, s_2 = position coordinates, and v_1, v_2 = linear velocities at 1 and 2, respectively.

(f) **Work of a moment** M required to rotate a body of mass moment of inertia I from the angular position 1 to the angular position 2 equals the change in rotary kinetic energy between these two positions (Fig. 3.05–6).

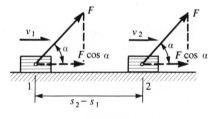

Fig. 3.05–5

$$\underbrace{(\theta_2 - \theta_1) \cdot (M \cos \alpha)}_{U_M} = \underbrace{\frac{I}{2}(\omega_2^2 - \omega_1^2)}_{\Delta T_\omega}$$

where U_M = work (Sec. 3.05–2a), ΔT_ω = change in rotary kinetic energy (Sec. 3.05–3d), θ_1, θ_2 = position angles, and ω_1, ω_2 = angular velocities at 1 and 2, respectively.

(g) **Force** F **required** to be applied in the direction of translatory motion of the body of mass m to increase its linear velocity from v_1 at s_1 to v_2 at s_2 is

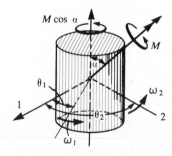

Fig. 3.05–6

$$F = \frac{m(v_2^2 - v_1^2)}{2(s_2 - s_1)} = \frac{m(v_2 - v_1)}{t_2 - t_1}$$

where s_1, s_2, v_1, v_2 have the same meaning as in (e) and t_1, t_2 = corresponding instants of time.

example:

An automobile of weight $W = 2{,}000$ kgf is moving at 50 km/h when the driver jams the brakes and skids rectilinearly to a full stop (Fig. 3.05–7). If the coefficient of kinetic friction $\mu_k = 0.6$, then from

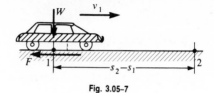

Fig. 3.05-7

$$F = \frac{m}{2}\frac{(v_2^2 - v_1^2)}{s_2 - s_1}$$

where $\quad F = -W \cdot \mu_k = -2{,}000 \times 0.6 = -1{,}200$ kgf, $\quad m = \dfrac{W}{g} = \dfrac{(2{,}000 \times 9.81)}{9.81} = 2{,}000$ kg, $\quad v_1 = 50 =$

$\dfrac{50{,}000}{3{,}600} = 13.9$ m/s, $v_2 = 0$, and $s_1 = 0$, the distance of skidding is

$$s = s_2 = \frac{-mv_1^2}{2F} = \frac{-2{,}000 \times 13.9^2}{-2 \times 1{,}200 \times 9.81} = 16.4 \text{ m}$$

and from

$$F = \frac{m(v_2 - v_1)}{t_2 - t_1}$$

where $t_1 = 0$, the time interval of skidding is

$$t = t_2 = \frac{-mv_1}{F} = \frac{-2{,}000 \times 13.9}{-1{,}200 \times 9.81} = 2.36 \text{ s}$$

(h) Moment M required to be applied about the axis and in the direction of the rotation of a body of mass moment of inertia I to increase its angular velocity from ω_1 to ω_2 is

$$M = \frac{I(\omega_2^2 - \omega_1^2)}{2(\theta_2 - \theta_1)} = \frac{I(\omega_2 - \omega_1)}{t_2 - t_1}$$

where $\theta_1, \theta_2, v_1, v_2 =$ same meaning as in (f) and $t_1, t_2 =$ corresponding instants of time.

(i) Work–kinetic energy equation of a rigid body in plane motion is the sum of relations (e) and (f); that is,

$$\underbrace{(s_2 - s_1) \cdot (F \cos \alpha_F)}_{\Delta U_F} + \underbrace{(\theta_2 - \theta_1) \cdot (M \cos \alpha_M)}_{\Delta U_M} = \underbrace{\frac{m}{2}(v_2^2 - v_1^2)}_{\Delta T_v} + \underbrace{\frac{I}{2}(\omega_2^2 - \omega_1^2)}_{\Delta T_\omega}$$

where α_F and $\alpha_M =$ direction angles of F and M, related to the direction of s and θ, respectively.

(4) Potential Energy

(a) Two types of potential energy frequently encountered in mechanics of solids are gravitational potential energy V_g and elastic potential energy V_e, also called *strain energy*. According to Sec. 3.05–3a, they are scalars measured in the same units as work.

(b) Gravitational potential energy V_g is the result of the position of a body of mass m in the gravitational field of the earth.

$$V_g = \text{weight} \times \text{height} = W \cdot h = m \cdot g \cdot h \qquad \left\{ \begin{array}{l} W \text{ in kgf} \\ m \text{ in kg} \\ h \text{ in m} \\ V_g \text{ in m-kgf} \end{array} \right\}$$

where $g = 9.806\ 65$ m/s^2 (acceleration due to gravity) and $h =$ height above ground level.

(c) **Linear strain energy** V_s is one form of the elastic potential energy V_e stored in a body during its linear deformation.

$$V_s = \tfrac{1}{2}k_s \cdot s^2 \qquad \left\{ \begin{array}{l} k_s \text{ in kgf/m} \\ s \text{ in m} \\ V_s \text{ in m-kgf} \end{array} \right\}$$

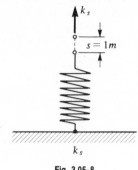

where k_s = linear stiffness defined as the force produced in an elastic body (Sec. 4.01) by a unit linear displacement and s = linear displacement (Fig. 3.05–8).

(d) **Angular strain energy** V_θ is another form of the elastic potential energy V_e stored in a body during its angular deformation.

$$V_\theta = \tfrac{1}{2}k_\theta \cdot \theta^2 \qquad \left\{ \begin{array}{l} k_\theta \text{ in kgf-m/rad} \\ \theta \text{ in rad} \\ V_\theta \text{ in m-kgf} \end{array} \right\}$$

k_s

Fig. 3.05–8

where k_θ = angular stiffness defined as the moment produced in an elastic body by a unit angular displacement and θ = angular displacement (Fig. 3.05–9).

(e) **Change in potential energy** due to the displacement of the body from one position to another is the negative of the work done by the acting forces on the body during this change in position.

$$V_g = -U_W \qquad V_s = -U_F \qquad V_\theta = -U_M$$

where U_W = work of gravity and U_F and U_M = work of force and of a moment, respectively.

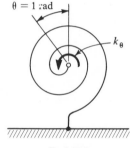

example:

The work of gravity on a body of mass 1,000 kg lifted to a height $h = 10$ m is

$$U_W = -h \cdot m \cdot g = -10 \times 1,000 \times 9.81 = -98,100 \text{ m-kgf}$$

and the gravitational potential energy of the body produced by this displacement is

$$V_g = -U_W = 98,100 \text{ m-kgf}$$

Fig. 3.05–9

(5) Conservation and Dissipation of Energy

(a) **Principle of conservation of mechanical energy.** For any isolated mechanical system the total energy E (kinetic + potential) remains constant (is conserved) at all times during the motion of the body; i.e., if at t_1 and t_2, respectively,

$$\underbrace{\underbrace{\frac{mv_1^2}{2}}_{T_{v1}} + \underbrace{\frac{I\omega_1^2}{2}}_{T_{\theta 1}}}_{T_1} + \underbrace{\underbrace{mgh_1}_{V_{g1}} + \underbrace{\frac{k_s s_1^2}{2}}_{V_{s1}} + \underbrace{\frac{k_\theta \theta_1^2}{2}}_{V_{\theta 1}}}_{V_1} = E \qquad \underbrace{\underbrace{\frac{mv_2^2}{2}}_{T_{v2}} + \underbrace{\frac{I\omega_2^3}{2}}_{T_{\theta 2}}}_{T_2} + \underbrace{\underbrace{mgh_2}_{V_{g2}} + \underbrace{\frac{k_s s_2^2}{2}}_{V_{s2}} + \underbrace{\frac{k_\theta \theta_2^2}{2}}_{V_{\theta 2}}}_{V_2} = E$$

then

$$T_1 + V_1 = T_2 + V_2 \qquad \text{or} \qquad \underbrace{(T_2 - T_1)}_{\Delta T} + \underbrace{(V_2 - V_1)}_{\Delta V} = 0 \qquad \text{or} \qquad \Delta T + \Delta V = 0$$

where ΔT = changes in kinetic energy and ΔV = changes in potential energy during the time interval $\Delta t = t_2 - t_1$.

example:

If a right circular cylinder of radius R, mass m, and mass moment of inertia I is released from point 1 on an inclined plane (Fig. 3.05–10) with $v_1 = 0$ and $\omega_1 = 0$, then its linear velocity v_2 and its angular velocity ω_2 when it rolls down without slipping to point 2 are computed from

$$\frac{mv_1^{2\,0}}{2} + \frac{I\omega_1^{2\,0}}{2} + mgh_1 = \frac{mv_2^2}{2} + \frac{I\omega_2^2}{2} + mgh_2^{\,0}$$

where h_1 is the vertical distance between 1 and 2.

Since slipping is prevented, $v = R\omega$ and

$$mgh_1 = \frac{mR^2\omega_2^2}{2} + \frac{I\omega_2^2}{2}$$

from which

$$\omega_2 = \sqrt{\frac{2mgh_1}{mR^2 + I}} \qquad \text{and} \qquad v_2 = R\sqrt{\frac{2mgh_1}{mR^2 + I}}$$

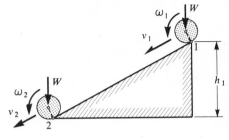

Fig. 3.05–10

(b) Dissipation of energy. If a part of the total energy of the system is transformed by friction into heat, this dissipated energy equals the change in the total energy ΔE.

$$\underbrace{(T_2 - T_1)}_{\Delta T} + \underbrace{(V_2 - V_1)}_{\Delta V} = \Delta E$$

where $\Delta E =$ work of friction, which is negative.

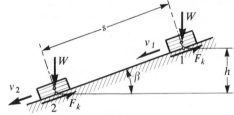

Fig. 3.05–11

example:

If a block of weight W slides from rest on an inclined plane of μ_k (Fig. 3.05–11) from point 1 at $h_1 = h$ to point 2 at $h_2 = 0$, then

$$\frac{mv_2^2}{2} - \frac{mv_1^{2\,0}}{2} + mgh_2^{\,0} - mgh_1 = -(mg \cos \beta)\mu_k s$$

from which the velocity of the block at 2 is

$$v_2 = \sqrt{2(gh - gs\mu_k \cos \beta)}$$

where $\beta =$ angle of the plane and $s \cos \beta = h \cot \beta$.

(6) Power

(a) Definition. Power is the time rate at which work is performed; it is a scalar.

$$\text{Power} = \frac{\text{work}}{\text{time}} = \frac{U}{t}$$

where U is either $U_F = s \cdot F_s$ (Sec. 3.05–1) or $U_M = \theta \cdot M_\theta$ (Sec. 3.05–2).

(b) Alternative definition. Power can also be defined as the product of cause (force or moment) and the velocity.

$$\text{Power} = \begin{cases} \text{force} \times \text{linear velocity} = F \cdot v \\ \text{moment} \times \text{angular velocity} = M \cdot \omega \end{cases}$$

where the respective velocity is acting in the direction of cause.

Dynamics of Rigid Bodies **101**

(c) SI unit of power is 1 watt (after James Watt), designated by the symbol W, and defined exactly as the work of 1 joule per second.

$$1 \text{ watt} = 1 \text{ W} = 1 \text{ joule/second} \qquad = 1 \text{ J/s}$$
$$= 1 \text{ meter} \times \text{newton/second} \qquad = 1 \text{ m-N/s}$$
$$= 0.102 \text{ meter} \times \text{kilogramf/second} = 0.102 \text{ m-kgf/s}$$

(d) Multiples and fractions of 1 watt frequently used in technology are

$$1 \text{ kilowatt} = 1 \text{ kW} = 10^3 \text{ W} = 102 \text{ m-kgf/s}$$
$$1 \text{ horsepower (MKS)} = 1 \text{ hp} = 735.5 \text{ W} = 75 \text{ m-kgf/s}$$

(e) FPS unit of power is 1 foot $\times$ poundf/second, designated by the symbol ft-lbf/sec, and defined exactly as 0.138 254 954 m-kgf/s.

(f) Multiples and fractions of 1 ft-lbf/sec frequently used in technology are

$$1 \text{ foot} \times \text{kip-force/second} = 1 \text{ ft-kipf/sec} = 10^3 \text{ ft-lbf/sec}$$
$$1 \text{ horsepower (FPS)} = 1 \text{ hp} = 550 \text{ ft-lbf/sec}$$

(g) Two basic conversion relations are

$$1 \text{ ft-lbf/sec} = 1.356 \text{ W} = 0.138 \text{ m-kgf/s}$$
$$1 \text{ m-kgf/s} = 9.81 \text{ W} = 7.23 \text{ ft-lbf/sec}$$
$$1 \text{ hp} \quad \text{(FPS)} = 745.7 \text{ W} = 1.014 \text{ hp} \quad \text{(MKS)}$$
$$1 \text{ hp} \quad \text{(MKS)} = 735.5 \text{ W} = 0.986 \text{ hp} \quad \text{(FPS)}$$

(h) Work in terms of power is the product of power and time, which leads to two commonly used units of work:

$$1 \text{ kilowatt} \times \text{hour} = 1 \text{ kilowatt-hour} = 1 \text{ kW-h}$$
$$1 \text{ horsepower} \times \text{hour} = 1 \text{ horsepower-hour} = 1 \text{ hp-h}$$

3.06 SIMPLE MACHINES

(1) Definition and Classification

(a) Machine is a device by which the point of application, magnitude, and/or direction of a force (or of forces) performing work are changed to gain practical advantage.

(b) Complex machines can be interpreted as systems of *simple machines*, which are

    (1) Lever   (4) Inclined plane

    (2) Wheel on axle   (5) Wedge

    (3) Free pulley   (6) Screw

(c) Operation of a machine is described by three parameters:

    (1) Input work U_i (work used to operate the machine).

    (2) Work loss U_f (work to overcome friction).

    (3) Output work U_0 (work done by the machine).

(d) Work-balance equation of a machine is

$$\underbrace{\text{Input work}}_{U_i} = \underbrace{\text{work loss}}_{U_f} + \underbrace{\text{output work}}_{U_0}$$

Hence the output work is always less than the input work.

(e) Efficiency of a machine is

$$\text{Eff.} = \frac{\text{output work}}{\text{input work}} = \frac{U_0}{U_i} = \frac{\text{output power}}{\text{input power}} = \frac{U_0/t}{U_i/t}$$

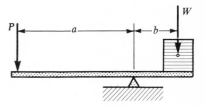

(f) Actual mechanical advantage of a machine is

$$\text{A.M.A.} = \frac{\text{output force}}{\text{input force}} = \frac{F_0}{F_i} = \text{force ratio}$$

(g) Ideal mechanical advantage of a machine is

$$\text{I.M.A.} = \frac{\text{displacement of input force}}{\text{displacement of output force}} = \frac{S_i}{S_0} = \text{displacement ratio}$$

Fig. 3.06–1

(2) Applications of the Lever

(a) Lever (Fig. 3.06–1) consists of a bar (straight, bent, or curved) pivoted at a point called the *fulcrum* and acted upon by the load W (output force) and the force P (input force).

$$Pa = Wb \qquad \text{A.M.A.} = \frac{W}{P}$$

example:

If $a = 4\,\text{m}$, $b = 2\,\text{m}$, and $W = 1{,}000\,\text{kgf}$, then

$$P = \frac{Wb}{a} = \frac{1{,}000 \times 2}{4} = 500\,\text{kgf} \qquad \text{A.M.A.} = \frac{1{,}000}{500} = 2$$

Fig. 3.06–2

(b) Wheel on an axle (Fig. 3.06–2) is the angular equivalent of the lever, where the moment of the wheel $P \cdot R$ is the input force moment and the moment of the axle $W \cdot r$ is the output moment and R, r are the radii of the wheel and of the axle, respectively.

$$PR = Wr \qquad \text{A.M.A.} = \frac{W}{P}$$

(3) Fixed and Free Pulleys

(a) Fixed pulley (Fig. 3.06–3) is another angular application of the principle of the lever, where the moment of the input force $P \cdot r$ equals the moment of the load $W \cdot r$, from which (Sec. 2.08–10)

$$P = W \qquad s_i = s_0 \qquad \text{A.M.A.} = 1$$

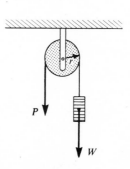

and the only advantage of this device is the change of direction of the input force.

Fig. 3.06–3

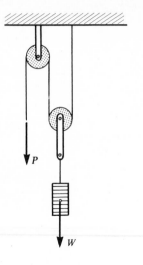 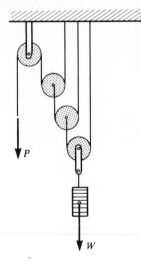 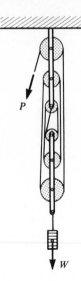

| Fig. 3.06–4 | Fig. 3.06–5 | Fig. 3.06–6 |

(b) One free pulley and one fixed pulley (Fig. 3.06–4) reduce the input force to

$$P = \frac{W}{2} \qquad s_i = 2s_0 \qquad \text{A.M.A.} = 2$$

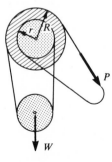

Fig. 3.06–7

(c) n free pulleys and one fixed pulley (Fig. 3.06–5) reduce the input force to

$$P = \frac{W}{2^n} \qquad s_i = 2^n s_0 \qquad \text{A.M.A.} = 2^n$$

(d) Block and tackle (Fig. 3.06–6) is a machine consisting of n free pulleys and n stationary pulleys which reduce the input force to

$$P = \frac{W}{2n} \qquad s_i = 2ns_0 \qquad \text{A.M.A.} = 2n$$

(e) Differential pulley (Fig. 3.06–7) consists of two pulleys of radius R and r, respectively, fastened together.

$$P = \frac{(R-r)W}{2R} \qquad s_i = 2\pi R \qquad s_0 = \pi(R-r) \qquad \text{A.M.A.} = \frac{2R}{R-r}$$

104 *Dynamics of Rigid Bodies*

(4) Applications of the Inclined Plane

(a) Inclined plane—tensile force (Fig. 3.06–8).

If the block of weight W is being moved up by a tensile force P, then

$$P = \frac{W \tan(\beta + \phi_k)}{\tan(\alpha + \phi_k)} \qquad s_i = \frac{s_0}{\sin \beta}$$

$$\text{A.M.A.} = \frac{\tan(\alpha + \phi_k)}{\tan(\beta + \phi_k)}$$

where ϕ_k = angle of kinetic friction.

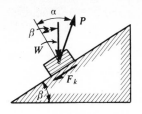

Fig. 3.06–8

(b) Inclined plane—compressive force (Fig. 3.06–9).

If the block of weight W is being moved up by a compressive force P, then

$$P = \frac{W \tan(\beta + \phi_k)}{\tan(\alpha - \phi_k)} \qquad s_i = \frac{s_0}{\sin \beta}$$

$$\text{A.M.A.} = \frac{\tan(\alpha - \phi_k)}{\tan(\beta + \phi_k)}$$

where ϕ_k = angle of kinetic friction.

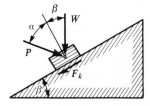

Fig. 3.06–9

(c) Unsymmetrical wedge–lifting (Fig. 3.06–10).

The force P required to lift the block of weight W is

$$P = W[\tan(\alpha + \phi_{k1}) + \tan(\beta + \phi_{k2})] \qquad s_i = \frac{s_0}{\tan \alpha + \tan \beta}$$

$$\text{A.M.A.} = \frac{1}{\tan(\alpha + \phi_{k1}) + \tan(\beta + \phi_{k2})}$$

where ϕ_{k1}, ϕ_{k2} = respective angles of kinetic friction.

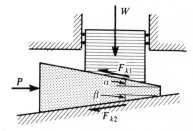

Fig. 3.06–10

(d) Unsymmetrical wedge–lowering (Fig. 3.06–11).

The force required to lower the block of weight W is

$$P = W[\tan(\phi_{k1} - \alpha) + \tan(\phi_{k2} - \beta)] \qquad s_i = \frac{s_0}{\tan \alpha + \tan \beta}$$

$$\text{A.M.A.} = \frac{1}{\tan(\phi_{k1} - \alpha) + \tan(\phi_{k2} - \beta)}$$

where ϕ_{k1}, ϕ_{k2} = respective angles of kinetic friction.

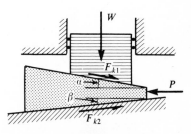

Fig. 3.06–11

(e) Screw–lifting (Fig. 3.06–12). The torque T required to raise the load W is

$$T = Pa = WR \tan(\alpha + \phi_k) \qquad s_i = \frac{2\pi R}{h} s_0$$

$$\tan \alpha = \frac{h}{2\pi R}$$

$$\text{A.M.A.} = \frac{a}{R \tan(\alpha + \phi_k)}$$

where $P = F_i$, $a = $ arm, and $\phi_k = $ angle of kinetic friction.

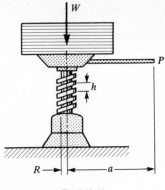

Fig. 3.06–12

(f) Screw–lowering (Fig. 3.06–12). The torque T required to lower the load W is

$$T = Pa = WR \tan(\phi_k - \alpha) \qquad s_i = \frac{2\pi R}{h} s_0$$

$$\tan \alpha = \frac{h}{2\pi R}$$

$$\text{A.M.A.} = \frac{a}{R \tan(\phi_k - \alpha)}$$

where P, a, and ϕ_k have the same meaning as in (e).

4

MECHANICS OF
DEFORMABLE SOLIDS

4.01 CHARACTERISTICS OF SOLIDS

(1) Structure of Solids

(a) Solid is an aggregate of matter which retains its shape (in absence of external constraints) for a length of time. Solids used in technology are called *materials*.

(b) Molecule is the physical limit of divisibility (by cutting, crushing, grinding, etc.) of a solid, and for inorganic materials the average size of a molecule is of the order 10^{-8} to 10^{-7} cm and its mass is of the order 10^{-24} to 10^{-21} g.

(c) Atom is the chemical limit of divisibility (by chemical decomposition) of a molecule and is composed of three types of elementary particles (present in the atom in different numbers) called *protons, neutrons,* and *electrons* (Sec. 7.01–1).

(2) Morphology of Solids

(a) Morphologically, solids are classified as crystalline or amorphous and as homogeneous or inhomogeneous.

(b) Crystalline solids are composed of crystals, each of which consists of atoms arranged in three-dimensional patterns called *Bravais lattices.* Steel, gold, and salts are typical examples of crystalline solids.

(c) Amorphous solids (also called *noncrystalline* solids) are not composed of repetitive, three-dimensional patterns of atoms. Sometimes they are called collectively *glasses.* Resins, wax, and glass are typical examples of amorphous solids.

(d) Homogeneous solids have the same properties in all their parts. Steel, clay, and glass are typical examples of homogeneous solids.

(e) Inhomogeneous solids have properties varying with their location in the solid. Certain soils and gravel deposits fall in this category.

(3) Causes and Effects

(a) Causes (stimuli) acting upon solids are classified as:

(α) *Primary causes,* which are the external forces (loads, reactions) and the internal forces (prestressing, weight, etc.)

(β) *Secondary causes,* which are changes in temperature and in moisture content

(b) Responses of solids to these causes are called the *effects,* and they are classified as *stresses* and *deformations.*

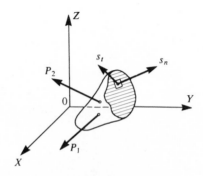

Fig. 4.01–1

(c) **Stress** in a solid is the intensity of force per unit area of its cross section, and it is a vector. According to its direction relative to the plane of action, a stress is designated as a *normal stress* s_n (normal to its plane of action) or as a *shearing stress* s_t (tangential to its plane of action) (Fig. 4.01–1).

(d) **Deformation** of a solid is the change in its form (change in volume and/or shape). The deformations are classified as elastic, plastic, elastoplastic, viscoelastic, and viscoplastic.

(e) **Mechanics of deformable solids (bodies)** is the systematic study of the stresses and deformations produced by the action of causes and of the cause-effect relations.

(f) **Rigid body (nondeformable body)**, encountered in Chaps. 2 and 3, is an analytical idealization which has no real equivalent in nature; all bodies are deformable to some extent.

(g) **Loading and unloading** define, respectively, the states of application and of removal of the causes of effects.

(4) Deformation Characteristics

(a) **Elasticity** is the property of the body of returning instantly and completely, after unloading, to its initial form (snapping back).

(b) **Plasticity** is the property of the body of retaining permanently and completely, after unloading, its deformed shape (complete permanent set).

(c) **Elastoplasticity** is the property of the body of retaining permanently but only partially, after unloading, its deformed shape (partially snapping back).

(d) **Viscoelasticity** is the property of the body of continuously increasing deformation (time-dependent creep) while loaded and of continuously retreating deformation after unloading (time-dependent relaxation), and of a return to the initial form after a certain interval of time (which can be infinite).

(e) **Viscoplasticity** is the property of the body of continuously increasing deformation (time-dependent creep) while loaded and of retention of the deformed shape after unloading.

4.02 MECHANICAL TESTS

(1) Classification

(a) **Mechanical tests** fall into six major categories: tensile test, compression test, hardness test, toughness test, fatigue test, and creep test.

(b) **Tensile and compression tests** investigate the capacity of materials to withstand static loads in tension and in compression, respectively.

(c) **Hardness test** is used to determine the resistance of materials to permanent deformation.

(d) **Toughness test,** also called the *impact* test, is used to determine the resistance of materials to shock-loading conditions.

(e) **Fatigue test** investigates the useful life span of materials under the action of cyclic loads (pulsating loads, reversing loads).

(f) **Creep test** is used to study the behavior of materials subjected to a sustained load for a long period of time.

(2) Tensile Test

(a) **Stress–deformation relation** must be determined from experiments and depends on the material. The most common test for the study of this relation is the tensile test performed at room temperature on a specimen (sample) of the material inserted in the testing machine.

(b) Specimen is usually a circular bar or a coupon of a plate with the profile shown in Fig. 4.02–1.

(c) Normal stress s_n is computed as the ratio of the tensile force P (applied at the ends of the specimen) to the initial cross-sectional area A_0.

$$s_n = \frac{P}{A_0}$$ $\begin{cases} s_n \text{ in kgf/cm}^2 \\ P \text{ in kgf} \\ A_0 \text{ in cm}^2 \end{cases}$

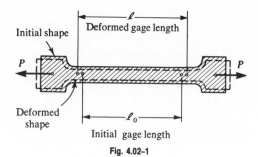

Initial shape

Deformed gage length

ℓ

P P

Deformed shape

Initial gage length

ℓ_0

Fig. 4.02–1

(d) Normal strain ϵ_n is the ratio of the elongation of the gage length $\Delta l = l - l_0$ to the initial gage length l_0 (Fig. 4.02–1).

$$\epsilon_n = \frac{l - l_0}{l_0}$$ $\begin{cases} \epsilon_n \text{ dimensionless} \\ l \text{ in cm} \\ l_0 \text{ in cm} \end{cases}$

(3) Tensile Stress-Strain Curve, General Case

(a) Testing of the specimen consists of applying a gradually increasing load and simultaneously reading the corresponding gage elongations from which the values of stress and strain are calculated.

(b) Graphical representation of these values is shown as the *stress-strain curve*, which is specific for each type of material (typical such curves are shown in Fig. 4.02–2). These curves define the behavior of the material in tension and reveal all or some of the following characteristics.

(c) Proportional limit (Fig. 4.02–2, point 1) is the stress up to which the ratio of stress and strain is constant.

(d) Elastic limit (Fig. 4.02–2, point 2) is the stress beyond which the material will not return to its original shape when unloaded and a permanent deformation will occur.

(e) Yield point (Fig. 4.02–2, point 3) is the stress at which there is an increase in strain with no increase in stress. Most materials do not possess this property.

(f) Ultimate strength (Fig. 4.02–2, point 4) is the maximum stress within the material before it begins to fail.

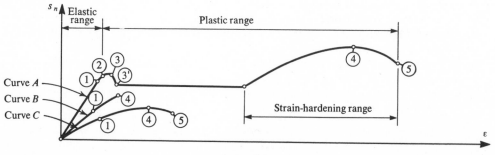

Fig. 4.02–2

(g) Breaking point is the stress at rupture (failure), which for some materials is lower than the ultimate stress.

(4) Tensile Stress-Strain Curve, Typical Cases

(a) **Ductile material with yield point.** Curve *A* of Fig. 4.02–2 rises first with constant slope up to the proportional limit (point 1), above which are located two yield points (points 3 and 3′) and a flat region in which the strain rises under the action of constant stress. With an additional increase in stress the curve begins to rise again and reaches its maximum (ultimate strength), after which the material flows and finally fails. During the transition from the ultimate strength to the breaking stress, the actual stress increases as the material forms a neck (Fig. 4.02–3), but since the stress is computed on the basis of the initial cross section, the breaking stress is somewhat lower than the ultimate stress. The stress-strain curve of mild steel is a typical *A* curve.

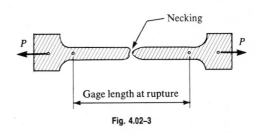

Fig. 4.02–3

(b) **Brittle material.** Curve *B* of Fig. 4.02–2 has neither a yield point nor a flat yield region, and after a straight-line rise ending with the proportional limit (point 1), the stress-strain curve terminates with point 4, which is the ultimate strength and also the breaking stress. Iron, brick, and rocks have stress-strain curves of this type.

(c) **Ductile material without yield point.** Curve *C* of Fig. 4.02–2 has, in addition to points 1 and 4, a distinct breaking point 5 and is typical for high strength steel and other metals.

(5) Observations from Tensile Test

(a) **Elastic region** is the straight-line portion of the stress-strain curve in which the stress is linearly proportional to the strain. Most engineering materials are or can be assumed to be elastic up to a certain limit.

(b) **Plastic region** is the portion of the stress-strain curve in which the strain increases under constant load.

(c) **Flexibility** is the property of the material to undergo large elastic deformations without rupture.

(d) **Ductility** is the property of a material to undergo large plastic deformations without rupture.

(e) **Brittleness** is the property of a material to fail after a very small deformation.

(f) **Toughness** is the property of a material to withstand large stresses and deformations before rupture (opposite of brittleness).

(6) Compression Test

(a) **For brittle materials** the tensile and compressive stress-strain curves are different, and their ultimate strengths in tension and in compression are also different. Figure 4.02–4 shows the difference of these curves for gray cast iron and concrete. Because of submicroscopic cracks, brittle materials are weak in tension but quite strong in compression.

(b) **For ductile materials** the tensile and compressive stress-strain curves are almost identical, and for technical purposes their ultimate strengths in tension and compression are assumed to be the same.

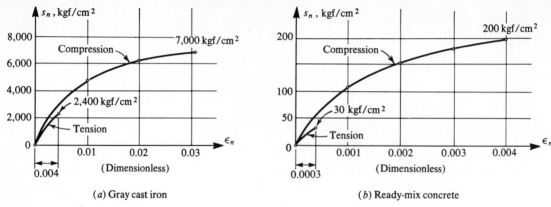

(a) Gray cast iron   (b) Ready-mix concrete

Fig. 4.02–4

4.03 LINEAR ELASTICITY, BASIC CASES

(1) Pure Tension and Compression

(a) Static equilibrium. If a slender bar of constant cross section acted upon by the axial tensile forces P applied at its ends (Fig. 4.03–1a) is intersected by a plane normal to its axis and separated into two parts (two free-body diagrams, Fig. 4.03–1b), then from the static equilibrium of either part,

$$P = s_n A$$

where $s_n A$ = resultant stress, s_n = normal stress, and A = area of the cross section. If P is a tensile force (Fig. 4.03–1a), s_n acts from the section (Fig. 4.03–1b). If P is a compressive force (Fig. 4.03–2a), s_n acts toward the section (Fig. 4.03–2b).

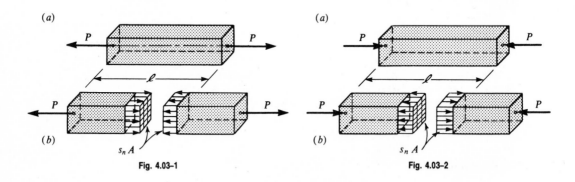

Fig. 4.03–1   **Fig. 4.03–2**

(b) Normal stress from (a) is then

$$s_n = \frac{P}{A} \qquad \left\{ \begin{array}{l} P \text{ in kgf} \\ A \text{ in cm}^2 \\ s_n \text{ in kgf/cm}^2 \end{array} \right\}$$

which is valid only if P is applied along the axis of the bar.

(c) Units of normal stress are

$$1 \text{ kilogram-force/centimeter}^2 = 1 \text{ kgf/cm}^2$$
$$1 \text{ pound-force/inch}^2 \qquad = 1 \text{ lbf/in.}^2$$

and their relations are

$$1 \text{ lbf/in.}^2 = 0.0703 \text{ kgf/cm}^2 \qquad 1 \text{ kgf/cm}^2 = 14.22 \text{ lbf/in.}^2$$

Units of the modulus of elasticity are identical (below).

(d) Hooke's law. Within the elastic range of homogeneous solids the ratio of the normal stress to the normal strain is constant; that is,

$$s_n = E\epsilon_n \qquad \begin{Bmatrix} E \text{ in kgf/cm}^2 \\ \epsilon_n \text{ dimensionless} \end{Bmatrix}$$

where E = constant of proportionality, called the *modulus of elasticity*, and ϵ_n = normal strain introduced in Sec. 4.02–2d.

(e) Axial deformation Δ_n is the change in length of the bar caused by the axial force P. In case of tension (Fig. 4.03–3), Δ_n = elongation. In case of compression (Fig. 4.03–4), Δ_n = contraction.

$$\Delta_n = l\epsilon_n = l\frac{s_n}{E} = \frac{Pl}{AE} \qquad \begin{Bmatrix} \Delta_n, l \text{ in cm} \\ P \text{ in kgf} \\ A \text{ in cm}^2 \end{Bmatrix}$$

where l = length of the bar.

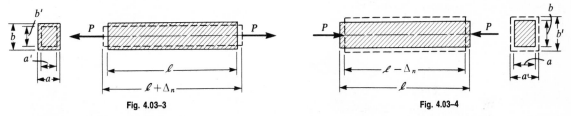

Fig. 4.03–3   Fig. 4.03–4

(f) Poisson's ratio. The elongation or contraction of the bar in a state of axial stress is always accompanied by the contraction or expansion of the cross section, respectively. The ratio of the normal strain ϵ_n to the lateral strain ϵ_a or ϵ_b is a constant called *Poisson's ratio* ν.

$$\nu = \frac{\epsilon_a}{\epsilon_n} = \frac{\epsilon_b}{\epsilon_n} \qquad \{\epsilon_a, \epsilon_b, \epsilon_n, \nu \text{ dimensionless}\}$$

Experiments show that Poisson's ratio for most engineering materials ranges from 0.0 to 0.5, with rubber approaching the top limit, cork the lower limit, and in the case of structural steel, $\nu = 0.3$.

(g) Lateral deformations (deformations of the cross section) in terms of (f) are

$$\Delta_a = a\epsilon_a = a\nu\epsilon_n = a\nu\frac{s_n}{E} = \nu\frac{Pa}{AE}$$

$$\Delta_b = b\epsilon_b = b\nu\epsilon_n = b\nu\frac{s_n}{E} = \nu\frac{Pb}{AE}$$

In Fig. 4.03–3,

$$a' = a - \Delta_a \qquad b' = b - \Delta_b$$

In Fig. 4.03–4,

$$a' = a + \Delta_a \qquad b' = b + \Delta_b$$

If a steel bar of $l = 10\,\text{m}$, $A = 10 \times 10\,\text{cm}^2$, $E = 2.1 \times 10^6\,\text{kgf/cm}^2$, and $\nu = 0.3$ is acted upon by the axial tensile force $P = 1.4 \times 10^5\,\text{kgf}$, then

$$s_n = \frac{P}{A} = \frac{1.4 \times 10^5}{10^2} = 1.4 \times 10^3\,\text{kgf/cm}^2$$

$$\Delta_n = \frac{Pl}{AE} = s_n \frac{l}{E} = \frac{1.4 \times 10^3 \times 10^3}{2.1 \times 10^6} = 0.7\,\text{cm} \qquad l' = 1000.7\,\text{cm}$$

$$\Delta_a = \Delta_b = \nu \frac{Pa}{AE} = \frac{\nu s_n a}{E} = \frac{0.3 \times 1.4 \times 10^3 \times 10}{2.1 \times 10^6} = 0.002\,\text{cm} \qquad a' = b' = 9.998\,\text{cm}$$

(2) Pure Shear

(a) Static equilibrium. If a short prismatic block acted upon by two tangential forces P applied along its vertical faces (Fig. 4.03–5a) is intersected by a vertical plane and separated into two parts (two free-body diagrams, Fig. 4.03–5b), then from the static equilibrium of either part,

$$P = \bar{s}_t A$$

where $\bar{s}_t A$ = resultant stress, $\bar{s}_t$ = average shearing stress, and A = area of the cross section.

(b) Average shearing stress from (a) is then

$$\bar{s}_t = \frac{P}{A} \qquad \left\{ \begin{array}{l} P \text{ in kgf} \\ A \text{ in cm}^2 \\ \bar{s}_t \text{ in kgf/cm}^2 \end{array} \right\}$$

which is in some cases only a poor approximation.

(c) Shearing stress distribution is not uniform over the cross section, and the maximum shearing stress on a given cross section is

$$s_{t,\text{max}} = \beta \frac{P}{A} \qquad \{s_{t,\text{max}} \text{ in kgf/cm}^2\}$$

where β = dimensionless shape factor.

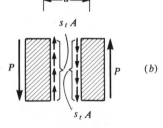

Fig. 4.03–5

(d) Shape factors for the most common cross sections are given below.

Cross section					
β	3/2	4/3	A/ht	3/2	$9c/4b$

(e) Units and conversion factors of $\bar{s}_t$ and $s_{t,\text{max}}$ are the same as those of $\bar{s}_n$ (Sec. 4.03–1c).

(f) Hooke's law. Within the elastic range of homogeneous solids the ratio of the shearing stress $\bar{s}_t$ to the shearing strain $\bar{\gamma}_t$ is constant; i.e.,

$$\bar{s}_t = G\bar{\gamma}_t \qquad \begin{cases} G \text{ in kgf/cm}^2 \\ \bar{\gamma}_n \text{ dimensionless} \end{cases}$$

where G = constant of proportionality, called the *modulus of rigidity*, and $\bar{\gamma}_t$ = average shearing strain defined below.

(g) Average shearing strain $\bar{\gamma}_t$ is the slope of detrusion of the faces of the block (Fig. 4.03–6).

$$\bar{\gamma}_t = \tan \phi = \frac{\Delta_t}{a} \qquad \begin{cases} \phi \text{ in rad} \\ \Delta_t \text{ in cm} \\ a \text{ in cm} \end{cases}$$

where Δ_t = shearing deformation.

(h) Shearing deformation Δ_t is the relative linear displacement of the external planes:

$$\Delta_t = a\bar{\gamma}_t = a\frac{\bar{s}_t}{G} = \frac{Pa}{AG} \qquad \begin{cases} P \text{ in kgf} \\ a \text{ in cm} \\ A \text{ in cm}^2 \end{cases}$$

where a = normal distance of these planes.

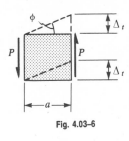

Fig. 4.03–6

(3) Volume Tension and Compression

(a) Volume stress s_v. If a cube of volume $V = a^3$ is acted upon by six forces of magnitude P, each normal to the respective face (Fig. 4.03–7), then the normal stress in the body on any plane is

$$s_v = \frac{P}{a^2} \qquad \{s_v \text{ in kgf/cm}^2\}$$

and the stress s_v is called *isotropic* (same in all directions).

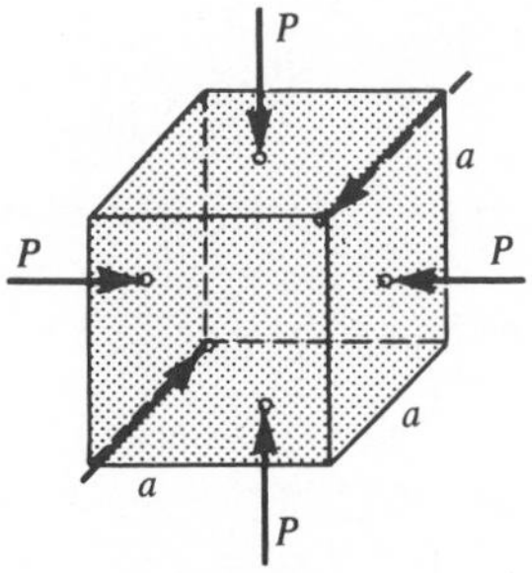

(b) Volume strain ϵ_v is defined as the ratio of the volume change ΔV to the initial volume V (Fig. 4.03–8):

$$\epsilon_v = \frac{\Delta V}{V} = \frac{a'^3 - a^3}{a^3} \qquad \{a, a' \text{ in cm}\}$$

where a = length of the edge of the cube and a' = length of the edge of the deformed cube.

(c) Hooke's law. Within the elastic range of homogeneous solids the ratio of the volume stress s_v to the volume strain ϵ_v is constant.

$$s_v = K\epsilon_v \qquad \begin{cases} s_v \text{ in kgf/cm}^2 \\ K \text{ in kgf/cm}^2 \\ \epsilon_v \text{ dimensionless} \end{cases}$$

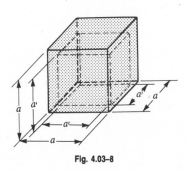

Fig. 4.03–7

Fig. 4.03–8

where K = constant of proportionality (which must be determined for a particular material by experiments) called the *bulk modulus*.

(d) Volume deformation ΔV is the change in volume V.

$$\Delta V = V\epsilon_v = V\frac{s_v}{K} = \frac{VP}{KA} = \frac{aP}{K} \qquad \begin{cases} a \text{ in cm} \\ P \text{ in kgf} \\ \Delta V \text{ in cm}^3 \end{cases}$$

where a = length of the edge of the cube.

(e) Relations of the modulus of elasticity E, modulus of rigidity G, bulk modulus K, and Poisson's ratio ν are given below.

$$E = 3(1-2\nu)K = 2(1+\nu)G = \frac{9KG}{3K+G}$$

$$G = \frac{E}{2(1+\nu)} = \frac{3(1-2\nu)K}{2(1+\nu)} = \frac{3KE}{9K-E} \qquad \left\{ \begin{array}{l} E,\ G,\ K \text{ in kgf/cm}^2 \\ \nu \text{ dimensionless} \end{array} \right\}$$

$$K = \frac{E}{3(1-2\nu)} = \frac{2(1+\nu)G}{3(1-2\nu)} = \frac{EG}{3(3G-E)}$$

example:

If for structural steel $E = 2.1 \times 10^6 \text{ kgf/cm}^2$ and $\nu = 0.3$, then

$$G = \frac{2.1 \times 10^6}{2(1+0.3)} = 0.807 \times 10^6 \text{ kgf/cm}^2$$

$$K = \frac{2.1 \times 10^6}{3(1-2\times 0.3)} = 1.750 \times 10^6 \text{ kgf/cm}^2$$

and ΔV of a cube of $a = 100 \text{ cm}$ in compression of $P = 100,000 \text{ kgf}$ is

$$\Delta V = \frac{100 \times 100,000}{1.750 \times 10^6} = 5.71 \text{ cm}^3$$

4.04 STATICS OF BEAMS

(1) Classification and Definitions

(a) Beam is a slender bar acted upon by forces and moments producing primarily bending.

(b) Bending of a bar is a combination of unilateral compression and tension during which the initially straight beam takes on the shape of a slightly curved beam (Fig. 4.04–1).

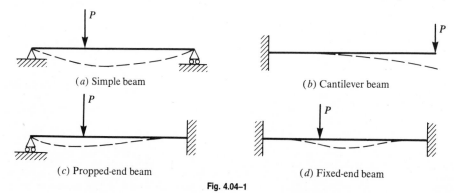

(a) Simple beam

(b) Cantilever beam

(c) Propped-end beam

(d) Fixed-end beam

Fig. 4.04–1

(c) Causes of this deformation are called the *loads*, and they are classified as concentrated loads, applied moments, and distributed loads (Fig. 4.04–2).

(d) Function of the beam is to carry loads, and to perform this function it must be supported. The forces developed by the loads at the point of support are called the *reactions*. Four most typical examples of beams are the simple beam (Fig. 4.04–1a), the cantilever beam (Fig. 4.04–1b), the propped-end beam (Fig. 4.04–1c), and the fixed-end beam (Fig. 4.04–1d).

(e) Planar beam is a straight bar of symmetrical cross section acted upon by forces in the plane of symmetry and by moments normal to this plane.

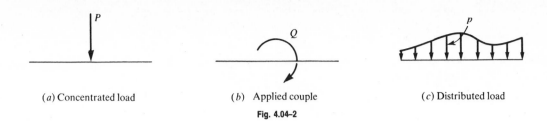

(a) Concentrated load   (b) Applied couple   (c) Distributed load

Fig. 4.04–2

(f) **Statically determinate beams.** If the reactions of a planar beam can be obtained from the conditions of static equilibrium only, the beam is said to be *statically determinate*. Analytical expressions for the reactions of six typical simple beams and cantilever beams are given in Tables 4.04–1 and 4.04–2, respectively.

(g) **Statically indeterminate beams.** If the reactions of a planar beam cannot be obtained from the equations of static equilibrium alone and the deformation conditions must be employed, the beam is said to be *statically indeterminate*. Analytical expressions for the reactions of six typical statically indeterminate beams are given in Tables 4.04–3 and 4.04–4.

(2) Stress Resultants

(a) **Internal effects.** If the planar beam of Fig. 4.04–3 is cut at a given point i and separated into two parts, two stress resultants occur at the centroid of the cut face of each part. They are the shearing force V and the bending moment M.

(b) **Shearing force**, or simply shear, V at a given section of the beam is equal to the algebraic sum of all forces and components of all forces acting parallel to and on one side of that section.

(c) **Bending moment** M at a given section of the beam is equal to the algebraic sum of all couples and of all static moments of all forces acting on one side of that section.

(d) **Signs of shear and bending moment** are governed by Fig. 4.04–4.

(e) **Stress resultant diagrams.** In general, the internal shear and bending moment vary from section to section, and the graphical representation of their variation is the shear diagram and the bending moment diagram, respectively. Examples of these diagrams are shown in Tables 4.04–1 through 4.04–4.

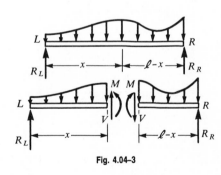

Fig. 4.04–3

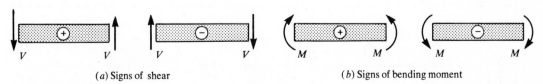

(a) Signs of shear   (b) Signs of bending moment

Fig. 4.04–4

TABLE 4.04–1 Simple Beam

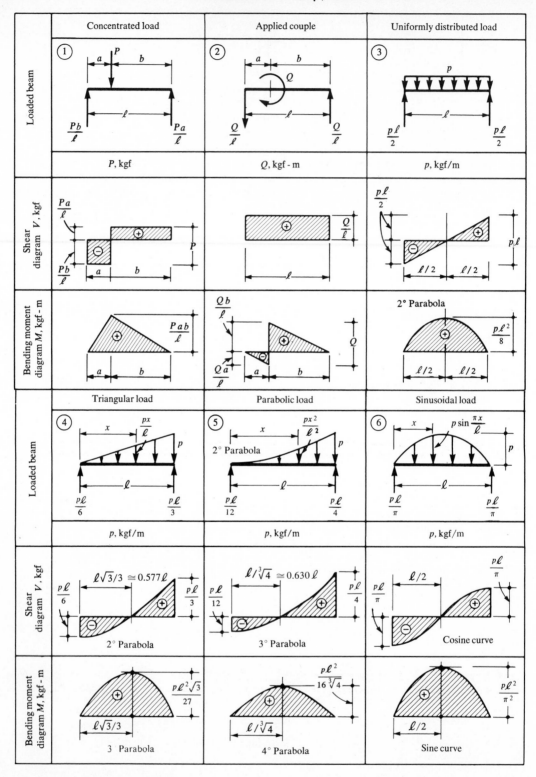

TABLE 4.04–2 Cantilever Beam

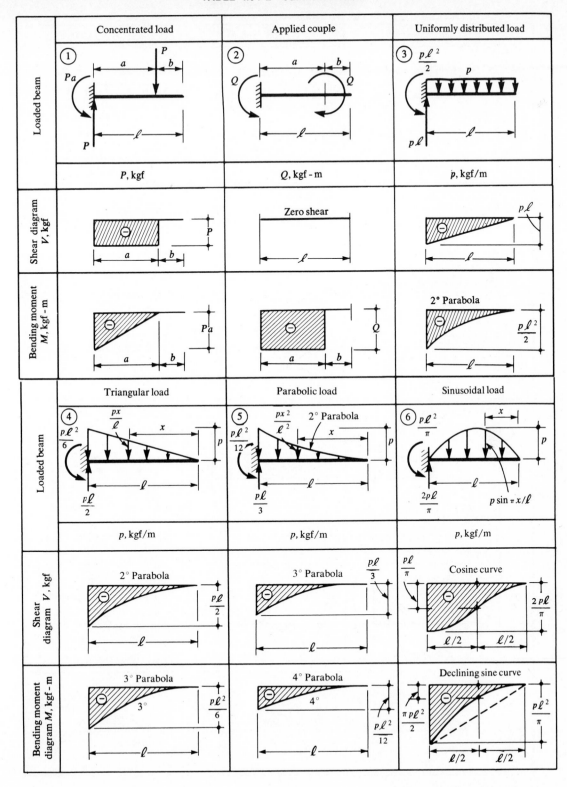

TABLE 4.04–3 Propped-End Beam

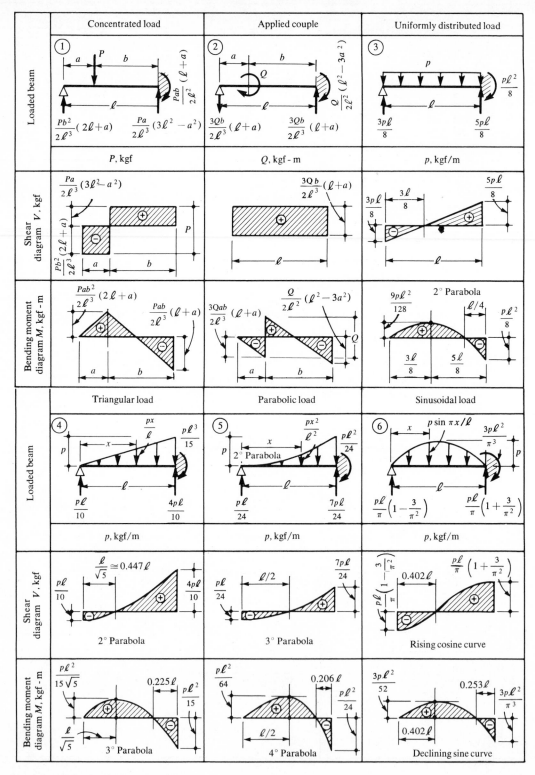

TABLE 4.04–4 Fixed-End Beam

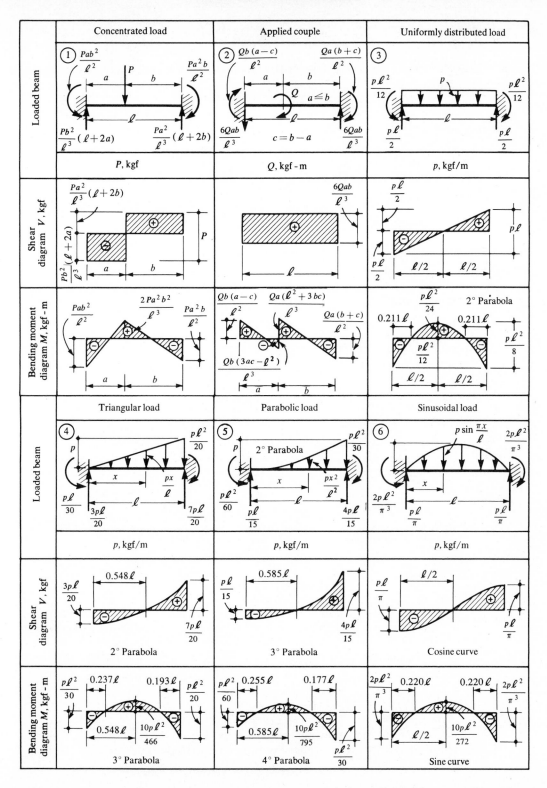

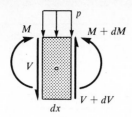 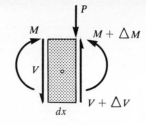 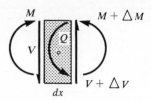

(a) Distributed load   (b) Concentrated load   (c) Applied couple

Fig. 4.04–5

(3) Differential and Difference Relations

(a) **Distributed load.** From the equilibrium of the differential element of Fig. 4.04–5a,

$$dV/dx = p \qquad dM/dx = -V$$

where p = intensity load (kgf/m), V = shear (kgf), M = bending moment (kgf-m). Thus, the slope of the shear diagram at any section of the beam equals the intensity of load at that section, and the slope of the bending moment diagram equals the negative value of shear at that section.

(b) **Concentrated load.** From the equilibrium of the differential element of Fig. 4.05–5b,

$$\Delta V = P \qquad \Delta M = 0$$

where Δ = difference in value, P = concentrated load (kgf). Thus, the sudden change in the shear diagram of the beam at the section of application of the concentrated load P equals the magnitude of that load.

(c) **Applied couple.** From the equilibrium of the differential element of Fig. 4.04–5c,

$$\Delta V = 0 \qquad \Delta M = -Q$$

where Q = applied couple (kgf-m). Thus, the sudden change in the bending moment diagram of the beam at the section of application of the applied couple equals the negative magnitude of that couple.

(4) Integral Relations

(a) **Distributed load.** The integration of the relationships in Sec. 4.04–3a between the limits x_1 and x_2 (the position coordinates of the two sections) yields

$$V_2 - V_1 = \int_{x_1}^{x_2} p \, dx \qquad M_2 - M_1 = - \int_{x_1}^{x_2} dV$$

where V_1, V_2, M_1, M_2 = the shear and bending moments of the respective sections. Thus, the change in shear between the two sections of the beam equals the area of the load diagram, and the change in bending moment between the two sections of the beam equals the negative of the area of the shear diagram between the sections.

(b) **Concentrated loads and applied couples.** From the relationships stated in (a) above it is apparent that in cases of n concentrated loads and/or applied couples,

$$V_2 - V_1 = \sum_{1}^{n} P_k \qquad M_2 - M_1 = - \sum_{1}^{n} Q_k$$

where V_1, V_2, M_1, M_2 have the same meaning as before and the right sides represent the sum of concentrated forces and applied couples, respectively.

(5) Extreme Values of Stress Resultants

(a) **Extreme values** of shear or bending moment occur at the sections at which the derivative of V or M equals zero.

$$dV/dx = p = 0 \qquad dM/dx = V = 0$$

(b) **Stress reversal.** The cross section of shearing stress resultant reversal (the section of change in sign of the shear) is also a location of extreme bending moment.

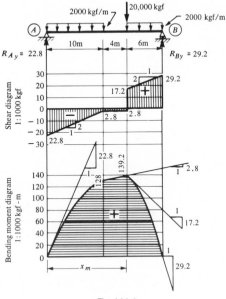

Fig. 4.04–6

example:

The reactions of the simple beam loaded as shown in Fig. 4.04–6 are:

$$R_{Ay} = 22{,}800 \text{ kgf} \qquad R_{By} = 29{,}200 \text{ kgf}$$

The shear equations in kgf are:

$$
\begin{aligned}
0 \leq x \leq 10 \text{ m} && V &= -22{,}800 + 2{,}000x \\
10 \text{ m} \leq x \leq 14 \text{ m} && V &= -2{,}800 \\
14 \text{ m} \leq x \leq 20 \text{ m} && V &= 17{,}200 + 2{,}000x
\end{aligned}
$$

The bending moment equations in kgf-m are:

$$
\begin{aligned}
0 \leq x \leq 10 \text{ m} && M &= 22{,}800x - 2{,}000x^2/2 \\
10 \text{ m} \leq x \leq 14 \text{ m} && M &= 22{,}800x - 2{,}000(10)(x - 5) \\
14 \text{ m} \leq x \leq 20 \text{ m} && M &= 22{,}800x - 2{,}000(10)(x - 5) - 20{,}000(x - 14) - 2{,}000(x - 14)^2/2
\end{aligned}
$$

The shear and bending moment diagrams based on these equations are shown in Fig. 4.04–6. The position of the cross section of zero shear (shear sign reversal)

$$x_m = 14 \text{ m}$$

gives the location of the maximum bending moment

$$M_{\max} = 139{,}200 \text{ kgf-m}$$

The slopes of the respective curves of these diagrams given by the equations of Sec. 4.04–3a are also indicated in Fig. 4.04–6.

4.05 BENDING OF HOMOGENEOUS BEAMS

(1) Symmetrical Bending in Elastic Range

(a) Bernoulli's hypothesis. The technical theory of the bending of straight beams is based on the assumption that the plane section of the beam remains plane during bending, which requires a linear variation of the normal strain ϵ_n from the maximum compressive strain on the top to the maximum tensile strain on the bottom, or vice versa. The cross section of the beam is assumed to be symmetrical with respect to the vertical axis Y, and all loads are applied in the vertical XY plane.

(b) Hooke's law. In the elastic range, the ratio of the normal stress to the normal strain is constant, and consequently the normal stress must also vary linearly from the maximum compressive stress on the top to the maximum tensile stress on the bottom of the beam, or vice versa.

(c) Neutral surface. As a consequence of (*a*) and (*b*) there must be a horizontal layer in the beam which neither deforms nor is in a state of stress and is therefore called the *neutral surface* (Fig. 4.05–1).

(d) Neutral axis is then the intersection of the neutral surface with a particular cross section, and it can be proved analytically that this axis passes through the centroid of that section and is normal to the plane of its symmetry.

(e) Bending stress s_n, which is normal to the cross section, is

$$s_n = \frac{My}{I} \qquad \left\{ \begin{array}{l} s_n \text{ in kgf/cm}^2 \\ M \text{ in kgf-cm} \\ y \text{ in cm} \\ I \text{ in cm}^4 \end{array} \right\}$$

where M = bending moment, y = position coordinate of stress measured from the neutral axis, and I = moment of inertia of the cross section about the neutral axis (Fig. 4.05–2).

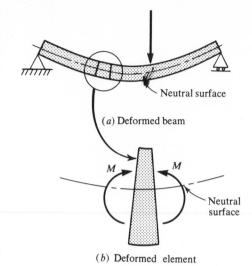

(*a*) Deformed beam

(*b*) Deformed element

Fig. 4.05–1

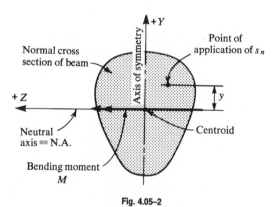

Fig. 4.05–2

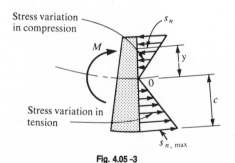

Fig. 4.05 –3

(f) Maximum bending stresses. The formula in (e) indicates that bending stress varies directly with the distance y (Fig. 4.05–3). If c is the distance to the remotest point of the cross section from the neutral axis, then

$$s_{n,\max} = \frac{Mc}{I} \qquad \left\{ \begin{array}{l} s_{n,\max} \text{ in kgf/cm}^2 \\ c \text{ in cm} \end{array} \right\}$$

which can be a tensile or compressive stress.

(g) Section modulus of the section is

$$Z = \frac{I}{c} \qquad \{Z \text{ in m}^3\}$$

in terms of which (f) becomes

$$s_{n,\max} = \frac{M}{Z}$$

(h) Table of I, Z, and $s_{n,\max}$ for the most commonly used cross sections is given below. More extensive tables of these values are given in Appendix A.

Cross section	Rectangle	Solid circle	Tube	Triangle
	①	②	③	④
I	$\dfrac{bh^3}{12}$	$\dfrac{\pi R^4}{4}$	$\dfrac{\pi(R^4 - r^4)}{4}$	$\dfrac{bh^3}{36}$
Z	$\dfrac{bh^2}{6}$	$\dfrac{\pi R^3}{4}$	$\dfrac{\pi(R^4 - r^4)}{4R}$	$\dfrac{bh^2}{24}$
$s_{n,\max}$	$\dfrac{6M}{bh^2}$	$\dfrac{4M}{\pi R^3}$	$\dfrac{4RM}{\pi(R^4 - r^4)}$	$\dfrac{24M}{bh^2}$

example:

If a rectangular section of $b = 12$ cm and $h = 20$ cm is acted upon by $M = 8 \times 10^5$ kgf-cm about the normal axis parallel to b (see table above), then

$$I = \frac{bh^3}{12} = \frac{12 \times 20^3}{12} = 8 \times 10^3 \text{ cm}^4 \qquad Z = \frac{bh^2}{6} = \frac{12 \times 20^2}{6} = 8 \times 10^2 \text{ cm}^3$$

$$s_{n,\max} = \frac{M}{Z} = \frac{8 \times 10^5}{8 \times 10^2} = 1{,}000 \text{ kgf/cm}^2$$

(2) Unsymmetrical Bending in Elastic Range

(a) Resolution of bending moment. If the cross section of the beam of Fig. 4.05–1 is symmetrical as shown in Fig. 4.05–4a or unsymmetrical as shown in Fig. 4.05–4b and the resulting bending moment produced by applied loads is skew by the angle β with respect to the centroidal axis Y, then the components of this bending moment become

$$M_y = M \cos \beta \qquad M_z = M \sin \beta \qquad \{M \text{ in kgf-cm}\}$$

and the state is called *unsymmetrical bending*.

(b) Bending stress on symmetrical section defined by Fig. 4.05–4a is

$$s_n = \frac{M_y z}{I_{yy}} - \frac{M_z y}{I_{zz}} \qquad \left\{ \begin{array}{l} s_n \text{ in kgf/cm}^2 \\ y, z \text{ in cm} \\ I_{yy}, I_{zz} \text{ in cm}^4 \end{array} \right\}$$

where M_y, M_z = bending moments due to loads about the respective centroidal axis, y, z = position coordinates of s_n on the section, and I_{yy}, I_{zz} = moments of inertia of the cross section about the respective centroidal axis.

(c) Neutral axis of the symmetrical section shown in Fig. 4.05–4a is given by the equation

$$\tan \alpha = \tan \beta \frac{I_{yy}}{I_{zz}} = \frac{z}{y}$$

where α = position angle of the neutral axis measured in the right-handed direction from the Y axis.

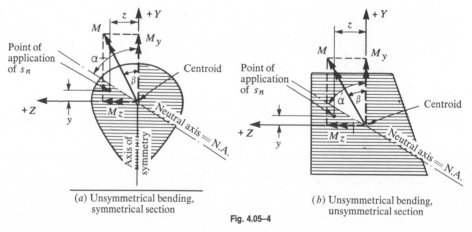

(a) Unsymmetrical bending, symmetrical section

(b) Unsymmetrical bending, unsymmetrical section

Fig. 4.05–4

(d) Bending stress on unsymmetrical section defined by Fig. 4.05–4b is

$$s_n = \frac{(M_y I_{zz} + M_z I_{yz})z}{I_{yy} I_{zz} - I_{yz}^2} - \frac{(M_y I_{yz} + M_z I_{yy})y}{I_{yy} I_{zz} - I_{yz}^2} \qquad \{I_{yz} \text{ in cm}^4\}$$

where I_{yz} = product of inertia of the section in the centroidal axis and the remaining symbols have the same meaning as in (b) above.

(e) Neutral axis of unsymmetrical section shown in Fig. 4.05–4b is given by the equation

$$\tan \alpha = \frac{I_{yz} + \tan \beta \, I_{yy}}{I_{zz} + \tan \beta \, I_{yz}} = \frac{z}{y}$$

where α = position angle of the same meaning as in (c) above.

(3) Symmetrical Bending in Plastic Range

(a) **Elastoplastic material.** If the material of the beam is elasto-plastic (Curve A, Fig. 4.02–2) so that the stress-strain curve can be represented by Fig. 4.05–5, then as the load is gradually being increased, the stress distribution diagram for a particular section of the beam will go through three distinct stages shown in Fig. 4.05–6, of which the last one is called the *full plastification* of the section. The extreme stress in all three cases is the *yield stress* $s_{n,y}$, defined in Sec. 4.02–3e as the yield point, which is particularly characteristic for structural steel.

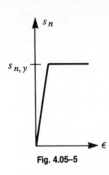

Fig. 4.05–5

(b) **Plastic bending theory** is again based on Bernoulli's hypothesis of a plane section (Sec. 4.05–1a), but as full plastification develops, the plastic strain becomes independent of the stress, which is constant over the section forming two stress blocks. One is acting over the tensile area A_T and is forming the tensile force T_P, and another is acting over the compressive area A_C and is forming the compressive force C_P.

$$T_P = s_{n,y} A_T \qquad C_P = s_{n,y} A_C$$

where the total area $A = A_T + A_C$ (in cm²).

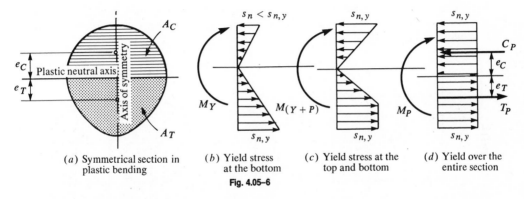

(a) Symmetrical section in plastic bending	(b) Yield stress at the bottom	(c) Yield stress at the top and bottom	(d) Yield over the entire section

Fig. 4.05–6

(c) **Position of neutral axis** in full plastic bending is located from the condition of static equilibrium of the horizontal forces requiring

$$T_P = C_P \qquad A_T = A_C = \frac{A}{2}$$

which indicates that the tensile and compressive area are equal.

(d) **Plastic moment** M_P on the section defined as the couple of T_P and C_P is

$$M_P = T_P e_T + C_P e_C = s_{n,y}(A_T e_T + A_C e_C) = s_{n,y} Z_P$$

where e_T, e_C = absolute values of the position coordinates of T_P and T_C, respectively, measured from the neutral axis of plastic bending, and Z_P = plastic section modulus (in cm³).

(e) **Shape factor** α_P is the ratio of the plastic moment M_P to the yield moment M_Y,

$$\alpha_P = \frac{M_P}{M_Y} = \frac{s_{n,y}(A_T e_T + A_C e_C)}{s_{n,y} I/y_{max}} = \frac{Z_P}{Z}$$

Several cases of plastic section modulus Z_P and of shape factor α_P are given in Appendix A.105.

4.06 SHEAR IN HOMOGENEOUS BEAMS

(1) Symmetrical Shear in Elastic Range

(a) **Warping of section.** Since the shearing stress s_t is neither constant nor linearly varying over the section, Bernoulli's hypothesis of the plane section is not satisfied in the case of combined shear and bending in beams, and the technical theory of bending is only a good approximation.

(b) **Shearing** s_t on the section of Fig. 4.06–1 at level $\overline{mm}$ is

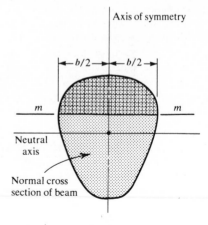

$$s_t = \frac{VQ}{Ib} = \frac{V}{R} \qquad \left\{\begin{array}{ll} s_t \text{ in kgf/cm}^2 & I \text{ in cm}^4 \\ V \text{ in kgf} & b \text{ in cm} \\ R \text{ in cm}^2 \end{array}\right\}$$

where V = shear of the section, Q = static moment of the shaded area about the neutral axis, I = moment of inertia of the section about the same axis, b = length $\overline{mm}$, and R = shear section modulus.

Fig. 4.06–1

(c) **Shearing stress variation** for the rectangular and circular section under a given shear is shown below. The maximum shearing stress $s_{t,max}$ always occurs on the neutral axis of elastic bending.

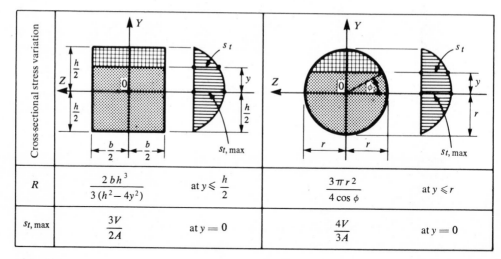

Cross-sectional stress variation		
R	$\dfrac{2bh^3}{3(h^2-4y^2)}$ at $y \leqslant \dfrac{h}{2}$	$\dfrac{3\pi r^2}{4\cos\phi}$ at $y \leqslant r$
$s_{t,max}$	$\dfrac{3V}{2A}$ at $y=0$	$\dfrac{4V}{3A}$ at $y=0$

example:

If a rectangular section of $b = 12$ cm and $h = 20$ cm is acted upon by $V = 1.2 \times 10^4$ kgf parallel to h, then

$$\left(\frac{Q}{b}\right)_{max} = \frac{b\left(\frac{h}{2}\right)\left(\frac{h}{4}\right)}{b} = \frac{12 \times 10 \times 5}{12} = 50 \text{ cm}^2$$

and

$$s_{t,max} = \frac{V}{I}\left(\frac{Q}{b}\right)_{max} = \frac{1.2 \times 10^4 \times 5 \times 10}{8 \times 10^3} = 75 \text{ kgf/cm}^2$$

where I is taken from the example of Sec. 4.05–1h.

(2) Unsymmetrical Shear in Elastic Range

(a) Resolution of shear. If the cross section of the beam of Fig. 4.05–1 is unsymmetrical as shown in Fig. 4.06–2 and the resulting shear produced by applied loads is skew by the angle β with respect to the centroidal axis Y, the components of this shear along the centroidal axes Z, Y are, respectively,

$$V_z = V \sin \beta \qquad V_y = V \cos \beta$$

and the state is called unsymmetrical shear.

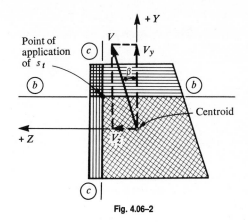

Fig. 4.06–2

(b) Shearing stresses on unsymmetrical section in Fig. 4.06–3 are

$$s_{t,z} = \frac{(Q_{cz}I_{yy} - Q_{cy}I_{yz})V_y}{(I_{yy}I_{zz} - I_{yz}{}^2)\bar{c}} + \frac{(Q_{cy}I_{zz} - Q_{cz}I_{yz})V_z}{(I_{yy}I_{zz} - I_{yz}{}^2)\bar{c}}$$

$$s_{t,y} = \frac{(Q_{bz}I_{yy} - Q_{by}I_{yz})V_y}{(I_{yy}I_{zz} - I_{yz}{}^2)\bar{b}} + \frac{(Q_{by}I_{zz} - Q_{bz}I_{yz})V_z}{(I_{yy}I_{zz} - I_{yz}{}^2)\bar{b}}$$

where V_z, V_y = components of shear defined in (b) above; $\bar{b}$, $\bar{c}$ = segments bb, cc respectively; Q_{bz}, Q_{by} = static moments of the shaded area limited by bb in Fig. 4.06–3a and b, respectively, about the centroidal axis denoted by the respective subscript; Q_{cz}, Q_{cy} = static moments of the shaded area limited by cc in Fig. 4.06–3a and b, respectively, about the centroidal axis denoted by the respective subscript; I_{zz}, I_{yy} = moments of inertia of the area of the cross section about the respective centroidal axis; I_{yz} = product of inertia of the area of the cross section in the centroidal axes; and $s_{t,z}$, $s_{t,y}$ = shearing stresses parallel to the respective centroidal axis and normal to the respective segment cc, bb as indicated in Fig. 4.06–3a and b.

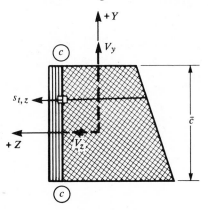

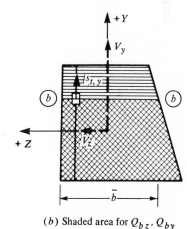

(a) Shaded area for Q_{cz}, Q_{cy} **Fig. 4.06 –3**   (b) Shaded area for Q_{bz}, Q_{by}

(c) Resulting shearing stress at a given point of the section is the vector sum of the shearing stresses in the respective case,

$$s_t = \sqrt{s_{t,z}{}^2 + s_{t,y}{}^2}$$

where the point of application of this resultant stress is the point of intersection of bb and cc.

4.07 ELASTIC CURVE OF HOMOGENEOUS BEAMS

(1) Geometry of Elastic Curve

(a) **Centroidal axis** of an initially straight bar acted upon by transverse loads takes the shape of a curve, called the *elastic* curve (Fig. 4.07–1).

(b) **Geometry of the elastic curve** is defined by the coordinate x, the vertical coordinate δ, called the *deflection of the beam*, and the angle θ, called the *slope*.

(c) **Flatness of the elastic curve** allows the following simplifications:

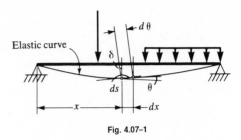

Fig. 4.07–1

$$\theta \cong \tan \theta \cong \sin \theta \qquad \cos \theta \cong 1 \qquad ds \cong dx$$

where ds = elemental length of the curve and the dx = projection of ds on the initial axis of the beam (X axis).

(2) Differential and Finite Relations

(a) **Beam functions.** The relations of shear V, bending moment M, slope θ, deflection δ, and variable intensity of load p are given by the following functions (Fig. 4.07–2a) called the *beam functions*.

$$\frac{d\delta}{dx} = \theta \qquad\qquad \frac{d^3\delta}{dx^3} = \frac{d^2\theta}{dx^2} = \frac{dM}{dx\,EI} = -\frac{V}{EI}$$

$$\frac{d^2\delta}{dx^2} = \frac{d\theta}{dx} = \frac{M}{EI} \qquad \frac{d^4\delta}{dx^4} = \frac{d^3\theta}{dx^3} = \frac{d^2M}{dx^2\,EI} = -\frac{dV}{dx\,EI} = \frac{p}{EI}$$

(b) **Sign convention** for V and M is given in Fig. 4.04–4. The signs of loads, slopes, and deflections are governed by the sign convention of analytic geometry.

(c) **Beam equations.** For a finite segment of a straight beam of symmetrical cross section acted upon by loads applied in the plane of symmetry as shown in Fig. 4.07–2b, the right end conditions in terms of the left end conditions and loads are

$$-V_R = -V_L + D_1$$

$$M_R = -V_L l + M_L + D_2$$

$$EI\theta_R = -\frac{V_L l^2}{2} + M_L l + EI\theta_L + D_3$$

$$\left. \begin{array}{l} V \text{ in kgf} \\ M \text{ in kgf-cm} \\ \theta \text{ in rad} \\ l,\,\delta \text{ in cm} \\ I \text{ in cm}^4 \\ E \text{ in kgf/cm}^2 \end{array} \right\}$$

$$EI\delta_R = -\frac{V_L l^3}{6} + \frac{M_L l^2}{2} + EI\theta_L l + EI\delta_L + D_4$$

where V_R = end shear at R, M_R = end bending moment at R, θ_R = end slope at R, δ_R = end deflection at R, l = length of segment, E = modulus of elasticity of beam material, I = moment of inertia of the cross section, D_1, D_2, D_3, D_4 = load functions defined in Sec. 4.07–2d below, and $V_L, M_L, \theta_L, \delta_L$ are the left end shear, bending moment, slope, and deflection, respectively.

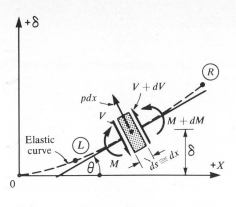
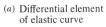

(a) Differential element of elastic curve

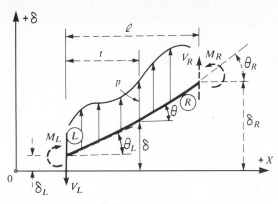

(b) Finite segment of elastic curve

Fig. 4.07–2

TABLE 4.07–1 Load Functions

Load	D_1	D_2	D_3	D_4
①	$-P$	$-Pb$	$-Pb^2/2$	$-Pb^3/6$
②	$-pb$	$-pb^2/2$	$-pb^3/6$	$-pb^4/24$
③	$-pb/2$	$-pb^2/6$	$-pb^3/24$	$-pb^4/120$
④	$-pb/3$	$-pb^2/12$	$-pb^3/60$	$-pb^4/360$
⑤	0	$-Q$	$-Qb$	$-Qb^2/2$

(d) **Load functions** introduced in the preceding section are the result of multiple integration of the differential equation of the beam,

$$\frac{d^4\delta}{dx^4} = \frac{p}{EI}$$

and their analytical forms are

$$D_1 = \int_0^l p\, dx \qquad\qquad D_3 = \int_0^l \int_0^x \int_0^x p\, dx\, dx\, dx$$

$$D_2 = \int_0^l \int_0^x p\, dx\, dx \qquad\qquad D_4 = \int_0^l \int_0^x \int_0^x \int_0^x p\, dx\, dx\, dx\, dx$$

The evaluation of these multiple integrals for the most important types of loads is recorded in Table 4.07–1.

(e) **Applications** of the beam equations introduced in (c) above are shown in Table 4.07–2 in symbolic form and in Tables 4.07–3 through 4.07–6 in specific form.

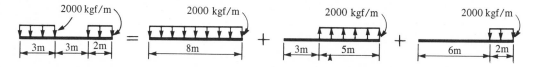

Fig. 4.07–3

(f) **Maximum deflection in a simple beam** acted upon by transverse loads of the same direction can always be approximated by the central deflection. The error resulting from this substitution cannot be numerically greater than 2.57 percent. The most common cases of central deflection δ and end slopes of simple beams are given in Tables 4.07–3 to 4.07–7.

(g) **Maximum deflection in a cantilever beam** acted upon by transverse loads of the same direction is always the deflection of the free end. The most common cases of end deflection δ_R and end slope θ_R are given in Tables 4.07–4 to 4.07–7.

(h) **Superposition.** Within the elastic range, the superposition of causes and effects is admissible, as shown below.

example:

The load functions for the load condition shown in Fig. 4.07–3 in terms of the relations given in Table 4.07–1 are

$$D_1 = -2000 \times 5 = -10^4 \text{ kgf}$$

$$D_2 = -2000(8^2 - 5^2 + 2^2)(100/2) = -4.3(10)^6 \text{ kgf-cm}$$

$$D_3 = -2000(8^3 - 5^3 + 2^3)(100^2/6) = -1.3(10)^9 \text{ kgf-cm}$$

$$D_4 = -2000(8^4 - 5^4 + 2^4)(100^3/24) = -2.9(10)^{11} \text{ kgf-cm}$$

TABLE 4.07–2 Application of Beam Equations

(a) Procedure. The application of the beam equations of Sec. 4.07–2c makes no distinction between statically indeterminate beams and statically determinate beams. There are always eight end conditions involved (V_L, M_L, θ_L, δ_L and V_R, M_R, θ_R, δ_R) of which four are unknown and four are zero. Thus four beam equations are adequate for the solution of any single-span beam problem.

(b) Simple beam

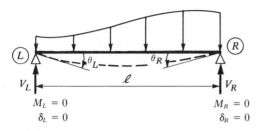

$$
\begin{bmatrix} V_L \\ V_R \\ EI\theta_L \\ EI\theta_R \end{bmatrix}
=
\begin{bmatrix} 0 & \dfrac{1}{l} & 0 & 0 \\ -1 & \dfrac{1}{l} & 0 & 0 \\ 0 & \dfrac{l}{6} & 0 & -\dfrac{1}{l} \\ 0 & -\dfrac{l}{3} & 0 & -\dfrac{1}{l} \end{bmatrix}
\begin{bmatrix} D_1 \\ D_2 \\ D_3 \\ D_4 \end{bmatrix}
$$

$M_L = 0 \qquad M_R = 0$
$\delta_L = 0 \qquad \delta_R = 0$

(c) Cantilever beam

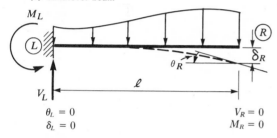

$$
\begin{bmatrix} V_L \\ M_L \\ EI\theta_R \\ EI\delta_R \end{bmatrix}
=
\begin{bmatrix} 1 & 0 & 0 & 0 \\ l & -1 & 0 & 0 \\ \dfrac{l^2}{2} & -l & 1 & 0 \\ \dfrac{l^3}{3} & -\dfrac{l^2}{2} & 0 & 1 \end{bmatrix}
\begin{bmatrix} D_1 \\ D_2 \\ D_3 \\ D_4 \end{bmatrix}
$$

$\theta_L = 0 \qquad V_R = 0$
$\delta_L = 0 \qquad M_R = 0$

(d) Propped-end beam

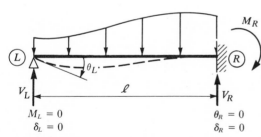

$$
\begin{bmatrix} V_L \\ EI\theta_L \\ V_R \\ M_R \end{bmatrix}
=
\begin{bmatrix} 0 & 0 & \dfrac{3}{l^2} & -\dfrac{3}{l^3} \\ 0 & 0 & \dfrac{1}{2} & -\dfrac{3}{2l} \\ -1 & 0 & \dfrac{3}{l^2} & -\dfrac{3}{l^3} \\ 0 & 1 & -\dfrac{3}{l} & \dfrac{3}{l^2} \end{bmatrix}
\begin{bmatrix} D_1 \\ D_2 \\ D_3 \\ D_4 \end{bmatrix}
$$

$M_L = 0 \qquad \theta_R = 0$
$\delta_L = 0 \qquad \delta_R = 0$

(e) Fixed-end beam

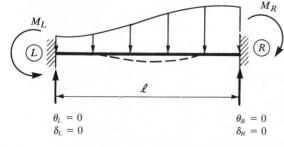

$$
\begin{bmatrix} V_L \\ M_L \\ V_R \\ M_R \end{bmatrix}
=
\begin{bmatrix} 0 & 0 & \dfrac{6}{l^2} & -\dfrac{12}{l^3} \\ 0 & 0 & \dfrac{2}{l} & -\dfrac{6}{l^2} \\ -1 & 0 & \dfrac{6}{l^2} & -\dfrac{12}{l^3} \\ 0 & 1 & -\dfrac{4}{l} & \dfrac{6}{l^2} \end{bmatrix}
\begin{bmatrix} D_1 \\ D_2 \\ D_3 \\ D_4 \end{bmatrix}
$$

$\theta_L = 0 \qquad \theta_R = 0$
$\delta_L = 0 \qquad \delta_R = 0$

TABLE 4.07-3 Elastic Curve—Concentrated Load

For notation used below refer to Sec. 4.07–2c. Special symbols used as abbreviations are:

$$\bar{\theta} = EI\theta \qquad \bar{\delta} = EI\delta \qquad m = \frac{a}{l} \qquad n = \frac{b}{l} \qquad u = \frac{x}{l} \qquad v = \frac{x'}{l}$$

For $x \le a$, $\langle x - a \rangle = 0$; for $x > a$, $\langle x - a \rangle = (x - a)$.

(a) Cantilever beam

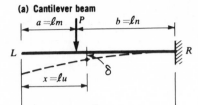

$$\bar{\theta}_L = \frac{Pb^2}{2} \qquad\qquad V_R = P$$

$$\bar{\delta}_L = -\frac{Plb^2(3 - n)}{6} \qquad M_R = -Pb$$

$$\bar{\delta} = \bar{\theta}_L x + \bar{\delta}_L - \frac{P\langle x - a \rangle^3}{6}$$

$$m = 0 \qquad \bar{\delta}_{max} = -\frac{Pl^3}{3} \qquad \text{at } u = 0$$

(b) Simple beam

$$\bar{\theta}_L = -\frac{Pl^2(1 + n)mn}{6} \qquad \bar{\theta}_R = \frac{Pl^2(1 + m)mn}{6}$$

$$V_L = -Pn \qquad\qquad V_R = Pm$$

$$\bar{\delta} = -\frac{V_L x^3}{6} + \bar{\theta}_L x - \frac{P\langle x - a \rangle^3}{6}$$

$$m = n \qquad \bar{\delta}_{max} = -\frac{Pl^3}{48} \qquad \text{at } u = \tfrac{1}{2}$$

$$m > n \qquad \bar{\delta}_{max} = -\frac{Pl^3 n u^3}{3} \qquad \text{at } u = \sqrt{\frac{1 - n^2}{3}}$$

$$m < n \qquad \bar{\delta}_{max} = -\frac{Pl^3 m v^3}{3} \qquad \text{at } v = \sqrt{\frac{1 - m^2}{3}}$$

(c) Propped-end beam

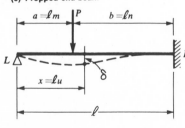

$$V_L = -\frac{Pn^2(2 + m)}{2} \qquad V_R = \frac{Pm(3 - m^2)}{2}$$

$$\bar{\theta}_L = -\frac{Pl^2 mn^2}{4} \qquad M_R = -\frac{Plm(1 - m^2)}{2}$$

$$\bar{\delta} = -\frac{V_L x^3}{6} + \bar{\theta}_L x - \frac{P\langle x - a \rangle^3}{6}$$

$$m < (\sqrt{2} - 1) \qquad \bar{\delta}_{max} = \frac{2M_R{}^3}{3V_R{}^2} \qquad \text{at } u = 1 + \frac{2M_R}{V_R l}$$

$$m > (\sqrt{2} - 1) \qquad \bar{\delta}_{max} = -\frac{Pl^3 mn^2 u}{6} \qquad \text{at } u = \sqrt{\frac{m}{2 + m}}$$

(d) Fixed-end beam

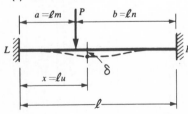

$$V_L = -Pn^2(1 + 2m) \qquad V_R = Pm^2(1 + 2n)$$

$$M_L = -Plmn^2 \qquad\qquad M_R = -Plm^2 n$$

$$\bar{\delta} = -\frac{V_L x^3}{6} + \frac{M_L x^2}{2} - \frac{P\langle x - a \rangle^3}{6}$$

$$m = n \qquad \bar{\delta}_{max} = -\frac{Pl^3}{192} \qquad \text{at } u = \tfrac{1}{2}$$

$$m > n \qquad \bar{\delta}_{max} = \frac{2M_L{}^3}{3V_L{}^2} \qquad \text{at } u = \frac{2M_L}{V_L l}$$

$$m < n \qquad \bar{\delta}_{max} = \frac{2M_R{}^3}{3V_R{}^2} \qquad \text{at } u = 1 + \frac{2M_R}{V_R l}$$

TABLE 4.07–4   Elastic Curve—Applied Couple

For the notation used below refer to Sec. 4.07–2c.   Special symbols used as abbreviations are:

$$\bar{\theta} = EI\theta \qquad \bar{\delta} = EI\delta \qquad m = \frac{a}{l} \qquad n = \frac{b}{l} \qquad u = \frac{x}{l} \qquad v = \frac{x'}{l}$$

For $x \le a$, $\langle x - a \rangle = 0$;   for $x > a$, $\langle x - a \rangle = (x - a)$.

(a) Cantilever beam

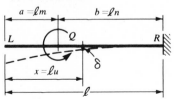

$$\bar{\theta}_L = Qln \qquad\qquad V_R = 0$$

$$\bar{\delta}_L = -\frac{Ql^2 n(2 - n)}{2} \qquad M_R = -Q$$

$$\bar{\delta} = \bar{\theta}_L x + \bar{\theta}_L - \frac{Q\langle x - a \rangle^2}{2}$$

$$m = 0 \qquad \bar{\delta}_{max} = -\frac{Ql^2}{2} \qquad \text{at } a = 0$$

(b) Simple beam

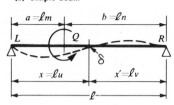

$$\bar{\theta}_L = -\frac{Ql(1 - 3n^2)}{6} \qquad \bar{\theta}_R = -\frac{Ql(1 - 3m^2)}{6}$$

$$V_L = -\frac{Q}{l} \qquad V_R = -\frac{Q}{l}$$

$$\bar{\delta} = -\frac{V_L x^3}{6} + \bar{\theta}_L x - \frac{Q\langle x - a \rangle^2}{2}$$

$$m = 0 \qquad \bar{\delta}_{max} = -\frac{Ql^2 v^3}{3} \qquad \text{at } v = \sqrt{\tfrac{1}{3}}$$

$$m > \sqrt{\tfrac{1}{3}} \qquad \bar{\delta}_{max} = -\frac{Ql^2 u^3}{3} \qquad \text{at } u = \sqrt{\frac{1 - 3n^2}{3}}$$

$$m < \sqrt{\tfrac{1}{3}} \qquad \bar{\delta}_{max} = -\frac{Ql^2 v^3}{3} \qquad \text{at } v = \sqrt{\frac{1 - 3m^2}{3}}$$

(c) Propped-end beam

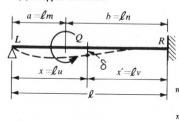

$$V_L = -\frac{3Q(1 - m^2)}{2l} \qquad V_R = -\frac{3Q(1 - m^2)}{2l}$$

$$\bar{\theta}_L = -\frac{Qln(3m - 1)}{4} \qquad M_R = \frac{Q(1 - 3m^2)}{2}$$

$$\bar{\delta} = -\frac{V_L x^3}{6} + \bar{\theta}_L x - \frac{Q\langle x - a \rangle^2}{2}$$

$$m > \tfrac{1}{3} \qquad \bar{\delta}_{max} = -\frac{Ql^2(1 - m^2)u^3}{2} \qquad \text{at } u = \sqrt{\frac{3m - 1}{3m + 3}}$$

$$m < \tfrac{1}{3} \qquad \bar{\delta}_{max} = \frac{2M_R{}^3}{3V_R{}^2} \qquad \text{at } u = 1 + \frac{2M_R}{V_R l}$$

(d) Fixed-end beam

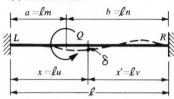

$$V_L = -\frac{6Qmn}{l} \qquad V_R = -\frac{6Qmn}{l}$$

$$M_L = -Qn(2 - 3n) \qquad M_R = Qm(2 - 3m)$$

$$\bar{\delta} = -\frac{V_L x^3}{6} + \frac{M_L x^2}{2} - \frac{Q\langle x - a \rangle^2}{2}$$

$$m = n \qquad \bar{\delta}_{max} = \mp\frac{Ql^2}{216} \qquad \text{at } u = \tfrac{1}{3}, \tfrac{2}{3}$$

$$m > \tfrac{1}{3} \qquad \bar{\delta}_{max} = \frac{2M_L{}^3}{3V_L{}^2} \qquad \text{at } u = \frac{2M_L}{V_L l}$$

$$m < \tfrac{1}{3} \qquad \bar{\delta}_{max} = \frac{2M_R{}^3}{3V_R{}^2} \qquad \text{at } u = 1 + \frac{2M_R}{V_R l}$$

Mechanics of Deformable Solids   **135**

TABLE 4.07-5 Elastic Curve—Uniformly Distributed Load

For the notation used below refer to Sec. 4.07-2c. Special symbols used as abbreviations are defined in Table 4.07-3. Also,

δ_c = central deflection

$$e = \frac{3V_R - \sqrt{9V_R{}^2 + 24pM_R}}{2p}$$

(a) Cantilever beam

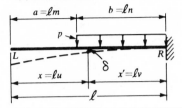

$a = \ell m$ $b = \ell n$

$x = \ell u$ $x' = \ell v$

ℓ

$$\bar\theta_L = \frac{pl^3 n^3}{6} \qquad\qquad V_R = pln$$

$$\bar\delta_L = -\frac{pl^4 n^3(4 - n)}{24} \qquad M_R = -\frac{pl^2 n^2}{2}$$

$$\bar\delta = \bar\theta_L x + \bar\delta_L - \frac{p\langle x - a\rangle^4}{24}$$

$$m = 0 \qquad \bar\delta_{max} = -\frac{pl^4}{8} \qquad \text{at } x = 0$$

(b) Simple beam

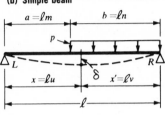

$a = \ell m$ $b = \ell n$

$x = \ell u$ $x' = \ell v$

ℓ

$$\bar\theta_L = -\frac{pl^3 n^2(2 - n^2)}{24} \qquad \bar\theta_R = \frac{pl^3 n^2(2 - n^2)}{24}$$

$$V_L = -\frac{pln^2}{2} \qquad\qquad V_R = \frac{pln(2 - n)}{2}$$

$$\bar\delta = -\frac{V_L x^3}{6} + \bar\theta_L x - \frac{p\langle x - a\rangle^4}{24}$$

$$m = 0 \qquad \bar\delta_{max} = -\frac{5pl^4}{384} \qquad\qquad \text{at } x = \frac{l}{2}$$

$$m < \tfrac{1}{2} \qquad \bar\delta_{max} \cong \bar\delta_c = -\frac{pl^4(5 - 12m^2 + 8m^4)}{384} \qquad \text{at } x = \frac{l}{2}$$

$$m > \tfrac{1}{2} \qquad \bar\delta_{max} \cong -\bar\delta_c = -\frac{pl^4 n^2(3 - 2n^2)}{96} \qquad \text{at } x = \frac{l}{2}$$

(c) Propped-end beam

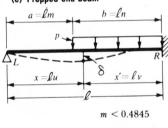

$a = \ell m$ $b = \ell n$

$x = \ell u$ $x' = \ell v$

ℓ

$$V_L = -\frac{pl}{8} n^3(4 - n) \qquad V_R = \frac{pln}{8}(8 - 4n^2 + n^3)$$

$$\bar\theta_L = -\frac{pl^3 n^3}{48}(4 - 3n) \qquad M_R = -\frac{pl^2}{8} n^2(2 - n)^2$$

$$\bar\delta = -\frac{V_L x^3}{6} + \bar\theta_L x - \frac{p\langle x - a\rangle^4}{24}$$

$$m = 0 \qquad \bar\delta_{max} = -\frac{pl^4}{185} \qquad \text{at } x = \frac{1 + \sqrt{33}}{16} l = 0.4215l$$

$$m < 0.4845 \qquad \bar\delta_{max} = \frac{V_R e^3}{6} + \frac{M_R e^2}{2} - \frac{pe^4}{24} \qquad \text{at } x = l - e$$

$$m \geq 0.4845 \qquad \bar\delta_{max} = \frac{V_L x^3}{6} + \bar\theta_L x \qquad \text{at } x = \sqrt{\frac{2\theta_L}{V_L}}$$

(d) Fixed-end beam

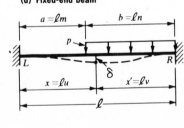

$a = \ell m$ $b = \ell n$

$x = \ell u$ $x' = \ell v$

ℓ

$$V_L = -\frac{pl}{2} n^3(2 - n) \qquad V_R = \frac{pl}{2} n(2 - 2n^2 + n^3)$$

$$M_L = -\frac{pl^2 n^3(4 - 3n)}{12} \qquad M_R = -\frac{pl^2 n^2(6 - 8n + 3n^2)}{12}$$

$$\bar\delta = -\frac{V_L x^3}{6} + \frac{M_L x^2}{2} - \frac{p\langle x - a\rangle^4}{24}$$

$$m = 0 \qquad \bar\delta_{max} = -\frac{pl^4}{384} \qquad \text{at } x = \frac{l}{2}$$

$$m < \sqrt{\tfrac{1}{3}} \qquad \bar\delta_{max} = \frac{Vle^3}{6} + \frac{M_R e^2}{2} - \frac{pe^4}{24} \qquad \text{at } x = l - e$$

$$m \geq \sqrt{\tfrac{1}{3}} \qquad \bar\delta_{max} = \frac{2M_L{}^3}{3V_L{}^2} \qquad \text{at } x = \frac{2M_L}{V_L}$$

TABLE 4.07–6   Elastic Curve—Triangular Load

For the notation used below refer to Sec. 4.07–2c.   Special symbols used as abbreviations are defined in Table 4.07–3.   Also, e and f are roots of the following quartic equations:

$$0 = -\frac{V_L e^2}{2} + \bar{\theta}_L - \frac{p(e-a)^4}{24b} \qquad 0 = -\frac{V_L f^2}{2} + M_L f - \frac{p(f-a)^4}{24b}.$$

(a) Cantilever beam

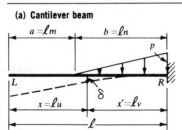

$a = \ell m$ $b = \ell n$

$x = \ell u$ $x' = \ell v$

$$\bar{\theta}_L = \frac{pb^3}{24} \qquad\qquad V_R = \frac{pb}{2}$$

$$\bar{\delta}_L = -\frac{plb^3(5-n)}{120} \qquad M_R = -\frac{pb^2}{6}$$

$$\bar{\delta} = \bar{\theta}_L x + \bar{\delta}_L - \frac{p(x-a)^5}{120b}$$

$$m = 0 \qquad \bar{\delta}_{max} = -\frac{pl^4}{30} \qquad \text{at } x = 0$$

(b) Simple beam

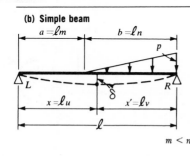

$a = \ell m$ $b = \ell n$

$x = \ell u$ $x' = \ell v$

$$\bar{\theta}_L = -\frac{pb^3(10-3n^2)}{360n} \qquad \bar{\theta}_R = \frac{pb^3(20-15n+3n^2)}{360n}$$

$$V_L = -\frac{pbn}{6} \qquad\qquad V_R = \frac{pb(3-n)}{6}$$

$$\bar{\delta} = -\frac{V_L x^3}{6} + \bar{\theta}_L x - \frac{p(x-a)^5}{120b}$$

$$m = 0 \qquad \bar{\delta}_{max} \cong -\frac{3pl^4}{460} \quad \text{at } x = l\sqrt{\frac{30-4\sqrt{30}}{30}} \cong 0.5193l$$

$$m < n \qquad \bar{\delta}_{max} \cong \bar{\delta}_c = -\frac{pl^4}{3840n}(25 - 50m + 5m^3 - 10m^5) \text{ at } x = \frac{l}{2}$$

$$m \geq n \qquad \bar{\delta}_{max} \cong \bar{\delta}_c = -\frac{pl^4}{480n}(5 - 2n^2) \qquad\qquad \text{at } x = \frac{l}{2}$$

(c) Propped-end beam

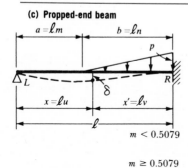

$a = \ell m$ $b = \ell n$

$x = \ell u$ $x' = \ell v$

$$V_L = -\frac{pbn^2(5-n)}{40} \qquad V_R = \frac{pb(20-5n^2+n^3)}{40}$$

$$\bar{\theta}_L = -\frac{pb^3(5-3n)}{240} \qquad M_R = -\frac{pb^2(20-15n+3n^2)}{120}$$

$$\bar{\delta} = -\frac{V_L x^3}{6} + \bar{\theta}_L x - \frac{p(x-a)^5}{120b}$$

$$m = 0 \qquad \bar{\delta}_{max} \cong -\frac{4pl^4}{1677} \qquad \text{at } x = l\sqrt{\frac{1}{5}} = 0.4472l$$

$$m < 0.5079 \qquad \bar{\delta}_{max} = -\frac{V_L e^3}{6} + \bar{\theta}_L e - \frac{p(e-a)^5}{120b} \qquad \text{at } x = e$$

$$m \geq 0.5079 \qquad \bar{\delta}_{max} = -\frac{V_L x^3}{6} + \bar{\theta}_L x \qquad\qquad \text{at } x = \sqrt{\frac{2\theta_L}{V_L}}$$

(d) Fixed-end beam

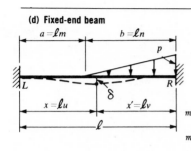

$a = \ell m$ $b = \ell n$

$x = \ell u$ $x' = \ell v$

$$V_L = -\frac{pbn^2(5-2n)}{20} \qquad V_R = \frac{pb(10-5n^2+2n^3)}{20}$$

$$M_L = -\frac{pb^2n(5-3n)}{60} \qquad M_R = -\frac{pb^2(10-10n+3n^2)}{60}$$

$$\bar{\delta}_L = -\frac{V_L x^3}{6} + \frac{M_L x^2}{2} - \frac{p(x-a)^5}{120b}$$

$$m = 0 \qquad \bar{\delta}_{max} \cong -\frac{pl^4}{764} \qquad \text{at } x \cong 0.5247l$$

$$m < 0.6039 \qquad \bar{\delta}_{max} = -\frac{V_L f^3}{6} + \frac{M_L f^2}{2} - \frac{p(f-a)^5}{120b} \qquad \text{at } x = f$$

$$m \geq 0.6039 \qquad \bar{\delta}_{max} = -\frac{V_L x^3}{6} + \frac{M_L x^2}{2} \qquad\qquad \text{at } x = \frac{2M_L}{V_L}$$

4.08 CONTINUOUS HOMOGENEOUS BEAMS

(1) Basic Relations

(a) **Continuous beam** is a slender bar supported at more than two points and acted upon by causes producing primarily bending. According to the number of spans, continuous beams are classified as two-, three-,..., and s-span beams. In general, they are statically indeterminate systems.

(b) **Statical indeterminacy** of a continuous beam is defined as

$$n = r - 3$$

where n = number of redundant moments over the supports and r = number of independent reactions.

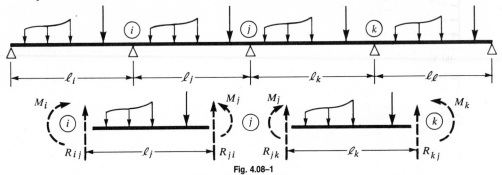

Fig. 4.08–1

(c) **Three-moment equation.** If an s-span continuous beam of constant cross section of moment of inertia I (in m^4) is acted upon by a system of transverse loads as shown in Fig. 4.08–1, the relation of the moments over any three consecutive supports i, j, k is given by the three-moment equation as

$$M_i l_j + 2M_j(l_j + l_k) + M_k l_k + 6EI(\tau_{ji} + \tau_{jk}) = 0$$

where l_j, l_k = length of span ij, jk, respectively (in m), M_i, M_j, M_k = unknown moments over the respective supports (in kgf-m), E = modulus of material of the beam (in kgf-m^2), and τ_{ji}, τ_{jk} = end slopes (in rad) of the respective simple beams ij, jk at j due to the transverse loads given for the most typical load conditions in Table 4.08–1.

TABLE 4.08–1 End Slopes of Simple Beam

τ_{LR}	Case	τ_{RL}	τ_{LR}	Case	τ_{RL}
$\dfrac{Pab\,(\ell+b)}{6\ell EI}$		$\dfrac{Pab\,(\ell+a)}{6\ell EI}$	$\dfrac{Q\,(\ell^2-3b^2)}{6\ell EI}$		$\dfrac{Q\,(\ell^2-3a^2)}{6\ell EI}$
$\dfrac{p\ell^3}{24\,EI}$		$\dfrac{p\ell^3}{24\,EI}$	$\dfrac{7p\ell^3}{360\,EI}$		$\dfrac{p\ell^3}{45\,EI}$

(2) Particular Cases

(a) **Moment coefficients.** Two-, three-, four-, and five-span continuous beams of equal spans of length l, of identical cross section, and of moment of inertia I, acted upon by general systems of transverse loads are considered in this section. Their moments over the supports are given as

$$M_1 = C_{11}m_1 + C_{12}m_2 + \cdots + C_{1j}m_j$$
$$M_2 = C_{21}m_1 + C_{22}m_2 + \cdots + C_{2j}m_j$$
$$\cdots\cdots\cdots\cdots\cdots\cdots\cdots\cdots\cdots\cdots$$
$$M_j = C_{j1}m_1 + C_{j2}m_2 + \cdots + C_{jj}m_j$$

where $C_{11}, C_{12}, \ldots, C_{jj}$ are the moment coefficients given in matrix form for each particular case below in Table 4.08–2.

(b) **Load coefficients** $m_1, m_2, \ldots, m_j$ are defined as

$$m_1 = \frac{3EI(\tau_{10} + \tau_{12})}{2l}$$
$$m_2 = \frac{3EI(\tau_{21} + \tau_{23})}{2l}$$
$$\cdots\cdots\cdots\cdots\cdots\cdots\cdots$$
$$m_j = \frac{3EI(\tau_{ji} + \tau_{jk})}{2l}$$

where $\tau_{10}, \tau_{12}, \ldots, \tau_{ji}, \tau_{jk}$ are the end slopes introduced in Sec. 4.08–1c. For a loadless span these end slopes equal zero.

TABLE 4.08–2 Moments in Equal-Span Continuous Beams

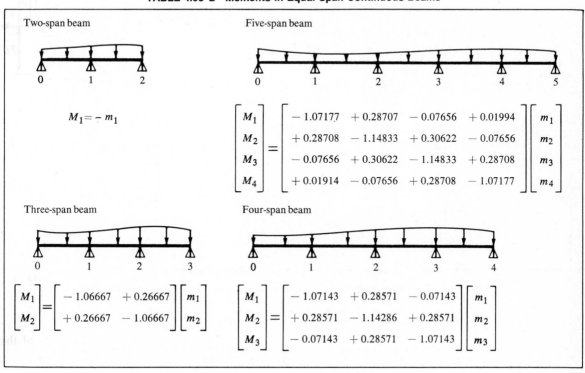

Two-span beam

$$M_1 = -m_1$$

Five-span beam

$$
\begin{bmatrix} M_1 \\ M_2 \\ M_3 \\ M_4 \end{bmatrix}
=
\begin{bmatrix}
-1.07177 & +0.28707 & -0.07656 & +0.01994 \\
+0.28708 & -1.14833 & +0.30622 & -0.07656 \\
-0.07656 & +0.30622 & -1.14833 & +0.28708 \\
+0.01914 & -0.07656 & +0.28708 & -1.07177
\end{bmatrix}
\begin{bmatrix} m_1 \\ m_2 \\ m_3 \\ m_4 \end{bmatrix}
$$

Three-span beam

$$
\begin{bmatrix} M_1 \\ M_2 \end{bmatrix}
=
\begin{bmatrix}
-1.06667 & +0.26667 \\
+0.26667 & -1.06667
\end{bmatrix}
\begin{bmatrix} m_1 \\ m_2 \end{bmatrix}
$$

Four-span beam

$$
\begin{bmatrix} M_1 \\ M_2 \\ M_3 \end{bmatrix}
=
\begin{bmatrix}
-1.07143 & +0.28571 & -0.07143 \\
+0.28571 & -1.14286 & +0.28571 \\
-0.07143 & +0.28571 & -1.07143
\end{bmatrix}
\begin{bmatrix} m_1 \\ m_2 \\ m_3 \end{bmatrix}
$$

4.09 BENDING OF SPECIAL BARS IN ELASTIC RANGE

(1) Symmetrical Bending of Curved Bars

(a) **Limitation of bending formula.** Whenever the ratio of depth h to radius of curvature R of the bar of symmetrical cross section shown in Fig. 4.09–1 exceeds 1/20, the technical theory of curved-bar bending must be introduced. The assumptions of this theory are the same as in Sec. 4.05–1, but the neutral axis and the centroidal axis do not coincide and the *stress distribution is hyperbolic*, not linear (Fig. 4.09–2).

(b) **Position of neutral axis** from the center of initial curvature is

$$r = \frac{R}{1 + B}$$

where $B = -\dfrac{1}{A}\displaystyle\int_{-c}^{d} \frac{\bar{y}\,dA}{R + \bar{y}}$

$$\begin{cases} A \text{ in cm}^2 \\ R, r, \bar{y}, c, d \text{ in cm} \\ B \text{ dimensionless} \end{cases}$$

in which A = area of cross section, R = initial radius of curvature, B = inertia function, $\bar{y}$ = vertical coordinate measured from the centroidal axis, y = vertical coordinate measured from the neutral axis, and c, d = vertical coordinates of the extreme points of the cross section measured from the centroidal axis.

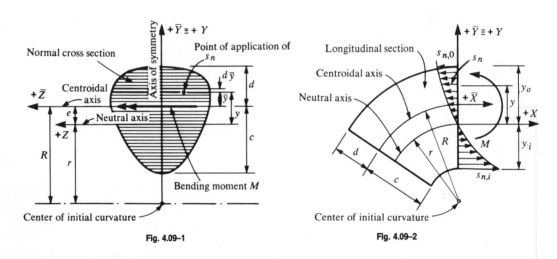

Fig. 4.09–1   Fig. 4.09–2

(c) **Bending stress normal** to the cross section at y is

$$s_n = \frac{M}{AR}\left[1 + \frac{y}{(R + y)B}\right]$$

$$\begin{cases} M \text{ in kgf-cm} \\ s_n \text{ in kgf/cm}^2 \end{cases}$$

where M = bending moment about the centroidal axis.

(d) **Extreme bending stresses** occur at the inner and outside fibers and are respectively

$$s_{n,i} = \frac{M}{AR}\left[1 + \frac{y_i}{(R + y_i)B}\right] \qquad s_{n,0} = -\frac{M}{AR}\left[1 + \frac{y_0}{(R + y_0)B}\right]$$

where y_i = negative coordinate of the inner fiber and y_0 = positive coordinate of the outside fiber, all measured from the neutral axis.

(e) Inertia functions B for the most commonly used sections are given below.

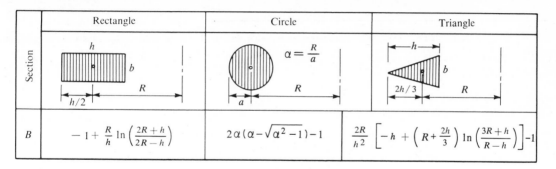

Section	Rectangle	Circle	Triangle
B	$-1 + \dfrac{R}{h} \ln\left(\dfrac{2R+h}{2R-h}\right)$	$2\alpha(\alpha - \sqrt{\alpha^2 - 1}) - 1$	$\dfrac{2R}{h^2}\left[-h + \left(R + \dfrac{2h}{3}\right)\ln\left(\dfrac{3R+h}{R-h}\right)\right] - 1$

For the Circle section: $\alpha = \dfrac{R}{a}$

TABLE 4.09 –1 Correction Factors K_i, K_o*

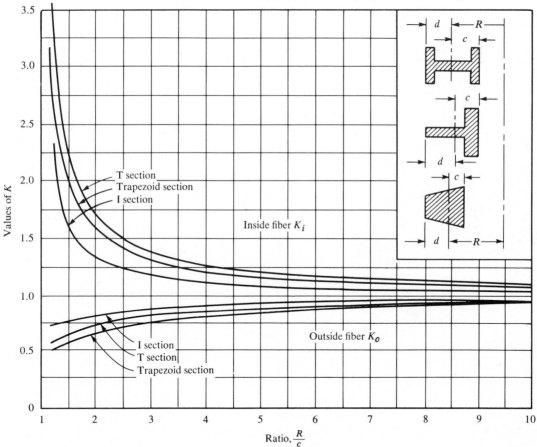

Values of K

Inside fiber K_i

Outside fiber K_o

T section
Trapezoid section
I section

I section
T section
Trapezoid section

Ratio, $\dfrac{R}{c}$

*B. J. Wilson and J. F. Querreau, "A Simple Method of Determining Stress in Curved Flexural Members," University of Illinois Experimental Station Bulletin 195, 1927.

(f) Correction factors.   If only the extreme values of the stress are desired, the straight-bar formulas of Sec. 4.05–1 can be modified by correction factors accounting for the bar's curvature, so that

$$s_{n,i} = K_i \frac{Mc}{I} \qquad s_{n,0} = -K_0 \frac{Md}{I}$$

where M = bending moment acting on the section about the centroidal axis, I = moment of inertia of the area of the section about the same axis, c, d = extreme vertical coordinates measured from the centroidal axis, and K_i, K_0 = correction factors, the value of which can be abstracted from the chart (Table 4.09–1) shown above.

(2) Symmetrical Bending of Beam-Columns

(a) **Beam-column** is a straight bar acted upon by a system of transverse loads and simultaneously by a compressive or tensile axial force. The bending moment in this beam on a symmetrical section given by the position coordinate x measured from the left end is

$$M = \mathrm{BM} \pm F\delta \qquad \left\{ \begin{array}{l} m, \text{BM in kgf-m} \\ F \text{ in kgf} \\ \delta \text{ in } m \end{array} \right\}$$

where BM = bending moment due to transverse loads, F = axial force, and δ = deflection of the beam at x.

(b) **Particular cases** of beam-columns acted upon by a compressive axial force are introduced in Tables 4.09–2 and 4.09–3. The symbols used in these tables are defined in Sec. 4.07–1, and the new equivalents used as abbreviations are:

$$\alpha = l\sqrt{\frac{F}{EI}} \qquad
\begin{array}{l} A_1 = \tan \alpha \\ A_2 = \sec \alpha \\ A_3 = \sin \alpha \end{array}
\qquad \beta = \frac{l}{2}\sqrt{\frac{F}{EI}} \qquad
\begin{array}{l} B_1 = \tan \beta \\ B_2 = \sec \beta \\ B_3 = \sin m\alpha \\ B_4 = \sin n\alpha \end{array}$$

TABLE 4.09–2 Cantilever Beam-Column

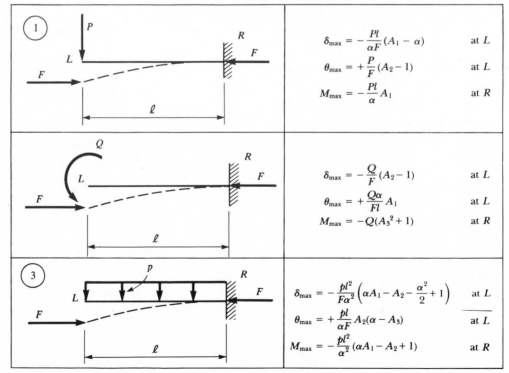

1

$$\delta_{max} = -\frac{Pl}{\alpha F}(A_1 - \alpha) \qquad \text{at } L$$

$$\theta_{max} = +\frac{P}{F}(A_2 - 1) \qquad \text{at } L$$

$$M_{max} = -\frac{Pl}{\alpha}A_1 \qquad \text{at } R$$

2

$$\delta_{max} = -\frac{Q}{F}(A_2 - 1) \qquad \text{at } L$$

$$\theta_{max} = +\frac{Q\alpha}{Fl}A_1 \qquad \text{at } L$$

$$M_{max} = -Q(A_3^2 + 1) \qquad \text{at } R$$

3

$$\delta_{max} = -\frac{pl^2}{F\alpha^2}\left(\alpha A_1 - A_2 - \frac{\alpha^2}{2} + 1\right) \qquad \text{at } L$$

$$\theta_{max} = +\frac{pl}{\alpha F}A_2(\alpha - A_3) \qquad \text{at } L$$

$$M_{max} = -\frac{pl^2}{\alpha^2}(\alpha A_1 - A_2 + 1) \qquad \text{at } R$$

TABLE 4.09–3 Simple Beam-Column

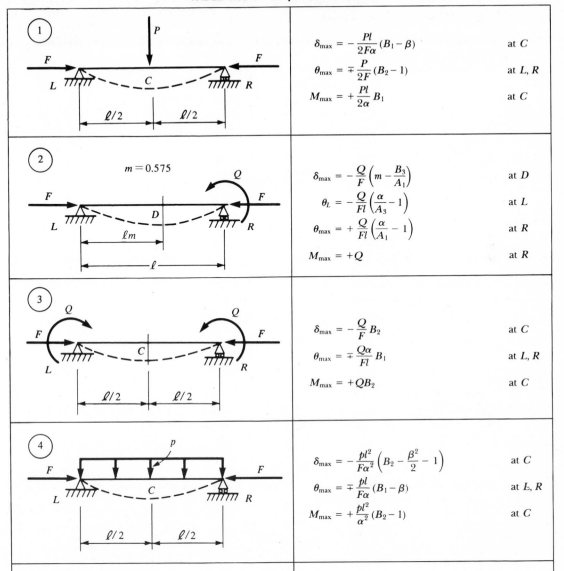
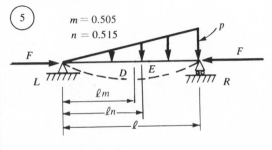

1

$$\delta_{\max} = -\frac{Pl}{2F\alpha}(B_1 - \beta) \qquad \text{at } C$$

$$\theta_{\max} = \mp\frac{P}{2F}(B_2 - 1) \qquad \text{at } L, R$$

$$M_{\max} = +\frac{Pl}{2\alpha}B_1 \qquad \text{at } C$$

2 $m = 0.575$

$$\delta_{\max} = -\frac{Q}{F}\left(m - \frac{B_3}{A_1}\right) \qquad \text{at } D$$

$$\theta_L = -\frac{Q}{Fl}\left(\frac{\alpha}{A_3} - 1\right) \qquad \text{at } L$$

$$\theta_{\max} = +\frac{Q}{Fl}\left(\frac{\alpha}{A_1} - 1\right) \qquad \text{at } R$$

$$M_{\max} = +Q \qquad \text{at } R$$

3

$$\delta_{\max} = -\frac{Q}{F}B_2 \qquad \text{at } C$$

$$\theta_{\max} = \mp\frac{Q\alpha}{Fl}B_1 \qquad \text{at } L, R$$

$$M_{\max} = +QB_2 \qquad \text{at } C$$

4

$$\delta_{\max} = -\frac{pl^2}{F\alpha^2}\left(B_2 - \frac{\beta^2}{2} - 1\right) \qquad \text{at } C$$

$$\theta_{\max} = \mp\frac{pl}{F\alpha}(B_1 - \beta) \qquad \text{at } L, R$$

$$M_{\max} = +\frac{pl^2}{\alpha^2}(B_2 - 1) \qquad \text{at } C$$

5 $m = 0.505$
$n = 0.515$

$$\delta_{\max} = -\frac{pl^2}{6F\alpha^2}\left[\alpha m^3 - m(6 - \alpha) + \frac{6B_3}{A_3}\right] \qquad \text{at } D$$

$$\theta_L = -\frac{pl}{6F\alpha}\left(\frac{1}{A_3} - \alpha - \frac{6}{\alpha}\right) \qquad \text{at } L$$

$$\theta_{\max} = \frac{pl}{3F\alpha_2}\left(\frac{3}{\alpha} - \alpha - \frac{3}{A_1}\right) \qquad \text{at } R$$

$$M_{\max} = +\frac{pl^2}{\alpha^2}\left(\frac{B_4}{A_3} - n\right) \qquad \text{at } E$$

Mechanics of Deformable Solids   143

4.10 TORSION OF HOMOGENEOUS BARS

(1) Classification and Definitions

(a) **Shaft** is a slender bar acted upon by forces and moments producing primarily torsion.

(b) **Torsion of a shaft** is developed by two twisting moments of equal magnitude M acting along its axis in opposite directions (Fig. 4.10–1), during which the surface lines parallel to its axis rotate through the angle γ_t (shearing strain), plane sections remain plane, and radii on these sections remain straight.

(c) **Causes of this deformation** may be moments distributed along the axes (Fig. 4.10–2a) or a system of several moments acting along this axis (Fig. 4.10–2b).

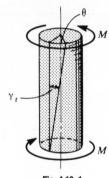

Fig. 4.10–1

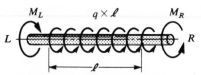

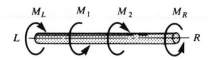

(a) Distributed twisting moment of intensity q (kgf-cm/cm) along the segment ℓ (cm)

(b) Two twisting moments M_1 and M_2 along the axis

Fig. 4.10–2

(d) **Function of the shaft** is to transmit the moment (or moments), and to perform this function it must be supported.   The torsional moments developed by the external causes at the points of support are called *reactive torques*.

example:

In Fig. 4.10–2b, if $M_R = 1{,}000{,}000$ kgf-cm, $M_1 = 200{,}000$ kgf-cm, and $M_2 = 400{,}000$ kgf-cm, then from the condition of static equilibrium of moments about the axis of the shaft,

$$M_L = 10^6 + 0.2 \times 10^6 - 0.4 \times 10^6 = 800{,}000 \text{ kgf-cm}$$

(e) **Statically determinate shaft** has one, and only one, unknown reactive moment (torque) since only one equation of static equilibrium is available.

(2) Stress Resultant

(a) **Internal effect.**   If the shaft of Fig. 4.10–3 is cut at a given point i and separated into two parts, one stress resultant, called the *internal torque T*, occurs at the centroid of the cut face of each part.

(b) **Internal torque (twisting moment)** T at a given section of the shaft is equal to the algebraic sum of all torsional moments acting on one side of this section.

(c) **Sign of torque** is governed by Fig. 4.10–4.

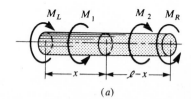

(a)

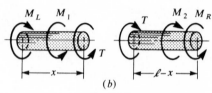

(b)

Fig. 4.10–3

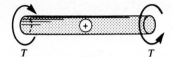

Fig. 4.10-4

(d) Torque diagram is the graphical representation of the variation of the internal torque.

(3) Torsional Stresses in Circular Shafts

(a) Circular shaft. The technical theory of torsion of circular shafts assumes that the plane cross section remains plane during the twist and that each section generates a distributed torque consisting of rings of shearing stress normal to the radius of the section and varying from zero at the center to their maximum at the circumference of the section (Fig. 4.10–5).

(b) Torsional stress s_t at the distance r from the center of the section is

$$s_t = \frac{Tr}{J} \qquad \begin{cases} T \text{ in kgf-cm} \\ r \text{ in cm} \\ J \text{ in cm}^4 \\ s_t \text{ in kgf/cm}^2 \end{cases}$$

where T = torque on the section and J = polar moment of inertia of the section about the axis of the shaft.

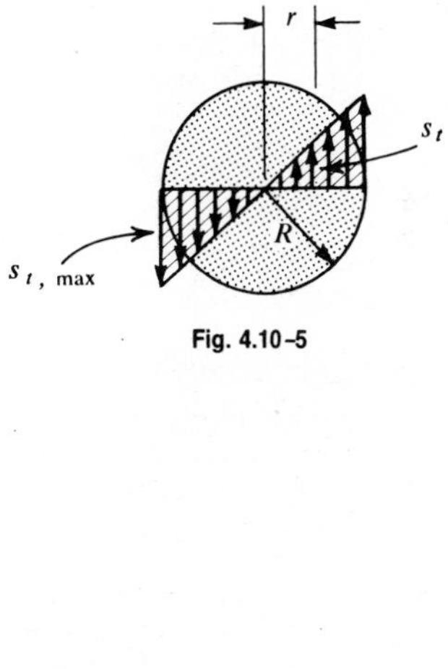

(c) Solid shaft. The polar moment of inertia of the cross section of a solid circular shaft of diameter D (radius R) is

$$J = \frac{\pi D^4}{32} = \frac{\pi R^4}{2} \qquad \begin{cases} J \text{ in cm}^4 \\ D, R \text{ in cm} \end{cases}$$

and the maximum torsional stress on this section is

$$s_{t,max} = \frac{16T}{\pi D^3} = \frac{2T}{\pi R^3} \qquad \begin{cases} s_{t,max} \text{ in kgf/cm}^2 \\ T \text{ in kgf-cm} \end{cases}$$

where T = torque on the section.

(d) Hollow shaft. The polar moment of inertia of the cross section of a hollow circular shaft of external diameter D (radius R) and of internal diameter d (radius r) is

$$J = \frac{\pi(D^4 - d^4)}{32} = \frac{\pi(R^4 - r^2)}{2} \qquad \{R, r, D, d \text{ in cm}\}$$

and the maximum torsional stress on this section is

$$s_{t,max} = \frac{16TD}{\pi(D^4 - d^4)} = \frac{2TR}{\pi(R^4 - r^4)} \qquad \begin{cases} s_{t,max} \text{ in kgf/cm}^2 \\ T \text{ in kgf-cm} \end{cases}$$

where T = torque on the section.

Fig. 4.10-5

(e) Horsepower transmitted by a shaft making n revolutions per minute under the torque T is

$$H = \frac{2\pi n T}{450,000}\,\text{hp} \qquad \text{(MKS system)} \qquad \{T \text{ in kgf-cm}\}$$

$$H = \frac{2\pi n T}{396,000}\,\text{hp} \qquad \text{(FPS system)} \qquad \{T \text{ in lbf-in.}\}$$

and the torque in terms of H (horsepower) is

$$T = \frac{71{,}620 H}{n} \qquad \text{(MKS system)} \qquad T = \frac{63{,}000 H}{n} \qquad \text{(FPS system)}$$

(f) Diameter of a circular solid shaft required to transmit H at n revolutions per minute with the torsional stress $s_{t,max}$ is

$$D = \sqrt[3]{\frac{365{,}000 H}{n \cdot s_{t,max}}} \quad \binom{\text{MKS system}}{D \text{ in cm}} \qquad D = \sqrt[3]{\frac{320{,}000 H}{n \cdot s_{t,max}}} \quad \binom{\text{FPS system}}{D \text{ in in.}}$$

example:

If $H = 500\,\text{hp}$, $n = 300\,\text{rev/min}$, and $s_{t,max} = 700\,\text{kgf/cm}^2$, then

$$D = \sqrt[3]{\frac{3.65 \times 10^5 \times 5 \times 10^2}{3 \times 10^2 \times 7 \times 10^2}} = 9.5\,\text{cm}$$

(4) Deformations of Circular Shafts

(a) Hooke's law. Within the elastic range of homogeneous solids the ratio of the torsional (shearing) stress s_t to the torsional (shearing) strain γ_t is constant; that is,

$$s_t = G\gamma_t \qquad \begin{Bmatrix} G \text{ in kgf/cm}^2 \\ \gamma_t \text{ dimensionless} \end{Bmatrix}$$

where G = modulus of rigidity introduced in Sec. 4.03–2f and related to E as stated in Sec. 4.03–3e.

(b) Torsional strain γ_t is the slope of the surface lines of the shaft produced by torsion (Fig. 4.10–6).

(c) Angle of twist θ of a circular shaft of length l within the elastic range is

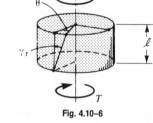

Fig. 4.10–6

$$\theta = l\frac{\gamma_t}{R} = l\frac{s_{t,max}}{GR} = \frac{Tl}{JG} \qquad \begin{Bmatrix} \theta \text{ in rad} & l \text{ in cm} \\ T \text{ in kgf-cm} & J \text{ in cm}^4 \end{Bmatrix}$$

where T = torque, J = polar moment of inertia of the cross section given in (c) and (d), and G = modulus of rigidity.

example:

If a circular solid steel shaft of $G = 8 \times 10^5\,\text{kgf/cm}^2$, $D = 10\,\text{cm}$, and $l = 300\,\text{cm}$ is subjected to a torque $T = 120{,}000\,\text{kgf-cm}$, then

$$s_{t,max} = \frac{16 T}{\pi D^3} = \frac{1.6 \times 10 \times 1.2 \times 10^5}{\pi \times 10^3} = 610\,\text{kgf/cm}^2$$

$$\theta = \frac{l s_{t,max}}{GR} = \frac{3 \times 10^2 \times 6.1 \times 10^2}{8 \times 10^5 \times 5} = 0.045\,\text{rad} = 2.58°$$

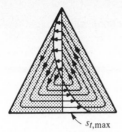

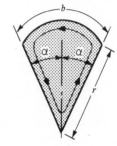

$s_{t,\text{max}}$

$s_{t,\text{max}}$

Fig. 4.10–7

(5) Torsion of Bars of Solid Noncircular Sections

(a) Solid sections of general shape.

As a general-shape section (Fig. 4.10–7) warps in a state of torsion, the simplified assumptions of the technical theory of torsion are not applicable and the problem becomes complicated. In general, the *angle of twist* and the *maximum torsional stress* can be expressed, respectively, as

$$\theta = \frac{Tl}{GK} \qquad s_t = \frac{T}{S} \qquad \left\{\begin{matrix} K \text{ in cm}^4 \\ S \text{ in cm}^3 \end{matrix}\right\}$$

where T = torque on the section, l = length of the bar, G = modulus of rigidity of the bar's material, K = torsional constant, and S = torsional section modulus (Tables 4.10–1 and 4.10–2).

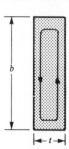
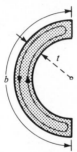

Fig. 4.10–8

(b) Simple narrow sections.

The torsional constant K and the section modulus of torsion in simple narrow sections such as those shown in Fig. 4.10–8 can be calculated by the following formulas:

Rectangle, $\frac{b}{t} > 3$

$$K = \frac{(b - 0.63t)t^3}{3} \qquad S = \frac{K}{t}$$

Circular rectangle, $\frac{b}{t} > 10$

$$K = \frac{bt^3}{3} \qquad S = \frac{K}{t}$$

Rectangle, $\frac{b}{t} > 10$

$$K = \frac{bt^3}{3} \qquad S = \frac{K}{t}$$

Circular sector, $2\alpha < \frac{\pi}{12}$

$$K = \frac{b^3(4r - 5b)}{48} \qquad S = \frac{b^2(4r - 5b)}{36}$$

In the circular sector the maximum stress occurs at $3r/4$ on the radial boundary.

(c) Composite narrow sections.

For sections composed of narrow rectangles of $\frac{b_i}{t_i} > 10$, such as those shown in Fig. 4.10–9,

$$K = \sum_{i=1}^{n} \frac{b_i t_i^3}{3} \qquad S = \frac{K}{t_{\text{max}}}$$

where t_{max} = largest t.

147

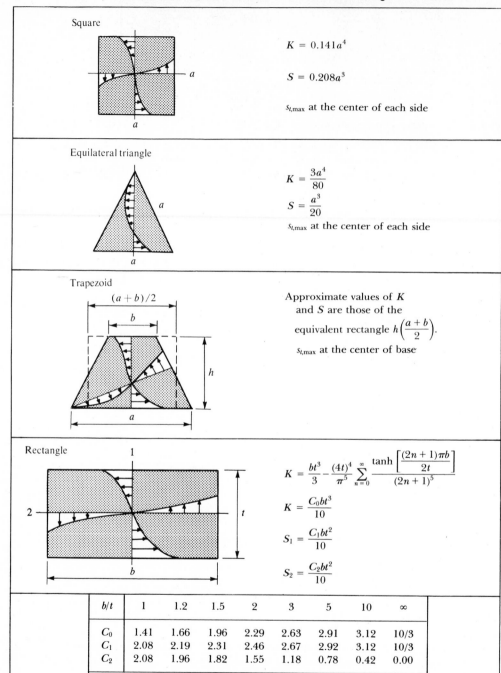

Square

$$K = 0.141a^4$$

$$S = 0.208a^3$$

$s_{t,\max}$ at the center of each side

Equilateral triangle

$$K = \frac{3a^4}{80}$$

$$S = \frac{a^3}{20}$$

$s_{t,\max}$ at the center of each side

Trapezoid

$(a + b)/2$

b

h

a

Approximate values of K
and S are those of the
equivalent rectangle $h\left(\dfrac{a + b}{2}\right)$.

$s_{t,\max}$ at the center of base

Rectangle

1

2

t

b

$$K = \frac{bt^3}{3} - \frac{(4t)^4}{\pi^5} \sum_{n=0}^{\infty} \frac{\tanh\left[\dfrac{(2n + 1)\pi b}{2t}\right]}{(2n + 1)^5}$$

$$K = \frac{C_0 bt^3}{10}$$

$$S_1 = \frac{C_1 bt^2}{10}$$

$$S_2 = \frac{C_2 bt^2}{10}$$

b/t	1	1.2	1.5	2	3	5	10	∞
C_0	1.41	1.66	1.96	2.29	2.63	2.91	3.12	10/3
C_1	2.08	2.19	2.31	2.46	2.67	2.92	3.12	10/3
C_2	2.08	1.96	1.82	1.55	1.18	0.78	0.42	0.00

$s_{t,\max 1}$ at the center of the longer side
$s_{t,\max 2}$ at the center of the shorter side

*For definition of symbols refer to Sec. 4.10–5a.

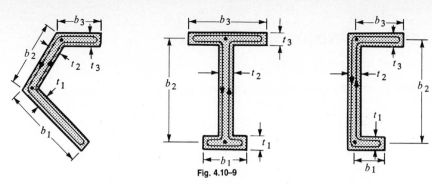

Fig. 4.10–9

TABLE 4.10–2 Torsional Constants of Solid Sections with Curvilinear Boundaries*

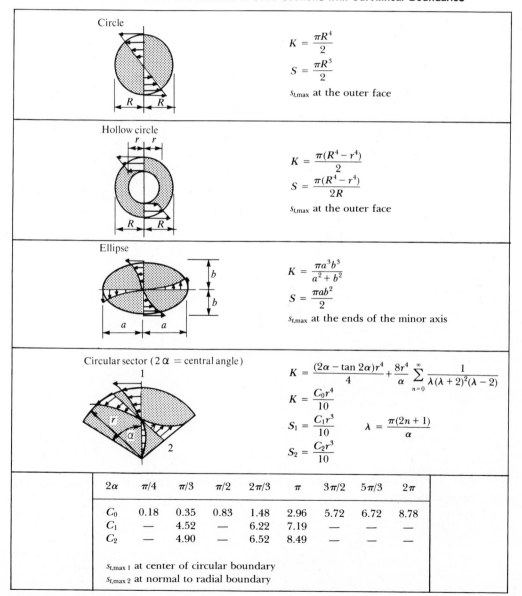

Circle

$$K = \frac{\pi R^4}{2}$$

$$S = \frac{\pi R^3}{2}$$

$s_{t,\max}$ at the outer face

Hollow circle

$$K = \frac{\pi(R^4 - r^4)}{2}$$

$$S = \frac{\pi(R^4 - r^4)}{2R}$$

$s_{t,\max}$ at the outer face

Ellipse

$$K = \frac{\pi a^3 b^3}{a^2 + b^2}$$

$$S = \frac{\pi a b^2}{2}$$

$s_{t,\max}$ at the ends of the minor axis

Circular sector (2α = central angle)

$$K = \frac{(2\alpha - \tan 2\alpha)r^4}{4} + \frac{8r^4}{\alpha} \sum_{n=0}^{\infty} \frac{1}{\lambda(\lambda + 2)^2(\lambda - 2)}$$

$$K = \frac{C_0 r^4}{10}$$

$$S_1 = \frac{C_1 r^3}{10} \qquad \lambda = \frac{\pi(2n + 1)}{\alpha}$$

$$S_2 = \frac{C_2 r^3}{10}$$

2α	$\pi/4$	$\pi/3$	$\pi/2$	$2\pi/3$	π	$3\pi/2$	$5\pi/3$	2π
C_0	0.18	0.35	0.83	1.48	2.96	5.72	6.72	8.78
C_1	—	4.52	—	6.22	7.19	—	—	—
C_2	—	4.90	—	6.52	8.49	—	—	—

$s_{t,\max 1}$ at center of circular boundary
$s_{t,\max 2}$ at normal to radial boundary

*For definition of symbols refer to Sec. 4.10–5a.

4.11 STRESSES AND STRAINS AT A POINT

(1) Variation of Stresses at a Point

(a) Notation. The states of stresses at the point i shown in Figs. 4.11–1 through 4.11–4 are defined in terms of the following symbols:

$$s_{nx}, s_{ny} = \text{normal stresses along } X, Y$$
$$s'_{nx}, s'_{ny} = \text{normal stresses along } X', Y'$$
$$s_{tx}, s_{ty} = \text{shearing stresses along } X, Y$$
$$s'_{tx}, s'_{ty} = \text{shearing stresses along } X', Y'$$

$$\alpha, \beta = \text{position angles of } X', Y'$$
$$\omega_a, \omega_b = \text{position angles of } A, B$$
$$\omega_c, \omega_d = \text{position angles of } C, D$$
$$\Delta = \sqrt{(s_{nx} - s_{ny})^2 + 4s_{ty}^2}$$

$$s_{na}, s_{nb} = \text{principal normal stresses along } A, B$$
$$s_{tc}, s_{td} = \text{extreme shearing stresses along } C, D$$
$$s_{nc}, s_{nd} = \text{equal normal stresses along } C, D$$

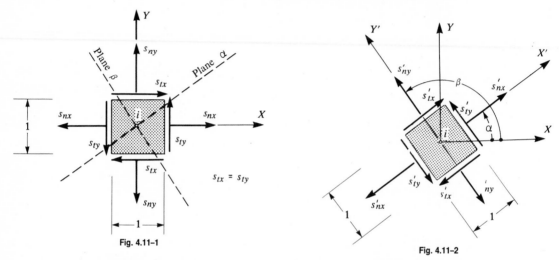

Fig. 4.11–1

Fig. 4.11–2

(b) Relationships between the stresses in the X, Y axes (Fig. 4.11–1) and those in the X', Y' axes (Fig. 4.11–2) are

$$s'_{nx} = \frac{s_{nx} + s_{ny}}{2} + \frac{s_{nx} - s_{ny}}{2} \cos 2\alpha + s_{ty} \sin 2\alpha$$

$$s'_{ny} = \frac{s_{nx} + s_{ny}}{2} - \frac{s_{nx} - s_{ny}}{2} \cos 2\alpha - s_{ty} \sin 2\alpha$$

$$s'_{tx} = s'_{ty} = -\frac{s_{nx} - s_{ny}}{2} \sin 2\alpha + s_{ty} \cos 2\alpha$$

where α is an arbitrary angle and $\beta = \pi/2 + \alpha$.

(c) Principal stress. For the given state of stress at i there exist two mutually perpendicular axes A, B (Fig. 4.11–3) along which the shearing stresses vanish and the normal stresses assume their extreme values. Such axes are called the *principal axes* of that given state and the normal stresses along these axes are denoted as the *principal stresses*, given analytically as

$$s_{na} = \frac{s_{nx} + s_{ny} + \Delta}{2} \qquad s_{nb} = \frac{s_{nx} + s_{ny} - \Delta}{2}$$

where $\omega_a = \frac{1}{2} \tan^{-1} \dfrac{2s_{ty}}{s_{nx} - s_{ny}}$ and $\omega_b = \pi/2 + \omega_a$.

(d) Extreme shearing stresses. At that given point i there are also two mutually perpendicular axes C, D (Fig. 4.11–4) along which the shearing stresses assume their extreme values and the normal stresses become of equal values. These axes bisect the angle of the A, B axes and their stress state is defined analytically as

$$s_{nc} = s_{nd} = \frac{s_{nx} + s_{ny}}{2} \qquad s_{tc} = s_{td} = \frac{\Delta}{2}$$

where $\omega_c = \frac{1}{2}\tan^{-1}\left(-\dfrac{s_{nx} - s_{ny}}{2s_{ty}}\right)$ and $\omega_d = \pi/2 + \omega_c$.

(e) Invariant relations of these stresses and angles are

$$s_{na} + s_{nb} = s_{nc} + s_{nd} = s_{nx} + s_{ny} = s'_{nx} + s'_{ny} = \text{constant}$$
$$\omega_b - \omega_a = \omega_d - \omega_c = \beta - \alpha = \pi/2$$

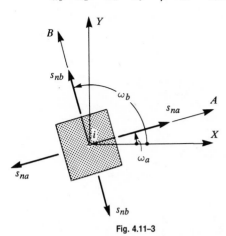

Fig. 4.11–3

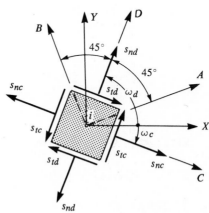

Fig. 4.11–4

(2) Variation of Strains at a Point

(a) Notation. The states of strains at the same point i shown in Figs. 4.11–5 through 4.11–8 are defined in terms of the following symbols:

$\varepsilon_{nx}, \varepsilon_{ny}$ = normal strains along X, Y
$\varepsilon'_{nx}, \varepsilon'_{ny}$ = normal strains along X', Y'
$\varepsilon_{tx}, \varepsilon_{ty}$ = shearing strains along X, Y
$\varepsilon'_{tx}, \varepsilon'_{ty}$ = shearing strains along X', Y'

α, β = position angles of X', Y'
ω_a, ω_b = position angles of A, B
ω_d, ω_c = position angles of D, C
$\Delta = \sqrt{(\varepsilon_{nx} - \varepsilon_{ny})^2 + \varepsilon_{ty}^2}$

$\varepsilon_{na}, \varepsilon_{nb}$ = principal normal strains along A, B
$\varepsilon_{tc}, \varepsilon_{td}$ = maximum shearing strains along C, D
$\varepsilon_{nc}, \varepsilon_{nd}$ = equal normal strains along C, D

(b) Relationships between the strains in the X, Y axes (Fig. 4.11–5) and those in the X', Y' axes (Fig. 4.11–6) are

$$\varepsilon'_{nx} = \frac{\varepsilon_{nx} + \varepsilon_{ny}}{2} + \frac{\varepsilon_{nx} - \varepsilon_{ny}}{2}\cos 2\alpha + \frac{\varepsilon_{ty}}{2}\sin 2\alpha$$

$$\varepsilon'_{ny} = \frac{\varepsilon_{nx} + \varepsilon_{ny}}{2} - \frac{\varepsilon_{nx} - \varepsilon_{ny}}{2}\cos 2\alpha - \frac{\varepsilon_{ty}}{2}\sin 2\alpha$$

$$\frac{\varepsilon'_{tx}}{2} = \frac{\varepsilon'_{ty}}{2} = -\frac{\varepsilon_{nx} - \varepsilon_{ny}}{2}\sin 2\alpha + \frac{\varepsilon_{ty}}{2}\cos 2\alpha$$

where α is an arbitrary angle and $\beta = \pi/2 + \alpha$.

(c) Principal strains. For the given state of strains at i there exist two mutually perpendicular axes A, B (Fig. 4.11–7) along which the shearing strains vanish and the normal strains assume their extreme values. Such axes are called the *principal axes* of that state and the normal strains along these axes are denoted as the *principal strains*, given analytically as

$$\varepsilon_{na} = \frac{\varepsilon_{nx} + \varepsilon_{ny} + \Delta}{2} \qquad \varepsilon_{nb} = \frac{\varepsilon_{nx} + \varepsilon_{ny} - \Delta}{2}$$

where $\omega_a = \frac{1}{2}\tan\dfrac{\varepsilon_{ty}}{\varepsilon_{nx} - \varepsilon_{ny}}$ and $\omega_b = \pi/2 + \omega_a$.

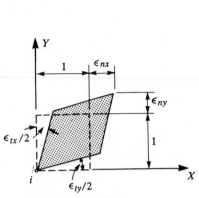

Fig. 4.11–5

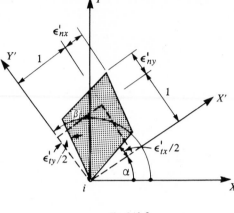

Fig. 4.11–6

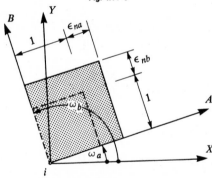

Fig. 4.11–7

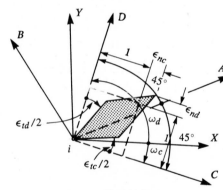

Fig. 4.11–8

(d) Extreme shearing strains. At that given point i there are also two mutually perpendicular axes C, D (Fig. 4.11–8) along which the shearing strains assume their extreme values and the normal strains become of equal values. These axes bisect the angle of the A, B axes, and their state is defined analytically as

$$\varepsilon_{nc} = \varepsilon_{nd} = \frac{\varepsilon_{nx} + \varepsilon_{ny}}{2} \qquad \varepsilon_{tc} = \varepsilon_{td} = \frac{\Delta}{2}$$

where $\omega_c = \frac{1}{2}\tan^{-1}\left(-\dfrac{\varepsilon_{nx} - \varepsilon_{ny}}{\varepsilon_{ty}}\right)$ and $\omega_d = \pi/2 + \omega_c$.

(e) Invariant relations of these strains and angles are

$$\varepsilon_{na} + \varepsilon_{nb} = \varepsilon_{nc} + \varepsilon_{nd} = \varepsilon_{nx} + \varepsilon_{ny} = \varepsilon'_{nx} + \varepsilon'_{ny} = \text{constant}$$
$$\omega_b - \omega_a = \omega_d - \omega_c = \alpha - \beta = \pi/2$$

(3) Mohr's Circle

(a) Representation. The stress and strain relations defined in Secs. 4.11–1 and 4.11–2 can be represented graphically by Mohr's circle. Since there is a perfect analogy between the stress and the strain equations, only two typical stress circle constructions are shown. All stresses are given in the XY plane and all constructions are carried out in the number sequence designated in the respective figures. All angles are measured from the given-state axes to the new-state axes in the counterclockwise direction.

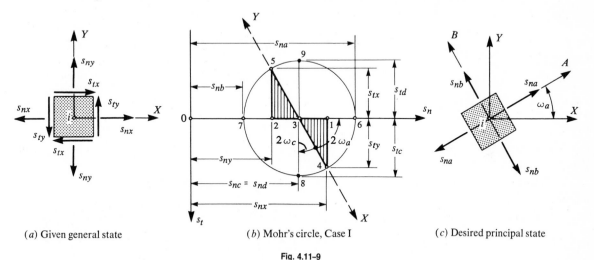

(a) Given general state

(b) Mohr's circle, Case I

(c) Desired principal state

Fig. 4.11–9

(b) Case I: From general state to principal state. The general state of stress is given in Fig. 4.11–9a. The corresponding Mohr's circle is constructed in Fig. 4.11–9b and the principal state derived from the circle construction is shown in Fig. 4.11–9c.

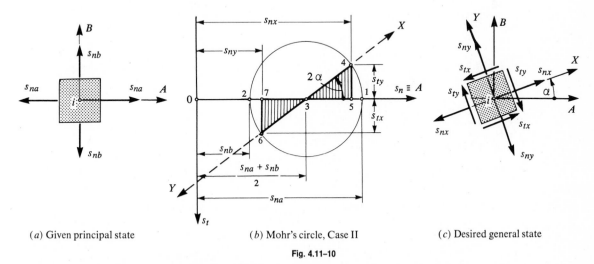

(a) Given principal state

(b) Mohr's circle, Case II

(c) Desired general state

Fig. 4.11–10

(c) Case II: From principal state to general state. The principal state of stress is given in Fig. 4.11–10a. The corresponding Mohr's circle is constructed in Fig. 4.11–10b and the desired general state is derived from the circle construction as shown in Fig. 4.11–10c.

4.12 COMPOUND STRESSES IN STRAIGHT HOMOGENEOUS BARS

(1) Superposition of Normal Stresses

(a) **Statement of problem.** A straight, short homogeneous bar of constant cross section is acted upon by a normal force applied eccentrically as shown in Fig. 4.12–1a.

(b) **Symbols** used in the geometric and stress relations are:

A = area of cross section (cm²) P = axial load along X (kgf)
M_y = bending moment about Y (kgf-cm) M_z = bending moment about Z (kgf-cm)
I_{yy} = moment of inertia of A about Y (cm⁴) I_{zz} = moment of inertia of A about Z (cm⁴)
k_{yy} = radius of gyration of A about Y (cm) k_{zz} = radius of gyration of A about Z (cm)
e = y coordinate of P (cm) f = z coordinate of P (cm)
s_n = normal stress along X (kgf-cm⁻²) α = direction angle of neutral axis

(c) **Sign convention.** x, y, z are positive in the positive direction of the coordinate axes, P is positive (negative) if causing tension (compression), M_y, M_z are positive if acting in the right-hand direction on the respective coordinate axis, and s_n is positive (negative) in tension (compression).

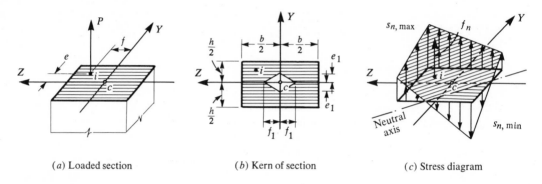

(a) Loaded section   (b) Kern of section   (c) Stress diagram

Fig. 4.12–1

(d) **Normal stress** in the elastic range at y, z on a cross section symmetrical with respect to the Y and Z axes (Fig. 4.12–1b) is

$$s_n = \frac{P}{A} + \frac{Pf}{I_{yy}} z - \frac{Pe}{I_{yz}} y = \frac{P}{A}\left(\frac{fz}{k_{yy}^2} - \frac{ey}{k_{zz}^2} + 1\right)$$

(e) **Equation of the neutral axis** is

$$\frac{ey}{k_{zz}^2} - \frac{fz}{k_{yy}^2} = 1 \qquad k_{yy} = \sqrt{\frac{I_{yy}}{A}} \qquad k_{zz} = \sqrt{\frac{I_{zz}}{A}}$$

and if this axis does not intersect the section, the section is in a state of tension or compression exclusively. For this state to exist the normal force must be applied in the region of the section called the *kern*. The extreme coordinates of the kern (Fig. 4.12–1c) are

$$e_1 = \frac{2k_{zz}^2}{h} \qquad f_1 = \frac{2k_{yy}^2}{b}$$

The most common cases of kerns are shown in Table 4.12–1, where $e_1 = e$ and $f_1 = f$.

TABLE 4.12–1 Extreme Coordinates of Kerns of Solid Symmetrical Sections*

(*a*) **Kern of rectangle**

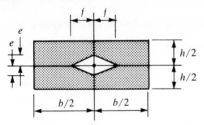

$e = h/6$ $f = b/6$

(*b*) **Kern of circle**

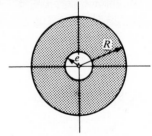

$e = R/4$

(*c*) **Kern of hollow circle**

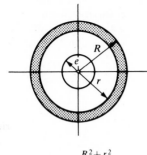

$$e = \frac{R^2 + r^2}{4R}$$

(*d*) **Kern of thin circular shell**

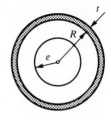

$e = R/2$

(*e*) **Kern of hollow rectangle**

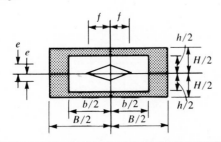

$$e = \frac{BH^3 - bh^3}{6\,(BH - bh)\,H}$$

$$f = \frac{HB^3 - hb^3}{6\,(BH - bh)\,B}$$

(*f*) **Kern of thin box shell**

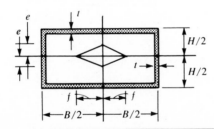

$$e = \frac{H\,(H + 3B)}{6\,(B + H)}$$

$$f = \frac{B\,(B + 3H)}{6\,(B + H)}$$

---

*For definition of symbols refer to Sec. 4.12–1e.

(2) Superposition of Tangential Stresses

(a) Statement of problem. A straight, homogeneous bar of symmetrical cross section with respect to the Y, Z axes is acted upon by a shearing force applied along the Y axis and a torque applied at the centroid of the section along the X axis as shown in Fig. 4.12–2.

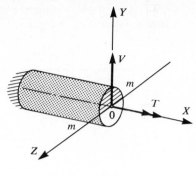

(b) Symbols used below are:

V = vertical shear (kgf)
T = torque (kgf-cm)
I = moment of inertia about Z (cm^4)
s_t = tangential stress (kgf/cm^2)

Fig. 4.12–2

(c) Sign convention. V is positive if acting in the direction of Y, T is positive if acting in the right-hand direction on X, and s_t is positive if acting in the direction of Y.

(d) Maximum tangential stress in the elastic range is the algebraic sum of the shearing stress due to V and the torsional stress due to T and occurs at the intersection of the Z axis with the contour of the section where both stresses are acting in the same direction. In general,

$$s_{t,\text{max}} = \frac{VQ}{It} + \frac{T}{S} \qquad \left\{ \begin{array}{l} S, Q \text{ in cm}^3 \\ t \text{ in cm} \end{array} \right\}$$

where Q = static moment of the shaded area of the section above the Z axis about this axis, S = torsional section modulus defined in Sec. 4.10–5a, and t = length of the segment mm in Fig. 4.12–2.

(e) Particular cases of this formula are given below.

Cross-sectional stress variation	Circle	Hollow circle	Rectangle *
	$t = 2R$	$\frac{t}{2}$	$b/2 \quad b/2$ $t = b$
$s_{t,\text{max}}$	$\dfrac{4V}{3\pi R^2} + \dfrac{2T}{\pi R^3}$	$\dfrac{4V}{3\pi (R^2 - r^2)} + \dfrac{2RT}{\pi (R^4 - r^4)}$	$\dfrac{3V}{2bh} + \dfrac{10T}{C_1 hb^2}$

*C_1 is given in Table 4.10–1.

5
MECHANICS OF FLUIDS AND SOILS

5.01 PROPERTIES OF FLUIDS

(1) Classification and Definitions

(a) **Matter (substance)** can be classified by the physical form of its occurrence (state, phase) as a solid, a liquid, or a gas (vapor).

(b) **Molecules of solids** at ordinary temperature and pressure have strong attraction for each other and remain in a fixed position relative to each other. Thus solids have a definite volume and shape and change this shape under the action of forces in much the same manner as springs (finite deformations).

(c) **Molecules of liquids** at ordinary temperature and pressure have small attraction for each other and drift relative to each other. Thus liquids have a definite volume but no definite shape, and when placed in a container they assume its shape but will not necessarily fill it (they can form a free surface).

(d) **Molecules of gases** at ordinary temperature and pressure have very small attraction for each other and their random drift is very large. Thus gases have neither a definite volume nor a definite shape, and when released in a container they assume its shape and fill the entire container (they do not form a free surface).

(e) **Fluids.** Because liquids and gases at ordinary temperature and pressure are unable to resist the action of a shear and continue to deform under its action as long as it exists, they are referred to as fluids.

(f) **Fluid mechanics**, defined as the branch of physics concerned with the investigation of effects of forces on fluids, is traditionally subdivided into two parts: statics and dynamics.

(g) **Statics of fluids** investigates the equilibrium of fluids under the action of stationary forces (forces which are in a fixed position on or in the fluid).

(h) **Dynamics of fluids** investigates the motion of fluids and the causes producing, sustaining, or opposing this motion.

(2) Mass–Volume Relationships

(a) **Mass density (unit mass)** of a fluid, designated by the symbol ρ (rho), is the mass of its unit volume.

$$\rho = \frac{\text{total mass}}{\text{total volume}} = \frac{m}{V} \qquad \left\{ \begin{array}{l} m \text{ in kg} \\ V \text{ in m}^3 \\ \rho \text{ in kg/m}^3 \end{array} \right\}$$

(b) **Units of mass density** are 1 g/cm^3, 1 kg/m^3, 1 kg/liter, 1 lb/in.^3, 1 lb/ft^3, 1 slug/ft^3.

(c) **Three basic conversion relations** are

$$1 \text{ g/cm}^3 = 0.0361 \text{ lb/in.}^3 \qquad 1 \text{ lb/in.}^3 = 27.68 \text{ g/cm}^3$$

$$1 \text{ kg/m}^3 = 0.0624 \text{ lb/ft}^3 \qquad 1 \text{ lb/ft}^3 = 16.02 \text{ kg/m}^3$$

$$1 \text{ lb/ft}^3 = 1/32.17 \text{ slug/ft}^3 \qquad 1 \text{ slug/ft}^3 = 32.17 \text{ lb/ft}^3$$

(d) **Mass density of distilled water** at 4°C is

$$1 \text{ g/cm}^3 = 1{,}000 \text{ kg/m}^3 = 1 \text{ kg/liter} = 62.428 \text{ lb/ft}^3 = 1.94 \text{ slug/ft}^3$$

(e) **Tables.** Mass densities of the most commonly used solids, liquids, and gases are given in Appendix A.

(f) **Specific volume** (v) is the inverse of mass density (is the volume occupied by a unit mass of the substance).

$$v = \frac{1}{\rho} = \frac{\text{total volume}}{\text{total mass}} = \frac{V}{m} \qquad \left\{ \begin{array}{l} v \text{ in m}^3/\text{kg} \\ V \text{ in m}^3 \\ m \text{ in kg} \end{array} \right\}$$

(3) Weight–Volume Relationships

(a) Weight density (unit weight) of a fluid, designated by the symbol γ (gamma), is the weight of its unit volume.

$$\gamma = \frac{\text{total weight}}{\text{total volume}} = \frac{W}{V} \qquad \left\{ \begin{array}{l} W \text{ in kgf} \\ V \text{ in m}^3 \\ \gamma \text{ in kgf/m}^3 \end{array} \right\}$$

(b) Units of weight density are $1\,\text{gf/cm}^3$, $1\,\text{kgf/m}^3$, $1\,\text{kgf/liter}$, $1\,\text{lbf/in.}^3$, $1\,\text{lbf/ft}^3$, $1\,\text{pd/ft}^3$.

(c) Three basic conversion relations are

$1\,\text{gf/cm}^3 = 0.0361\,\text{lbf/in.}^3 \qquad 1\,\text{lbf/in.}^3 = 27.68\,\text{gf/cm}^3$

$1\,\text{kgf/m}^3 = 0.0624\,\text{lbf/ft}^3 \qquad 1\,\text{lbf/ft}^3 = 16.02\,\text{kgf/m}^3$

$1\,\text{lbf/ft}^3 = 32.17\,\text{pd/ft}^3 \qquad 1\,\text{pd/ft}^3 = 1/32.17\,\text{lbf/ft}^3$

where pd = poundal.

(d) Weight density of distilled water at 4°C is

$1\,\text{gf/cm}^3 = 1{,}000\,\text{kgf/m}^3 = 1\,\text{kgf/liter} = 62.428\,\text{lbf/ft}^3 = 2{,}008.6\,\text{pd/ft}^3$

(e) Tables. Weight densities of the other most commonly used solids, liquids, and gases are given in Appendix A.

(4) Specific Gravity

(a) Specific gravity of a solid or fluid, designated by the symbol G, is the ratio of its weight to the weight of an equal volume of standard substance and is a dimensionless number, independent of the system of measure (same number in the SI and FPS systems).

$$G = \frac{\text{total weight of substance}}{\text{weight of equal volume of standard substance}}$$

$$= \frac{\text{weight density of substance}}{\text{weight density of standard substance}}$$

(b) Standard substance. Solids and liquids are referred to distilled water at 4°C as standard, while gases are often referred to air at 0°C at 1 atmosphere pressure as standard.

example:

If for gold $\gamma = 19{,}300\,\text{kgf/m}^3 = 1{,}203\,\text{lbf/ft}^3$ and for standard water $\gamma = 1{,}000\,\text{kgf/m}^3 = 62.428\,\text{lbf/ft}^3$, then the specific gravity of gold is

$G = 19{,}300/1{,}000 = 1{,}203/64{,}428 = 19.3$

(c) Densities in terms of G are

$$\rho = G \cdot \rho_s \qquad \gamma = G \cdot \gamma_s$$

where ρ_s and γ_s = mass density and weight density of the standard substance, respectively.

example:

If for gasoline $G = 0.68$ and for standard water $\rho_s = 1.94\,\text{slug/ft}^3 = 1{,}000\,\text{kg/m}^3$ and $\gamma_s = 62.43\,\text{lbf/ft}^3 = 1{,}000\,\text{kgf/m}^3$, then

$\rho = 0.68 \times 1.94 = 1.32\,\text{slug/ft}^3 \qquad \gamma = 0.68 \times 62.43 = 42.45\,\text{lbf/ft}^3$

$\quad\, = 0.68 \times 1{,}000 = 680\,\text{kg/m}^3 \qquad\quad = 0.68 \times 1{,}000 = 680\,\text{kgf/m}^3$

(5) Viscosity

(a) **Definition.** Viscosity is the resistance of fluid to shearing action which takes place when an attempt is made to slide one plate over another with an intervening layer of fluid between them.

(b) **Newton's law of viscosity** states that the shearing stress τ (tau) developed by this motion between the plate and the fluid is linearly proportional to the angular velocity du/dy (Fig. 5.01–1).

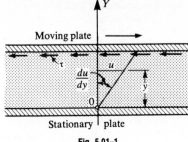

Fig. 5.01–1

$$\tau = \mu \frac{du}{dy} \qquad \left\{ \begin{array}{l} \tau \text{ in kgf/cm}^2 \\ u \text{ in cm/s} \\ y \text{ in cm} \\ \mu \text{ in (kgf-s)/cm}^2 \end{array} \right\}$$

where u = linear velocity of the layer of fluid, y = vertical coordinate of the layer, and μ (mu) = absolute viscosity (also called the *dynamic viscosity*), which must be determined for a particular fluid by experiments.

(c) **Absolute (dynamic) viscosity** μ (mu) depends on pressure and temperature, but the influence of pressure is so small that it can be neglected (except at extremely high pressures). The viscosity of liquids decreases with an increase in temperature; for gases the opposite is true.

example:

Molasses, oils, and tars are examples of highly viscous liquids, while water and alcohol have very low viscosities.

(d) **SI unit of absolute viscosity** is 1 poise (after Jean L. Poiseuille).

$$1 \text{ poise} = \frac{g}{cm\text{-}s} = \frac{dyne\text{-}s}{cm^2} = \frac{N\text{-}s}{10\,m^2}$$

(e) **FPS unit of absolute viscosity** is 1 reyn (after Osborne Reynolds).

$$1 \text{ reyn} = \frac{12 \text{ slugs}}{\text{in.-sec}} = \frac{lbf\text{-sec}}{in.^2} = \frac{144 \, lbf\text{-sec}}{ft^2}$$

(f) **Conversion relations** are

$$1 \text{ poise} = 100 \text{ centipoises} = 1 \text{ dyne-s/cm}^2$$

$$1 \frac{kgf\text{-}s}{m^2} = 98.1 \text{ poises} = \frac{0.205 \, lbf\text{-sec}}{ft^2}$$

$$1 \frac{lbf\text{-sec}}{ft^2} = 478 \text{ poises} = \frac{4.88 \, kgf\text{-}s}{m^2}$$

(g) **Kinematic viscosity** ν (nu) is the ratio of dynamic viscosity μ to mass density ρ of the same fluid.

$$\nu = \frac{\mu}{\rho} \qquad \left\{ \begin{array}{l} \mu \text{ in kg/cm-s} \\ \rho \text{ in kg/cm}^3 \\ \nu \text{ in cm}^2/s \end{array} \right\}$$

where μ is defined in (b). For distilled water at 20°C, $\nu = 0.01 \text{ cm}^2/\text{s}$.

(h) SI unit of kinematic viscosity is $1 \text{ m}^2/\text{s}$, expressed as

$$1 \text{ m}^2/\text{s} = \frac{10^4 \text{ cm}^2}{\text{s}} = 10^4 \text{ stokes} = 10^4 \text{ S} = 10^6 \text{ cS}$$

where $1 \text{ S} = \text{cm}^2/\text{s} = 1 \text{ stoke}$ (after George Gabriel Stokes) and $1 \text{ cS} = \text{S}/100 = 1$ centistoke.

(i) FPS unit of kinematic viscosity is $1 \text{ ft}^2/\text{sec}$, related to the SI units as

$$1 \text{ ft}^2/\text{sec} = (30.48)^2 \text{ cm}^2/\text{s} = (30.48)^2 \text{ S} = 0.0929 \text{ m}^2/\text{s}$$

example:

If the absolute viscosity of a liquid is $\mu = 0.04$ poise and its specific gravity $G = 0.80$, then

$$\nu = \frac{\mu}{\rho} = \frac{\mu}{G \cdot \rho_s} = \frac{0.04}{0.8 \times 1.0} = 0.05 \text{ S} = 5 \times 10^{-6} \text{ m}^2/\text{s} = \frac{0.04/478}{0.8 \times 1.935} = 5.4 \times 10^{-5} \text{ ft}^2/\text{sec}$$

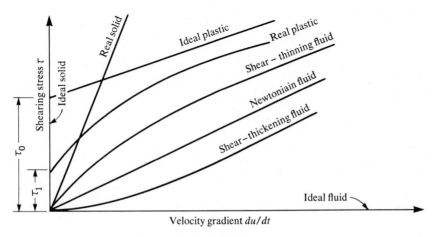

Fig. 5.01–2

(j) Newtonian fluids. If the shearing stress–velocity gradient relation is governed by the equation of Sec. 5.1–5b, the graph of this relation is a straight line (Fig. 5.01–2) and the fluid is called a *Newtonian fluid*. Other fluids and solids are designated as non-Newtonian.

(k) Non-Newtonian fluids and solids fall into the following categories:

Ideal plastics $\tau = \tau_0 + \mu \dfrac{du}{dt}$

Real plastics $\tau = \tau_1 + \mu \left(\dfrac{du}{dt}\right)^{1/n}$

Shear-thickening fluids $\tau = \mu \left(\dfrac{du}{dt}\right)^{n}$ $\Big\} \; n > 1$

Shear-thinning fluids $\tau = \mu \left(\dfrac{du}{dt}\right)^{1/n}$

(l) Ideal solids and fluids have shearing stress independent of the velocity gradient and velocity gradient independent of the shearing stress (Fig. 5.01–2).

(6) Pressure

(a) Definition. Pressure, designated by the symbol p, is the normal compressive force per unit area.

$$p = \frac{\text{force acting normal over an area}}{\text{area over which the force is distributed}} = \frac{F}{A} \qquad \left\{ \begin{array}{l} F \text{ in kgf} \\ A \text{ in m}^2 \\ p \text{ in kgf/m}^2 \end{array} \right\}$$

(b) SI unit of pressure is 1 pascal (after Blaise Pascal), designated by the symbol Pa, and defined as the pressure of 1 newton per square meter.

$$1 \text{ pascal} = 1 \text{ Pa} = 1 \text{ N/m}^2 = 10 \text{ dynes/cm}^2 = 0.10197 \text{ kgf/m}^2$$

(c) Technical atmosphere (at) is the pressure of 1 kilogram-force per square centimeter at latitude 45°, sea level, and 0°C.

$$1 \text{ technical atmosphere} = 1 \text{ at} = 1 \text{ kgf/cm}^2$$

(d) Bar (b) is the pressure of 10 newtons per square centimeter.

$$1 \text{ bar} = 1 \text{ b} = 10 \text{ N/cm}^2 = 1.0197 \text{ kgf/cm}^2$$

(e) Torr (torr) is the pressure of 1 mm of mercury per square centimeter at latitude 45°, sea level, and 0°C.

$$1 \text{ torr} = 1.333 \times 10^{-2} \text{ N/cm}^2 = 1.3596 \times 10^{-3} \text{ kgf/cm}^2$$

(f) FPS unit of pressure is 1 pound-force per square foot, designated by the symbol lbf/ft^2.

(g) Physical atmosphere (standard atmosphere, atm) is the pressure of 760 mm of mercury per square centimeter at latitude 45°, sea level, and 0°C.

$$
\begin{aligned}
1 \text{ physical atmosphere} = 1 \text{ atm} &= 10.1325 \text{ N/cm}^2 = 1.0332 \text{ kgf/cm}^2 = 14.69 \text{ lbf/in.}^2 \\
&= 76 \text{ cm of mercury/cm}^2 \text{ at } 0°\text{C} = 1{,}033 \text{ cm of water/cm}^2 \text{ at } 4°\text{C} \\
&= 29.92 \text{ in. of mercury/in.}^2 \text{ at } 0°\text{C} = 406.8 \text{ in. of water/in.}^2 \text{ at } 4°\text{C}
\end{aligned}
$$

(h) Commonly used conversion factors are

$$1 \text{ Pa} = 9.8692 \times 10^{-6} \text{ atm} = 1.0197 \times 10^{-5} \text{ kgf/cm}^2 = 1.4504 \times 10^{-4} \text{ lbf/in.}^2$$

$$1 \text{ kgf/cm}^2 = 9.6784 \times 10^{-1} \text{ atm} = 9.8066 \times 10^4 \text{ Pa} = 1.4223 \times 10 \text{ lbf/in.}^2$$

$$1 \text{ lbf/in.}^2 = 6.8046 \times 10^{-2} \text{ atm} = 6.8948 \times 10^3 \text{ Pa} = 7.0300 \times 10^{-2} \text{ kgf/cm}^2$$

$$1 \text{ lbf/ft}^2 = 4.7254 \times 10^{-4} \text{ atm} = 4.7883 \times 10 \text{ Pa} = 4.8827 \text{ kgf/m}^2$$

(i) Vapor pressure is the pressure at which a liquid boils and is in equilibrium with its own vapor. The vapor pressure increases with temperature.

(j) Bulk modulus of elasticity K is the measure of compressibility of a liquid and is defined as the ratio of the change in unit pressure to the change in unit volume.

$$K = \frac{\Delta p}{\Delta V / V} \qquad \left\{ \begin{array}{l} K \text{ in kgf/m}^2 \\ p \text{ in kgf/m}^2 \\ V \text{ in m}^3 \end{array} \right\}$$

where V = volume. Bulk moduli of selected liquids are given in Appendix A.

5.02 STATICS OF FLUIDS

(1) Static Fluid Pressure

(a) Fluid statics investigates the equilibrium of fluids at rest with respect to the earth or at rest with respect to a frame of reference moving in uniform rectilinear motion with respect to the earth.

(b) Conditions of static equilibrium require that the acceleration of the fluid be zero in all directions and consequently that the sum of all forces acting on the fluid and on any particle of it be zero in every direction.

(c) Static pressure at any given point in a fluid at rest is the same in all directions and acts normal to any plane. Consequently, on the same horizontal plane in the fluid the static pressures are equal (uniformly distributed compression).

(d) Pascal's principle states that the change in pressure on any part of a confined fluid introduces the same change in pressure in all parts of the fluid.

(e) Pressure variation with depth within a homogeneous liquid (Fig. 5.02–1) is

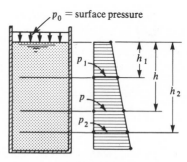

$$p = p_0 + \gamma h \qquad \left\{ \begin{array}{l} p \text{ in kgf/m}^2 \\ p_0 \text{ in kgf/m}^2 \\ \gamma \text{ in kgf/m}^3 \\ h \text{ in m} \end{array} \right\}$$

Fig. 5.02–1

where p_0 = surface pressure, γ = weight density of liquid, h = depth below the surface, and p = absolute pressure.

(f) Atmospheric pressure is the weight of a unit column (of base area 1 cm^2) of air above sea level extending to the top of the atmosphere (approximately 30 km) and is called the physical atmosphere (atm), defined in Sec. 5.01–5g.

(g) Free surface pressure p_s under standard conditions is the pressure of 1 atm defined in (f) above.

$$p_s = 1 \text{ atm} = 101{,}325 \text{ N/m}^2 = 10{,}332 \text{ kgf/m}^2$$
$$= 2{,}116 \text{ lbf/ft}^2 = 14.69 \text{ lbf/in.}^2$$

(h) Gage pressure is the difference between the absolute pressure and the atmospheric pressure defined in (f) and (g) above. The gage pressure is positive above 1 atm and negative below 1 atm.

(i) Difference in pressure Δp between two points at levels h_1 and h_2 in the given liquid (Fig. 5.02–1) is

$$\Delta p = p_2 - p_1 = \gamma(h_2 - h_1) \qquad \{\Delta p \text{ in kgf/m}^2\}$$

where γ, h_2, h_1, p_2, p_1 are defined in (e).

(j) Pressure head h, defined as the height of the column of homogeneous liquid that produces a given pressure p, is

$$h = \frac{p}{\gamma} \qquad \{h \text{ in m}\}$$

where γ = weight density of the liquid.

(k) Hydrostatic paradox states that the total force P exerted by a fluid of weight density γ and of head h on the area A is independent of the shape of the container (Fig. 5.02–2).

$$P = A\gamma h \qquad \left\{ \begin{array}{l} P \text{ in kgf} \\ A \text{ in m}^2 \\ \gamma \text{ in kgf/m}^3 \\ h \text{ in m} \end{array} \right\}$$

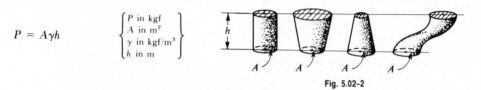

Fig. 5.02–2

(l) Free-surface liquid levels in two communicating vessels must lie in the same horizontal plane because they are acted upon by the same surface pressure p_s (Fig. 5.02–3).

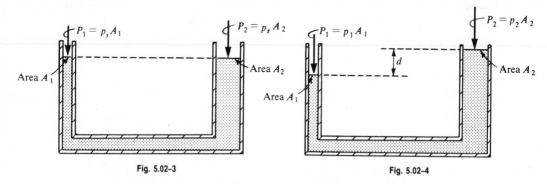

Fig. 5.02–3 Fig. 5.02–4

(m) Compressed-surface liquid levels in two communicating vessels (Fig. 5.02–4) of surface areas A_1 and A_2 acted upon by compressive forces P_1 and P_2, respectively, differ by

$$d = \frac{P_1/A_1 - P_2/A_2}{\gamma} = \frac{p_1 - p_2}{\gamma} \qquad \left\{ \begin{array}{l} P_1, P_2 \text{ in kgf} \\ A_1, A_2 \text{ in m}^2 \\ \gamma \text{ in kgf/m}^3 \\ d \text{ in m} \end{array} \right\}$$

where γ = weight density of liquid.

(2) Measurement of Pressure

(a) Barometer is the pressure gage used to measure atmospheric pressure. Two types of barometer are shown diagramatically in Fig. 5.02–5. The closed-end tube is filled with a liquid (usually mercury) and inverted in the position shown. If the tube is sufficiently long, there will be an empty space at the top called a *vacuum* (producing zero pressure). Then the pressure at the free surface (below) becomes

$$P_{atm} = \gamma_{mercury}h$$

where

$$\gamma_{mercury} = 850 \text{ lbf/ft}^3 = 13,600 \text{ kgf/m}^3$$

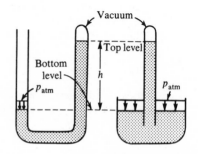

Fig. 5.02–5

and for standard conditions $p_{atm} = 1$ atm, $h = 0.76$ m.

(b) **Simple manometer** is the pressure gage used to measure the unknown pressure or the pressure difference in a fluid system. It is essentially a U tube filled with a reference liquid (called *manometric liquid*), the end levels of which are exposed to the atmospheric pressure and to the fluid system pressure, respectively. The difference of these two levels is used for the calculation of the gage pressure p_g.

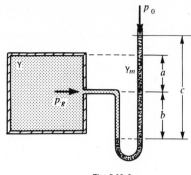

$$p_g = c\gamma_m - b\gamma$$

$$a = \frac{c\gamma_m}{\gamma} - b$$

Fig. 5.02–6

where γ_m = weight density of manometric fluid, γ = weight density of investigation fluid, and a, b, c = heads shown in Fig. 5.02–6.

(c) **Differential manometer** consists of two connected U tubes and is used to measure the pressure difference between two levels in a fluid system (Fig. 5.02–7).

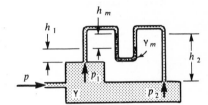

$$\Delta p_g = p_2 - p_1 = (h_2 - h_1)\gamma + h_m(\gamma_m - \gamma)$$

where Δp_g = difference in gage pressure, p_1, p_2 = gage pressures, γ_m = weight density of manometric fluid, γ = weight density of investigated fluid, and h_1, h_2, h_m = heads shown in Fig. 5.02–7.

Fig. 5.02–7

(d) **Bourdon gage** consists of a sealed spiral of flat metal tubing (Fig. 5.02–8). As the pressure fills this tube, the tube tends to unwind and straighten out. This effect is communicated to a pointer by a pinion gear. The dial on which the pointer indicates the pressure is calibrated with respect to a given standard.

Fig. 5.02–8

(e) **Hand gage** used to measure the pressure in automobile tires consists of a tube with a piston attached to a spring. As the air pressure pushes against the piston the spring contracts and the calibrated scale attached to the piston indicates the gage pressure of the tire.

(3) Pressure on Plane Areas

(a) **Resultant of pressure** on a plane area A by a liquid of weight density γ (Fig. 5.02–9) is perpendicular to the plane of A and its magnitude is

$$P = Ah_c\gamma \qquad \begin{cases} A \text{ in m}^2 & P \text{ in kgf} \\ h_c \text{ in m} & \gamma \text{ in kgf/m}^3 \end{cases}$$

where A = area subjected to pressure and h_c = vertical distance of the centroid of A from the free surface of the liquid.

(b) **Center of pressure** m is the point of application of the resultant of pressure P. For horizontal plane areas this center coincides with the centroid of the respective area. For inclined or vertical plane areas the center of pressure is at some distance below the centroid of the respective area (Fig. 5.02–9).

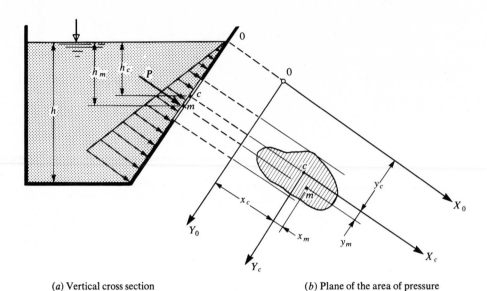

(a) Vertical cross section   (b) Plane of the area of pressure

Fig. 5.02–9

(c) **Coordinates of center of pressure** m referred to the centroidal axes X_c, Y_c of the pressure area A in Fig. 5.02–9 are

$$x_m = \frac{I_{cxy}}{y_c A} \qquad y_m = \frac{I_{cxx}}{y_c A} \qquad \left\{ \begin{array}{l} x_m,\, y_m,\, y_c \text{ in m} \\ A \text{ in m}^2 \\ I_{cxx},\, I_{cxy} \text{ in m}^4 \end{array} \right\}$$

where A = pressure area, I_{cxx} = moment of inertia of A about the X_c axis, I_{cxy} = product of inertia of A with respect to the X_c and Y_c axis, and y_c = y coordinate of the centroid of A.

example:

Referring to the pressure areas 1 and 2 on the walls of the water tank of Fig. 5.02–10,

$A_1 = 3 \times 6 = 18\ \text{m}^2$ $\qquad\qquad$ $A_2 = (4 \times 6)/2 = 12\ \text{m}^2$

$h_{c,1} = 10 + 3 = 13\ \text{m}$ $\qquad\qquad$ $h_{c,2} = 10 + \frac{2}{3}(0.707 \times 6) = 12.83\ \text{m}$

$y_{c,1} = 10 + 3 = 13\ \text{m}$ $\qquad\qquad$ $y_{c,2} = \frac{10}{0.707} + 4 = 18.14\ \text{m}$

$I_{cxx,1} = \dfrac{3 \times 6^3}{12} = 54\ \text{m}^4$ $\qquad\qquad$ $I_{cxx,2} = \dfrac{4 \times 6^3}{36} = 24\ \text{m}^4$

$I_{cxy,1} = 0$ $\qquad\qquad\qquad\qquad$ $I_{cxy,2} = 0$

$P_1 = 18 \times 13 \times 1{,}000 = 234{,}000\ \text{kgf}$ $\qquad$ $P_2 = 12 \times 12.83 \times 1{,}000 = 153{,}960\ \text{kgf}$

$x_{m,1} = 0 \qquad y_{m,1} = \dfrac{54}{13 \times 18} = 0.23\ \text{m}$ $\qquad$ $x_{m,2} = 0 \qquad y_{m,2} = \dfrac{24}{18.14 \times 13} = 0.1\ \text{m}$

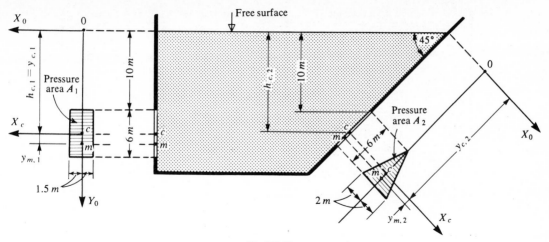

Fig. 5.02-10

(4) Pressure on Curved Surfaces

(a) Horizontal component P_H of the resultant of pressure P on a symmetrical surface S by liquid of weight density γ (Fig. 5.02–11a) equals the pressure force exerted on the vertical projection of that surface (Fig. 5.02–11b) and acts through the pressure center of this projection.

$$P_H = A_V h_V \gamma \qquad \begin{cases} A_V \text{ in m}^2 & P_H \text{ in kgf} \\ h_V \text{ in m} & \gamma \text{ in kgf/m}^3 \end{cases}$$

where A_V = area of the vertical projection of surface and h_V = vertical coordinate of the centroid of A_V.

(b) Vertical coordinate of the center of pressure is

$$h_m = h_V + \frac{I_{Vxx}}{h_V A_V} \qquad \begin{cases} h_V \text{ in m} \\ I_{Vxx} \text{ in m}^4 \end{cases}$$

where h_V = vertical coordinate of the centroid c_V of A_V and I_{Vxx} = moment of inertia of A_V about its centroidal axis parallel to the free surface.

(c) Vertical component P_V of the resultant of pressure P on a symmetrical surface S by a liquid of weight density γ (Fig. 5.02–11a) equals the weight of the volume of the liquid above this surface and acts at the centroid of this volume.

$$P_V = V\gamma \qquad \begin{cases} V \text{ in m}^3 & \gamma \text{ in kgf/m}^3 \\ P_V \text{ in kgf} \end{cases}$$

where V = volume of the liquid above the surface (shaded area in Fig. 5.02–11a).

(d) Resultant of P_H **and** P_V is

$$P = \sqrt{P_H^2 + P_V^2}$$

and its direction is given by

$$\tan \alpha = \frac{P_V}{P_H}$$

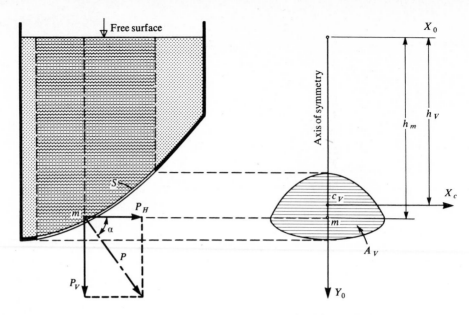

(a) Vertical cross section in plane of symmetry   (b) Projected pressure area A_V

Fig. 5.02–11

(e) Unsymmetrical surface. If the surface of Fig. 5.02–11 is unsymmetrical, the resultant of pressure can be represented by three cartesian components P_x, P_y, P_z parallel to the respective X and Y axes which are placed in the plane of the free surface of the liquid and the Z axis acting normal to this free surface.

$$P_x = A_{yz}h_{Vx}\gamma$$
$$P_y = A_{zx}h_{Vy}\gamma$$
$$P_z = V\gamma$$

$$\left.\begin{array}{l} A_{yz}, A_{zx} \text{ in m}^2 \\ h_{Vx}, h_{Vy} \text{ in m} \\ \gamma \text{ in kgf/m}^3 \\ P_x, P_y, P_z \text{ in kgf} \end{array}\right\}$$

where A_{yz} = projection of the surface in the YZ vertical plane, A_{zx} = projection of the surface in the ZX vertical plane, V = volume of the liquid above the surface, h_{Vx} = vertical coordinate of the centroid of A_{yz}, and h_{Vy} = the vertical coordinate of the centroid of A_{zx}.

(f) Location of P_x, P_y is given by the coordinates x_{Vx}, h_{Vx} and y_{Vy}, h_{Vy} of the center of pressure of A_{yz}, A_{zx}, respectively, computed analogically to those of Sec. 5.02–3c, and P_z acts vertically through the centroid of V (as in Sec. 5.02–4b) given by x_{cz}, y_{cz}, z_{cz}.

(g) Resultant of unsymmetrical pressure is a force,

$$P = \sqrt{P_x^2 + P_y^2 + P_z^2} \qquad \{P, P_x, P_y, P_z \text{ in kgf}\}$$

and a moment

$$M = \sqrt{M_x^2 + M_y^2 + M_z^2} \qquad \{M, M_x, M_y, M_z \text{ in kgf-m}\}$$

where M_x = static moment of P_y and P_z about the X axis, M_y = static moment of P_z and P_x about the Y axis, and M_z = static moment of P_x and P_y about the Z axis.

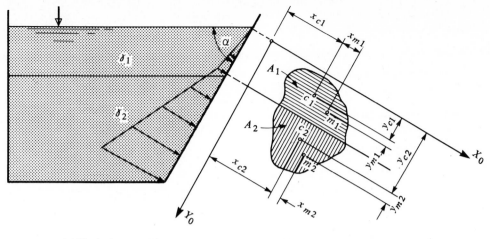

(*a*) Vertical cross section   (*b*) Plane of the area of pressure

Fig. 5.02–12

(5) Pressure on Plane Areas in Layered Fluids

(a) **Resultants of pressure** produced by layered liquids of weight densities γ_1, γ_2 on the respective areas A_1, A_2 in Fig. 5.02–12 are

$$P_1 = p_{c1}A_1 \qquad P_2 = p_{c2}A_2$$

where p_{c1}, p_{c2} = pressures at the centroids of these areas.

(b) **Centers of pressure** of these areas are given by

$$x_{m1} = \frac{\gamma_1 \sin \alpha I_{c1xy}}{P_1} \qquad y_{m1} = \frac{\gamma_1 \sin \alpha I_{c1xx}}{P_1}$$

$$x_{m2} = \frac{\gamma_2 \sin \alpha I_{c2xy}}{P_2} \qquad y_{m2} = \frac{\gamma_2 \sin \alpha I_{c2xx}}{P_2}$$

where α = angle of plane inclination, I_{c1xx}, I_{c2xx} = moments of inertia of the respective areas about their centroidal axes X_{c1}, X_{c2}, respectively, and I_{c1xy}, I_{c2xy} = products of inertia of the same areas in their centroidal axes.

(c) **Total resultant of pressure** on the area

$$A = A_1 + A_2$$

is

$$P = P_1 + P_2$$

where P_1, P_2 and A_1, A_2 are those defined in (*a*) above.

(d) **Center of total pressure** is given by

$$x_m = n_1(x_{c1} + x_{m1}) + n_2(x_{c2} + x_{m2})$$

$$y_m = n_1(y_{c1} + y_{m1}) + n_2(y_{c2} + y_{m2})$$

where $n_1 = P_1/P$ and $n_2 = P_2/P$.

(6) Membrane Stresses in Thin-Walled Containers

(a) Membrane state. When the walls of a container of revolution are relatively thin (one-tenth or less of the radius of curvature of the wall), the geometry of the wall shows no abrupt changes, and the walls are acted upon by distributed loads of constant or smoothly varying axially symmetrical intensity, the container is called a *thin shell* and the general state of stress in this shell reduces to the *membrane state* represented by Fig. 5.02–13.

(b) Stress and displacement equations of the membrane states in three typical thin shells are given below. The symbols used in these particular cases are:

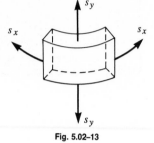

Fig. 5.02–13

s_x, s_y = normal stresses (kgf/cm²)
E = modulus of elasticity (kgf/cm²)
ν = Poisson's ratio (dimensionless)
h = height (cm)
θ = position angle (rad)
Δ_R = radial displacement (cm)
p = intensity of pressure (kgf/cm²)
R = radius of curvature (cm)
t = wall thickness (cm)
L = shell length (cm)
α = extreme position angle (rad)
γ = weight density (kgf/cm³)

All shells considered below are simply supported by tangential reactions of constant intensity q (kgf/cm) called the *line reactions*.

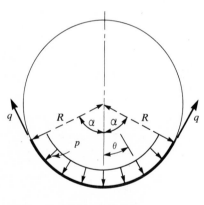

(a) Spherical shell, constant pressure

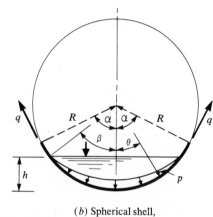

(b) Spherical shell, static pressure

Fig. 5.02–14

(c) Spherical shell of constant radius R and thickness t is shown in Fig. 5.02–14.
Loading: External or internal pressure $p = \pm$ constant.

$$s_x = \frac{pR}{2t} \qquad s_y = \frac{pR}{2t}$$

$$q = \frac{pR}{2} \qquad \Delta_R = \frac{pR^2}{2Et}(1-\nu)$$

Loading: Internal static pressure due to liquid of weight density γ, $p = \gamma(R - R\cos\theta - h)$

Above the liquid level,

$$s_x = -\frac{\gamma(3R-h)h^2}{6tR\sin^2\theta}$$

$$s_y = +\frac{\gamma(3R-h)h^2}{6tR\sin^2\theta}$$

At the edge,

$$q = \frac{\gamma(3R-h)h^2}{6R\sin^2\alpha}$$

Below the liquid level,

$$s_x = \frac{\gamma R^2}{6t}\left(\frac{3h}{R} - 5 + \frac{2(3+2\cos\theta)\cos\theta}{1+\cos\theta}\right)$$

$$s_y = \frac{\gamma R^2}{6t}\left(\frac{3h}{R} - 1 + \frac{2\cos\theta}{1+\cos\theta}\right)$$

At the liquid level,

$$\Delta_R = \frac{R\sin\beta}{E}(s_x - \nu s_y)$$

where Δ_R is measured along the liquid level ($\theta = \beta$).

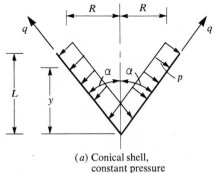

(a) Conical shell,
constant pressure

Fig. 5.02–15

(b) Conical shell,
static pressure

(d) Conical shell of vertex angle α and constant thickness t is shown in Fig. 5.02–15.
Loading: External or internal pressure $p = \mp$constant.

$$s_x = \frac{pR}{t\sin\alpha} \qquad s_y = \frac{pR}{2t\sin\alpha}$$

$$q = \frac{pR}{2\sin\alpha} \qquad \Delta_R = \frac{pR}{E\sin\alpha}\left(1 - \frac{\nu}{2}\right)$$

Loading: Internal static pressure due to liquid of weight density γ, $p = -\gamma(h-y)$

Above the liquid level,

$$s_x = 0$$

$$s_y = \frac{\gamma h^3 \sin^2\alpha}{6Rt\cos^3\alpha}$$

At the edge,

$$q = \frac{\gamma h^2 \sin\alpha}{6L\cos^2\alpha}$$

Below the liquid level,

$$s_x = \frac{\gamma R}{t}\left(\frac{h}{\cos\alpha} - \frac{R}{\sin\alpha}\right)$$

$$s_y = \frac{\gamma R}{2t}\left(\frac{3h}{\cos\alpha} - \frac{2R}{\sin\alpha}\right)$$

At the liquid level,

$$\Delta_R = -\frac{L\tan\alpha}{E}\nu s_y$$

where Δ_R is measured along the liquid level ($y = h$).

(e) **Cylindrical shell** of constant radius R, length L, and thickness t, closed at both ends, is shown in Fig. 5.02–16.

Loading: External or internal pressure $p = \mp$ constant.

$$s_x = \frac{pR}{t} \qquad\qquad s_y = \frac{pR}{2t}$$

$$q = 0 \qquad\qquad \Delta_R = \frac{pR^2}{Et}$$

Loading: Internal static pressure due to liquid of weight density γ

$$p = -\gamma(h - y)$$

Above the liquid level,   Below the liquid level,

$$s_x = 0 \qquad\qquad\qquad\qquad s_x = -\frac{pR}{t}$$

$$s_y = 0 \qquad\qquad\qquad\qquad s_y = 0$$

$$\Delta_R = 0 \qquad\qquad\qquad\qquad \Delta_R = -\frac{pR^2}{Et}$$

where Δ_R is measured along the radius at level y.

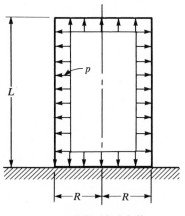 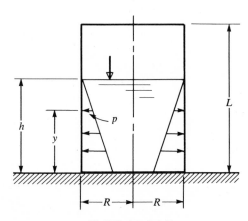

(*a*) Cylindrical shell,   **Fig. 5.02–16**   (*b*) Cylindrical shell,
constant pressure   static pressure

(7) Buoyancy and Flotation

(a) **Archimedes' principle.** A body, floating or immersed in a liquid, is acted upon by an upward force, called the *buoyant force*, which equals the weight of the displaced fluid.

(b) **Center of buoyancy** is the point through which the buoyant force acts, and it is the center of gravity of the volume of displaced liquid. If the body is homogeneous and wholly immersed, its center of gravity coincides with the center of buoyancy.

(c) **Position of a body in a liquid** depends on the relations of its weight W to the buoyant force B. If

$$W < B \qquad \text{body floats}$$
$$W = B \qquad \text{body remains suspended}$$
$$W > B \qquad \text{body sinks}$$

Depth of flotation of a homogeneous solid of volume V_1 and weight density γ_1 in a liquid of weight density γ_2 is the depth of the lowest point of the solid below the surface of the liquid. Five typical cases of flotation are given in Fig. 5.02–17 below, where

R = radius (m) $\qquad$ $n = \gamma_1/\gamma_2$ (dimensionless)
h = height (m) $\qquad$ $\phi = \cos^{-1}(1 - 2n)$ (rad)
f = depth of flotation (m) $\qquad$ ω = central angle (rad)

The central angle ω is the root of the transcendental equation,

$$\omega - \sin\omega = 2\pi n$$

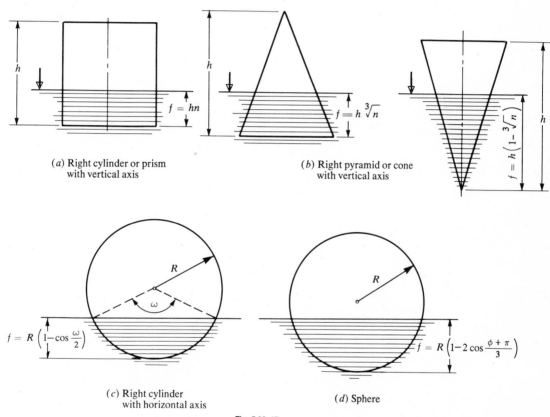

(*a*) Right cylinder or prism with vertical axis $\qquad$ $f = hn$

(*b*) Right pyramid or cone with vertical axis $\qquad$ $f = h\sqrt[3]{n}$

$f = h\left(1 - \sqrt[3]{n}\right)$

(*c*) Right cylinder with horizontal axis $\qquad$ $f = R\left(1 - \cos\dfrac{\omega}{2}\right)$

(*d*) Sphere $\qquad$ $f = R\left(1 - 2\cos\dfrac{\phi + \pi}{3}\right)$

Fig. 5.02–17

example:

A wooden block of length 10 m and of cross-sectional area 100 cm² is to be used as a buoy projecting 1 m above the water surface. Determine the volume of a concrete block which will hold this buoy in the given position. $\gamma_{\text{water}} = 1{,}000\ \text{kgf/m}^3$, $\gamma_{\text{concrete}} = 2{,}300\ \text{kgf/m}^3$, and $\gamma_{\text{wood}} = 500\ \text{kgf/m}^3$.

Total volume $= 10 \times \dfrac{100}{10^4} + (V_{\text{concrete}} = x) = V$ $\qquad$ (in m³)

Total weight $= 0.1 \times 500 + 2{,}300x = W$ $\qquad$ (in kgf)

Buoyant force $= 0.09 \times 1{,}000 + 1{,}000x = B$ $\qquad$ (in kgf)

By Archimedes' principle $W = B$ or $50 + 2{,}300x = 90 + 1{,}000x$ from which $x = 40/1{,}300 = 0.03\ \text{m}^3$.

(e) **Stable equilibrium** of a floating body (Fig. 5.02–18a) occurs when the center of gravity of the body lies below the center of buoyancy (since any small rotation will set up a restoring couple, returning the body to its initial position).

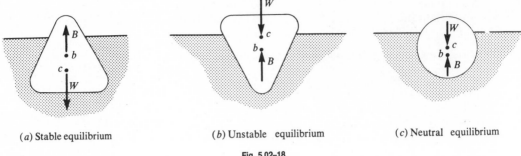

(a) Stable equilibrium   (b) Unstable equilibrium   (c) Neutral equilibrium

Fig. 5.02–18

(f) **Unstable equilibrium** of a floating (Fig. 5.02–18b) occurs when the center of gravity of the body lies above the center of buoyancy (since any small rotation will set up an overturning couple, rotating the body into a new position of static equilibrium).

(g) **Neutral equilibrium** (Fig. 5.02–18c) occurs for all floating right cylinders, and spheres (since their weight and buoyancy force are always collinear).

(h) **Metacenter** M of a floating body (which was rotated by θ about an axis parallel to the surface of liquid) is the point of intersection of the buoyant force and the initial vertical center line of the body (Fig. 5.02–19).

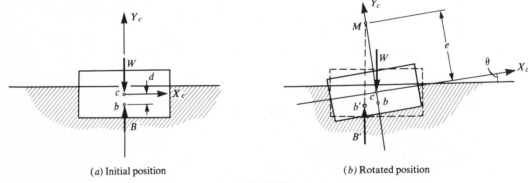

(a) Initial position   (b) Rotated position

Fig. 5.02–19

(i) **Metacentric height** e is the distance $\overline{cM}$, and for a small θ,

$$e = \frac{I}{V'} \mp d \qquad \left\{ \begin{array}{ll} I \text{ in } m^4 & d, e \text{ in } m \\ V' \text{ in } m^3 & \end{array} \right\}$$

where I = moment of inertia of the horizontal cross section of the body at the liquid surface about its centroidal axis parallel to the axis of rotation, V' = volume of displaced liquid, and $d = \overline{bc}$. The minus sign is used if c is above b, the plus sign if c is below b.

(j) **Angular stability of ships.** In surface vessels, the center of gravity is usually above the center of buoyancy, and yet they are in a position of stable equilibrium as long as the righting couple produced by W and B is acting against rotation (Fig. 5.02–19b). A large metacentric height indicates a large righting couple and a rapid return from the heeling-over position. If the couple of W and B is acting in the direction of rotation, the vessel will capsize.

(k) Weight density γ of a body of unknown volume V can be determined by means of the weight W of the body in the atmosphere and of the weight W' of the same body when submerged in a testing liquid of known weight density γ_t (Fig. 5.02–20).

$$\gamma = \gamma_t \frac{W}{W - W'} \qquad \left\{\begin{array}{l} \gamma, \gamma_t \text{ in kgf/m}^3 \\ W, W' \text{ in kgf} \end{array}\right\}$$

(l) Volume V of the same body is then

$$V = \frac{W - W'}{\gamma_t} \qquad \{V \text{ in m}^3\}$$

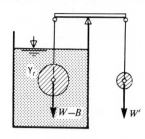

Fig. 5.02–20

(m) Hydrometer is an instrument which uses the principle of buoyancy to determine the specific gravity of liquids. It consists of a slender calibrated stem and a bulb filled with weights. In distilled water of $G_0 = 1.00$, it floats as shown in Fig. 5.02–21a, marking 1.00 on its scale. In other liquids, it floats as shown in Fig. 5.02–21b, marking the specific gravity of that liquid.

(8) Cohesion, Adhesion, and Capillarity

(a) Adhesion is the attraction which exists between the molecules of two different substances, and in liquids it is the degree of attraction between the liquid and the walls or the surface of the container.

(b) Cohesion is the attraction which exists between the molecules of two like substances, and in liquids it is the prime cause of the surface tension described below.

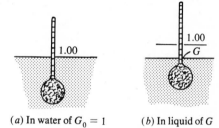

(a) In water of $G_0 = 1$ (b) In liquid of G

Fig. 5.02–21

(c) Surface tension is the stress in the interface of two immiscible fluids (as if there were a film under tension).

(d) Wetting and nonwetting fluids. When a liquid comes into contact with the surface of a solid or with another fluid, one of two distinct cases may arise. If the forces between the molecules of the liquid are less than the forces between the molecules of the liquid and the molecules of the surface, the liquid spreads (Fig. 5.02–22a) and is called a *wetting fluid*. If the opposite is true, the liquid forms a drop and is called a *nonwetting fluid* (Fig. 5.02–22b). Water is a typical wetting fluid and mercury is a typical nonwetting fluid.

(a) Wetting fluid   (b) Nonwetting fluid

Fig. 5.02–22

(e) Capillarity. When a small-diameter tube is placed vertically in a container with liquid, the liquid surface in the tube either rises (wetting fluid) or depresses (nonwetting fluids) relative to the level of the liquid in the container (Fig. 5.02–23).

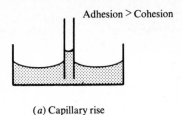

Adhesion > Cohesion

(a) Capillary rise

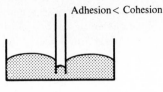

Adhesion < Cohesion

(b) Capillary depression

Fig. 5.02–23

(f) Meniscus is the form of the capillary surface, which is concave for the wetting fluids (Fig. 5.02–23a) and convex for the nonwetting fluids (Fig. 5.02–23b).

(g) Capillary rise and depression in a tube of radius r for a liquid of weight density γ (Fig. 5.02–24) is

$$h = \frac{2\sigma \cos \alpha}{r\gamma} \qquad \left\{ \begin{array}{l} h \text{ in cm} \\ \sigma \text{ in gf/cm} \\ r \text{ in cm} \\ \gamma \text{ in gf/cm}^3 \end{array} \right\}$$

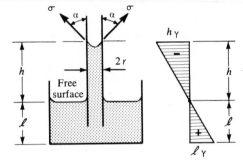

where σ = surface tension and α = angle of contact. The derivation of this equation is based on the static equilibrium of the suspended column of fluid in the tube and the vertical component of the surface tension resultant.

Fig. 5.02–24

(h) Capillarity of water in a glass tube of radius r at atmospheric pressure and at 20°C has $\sigma = 0.074 \text{ gf/cm}$ and $h = 0.148/r$ cm (capillary rise).

(i) Capillarity of mercury in a glass tube of radius r at atmospheric pressure and at 20°C has $\sigma = 0.705 \text{ gf/cm}$ and $h = -4.61/r$ cm (capillary depression).

(j) Vapor pressure and surface tension. The vapor pressure p_v is the pressure at which the liquid boils and is in equilibrium with its own vapor. The relationship between the surface tension and the vapor pressure of pure water in contact with air at a given temperature is shown below.

T, °C	σ, gf/cm	p_v, kgf/cm^2	T, °C	σ, gf/cm	p_v, kgf/cm^2
0	7.69 (−02)	6.230 (−03)	50	6.92 (−02)	1.259 (−01)
10	7.57 (−02)	1.252 (−02)	60	6.75 (−02)	2.031 (−01)
20	7.40 (−02)	2.383 (−02)	70	6.56 (−02)	3.177 (−01)
30	7.26 (−02)	4.326 (−02)	60	6.38 (−02)	4.828 (−01)
40	7.09 (−02)	7.520 (−02)	90	6.20 (−02)	7.149 (−01)
			100	6.00 (−02)	1.033 (+00)

Conversion: 1 gf/cm = 9.806 650 (−01) N/m = 5.599 742 (−04) lbf/in
1 kgf/cm^2 = 9.806 650 (+04) Pa = 1.422 334 (+01) lbf/in^2

5.03 DYNAMICS OF FLUIDS

(1) Classification and Definitions

(a) **Motion of fluid mass**, called *fluid flow*, is described as a continuous change in the position of all or of some particles of the mass.

(b) **Kinematics and kinetics** of fluid particles are defined analogically as their counterparts in Sec. 3.01–1, but fluid flow is complex and not always subject to exact mathematical analysis.

(c) **Basic laws** describing the complete motion of a fluid are in general unknown, and recourse to experimentation is required.

(d) **Kinematic quantities** describing the motion of a fluid particle are the displacement, velocity, and acceleration, all of which are functions of time or constants.

(e) **Pathline** is a continuous real line traced by the moving particle in a given interval of time. Pathlines of a fluid system may cross one another and represent the histogram of motion.

(f) **Streamline** is a continuous imaginary line tangent everywhere to the respective velocity vector. Streamlines of a fluid system never cross one another and represent an instantaneous flow pattern (at a given instant) which at another instant may be completely different.

(g) **Streakline** is the locus of fluid particles which have passed through a certain point. A streakline can be produced experimentally by the continuous release of a marked substance (dye, smoke).

(h) **Timeline** is a line formed by a set of fluid particles at a given instant, and like the streamline is an instantaneous line.

(i) **Fluid flow analysis** is based on the three conservation laws of mechanics:
   (α) Conservation of mass (law of continuity)
   (β) Conservation of momentum (Newton's second law)
   (γ) Conservation of energy (first law of thermodynamics)

(j) **Methods of analysis** are classified as integral analysis (control volume method), differential analysis (elemental system method), and dimensional analysis (experimental method).

(2) Types of Fluid Flow

(a) **Fluid flows** are classified as steady or unsteady, uniform or nonuniform, laminar or turbulent.

(b) **Steady flow** at a point occurs when the parameters of flow (velocity, density, pressure) are independent of time and of temperature (do not change, are constant) at that point.

(c) **Unsteady flow** at a point occurs when some or all of the parameters listed in (b) vary with time and/or with temperature at that point.

(d) **Uniform flow** at an instant in a field occurs when the velocity vector is of constant magnitude and direction at all points of that field.

(e) **Nonuniform flow** at an instant in a field occurs when the velocity vector varies in magnitude and/or direction from point to point in that field.

(f) **Turbulent flow** occurs when the fluid particles move in very irregular (chaotic) pathlines, causing losses of energy approximately proportional to the square of flow velocity. The development of this flow requires a high velocity and large flow profiles.

(g) **Laminar flow** occurs when the fluid particles move in smooth, orderly streams (laminas), causing energy losses directly proportional to flow velocity. The development of this flow requires high viscosity, low velocity, and small flow profiles.

(3) Fluid in Rigid-Body Translation

(a) Relative translatory equilibrium.
A fluid mass subjected to translation at constant linear acceleration is in a relative translatory equilibrium; i.e., there is no relative motion between fluid particles.

(b) Horizontal translation.
If the vessel containing a liquid of weight density γ translates horizontally with constant acceleration a (Fig. 5.03–1), the initially horizontal surface of this liquid becomes an inclined plane of slope

$$\tan \theta = \frac{a}{g} \qquad \left\{ \begin{array}{l} a, g \text{ in m/s}^2 \\ \theta \text{ in rad} \end{array} \right\}$$

and the pressure on the bottom of the vessel is

$$p = \gamma\left(h + \frac{la}{2g} - \frac{xa}{g}\right) \qquad \left\{ \begin{array}{l} p \text{ in kgf/m}^2 \\ \gamma \text{ in kgf/m}^3 \\ h, l, x \text{ in m} \end{array} \right\}$$

where $g = 9.81$ m/s², h = initial level of the free surface measured from the bottom, l = length of vessel, and x = horizontal coordinate.

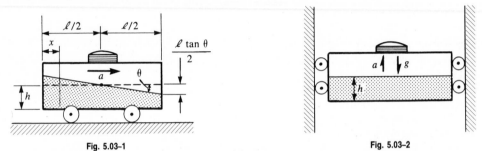

Fig. 5.03–1	Fig. 5.03–2

(c) Vertical motion.
If the vessel containing a liquid of weight density γ moves vertically with constant acceleration a (Fig. 5.03–2), the initially horizontal surface remains horizontal and the pressure p at any level y below the free surface is

$$p = \gamma y\left(1 \pm \frac{a}{g}\right) \qquad \left\{ \begin{array}{ll} p \text{ in kgf/m}^2 & y \text{ in m} \\ \gamma \text{ in kgf/m}^3 & a \text{ in m/s}^2 \end{array} \right\}$$

where $g = 9.81$ m/s², $(-)$ is used for the downward motion, and $(+)$ for the upward motion.

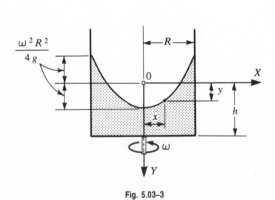 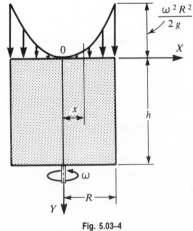

Fig. 5.03–3	Fig. 5.03–4

(4) Fluid in Rigid-Body Rotation

(a) Relative rotatory equilibrium. A fluid mass subjected to a constant angular speed ω with respect to its vertical axis of symmetry is in a relative rotatory equilibrium; that is, there is no relative motion between particles.

(b) Axial rotation—open vessel. If an open cylindrical tank of radius R containing a liquid of weight density γ rotates with a constant angular speed ω (Fig. 5.03–3), the initially horizontal surface becomes a rotational concave paraboloid with vertex at $y = \omega^2 R^2/4g$ below the initial surface, and any vertical plane through the axis of rotation cuts this surface in a parabola, defined by

$$y = \frac{\omega^2 R^2}{4g} - \frac{\omega^2 x^2}{2g} \qquad \begin{Bmatrix} x,\ y,\ R \text{ in m} \\ \omega \text{ in rad/s} \end{Bmatrix}$$

and the pressure on the bottom of the cylinder is

$$p = \gamma\left(h - \frac{\omega^2 R^2}{4g} + \frac{\omega^2 x^2}{2g}\right) \qquad \begin{Bmatrix} p \text{ in kgf/m}^2 & h \text{ in m} \\ \gamma \text{ in kgf/m}^3 & \end{Bmatrix}$$

where $g = 9.81$ m/s², h = initial level of free surface, y = vertical coordinate from free surface, and x = horizontal distance from the axis of rotation.

(c) Axial rotation—closed vessel. If a closed cylinder of radius R is completely filled with a liquid of weight density γ under pressure p_0 and rotates with a constant angular speed ω (Fig. 5.03–4), the pressure on its bottom is

$$p = p_0 + \gamma\left(h + \frac{\omega x^2}{2g}\right) \qquad \begin{Bmatrix} p,\ p_0 \text{ in kgf/m}^2 & \omega \text{ in rad/s} \\ \gamma \text{ in kgf/m}^3 & h,\ x \text{ in m} \end{Bmatrix}$$

where the symbols have the same meaning as in (b).

(5) Basic Equations of Fluid Flow

(a) Assumptions. Basic equations of fluid flow are based on the following assumptions:

(α) *Steady, frictionless flow* in a stream tube
(β) *Velocity normal* to the flow profile cross section of the stream tube
(γ) *Velocity and density constant* over the instantaneous cross section
(δ) *One inlet and one outlet* to the volume of the tube (Fig. 5.03–5)

(b) Rate of flow (discharge) Q is defined as the volume of fluid that passes a given cross section per unit of time (Fig. 5.03–5).

$$Q = A\bar{v} \qquad \begin{Bmatrix} Q \text{ in m}^3/\text{s} \\ A \text{ in m}^2 \\ \bar{v} \text{ in m/s} \end{Bmatrix}$$

Fig. 5.03–5

where A = area of cross section and $\bar{v}$ = average velocity normal to the cross section.

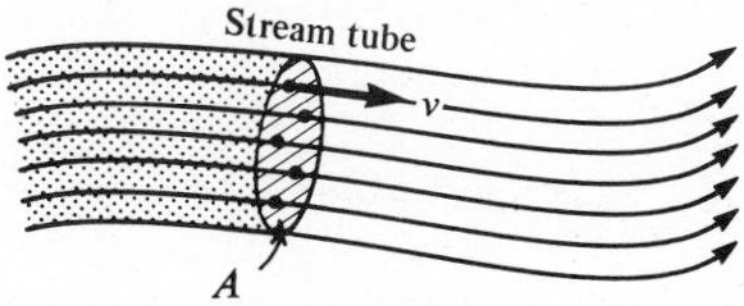

(c) Units of Q are

$$1 \text{ m}^3/\text{s} = 35.31 \text{ ft}^3/\text{sec} \qquad 1 \text{ ft}^3/\text{sec} = 0.0283 \text{ m}^3/\text{s}$$

$$1 \text{ liter/s} = 0.03531 \text{ ft}^3/\text{sec} \qquad 1 \text{ gallon/s} = 0.1337 \text{ ft}^3/\text{sec}$$

$$= 0.2642 \text{ gallon/sec} \qquad = 3.785 \text{ liters/s}$$

where 1 gallon = 1 U.S. liquid gallon.

(d) Conservation of mass. The mass of fluid passing in one unit of time through all cross sections of a stream tube is constant.

$$Q_m = \rho_1 A_1 \bar{v}_1 = \rho_2 A_2 \bar{v}_2 = \cdots = \rho_n A_n \bar{v}_n = \text{constant} \qquad \left\{ \begin{array}{l} Q_m \text{ in kg/s} \\ \rho \text{ in kg/m}^3 \\ A \text{ in m}^2 \\ \bar{v} \text{ in m/s} \end{array} \right\}$$

where Q_m = mass of fluid passing through any cross section of the tube per unit of time, ρ = mass density, A = area of cross section, and $\bar{v}$ = average flow velocity.

(e) Conservation of volume. If $\rho_1 = \rho_2 = \cdots = \rho_n$, the fluid is said to be *incompressible*, and the volume of fluid passing per unit time through all cross sections is constant.

$$Q = A_1 \bar{v}_1 = A_2 \bar{v}_2 = \cdots = A_n \bar{v}_n = \text{constant} \qquad \{Q \text{ in m}^3/\text{s}\}$$

example:

In Fig. 5.03–6, if $A_1 = 0.1\,\text{m}^2$, $A_2 = 0.3\,\text{m}^2$, $\gamma_1 = \gamma_2 = 1{,}000\,\text{kgf/m}^3$, and $\bar{v}_1 = 2\,\text{m/s}$, then

$$\bar{v}_2 = \frac{A_1 \bar{v}_1}{A_2} = \frac{0.1 \times 2}{0.3} = 0.67\,\text{m/s}$$

and

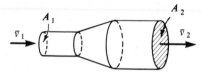

Fig. 5.03–6

$$Q_{m1} = \rho A_1 \bar{v}_1 = 1{,}000 \times 0.1 \times 2 = 200\,\text{kg/s}$$

$$Q_{m2} = \rho A_2 \bar{v}_2 = 1{,}000 \times 0.3 \times 0.67 = 200\,\text{kg/s}$$

(f) Total energy possessed by a moving fluid consists of the potential energy E_p and the kinetic energy E_k; for incompressible fluids the total energy per unit volume is constant (Bernoulli's theorem).

$$E = p + \gamma h + \frac{\gamma v^2}{2g} = \text{constant} \qquad \{E \text{ in (m-kgf)/m}^3\}$$

where p = absolute pressure on the unit volume, h = vertical coordinate of the unit volume (called *head*) related to a given datum, γ = weight density of fluid, v = velocity of fluid, and $g = 9.81\,\text{m/s}^2$.

(g) Conservation of energy. For an isolated fluid flow (no loss of head, no friction) the change in total energy between two cross sections of the stream tube is zero.

$$p_2 - p_1 + \gamma(h_2 - h_1) + \frac{\gamma(v_2^2 - v_1^2)}{2g} = 0$$

where the subscripts refer to the cross sections of Fig. 5.03–7.

example:

In Fig. 5.03–7, $h_3 = 10\,\text{m}$, $h_2 = 6\,\text{m}$, $h_1 = 2\,\text{m}$, and $\gamma = 1{,}000\,\text{kgf/m}^3$, then from

$$E_3 - E_2 = 0, \qquad v_2 = \sqrt{2g(h_3 - h_2)} = 8.9\,\text{m/s}$$

$$E_3 - E_1 = 0 \qquad v_1 = \sqrt{2g(h_3 - h_1)} = 12.5\,\text{m/s}$$

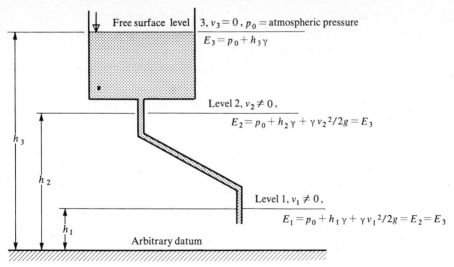

Free surface level | 3, $v_3 = 0$, p_0 = atmospheric pressure
$E_3 = p_0 + h_3\gamma$

Level 2, $v_2 \neq 0$,
$E_2 = p_0 + h_2\gamma + \gamma v_2{}^2/2g = E_3$

Level 1, $v_1 \neq 0$,
$E_1 = p_0 + h_1\gamma + \gamma v_1{}^2/2g = E_2 = E_3$

Arbitrary datum

Fig. 5.03–7

(h) Horizontal flow. If the velocity $\bar{v}_1$ and the pressure p_1 at 1 in the stream tube of Fig. 5.03–8 are known, their counterparts at 2 are

$$\bar{v}_2 = \frac{A_1\bar{v}_1}{A_2} = n\bar{v}_1$$

$$p_2 = p_1 - \frac{\gamma(\bar{v}_2{}^2 - \bar{v}_1{}^2)}{2g} = p_1 - \frac{\gamma\bar{v}_1{}^2(n^2-1)}{2g}$$

where A_1, A_2 = areas of cross section, p_1, p_2 = pressures, $\bar{v}_1$, $\bar{v}_2$ = velocities, γ = weight density, $g = 9.81$ m/s², and $n = A_1/A_2$.

Fig. 5.03–8

(i) Curvilinear flow. If the velocity $\bar{v}_1$, the pressure p_1, and the relative head h_1 at 1 of Fig. 5.03–9 are known, then at the relative head h_2 their counterparts are

$$\bar{v}_2 = \frac{A_1\bar{v}_1}{A_2} = n\bar{v}_1$$

$$p_2 = p_1 - \gamma(h_2 - h_1) - \frac{\gamma(\bar{v}_2{}^2 - \bar{v}_1{}^2)}{2g} = p_1 - \gamma h - \frac{\gamma\bar{v}_1{}^2(n^2-1)}{2g}$$

where A_1, A_2, $\bar{v}_1$, $\bar{v}_2$, p_1, p_2, n, γ, and g are the same as in (h) and $h = h_2 - h_1$.

(j) Total head H is the sum of the potential head (E_p/γ) and of the kinetic head (E_k/γ).

$$H = \frac{p}{\gamma} + h + \frac{v^2}{2g} \qquad \{H \text{ in m}\}$$

where E_p, E_k, and their units are those defined in (f).

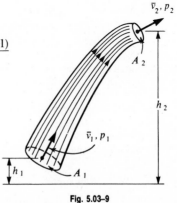

Fig. 5.03–9

(k) Compression work U done by a piston forcing the volume of liquid V into a cylinder against an opposing pressure p is

$$U = Vp \qquad \left\{ \begin{array}{l} U \text{ in m-kgf} \\ V \text{ in m}^3 \\ p \text{ in kgf/m}^2 \end{array} \right\}$$

(l) Lifting work U done by a piston of cross-sectional area A moving the volume of liquid V along the total head H is

$$U = (V\gamma)H = (Av\gamma)Ht$$

where H, V, γ are those of (i) and (j), and v = velocity of flow, t = time of lift. For units and conversion factors of work, refer to Sec. 3.05–1.

(m) Power of the isolated fluid flow of (f) at a given cross section of area A and total head H is

$$\text{Power} = AH\gamma v = Q\gamma H \qquad \{\text{power in m-kgf/s}\}$$

where γ = weight density of fluid and v = velocity through A. For units and conversion factors of power, see Sec. 3.05–6.

example:

The power of a water pump of lift $H = 25\,\text{m}$ and discharge $Q = 2\,\text{m}^3/\text{s}$ is

$$QH\gamma = 2 \times 25 \times 1{,}000 = 50{,}000 \text{ m-kgf/s}$$
$$= 50{,}000 \times 9.81 = 490{,}000 \text{ W}$$
$$= 490{,}000 \times 1.359 \times 10^{-3} = 667 \text{ hp} \qquad \text{(MKS)}$$

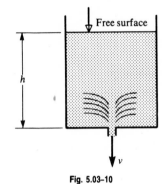

Fig. 5.03–10

(6) Orifices

(a) Small base discharge. For a small orifice of area A (Fig. 5.03–10), the discharge velocity is

$$v = C_v \sqrt{2hg}$$

and the discharge is

$$Q = C_Q A \sqrt{2hg}$$

where C_v = coefficient of velocity (0.95 to 0.99) and C_Q = coefficient of discharge (for sharp-edged orifices, 0.60 to 0.67; for well-rounded orifices, 0.90 to 0.97).

(b) Small lateral discharge. For a small orifice of area A (Fig. 5.03–11), the discharge velocity is

$$v = C_v \sqrt{2h_1 g}$$

the discharge is

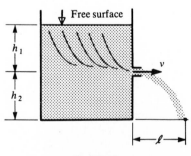

Fig. 5.03–11

$$Q = C_Q A \sqrt{2h_1 g}$$

and the horizontal length of the jet is $l = 2\sqrt{h_1 h_2}$ where C_v and C_Q are those of (a).

(c) Large lateral discharge. For a large rectangular orifice of width b and depth $h_2 - h_1$ (Fig. 5.03–12), the average discharge velocity is

$$\bar{v} = \frac{2C_v(h_2^{2/3} - h_1^{2/3})\sqrt{2g}}{3(h_2 - h_1)}$$

and the discharge is

$$Q = 2/3\, C_Q b(h_2^{2/3} - h_1^{2/3})\sqrt{2g}$$

where C_Q is the same as in (a).

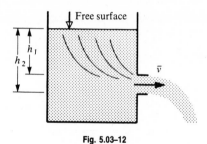

Fig. 5.03–12

(d) Small lateral discharge—excess pressure. For the small orifice of area A (Fig. 5.03–13), the discharge velocity of liquid under surface pressure p (gage pressure) is

$$v = C_v\sqrt{2g\left(h_1 + \frac{p}{\gamma}\right)}$$

the discharge is

$$Q = C_Q A\sqrt{2g\left(h_1 + \frac{p}{\gamma}\right)}$$

and the jet length is

$$l = 2\sqrt{\left(h_1 + \frac{p}{\gamma}\right)h_2}$$

where C_v, C_Q are those of (a) and $\gamma =$ weight density of liquid.

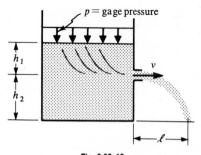

Fig. 5.03–13

(e) Change in cross section. For the orifice of Fig. 5.03–14, the discharge velocity is

$$v = C_v\sqrt{\frac{2gh_1}{1 - (nC_Q)^2}}$$

the discharge is

$$Q = C_Q A_1\sqrt{\frac{2gh_1}{1 - (nC_Q)^2}}$$

and the jet length is

$$l = 2\sqrt{\frac{h_1 h_2}{1 - (nC_Q)^2}}$$

where C_v, C_Q are those of (a) and $n = A_1/A_2$.

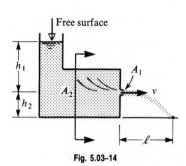

Fig. 5.03–14

(f) General case arises when the liquid approaches the orifice with velocity v_1 under pressure p_1 and head h_1 (Fig. 5.03–15). Then the discharge velocity at point 2 throughout the orifice of area A is

$$v_2 = C_v \sqrt{2g\left(\frac{p_1}{\gamma} - \frac{p_2}{\gamma} + \frac{v_1^{\,2}}{2g} - h_1\right)}$$

the discharge is

$$Q = C_Q A \sqrt{2g\left(\frac{p_1}{\gamma} - \frac{p_2}{\gamma} + \frac{v_1^{\,2}}{2g} - h_1\right)}$$

and the jet length is

$$l = 2\sqrt{\left(\frac{p_1}{g} - \frac{p_2}{g} + \frac{v_1^{\,2}}{2g} - h_1\right)h_2}$$

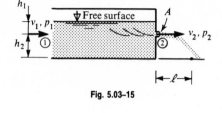

Fig. 5.03–15

where p_2 = pressure of exit medium, γ = weight density of liquid, and C_v, C_Q are those of (a).

(g) Lowering of head. Theoretically, Bernoulli's equation applies only to steady flow, but for slow drainage of a vessel from head h_1 to h_2 through a small orifice of area A_1, the average discharge velocity (Fig. 5.03–16) is

$$\bar{v} = \frac{C_v(\sqrt{gh_1} + \sqrt{gh_2})}{\sqrt{2}}$$

and the time required to lower the liquid level from h_1 to h_2 is

$$t = \frac{2A_2}{C_Q A_1}\left(\sqrt{\frac{h_1}{2g}} - \sqrt{\frac{h_2}{2g}}\right)$$

where A_1 = area of horizontal cross section of the vessel, A_2 = area of orifice, and C_v, C_Q are those of (a).

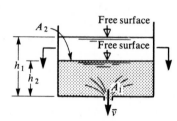

Fig. 5.03–16

5.04   FLUID FLOW IN PIPES

(1) Classification and Definitions

(a) **Pipe flow** is defined as fluid flow in a circular closed conduit whose diameter is small compared to its length and which is entirely filled with fluid.

(b) **Assumptions.** The fluid considered is an incompressible real fluid of known constant density and viscosity and is in a state of steady flow, which can be laminar or turbulent.

(c) **Reynolds number**, Re, specific for the given system is defined as the ratio of the inertial force to the viscous force:

$$\text{Re} = \frac{\bar{v}\rho d}{\mu} = \frac{\bar{v}d}{\nu} \qquad \begin{cases} \text{Re dimensionless} & d \text{ in m} \\ \bar{v} \text{ in m/s} & \rho \text{ in kg/m}^3 \\ \mu \text{ in kgf-s/m}^2 & \nu \text{ in m}^2/\text{s} \end{cases}$$

where $\bar{v}$ = average velocity of fluid flow, d = diameter of pipe, ρ = mass density of fluid, μ = absolute viscosity of fluid, and ν = kinematic viscosity of fluid (both μ and ν were defined in Sec. 5.01–5).

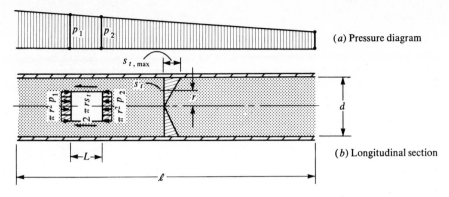

(a) Pressure diagram

(b) Longitudinal section

Fig. 5.04–1

(d) **Shearing stress** variation at a given cross section of the pipe is linear (Fig. 5.04–1); that is,

$$s_t = \frac{(p_1 - p_2)r}{2L} \qquad \left\{ \begin{array}{l} s_t,\ p_1,\ p_2 \text{ in kgf/m}^2 \\ r,\ L \text{ in m} \end{array} \right\}$$

where p_1, p_2 = fluid pressures, r = radial position of s_t, and L = length of the pipe element.

(e) **Maximum shearing stress** occurs at the interface of the fluid and the pipe ($r = d/2$), where

$$s_{t,\max} = \frac{(p_1 - p_2)d}{4L} = f\frac{\rho \bar{v}^2}{8} \qquad \left\{ \begin{array}{l} f \text{ dimensionless} \\ \rho \text{ in kg/m}^3 \\ \bar{v} \text{ in m/s} \end{array} \right\}$$

where f = friction factor (see Secs. 5.04–2 and 5.04–3), ρ = mass density of fluid, and $\bar{v}$ = average velocity of fluid.

(f) **Average velocity of flow** is, from (e),

$$\bar{v} = \sqrt{\frac{2(p_1 - p_2)d}{f\rho L}} \qquad \{\bar{v} \text{ in m/s}\}$$

(g) **Discharge** Q in a pipe of flow cross section A is

$$Q = A\sqrt{\frac{2(p_1 - p_2)d}{f\rho L}} \qquad \left\{ \begin{array}{l} Q \text{ in m}^3\text{/s} \\ A \text{ in m}^2 \end{array} \right\}$$

(h) **Loss of head due to friction** is

$$h_L = f\frac{L\bar{v}^2}{2gd} \qquad \{h_L \text{ in m}\}$$

where f, L, $\bar{v}$, and d are those defined above and $g = 9.81 \text{ m/s}^2$.

(i) **Diameter of pipe** d required to yield a discharge Q is

$$d = \sqrt[5]{\frac{fL\rho}{2(p_1 - p_2)}\left(\frac{4Q}{\pi}\right)^2} = \sqrt[5]{\frac{fL}{2gh}\left(\frac{4Q}{\pi}\right)^2} \qquad \{d \text{ in m}\}$$

where $h = \dfrac{p_1 - p_2}{g\rho}$ = pressure head.

(2) Laminar Flow in Pipes

(a) **Velocity distribution** of laminar flow follows the parabolic pattern of Fig. 5.04–2, and the maximum velocity is twice the average velocity $\bar{v}$ given in Sec. 5.04–1f.

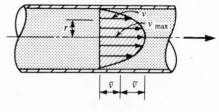

$$v_{max} = 2\bar{v} \qquad \{v_{max} \text{ in m/s}\}$$

Fig. 5.04–2

(b) **Reynolds number** of laminar flow in circular pipes (Sec. 5.04–1c) is

$$\text{Re} \leqslant 2{,}000 \qquad \{\text{Re dimensionless}\}$$

(c) **Loss of head due to friction** in laminar pipe flow (Hagen–Poiseuille equation) is

$$h_L = \frac{32\mu L\bar{v}}{\rho g d^2} = \frac{32\mu L\bar{v}}{\gamma d^2} = \frac{32\nu L\bar{v}}{g d^2} \qquad \begin{cases} h_L, d, L \text{ in m} & v \text{ in m/s} \\ \mu \text{ in kgf-s/m}^2 & \nu \text{ in m}^2\text{/s} \\ \gamma \text{ in kgf/m}^3 & \rho \text{ in kg/m}^3 \end{cases}$$

where μ = absolute viscosity (Sec. 5.01–5), ν = kinematic viscosity (Sec. 5.01–5), L = length of pipe, $\bar{v}$ = average velocity, ρ = mass density of fluid, g = 9.81 m/s², and d = diameter of pipe.

(d) **Darcy–Weisbach formula** for loss of head is the formula of Sec. 5.04–1h, where the friction factor $f = 64/\text{Re}$, that is,

$$h_L = f\frac{L\bar{v}^2}{2gd} = \frac{64L\bar{v}^2}{2gd\,\text{Re}} \qquad \{h_L \text{ in m}\}$$

where Re = Reynolds number (Sec. 5.04–1c).

example:

If oil of γ = 900 kgf/m³ and μ = 0.005 kgf-s/m² flows with $\bar{v}$ = 1 m/s through a 10-cm pipe of length 100 m, then ρ = 900 kg/m³, μ = 0.005 × 9.81 kg/m-s, and

$$\text{Re} = \frac{1 \times 900 \times 0.1}{9.81 \times 0.005} = 1{,}835 \qquad \text{(laminar flow)}$$

$$h_L = \frac{64 \times 100 \times 1^2}{2 \times 9.81 \times 0.1 \times 1{,}835} = 1.77 \text{ m}$$

and the loss in pressure is

$$\Delta p = h_L \gamma = 1600 \text{ kgf/m}^2 = 0.16 \text{ kgf/cm}^2$$

(3) Turbulent Flow in Pipes

(a) **Velocity distribution** of turbulent flow is more uniform and follows the pattern of Fig. 5.04–3. The maximum velocity is

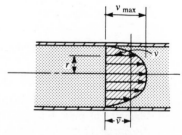

$$v_{max} = (1 + 1.33\sqrt{f})\bar{v} \qquad \begin{cases} v_{max}, \bar{v} \text{ in m/s} \\ f \text{ dimensionless} \end{cases}$$

Fig. 5.04–3

where f = friction factor (see below) and $\bar{v}$ = average velocity.

(b) **Friction factor** for turbulent flow is an experimental value which can be estimated from the chart of Appendix A.65 (see example below).

(c) Loss of head due to friction in turbulent flow (Darcy–Weisbach formula) is given by the formula of Sec. 5.04–1h, where f is taken from the chart of Appendix A.65.

$$h_L = f\frac{L\bar{v}^2}{2gd} \qquad \left\{ \begin{array}{l} L,\ d \text{ in m} \\ v \text{ in m/s} \\ f(\epsilon) \text{ dimensionless} \end{array} \right\}$$

where $f = f(\epsilon)$ = friction factor mentioned above.

(d) Roughness values ϵ (in m) used in the chart of Appendix A.65 are also listed in the same table.

example:

If oil of $\gamma = 900\ \text{kgf/m}^3$ and $\mu = 0.005\ \text{kgf-s/m}^2$ flows with $\bar{v} = 4$ m/s through a 10-cm pipe of length 100 m and of $\epsilon/d = 0.0002$, then

$$\text{Re} = \frac{4 \times 900 \times 0.1}{9.81 \times 0.005} = 7{,}340 \qquad f = 0.035 \text{ (from chart)}$$

$$h_L = \frac{0.035 \times 100 \times 4^2}{2 \times 9.81 \times 0.1} = 28.5 \text{ m}$$

and the loss of pressure is

$$\Delta p = h_L\gamma = 28.5 \times 900 = 25{,}700\ \text{kgf/m}^2 = 2.57\ \text{kgf/cm}^2$$

(e) Alternative formula for f in smooth pipes for Re between 3,000 and 100,000 is

$$f = \frac{0.316}{\sqrt[4]{\text{Re}}}$$

known as *Blasius' formula*, where Re = Reynolds number as given in Sec. 5.04–1c.

example:

For the flow considered in example (d),

$$f = \frac{0.316}{\sqrt[4]{7{,}340}} = 0.034$$

which is in a good agreement with the value obtained from the chart.

(f) For water at standard temperature (between 15 and 25°C), the average values of f for various sizes of pipes and different velocities are given in Appendices A.63 and A.64.

(4) Minor Losses in Pipes

(a) Sudden enlargement. The loss of head due to a sudden enlargement of cross section (Fig. 5.04–4) is

$$h_L = \frac{(\bar{v}_1 - \bar{v}_2)^2}{2g}$$

where $\bar{v}_1$, $\bar{v}_2$ = average velocity before and after enlargement and $g = 9.81\ \text{m/s}^2$.

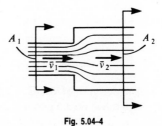

Fig. 5.04–4

(b) Sudden contraction. The loss of head due to a sudden contraction of cross section (Fig. 5.04–5) is

$$h_L = K_1 \frac{\bar{v}_2^{\,2}}{2g}$$

where K_1 = contraction coefficient given below and $\bar{v}_2$ = average velocity in the smaller pipe.

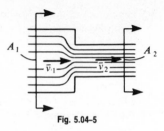

Fig. 5.04–5

A_2/A_1	0.1	0.2	0.3	0.4	0.5	0.6	0.7	0.8	0.9	1.0
K_1	0.36	0.34	0.31	0.27	0.22	0.16	0.10	0.05	0.02	0

(c) Gradual enlargement. The loss of head due to a gradual conical enlargement (Fig. 5.04–6) is

$$h_L = K_2 \frac{\bar{v}_1^{\,2}}{2g}$$

where K_2 = experimental coefficient given below and $\bar{v}_1$ = average velocity in the smaller pipe.

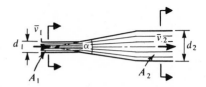

Fig. 5.04–6

	α \ d_1/d_2	0.1	0.2	0.3	0.4	0.5	0.6	0.7	0.8	0.9
	10°	0.04	0.04	0.04	0.04	0.04	0.04	0.03	0.02	0.01
	30°	0.16	0.16	0.16	0.16	0.16	0.15	0.13	0.10	0.06
K_2	60°	0.49	0.49	0.48	0.48	0.46	0.43	0.37	0.27	0.16
	90°	0.64	0.63	0.63	0.62	0.60	0.55	0.49	0.38	0.20
	120°	0.72	0.72	0.71	0.70	0.67	0.62	0.54	0.43	0.24

Source: H. W. King, "Handbook of Hydraulics," 4th ed., pp. 6–16, McGraw-Hill, New York, 1954.

(d) Loss due to bends of the pipe (Fig. 5.04–7) is

$$h_L = K_3 \frac{\bar{v}^{\,2}}{2g}$$

where K_3 = experimental constant given below and $\bar{v}$ = average velocity.

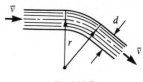

Fig. 5.04–7

d/r	0.1	0.2	0.3	0.4	0.5	0.6	0.7	0.8	0.9
K_3	0.310	0.185	0.166	0.180	0.190	0.220	0.280	0.285	0.290

(e) Importance of losses. Since the friction factor f (Sec. 5.04–3c) cannot be estimated closer than within ± 5 percent, the losses of (a) through (d) may be disregarded if they are equal to or less than 5 percent of the total head.

(5) Large Losses in Pipes

(a) Open valves, elbows, and tees.
The loss of head due to these special conditions is calculated by the same formula as used in Sec. 5.04–4d but in terms of the experimental constant K_4 given below (1 in = 2.54 cm).

Nominal diameter, in	Screwed valves				Flanged valves				
	$\frac{1}{2}$	1	2	4	1	2	4	8	20
Valves (fully open):									
Globe	14	8.2	6.9	5.7	13	8.5	6.0	5.8	5.5
Gate	0.30	0.24	0.16	0.11	0.80	0.35	0.16	0.07	0.03
Swing check	5.1	2.9	2.1	2.0	2.0	2.0	2.0	2.0	2.0
Angle	9.0	4.7	2.0	1.0	4.5	2.4	2.0	2.0	2.0
Elbows:									
45° regular	0.39	0.32	0.30	0.29					
45° long radius					0.21	0.20	0.19	0.16	0.14
90° regular	2.0	1.5	0.95	0.64	0.50	0.39	0.30	0.26	0.21
90° long radius	1.0	0.72	0.41	0.23	0.40	0.30	0.19	0.15	0.10
180° regular	2.0	1.5	0.95	0.64	0.41	0.35	0.30	0.25	0.20
180° long radius					0.40	0.30	0.21	0.15	0.10
Tees:									
Line flow	0.90	0.90	0.90	0.90	0.24	0.19	0.14	0.10	0.07
Branch flow	2.4	1.8	1.4	1.1	1.0	0.80	0.64	0.58	0.41

SOURCE: Adapted from F. M. White, "Fluid Mechanics," p. 354, McGraw-Hill, New York, 1979.

(b) Partially open valves.
Enormous losses occur due to a partially open valve. In those cases the constant K_4 introduced above must be magnified by n, the value of which is given below.

Type	Constant n			
	Valve, open	Closed, 25%	Closed, 50%	Closed, 75%
Globe valve	1.0	1.5–2.0	2.0–3.0	6.0–8.0
Gate valve	1.0	3.0–5.0	12–22	70–120

SOURCE: F. M. White, "Fluid Mechanics," p. 355, McGraw-Hill, New York, 1979.

(6) Noncircular Section

(a) Equivalent diameter.
Most of the formulas for head loss introduced in the preceding sections for circular pipes may be applied to noncircular ducts if the diameter d in the respective formulas is replaced by the equivalent diameter d_e, defined as

$$d_e = 4r_H = 4 \frac{\text{flow area}}{\text{wetted perimeter}} = \frac{4A}{s}$$

where r_H = hydraulic radius of a duct flow.

(b) Special cases of d_e are given below:

For square of side a: $\qquad\qquad\qquad\qquad d_e = a$

For rectangle of sides a, b: $\qquad\qquad d_e = \dfrac{2ab}{a+b}$

For annulus of diameters $d_0 > d_i$: $\qquad d_e = d_0 - d_i$

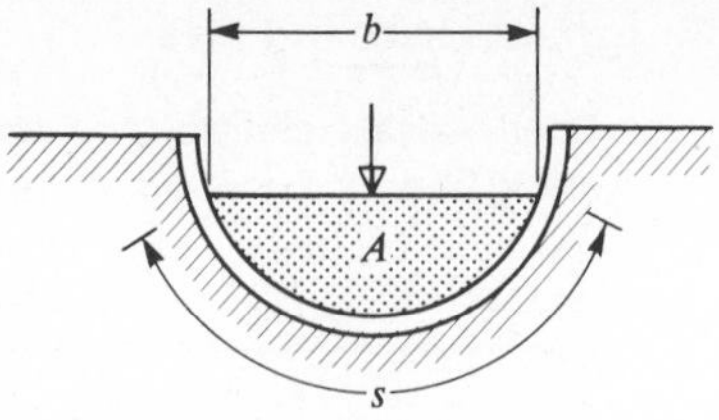

Fig. 5.05–1

5.05 FLUID FLOW IN OPEN CHANNELS

(1) Classification and Definitions

(a) Open channel is a conduit through which the liquid flows with a free surface subjected to only atmospheric pressure. Open channels are either natural (streams, rivers, etc.) or artificial (canals, ditches, etc.).

(b) Cause of flow in open channels is gravity (weight of the liquid), and consequently the slope of the channel is the most important parameter of this flow.

(c) Hydraulic radius r_H of an open-channel flow is the ratio of the area A of the flow cross section to the wetted perimeter of the channel (Fig. 5.05–1).

$$r_H = \frac{A}{s} \qquad \left\{ \begin{matrix} A \text{ in m}^2 \\ s, r_H \text{ in m} \end{matrix} \right\}$$

(d) Hydraulic depth d_H of an open-channel flow is the ratio of the area A of flow cross section to the width b of the cross section at the liquid surface (Fig. 5.05–1).

$$d_H = \frac{A}{b} \qquad \left\{ \begin{matrix} A \text{ in m}^2 \\ b \text{ in m} \\ d_H \text{ in m} \end{matrix} \right\}$$

(e) Reynolds number Re of an open-channel flow is

$$\mathrm{Re} = \frac{4 r_H \bar{v}}{\nu} = \frac{4 A \bar{v}}{\nu s} \qquad \left\{ \begin{matrix} \text{Re dimensionless} & A \text{ in m}^2 \\ r_H, s \text{ in m} & \bar{v} \text{ in m/s} \\ \nu \text{ in m}^2/\text{s} \end{matrix} \right\}$$

where $\bar{v}$ = average velocity, ν = kinematic viscosity (Sec. 5.01–5g), and r_H, A, s are defined in (c) and (d).

(f) Froude number N_F of an open-channel flow is

$$N_F = \frac{\bar{v}}{\sqrt{g d_H}} \qquad \left\{ \begin{matrix} N_F \text{ dimensionless} \\ v \text{ in m/s} \\ d_H \text{ in m} \end{matrix} \right\}$$

where $\bar{v}$ = average velocity, d_H = hydraulic depth defined in (d), and $g = 9.81$ m/s^2.

(g) Hydraulic slope (slope of the energy line) is

$$S_H = \frac{h_L}{L} \qquad \left\{ \begin{matrix} S_H \text{ dimensionless} \\ h_L, L \text{ in m} \end{matrix} \right\}$$

where h_L = loss of head and L = length of channel measured along the bottom of the channel (Fig. 5.05–2).

(h) Slope of channel (Fig. 5.05–2) is

$$S_0 = \tan\theta = \frac{y_1 - y_2}{\sqrt{L^2 - (y_1 - y_2)^2}} \qquad \left\{ \begin{matrix} S_0 \text{ dimensionless} \\ L, y_1, y_2 \text{ in m} \\ \theta \text{ in rad} \end{matrix} \right\}$$

(i) Slope of surface of liquid in an open channel (Fig. 5.05–2) is

$$S_l = S_0 + \frac{d_1 - d_2}{\sqrt{L^2 - (y_1 - y_2)^2}} \qquad \left\{ \begin{matrix} S \text{ dimensionless} \\ d_1, d_2 \text{ in m} \end{matrix} \right\}$$

where d_1, d_2 = depth of flow at 1, 2, respectively, and S_0, L, y_1, y_2 are those of (h).

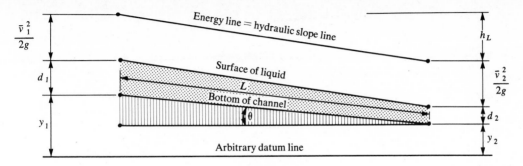

Fig. 5.05–2

(2) Types of Fluid Flow

(a) **Fluid flows in open channels** are classified as steady or unsteady, uniform or nonuniform, laminar, transitional, or turbulent.

(b) **Laminar flow** occurs at values of the Reynolds number below 4,000, and a transition from laminar to turbulent flow extends approximately from Re = 4,000 to Re = 11,000.

(c) **Turbulent flow** occurs for Re = 11,000, and most open-channel flows are turbulent.

(d) **Steady flow** occurs if the rate of flow passing a given section is constant, and the flow becomes unsteady when the rate of flow at that section varies with time.

(e) **Uniform flow** occurs when the parameters of flow are the same at all sections of the channel, and if they change, their changes must occur simultaneously at all sections; otherwise the flow is nonuniform.

(f) **Froude number classification** (Sec. 5.05–1f) defines the velocity of flow as:

    (α) *Subcritical* ($N_F < 1$), also described as *tranquil*, at which the effect of gravitational forces overrides the effect of inertial forces and the downstream conditions affect the upstream conditions

    (β) *Critical* ($N_F = 1$), which equals the velocity of small gravity waves

    (γ) *Supercritical* ($N_F > 1$), also called *rapid* or *torrential*, at which the effect of inertial forces exceeds the effect of gravitational forces and the downstream conditions cannot affect the upstream conditions

(g) **Velocity of flow** in an open channel varies from zero at the wetted perimeter to a maximum or near maximum at the free surface. The vertical distribution of velocity is assumed to be parabolic in laminar flow and logarithmic in turbulent flow.

(h) **Specific energy of flow** is defined as the vertical distance between the bottom of the channel and the energy line (Fig. 5.05–2).

$$E_s = d + \frac{\bar{v}^2}{2g} \qquad \{E_s \text{ in m}\}$$

where d, $\bar{v}$, and g are the same as before.

(i) **Two-dimensional flow rate** q for a wide rectangular channel is defined as

$$q = \frac{Qd}{A} \qquad \left\{ \begin{array}{l} Q \text{ in m}^3/\text{s} \\ q \text{ in m}^2/\text{s} \\ A \text{ in m}^2 \end{array} \right\}$$

where Q = total flow rate and A = flow cross section.

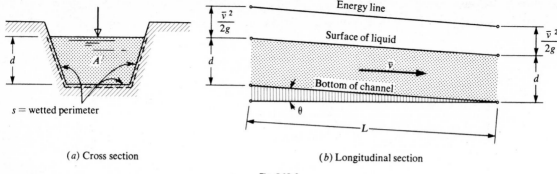

(a) Cross section $\qquad\qquad\qquad\qquad$ *(b)* Longitudinal section

Fig. 5.05–3

(j) Critical depth d_c for a constant q in the same channel occurs when the specific energy is minimum.

$$d_c = \sqrt[3]{q^2/g} = \tfrac{2}{3}E_c = \bar{v}_c^2/g$$

where E_c = minimum specific energy and $\bar{v}_c$ = critical velocity.

(3) Steady Uniform Flow

(a) Conditions of a steady uniform flow are: the discharge Q, the flow area A, the depth d, and the slope of the channel bottom S_0 are all constant; the slope of liquid surface S_l and the slope of the energy line S_H are parallel to the bottom of the channel (Fig. 5.05–3).

$$S_0 = S_l = S_H = \text{constant} \qquad\qquad \{\text{all dimensionless}\}$$

(b) Average velocity of the flow is then given by the Chézy formula as

$$\bar{v} = C\sqrt{r_H S_0} \qquad\qquad \begin{cases} \bar{v} \text{ in m/s} \\ r_H \text{ in m} \\ C \text{ in } \sqrt{\text{m/s}^2} \end{cases}$$

where r_H = hydraulic radius (Sec. 5.05–1c), $S_0 = \tan\theta \cong \theta$, and C = coefficient computed by one of the formulas given below.

(c) Manning formula gives the value of C as

$$C = \frac{\sqrt[6]{r_H}}{n} \qquad\qquad \{C \text{ in } \sqrt{\text{m/s}^2}, r_H \text{ in m}\}$$

$$C = \frac{1.486}{n}\sqrt[6]{r_H} \qquad\qquad \{C \text{ in } \sqrt{\text{ft/sec}^2}, r_H \text{ in ft}\}$$

where the first formula is in SI units, the second formula is in FPS units, and n = roughness factor of the lining given in *(e)* below.

(d) Bazin formula gives the value of C as

$$C = \frac{87}{1 + m/\sqrt{r_H}} \qquad\qquad \{C \text{ in } \sqrt{\text{m/s}^2}, r_H \text{ in m}\}$$

$$C = \frac{157.6}{1 + m/\sqrt{r_H}} \qquad\qquad \{C \text{ in } \sqrt{\text{ft/sec}^2}, r_H \text{ in ft}\}$$

where the first formula is in SI units, the second formula is in FPS units, and m = roughness factor of the lining, given in *(e)* below.

(e) Average values of n and m.

Lining	n	m	Lining	n	m
Smooth concrete	0.010	0.11	Corrugated metal	0.022	1.40
Rough concrete	0.015	0.40	Earth, well maintained	0.023	1.54
Planed timber	0.012	0.20	Earth, average condition	0.027	2.30
Cast iron	0.013	0.27	Rocks, solid walls	0.040	3.50
Brick	0.016	0.60	Gravels, river bed	0.030	3.00

(f) Discharge Q is

$$Q = \bar{v}A \qquad \left\{ \begin{array}{l} Q \text{ in m}^3/\text{s} \\ \bar{v} \text{ in m/s} \\ A \text{ in m}^2 \end{array} \right\}$$

where $\bar{v}$ = average velocity given in (b) and A = area of the cross section of flow.

(g) Loss of head h_L per length L is

$$h_L = \frac{L}{r_H}\left(\frac{\bar{v}}{C}\right)^2 \qquad \{h_L \text{ in m}\}$$

where r_H, $\bar{v}$, and C are defined in (b) to (d).

(h) Most efficient cross section has the least wetted perimeter. The semicircular cross section is the most efficient of all sections. The most efficient rectangular cross section has the bottom twice the depth, and the most efficient trapezoidal cross section is half of a hexagon.

example:

If the open channel of Fig. 5.05–4 has the bottom slope $S_0 = 0.0009$ and the roughness factor $n = 0.022$, then

$$A = 12 \times 8 + 6 \times 8 = 144 \text{ m}^2$$

$$s = 2\sqrt{6^2 + 8^2} + 12 = 32 \text{ m} \qquad r_H = \frac{A}{s} = 4.5 \text{ m}$$

$$C = \frac{\sqrt[6]{4.5}}{0.022} = 58.4\sqrt{\text{m/s}^2}$$

$$v = 58.4 \times 4.5^{1/2} \times 0.0009^{1/2} = 3.72 \text{ m/s}$$

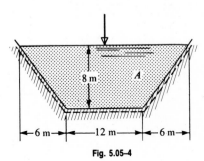

Fig. 5.05–4

By (b),

$$Q = 3.72 \times 144 = 535 \text{ m}^3/\text{s}$$

The Froude number (Sec. 5.05–1f) of this flow is

$$N_F = \frac{3.72}{(9.81 \times 8)^{1/2}} = 0.42 < 1$$

and the critical depth (Sec. 5.05–2j) is

$$d_c \cong \sqrt[3]{\frac{(535 \times 8/144)^2}{9.81}} = 4.5 \text{ m} < 8 \text{ m}$$

which shows that the flow is subcritical.

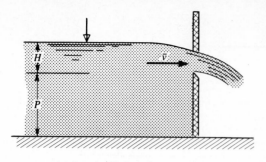

(a) Longitudinal section

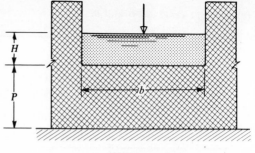

(b) Cross section

Fig. 5.05–5

(4) Weirs

(a) Weir is a barrier in an open channel used to measure and control the discharge. The edge over which the liquid flows is called the *crest*, and the overflow is called the *nappe*. The nappe discharging into the air is called the *free* discharge. If the discharge is partially under the surface of liquid, the weir is said to be *submerged* (drowned).

(b) Theoretical free discharge Q_{theor} for the rectangular weir of Fig. 5.05–5 is

$$Q_{\text{theor}} = \frac{2}{3} b \sqrt{2g} \left[\left(H + \frac{\bar{v}^2}{2g} \right)^{3/2} - \left(\frac{\bar{v}^2}{2g} \right)^{3/2} \right] \qquad \begin{cases} H, b \text{ in m} \\ \bar{v} \text{ in m/s} \\ Q_{\text{theor}} \text{ in m}^3/\text{s} \end{cases}$$

where b = effective width of the weir, H = head on the weir, $\bar{v}$ = average approach velocity, and $g = 9.81 \text{ m/s}^2$.

(c) Approximate theoretical free discharge $\bar{Q}_{\text{theor}}$ is

$$\bar{Q}_{\text{theor}} = \frac{2b\sqrt{2g}}{3} H^{3/2} \qquad \{\bar{Q}_{\text{theor}} \text{ in m}^3/\text{s}\}$$

where $\bar{v}^2/2g$ in the formula of (b) is assumed to be negligibly small.

(d) Approximate actual free discharge $\bar{Q}$ is

$$\bar{Q} = KbH^{3/2} \qquad \{\bar{Q} \text{ in m}^3/\text{s}\}$$

where

$$K = 1.80 + 0.221 \frac{H}{P} \qquad \{K \text{ in } \sqrt{\text{m/s}^2}\}$$

is von Mises' correction factor and b, H, P are the segments defined by Fig. 5.05–5.

(e) Theoretical submerged discharge Q_{theor} for the rectangular weir of Fig. 5.05–6 is

$$Q_{\text{theor}} = bH \sqrt{2g \left(d + \frac{\bar{v}^2}{2g} \right)} \qquad \{b, h, d \text{ in m}\}$$

where bH = cross-sectional area of weir and d = difference in water tables.

(f) Approximate actual submerged discharge $\bar{Q}$ is

$$\bar{Q} = KbH\sqrt{d} \qquad \{\bar{Q} \text{ in m}^3/\text{s}\}$$

where K = correction factor (2.8 to 3.0) and b, h, d are the segments defined by Fig. 5.05–6.

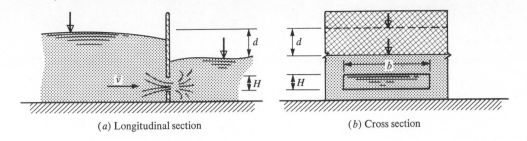

(a) Longitudinal section   (b) Cross section

Fig. 5.05–6

(g) Weirs of special shapes are shown in Fig. 5.05–7. Their approximate actual free discharge is given by the general formula

$$\bar{Q} = KA\sqrt{H} \qquad \{\bar{Q} \text{ in m}^3/\text{s}\}$$

where A = area of the cross section of flow through the weir, B = effective channel width, b = effective weir width, H = head on the weir, and K = specific correction factor given below each case.

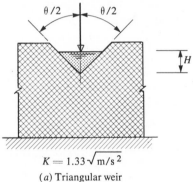

$$K = 1.33\sqrt{\text{m/s}^2}$$

(a) Triangular weir

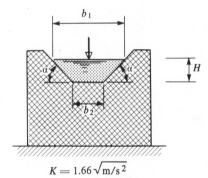

$$K = 1.66\sqrt{\text{m/s}^2}$$

(b) Trapezoidal weir

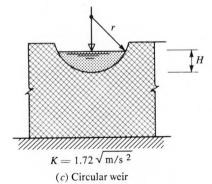

$$K = 1.72\sqrt{\text{m/s}^2}$$

(c) Circular weir

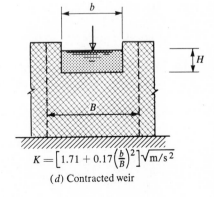

$$K = \left[1.71 + 0.17\left(\frac{b}{B}\right)^2\right]\sqrt{\text{m/s}^2}$$

(d) Contracted weir

Fig. 5.05–7

5.06 TURBOMACHINERY

(1) Action of Fluid Jets

(a) **Momentum equation** for a free body of incompressible fluid states that the resultant force F produced by the change in linear momentum of the fluid mass is

$$F = \rho Q(\bar{v}_2 - \bar{v}_1) \qquad \begin{cases} F \text{ in kgf} & Q \text{ in m}^3/\text{s} \\ \rho \text{ in kg/m}^3 & \bar{v}_1, \bar{v}_2 \text{ in m/s} \end{cases}$$

where ρ = mass density of fluid, Q = discharge of liquid per unit of time, and $\bar{v}_1$, $\bar{v}_2$ = average velocity of the fluid mass at t_1, t_2, respectively.

(b) **Reaction of a jet** flowing through an orifice of a tank (Fig. 5.06–1) is

$$R = 2C_Q A \gamma h \qquad \begin{cases} R \text{ in kgf} & \gamma \text{ in kgf/m}^3 \\ A \text{ in m}^2 & h \text{ in m} \\ C_Q \text{ dimensionless} \end{cases}$$

where A = area of jet cross section, γ = weight density of liquid, h = head at orifice, and C_Q = coefficient of discharge (Sec. 5.03–6).

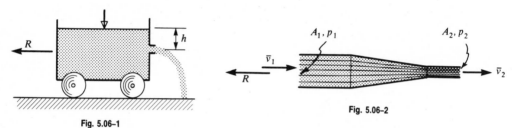

Fig. 5.06–1

Fig. 5.06–2

(c) **Kinetic energy of a jet** of cross-sectional area A, mass density ρ, and average velocity $\bar{v}$ is

$$E_k = \frac{A\rho\bar{v}^2}{2}(1) = A\gamma\bar{h}_v(1) \qquad \begin{cases} E_k \text{ in m-kgf} & \rho \text{ in kg/m}^3 \\ A \text{ in m}^2 & \gamma \text{ in kgf/m}^3 \\ \bar{v} \text{ in m/s} & h_v \text{ in m} \end{cases}$$

where γ = weight density of fluid, $\bar{h}_v = \bar{v}^2/2g$ = velocity head, and (1) stands for one unit of length of the jet.

(d) **Pipeline contraction.** The reaction of fluid of weight density γ on the pipeline contraction of Fig. 5.06–2 is

$$R = \frac{A_1\gamma\bar{v}_1}{g}(\bar{v}_2 - \bar{v}_1) + A_2 p_2 - A_1 p_1$$

where A_1, A_2 = cross-sectional areas of flow, $\bar{v}_1$, $\bar{v}_2$ = average velocities, and p_1, p_2 = fluid pressures.

(e) **Fixed vane.** When a free jet of fluid of weight density γ and average velocity $\bar{v}$ strikes the fixed blade (vane) of Fig. 5.06–3, the reaction of the blade is

$$R = \frac{A\gamma\bar{v}^2}{g}\sqrt{2(1 - \cos\theta)} \qquad \begin{cases} R \text{ in kgf} & \bar{v} \text{ in m/s} \\ A \text{ in m}^2 & \theta \text{ in rad} \\ \gamma \text{ in kgf/m}^3 \end{cases}$$

and the cartesian components of R are

$$R_x = \frac{A\gamma\bar{v}^2}{g}(1 - \cos\theta) \qquad R_y = \frac{A\gamma\bar{v}^2}{g}\sin\theta \qquad \{R_x, R_y \text{ in kgf}\}$$

where A = cross-sectional area of jet, θ = angular deviation of flow, and g = 9.81 m/s^2.

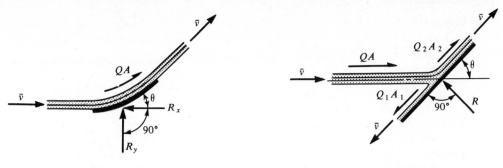

Fig. 5.06–3   Fig. 5.06–4

(f) Fixed inclined plane.   When a free jet of fluid of weight density γ and average velocity $\bar{v}$ strikes the fixed inclined plane of Fig. 5.06–4, then the flow distribution of the jet discharge

$$\rho Q = \frac{A\gamma}{g}\bar{v} \qquad \left\{\begin{array}{ll} Q \text{ in m}^3\text{/s} & A \text{ in m}^2 \\ \gamma \text{ in kgf/m}^3 & \rho \text{ in kg/m}^3 \\ \bar{v} \text{ in m/s} & \end{array}\right\}$$

is

$$\rho Q_1 = \frac{A_1\gamma}{g}\bar{v} \qquad \rho Q_2 = \frac{A_2\gamma}{g}\bar{v} \qquad Q_1 = Q(1-\cos\theta)/2 \qquad Q_2 = Q(1+\cos\theta)/2$$

where A, A_1, A_2 = cross-sectional areas of the respective flows and the reaction of the plate is

$$R = Q\rho\bar{v}\sin\theta \qquad\qquad \{R \text{ in kgf} \quad \theta \text{ in rad}\}$$

(g) Moving vane.   When a free jet of fluid of weight density γ and average velocity $\bar{v}$ strikes a blade (vane) moving with velocity $\bar{v}_0$, the reaction of the blade (Fig. 5.06–5) is

$$R = \frac{A\gamma(\bar{v}-\bar{v}_0)^2}{g}\sqrt{2(1-\cos\theta)} \qquad \left\{\begin{array}{ll} R \text{ in kgf} & \bar{v},\ \bar{v}_0 \text{ in m/s} \\ A \text{ in m}^2 & \theta \text{ in rad} \\ \gamma \text{ in kgf/m}^3 & \end{array}\right\}$$

and the cartesian components of R are

$$R_x = \frac{A\gamma(\bar{v}-\bar{v}_0)^2}{g}(1-\cos\theta) \qquad R_y = \frac{A\gamma(\bar{v}-\bar{v}_0)^2}{g}\sin\theta \qquad \{R_x,\ R_y \text{ in kgf}\}$$

where A = cross-sectional area of jet and θ = angular deviation.

(h) Power exerted on a moving vane by a jet of fluid of weight density γ and velocity $\bar{v}$ is

$$R_x\bar{v}_0 = \frac{A\gamma\bar{v}_0(\bar{v}-v_0)^2}{g}(1-\cos\theta) \qquad \{\text{power in m-kgf/s}\}$$

where the symbols are those used in (g).

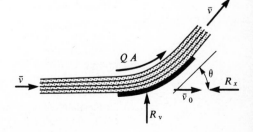

Fig. 5.06–5

(2) Classification and Definitions

(a) Turbomachine (rotating machine) is a machine which adds energy to the fluid system (pump) or extracts energy from the fluid system (turbine) by means of a rotating system of blades rigidly attached to a shaft within the machine.

(b) Rotating system of a pump is called an *impeller* and that of a turbine, a *runner*. In a pump, energy is transferred from the impeller to the fluid, and in a turbine, energy is transferred from the fluid to the runner.

(c) Velocities. If the fluid under steady flow condition enters the curved channel formed by the rotating vane at radius R_1 and leaves the vane at radius R_2 (Fig. 5.06–6), then the absolute linear velocities of the fluid at R_1 and R_2 are, respectively,

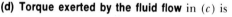

Fig. 5.06–6

Direction of rotation

$$v_1 = \sqrt{v_{f1}^2 + v_{b1}^2 + 2v_{f1}v_{b1}\cos\alpha_1}$$

$$v_2 = \sqrt{v_{f2}^2 + v_{b2}^2 + 2v_{f2}v_{b2}\cos\alpha_2}$$

$$\left\{ \begin{array}{l} v \text{ in m/s} \\ \alpha \text{ in rad} \end{array} \right\}$$

where v_{f1}, v_{f2} = relative linear velocities of fluid at R_1, R_2, respectively, v_{b1}, v_{b2} = linear velocities of vane at R_1, R_2, respectively, and α_1, α_2 = angles of vane with the tangent at R_1, R_2, respectively.

(d) Torque exerted by the fluid flow in (c) is

$$T = Q\rho(v_{f1}R_1\cos\alpha_1 - v_{f2}R_2\cos\alpha_2)$$

$$\left[\begin{array}{ll} T \text{ in kgf-m} & Q \text{ in m}^3\text{/s} \\ \rho \text{ in kg/m}^3 & v \text{ in m/s} \\ R \text{ in m} & \alpha \text{ in rad} \end{array} \right]$$

where Q = rate of fluid flow through the machine, ρ = mass density of fluid, and v_{f1}, v_{f2}, R_1, R_2, α_1, α_2 are defined in (c).

(e) Power of the fluid system equals the torque T given by (d) times the angular velocity of the shaft ω.

$$\text{Power} = Q\rho(v_{f1}v_{b1}\cos\alpha_1 - v_{f2}v_{b2}\cos\alpha_2) \qquad \{\text{power in m-kgf/s}\}$$

where $v_{b1} = R_1\omega$ and $v_{b2} = R_2\omega$.

(f) Energy transfer E of unit weight of fluid is

$$E = \frac{\gamma}{2g}[(v_1^2 - v_2^2) + (v_{f1}^2 - v_{f2}^2) + (v_{b1}^2 - v_{b2}^2)] \qquad \{E \text{ in m-kgf}\}$$

where the first term $\gamma(v_1^2 - v_2^2)/2g$ = change in kinetic energy of the fluid, the second term $\gamma(v_{f1}^2 - v_{f2}^2)/2g$ = pressure change produced by the centrifugal effect, and the third term $\gamma(v_{b1}^2 - v_{b2}^2)/2g$ = pressure change due to relative kinetic energy. E is positive for pumps and negative for turbines.

(g) Energy classification. Turbomachines are classified according to the energy transfer function as impulse machines [last two terms in (f) are zero] and as reaction machines [last two terms in (f) are not zero].

(h) Flow classification. According to the type of flow they are also classified as tangential-flow machines, radial-flow machines, and axial-flow machines.

(3) Performance Characteristics

(a) **Notation.** The performance characteristics of *centrifugal pumps* and *turbines* are defined below in terms of the following symbols:

D = diameter of rotating element (m)

H = head (m)

N = speed of rotation (rpm)

P = power (W)

Q = discharge (m³/s)

T = shaft torque (kgf-m)

γ = weight density of liquid (kgf/m³)

ω = angular speed (rad/s)

(b) **Power characteristics** are

$$P_W = \gamma Q H \qquad P_B = \omega T \qquad \eta = P_W/P_B$$

where P_W = water power, P_B = brake power, and η = efficiency.

(c) **Efficency** η is composed of three parts,

$$\eta_V = \frac{Q}{Q + Q_L} \qquad \eta_H = \frac{H_S - H_F}{H_S} \qquad \eta_M = \frac{P_B - P_F}{P_B}$$

so that

$$\eta = \eta_V \cdot \eta_H \cdot \eta_M$$

where η_V = volumetric efficiency, η_H = hydraulic efficiency, η_M = mechanical efficiency, Q_L = loss of fluids due to leakage, H_S = supplied head, H_F = loss of head due to friction and lack of match, and P_F = loss of power due to friction.

(d) **Dimensionless speed factor** in SI units is

$$\phi = \frac{0.226r\omega}{\sqrt{H}} = \frac{0.0236rN}{\sqrt{H}} \qquad \left\{ \begin{array}{l} r \text{ in m} \\ \omega \text{ in rad/s} \\ N \text{ in rpm} \\ H \text{ in m} \end{array} \right\}$$

and in FPS units is

$$\phi = \frac{0.125r\omega}{\sqrt{H}} = \frac{0.0131rN}{\sqrt{H}} \qquad \left\{ \begin{array}{l} r \text{ in ft} \\ \omega \text{ in rad/sec} \\ N \text{ in rpm} \\ H \text{ in ft} \end{array} \right\}$$

where r = radius of rotating element.

(e) **Specific speed** in SI and FPS units is

$$N_S = N\frac{\sqrt{Q}}{H^{3/4}} = N\frac{\sqrt{P_W}}{H^{5/4}}$$

where P_W is the same as defined in (*b*) above.

(f) **Unit factors** are

$$N_U = \frac{D_1 N}{\sqrt{H}} \qquad Q_U = \frac{Q}{D_1^2\sqrt{H}} \qquad P_U = \frac{P}{D_1^2\sqrt{H}}$$

where N_U = unit speed, Q_U = unit discharge, P_U = unit power, and D_1 = diameter of the rotating element in inches or centimeters, the FPS or SI units, respectively.

5.07 PROPERTIES OF SOILS

(1) Definitions

(a) Soil is a system of solid particles derived from rocks by physical and chemical disintegration and may include water, air (gases), and organic matter.

(b) Visual classification of soils identifies the following basic types: boulders (>25 cm in size), cobbles (5 to 25 cm in size), pebbles (4 mm to 5 cm), gravels (2 mm to 15 cm), coarse sand (0.60 to 2.00 mm), medium sand (0.20 to 0.60 mm), fine sand (0.05 to 0.20 mm), silt (0.005 to 0.050 mm), and clay (less than 0.005 mm).

(c) Graphical symbols of rocks and soils introduced above and of three additional types unsuitable as foundation materials are shown in Fig. 5.07–1 below. *Mud* is defined as a sticky mixture of earth and water, *peat* is partly decayed organic matter, and *loam* is a mixture of sand, silt, and/or clay, called *humus*.

Bedrock	Broken rock	Boulders	Gravels
Sand	Silt	Stiff clay	Soft clay
Silt and clay	Mud	Peat	Loam

Fig. 5.07–1

(d) Volume and weight relationships of a soil sample of volume V (m³) and weight W (kgf) shown in Fig. 5.07–2 are defined analytically in Table 5.07–1 in terms of the following symbols:

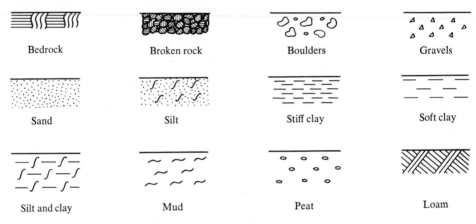

Fig. 5.07–2

e = void ratio (dimensionless)
n = porosity (%)
w = water content (%)
γ = weight density (kgf/m³)
G = specific gravity (dimensionless)
S = degree of saturation (dimensionless)
V = volume (m³)
W = weight (kgf)

The subscripts of these symbols are: d = dry, g = gases (air), s = solids, sat = saturated, v = voids, w = water, 0 = distilled water.

TABLE 5.07–1

(a) Volume

$$V = V_s + V_w + V_g \cong V_s + V_v$$

(e) Weight

$$W = W_s + W_w + W_g \cong W_s + W_w$$

(b) Porosity

$$n = \frac{V_v}{V} = \frac{V_v}{V_s + V_v} = \frac{e}{1 + e}$$

(f) Void ratio

$$e = \frac{V_v}{V_s} = \frac{V_v}{V - V_v} = \frac{n}{1 - n}$$

(c) Degree of saturation

$$S = \frac{V_w}{V_v}\,(\%)$$

If $V_w = 0$, $S = 0\%$ (dry soil)

$V_w < V_v/4$, $S < 25\%$ (humid soil)

$V_v/4 < V_w < V_v/2$, $26\% < S < 50\%$ (damp soil)

$V_v/2 < V_w < 3V_v/4$, $51\% < S < 75\%$ (moist soil)

$3V_v/4 < V_w < V_v$, $76\% < S < 99\%$ (wet soil)

$V_w = V_v$, $S = 100\%$ (saturated soil)

(g) Weight density

$$\gamma = \frac{W}{V}, \quad \gamma_s = \frac{W_s}{V_s}$$

$$\gamma_d = \frac{W_s}{V}, \quad \gamma_w = \frac{W_w}{V_w}$$

$$\gamma_{sat} = \frac{W_s + W_w}{V}$$

(h) Water content

$$w = \frac{W_w}{W_s}\,(\%)$$

(d) Specific gravity

$$G = \frac{\gamma}{\gamma_0}, \quad G_s = \frac{\gamma_s}{\gamma_0} \qquad G_d = \frac{\gamma_d}{\gamma_0}, \quad G_w = \frac{\gamma_w}{\gamma_0}$$

(i) Properties of soils in natural state*

Soil		n, %	e	w, %	γ_d N-m^{-3}	γ_d kgf-m^{-3}	γ_d lbf-ft^{-3}	γ_{sat} N-m^{-3}	γ_{sat} kgf-m^{-3}	γ_{sat} lbf-ft^{-3}
Uniform sand	Loose	46	0.85	32	14,000	1,430	90	18,500	1,890	118
	Dense	34	0.51	19	17,200	1,750	109	20,500	2,090	130
Mixed-grained sand	Loose	40	0.67	25	15,600	1,590	99	19,500	1,990	124
	Dense	30	0.43	16	18,200	1,860	116	21,200	2,160	135
Glacial mixed-grain till		20	0.25	9	20,800	2,120	132	22,800	2,320	145
Glacial clay	Soft	55	1.20	45	—	—	—	17,400	1,770	110
	Stiff	37	0.60	22	—	—	—	20,300	2,070	129
Soft clay	Slightly organic	66	1.90	70	—	—	—	15,500	1,580	98
	Very organic	75	3.00	110	—	—	—	14,000	1,430	89
Soft bentonite		84	5.20	194	—	—	—	12,500	1,270	80

*K. Terzaghy and R. B. Peck, "Soil Mechanics in Engineering Practice," p. 28, Table 6.3, Wiley, New York, 1967.

(e) **Structure** of soil is the arrangement of these three components and appears in three categories: *cohesionless soils, cohesive soils,* and *skeletal soils.* Cohesionless soil when unconfined has little or no strength when air-dried and has little or no cohesion when submerged. The opposite is true for cohesive soil. Finally, skeletal soil is a system of large grains held together by a binder.

(f) **Consistency** of cohesive soil denotes the ease with which the soil can be deformed and is denoted by such terms as soft, medium, stiff, and hard. The measure of consistency is the intensity of load at which a sample of unconfined soil fails in a simple compression test; consistency q_u = unconfined compressive strength of an undisturbed soil sample. The ranges of soil consistency are given below.

Description	Consistency q_u		Description	Consistency q_u	
	kgf-m^{-2}	kN-m^{-2}		kgf-m^{-2}	kN-m^{-2}
Very soft	<2,500	<24.5	Stiff	10,000–20,000	98.0–196.0
Soft	2,500–5,000	24.5–49.0	Very stiff	20,000–40,000	196.0–392.0
Medium	5,000–10,000	49.0–98.0	Hard	>40,000	>392.0
1 kN-m^{-2} = 20.9 lbf-ft^{-2}					

(g) **Atterberg's limits** are used to measure the general consistency of fine-grain soils; they are:
Liquid limit, defined as the water content w_L at which the soil ceases to behave as a liquid and begins to behave as a paste.
Plastic limit, defined as the water content w_P at which the soil ceases to behave as a paste and begins to crumble when rolled out into a cylinder of $\frac{1}{8}$-in (3.175 mm) diameter.
Shrinkage limit, defined as the water content w_S at which the reduction in water will not cause a decrease in volume of the soil sample.

(h) **Atterberg's indices** based on the limits defined above are as follows:
Plasticity index I_P is the difference between the liquid limit and the plastic limit.

$$I_P = w_L - w_P$$

Shrinkage index I_S is the difference between the plastic limit and the shrinkage limit.

$$I_S = w_P - w_S$$

Liquid index I_L is the ratio of the difference of the water content of the soil w and the plastic limit to the plasticity index.

$$I_L = (w - w_P)/I_P$$

(i) **Atterberg's system** showing the relationships of states, limits, and indexes is given below.

Liquid state		
	Liquid limit w_L	
Plastic state		Plasticity index $I_P = w_L - w_P$
	Plastic limit w_P	
Semisolid state		Shrinkage index $I_S = w_P - w_S$
	Shrinkage limit w_S	
Solid state		

(j) Sensitivity of soil is measured by the ratio of the unconfined compressive strength of an undisturbed soil sample to the unconfined strength of a remolded sample of the same soil; it is called the *degree of sensitivity* S_t.

$$S_t = \frac{q_u \text{ (undisturbed)}}{q_u \text{ (remolded)}}$$

For most clays, $S_t = 2$ to 4.

Description	S_t
Insensitive	<2
Moderately sensitive	2–4
Sensitive	4–8
Extra sensitive	8–16
Quick	>16

(2) Classification Systems

(a) Grain-size classification systems are based on sieve analysis accomplished by means of standard sieves, the sizes of which are given in Appendix A.29. Six classification charts are given in the same table.

(b) Textural classification is based on the percentages of size groups contained in the sample. The classification chart is shown in Fig. 5.07–3.*

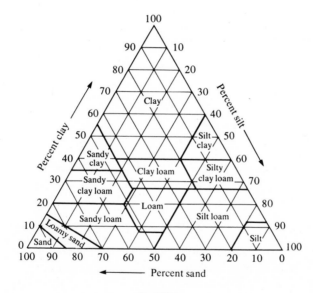

Fig. 5.07–3

(c) AASHO classification system is based on the chart shown in Appendix A.30. This system, devised by the American Association of State Highway Officials,† divides soils into four major groups: granular soils (A-1, A-2), fine sands (A-3), clays and silts (A-4, A-5, A-6, A-7), and organic soils (A-8).

*M. Peech et al., "Methods of Soil Analysis for Soil-Fertility Investigation," U.S. Department of Agriculture Circular 757, 1947.

†"Report of Committee on Classification of Materials for Subgrades and Granular Type Roads," *Proc. Highway Res. Bd.*, Vol. 25, Washington, D.C., 1945, pp. 375–388; discussion pp. 388–392.

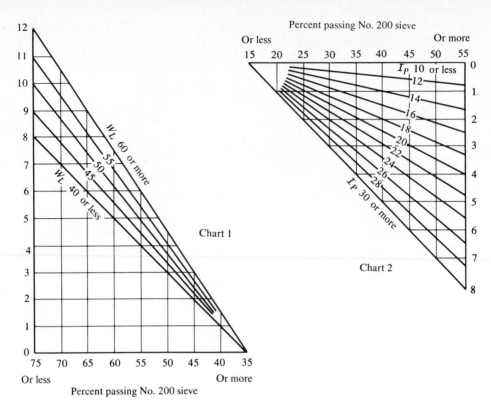

Fig. 5.07-4

(d) Group index is a supplementary criterion of the AASHO classification defined by the empirical formula

$$I_G = 0.2a + 0.005ac + 0.01bd$$

where a = portion of percentage passing a no. 200 sieve greater than 35 percent and not exceeding 75 percent, expressed as a positive whole number from 1 to 40; b = portion of percentage passing a no. 200 sieve greater than 15 percent and not exceeding 55 percent, expressed as a positive whole number from 1 to 40, c = portion of the numerical liquid limit w_L greater than 40 and not exceeding 60, expressed as a positive whole number from 1 to 20; and d = portion of the numerical plasticity index I_P greater than 10 and not exceeding 30, expressed as a positive whole number from 1 to 20. This index can also be obtained from the charts of Fig. 5.07–4 as the sum of the vertical readings in charts 1 and 2. The group index rating is shown schematically below.

Group index	0	1	2	3–4	5–9	10–20
Rating	Excellent	Very Good	Good	Fair	Poor	Very Poor

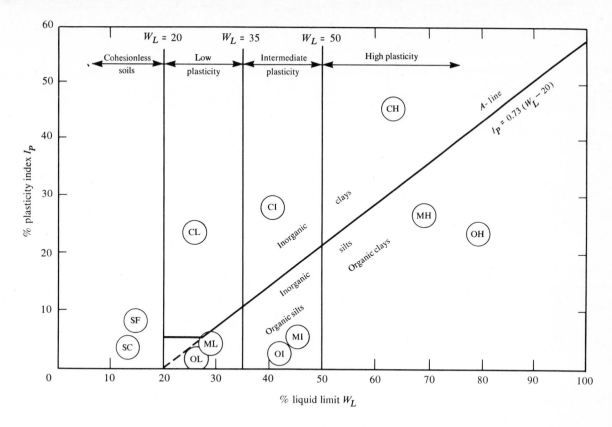

Fig. 5.07–5

(e) **Unified soil classification system** is the most widely used classification system, developed by A. Casagrande[*] at the request of the U.S. Army Corps of Engineers in cooperation with the U.S. Bureau of Reclamation. The parameters of this classification are soil components, soil groups, identification procedures, values of subgrade, subbase, base, frost susceptibility, volume change, drainage characteristics, dry weight density, California bearing ratio, and modulus of subgrade. The symbols used in this system are listed in Appendix A.31, the descriptive classification chart is shown in Appendix A.32*a*, and the corresponding chart of design values is shown in Appendix A.32*b*.

(f) **Plasticity chart** of Fig. 5.07–5 included in this classification system is used to separate the more claylike materials from those that are siltlike.

(g) **California bearing ratio (CBR)** is the ratio of the pressure required to penetrate a soil mass with a 3-in^2 circular piston at the rate of 0.05 in/min to the pressure required to penetrate a standard material with the same piston at the same rate.

(h) **Modulus of subgrade** k is the ratio of the load per unit area of a mass of soil to the corresponding settlement of the surface. It is determined as the slope of the secant, drawn between the point corresponding to zero settlement and the point of 0.05-in settlement on the load settlement curve derived by the plate load test using the 30-in diameter plate (1 in = 2.54 cm).

[*]A. Casagrande, "Classification and Identification of Soils," *Trans. ASCE*, Vol. 113, 1948, p. 901.

(3) Permeability

(a) Definition. The permeability (fluid conductivity) k is the facility with which water is able to travel through the soil. The numerical value of k for a particular soil is an experimental constant. For laminar flow in fully saturated soils of grain size less than 1 mm, the total discharge is given by Darcy's formula as

$$Q = vA = kiA \qquad \begin{cases} v \text{ in m/s} \\ i \text{ dimensionless} \\ k \text{ in m/s} \\ A \text{ in m}^2 \end{cases}$$

where v = discharge velocity, i = hydraulic gradient, k = permeability, and A = area of flow profile. The value of k varies from 100 cm/s for clean gravels to 10^{-9} cm/s for homogeneous clay.

(b) Casagrande's chart. The relationship between the permeability and other characteristics of soils with the different methods available for the determination of k is given in the chart below. More elaborate charts are given in Appendix A.33.

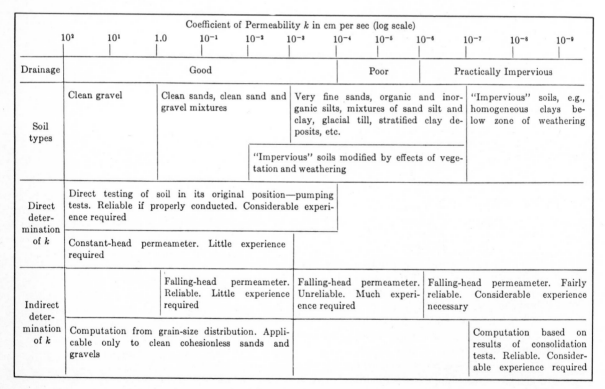

SOURCE: A. Casagrande and R. E. Fadum, "Notes for Soil Testing for Engineering Purposes," Harvard University Graduate School of Engineering, 1940, pub. 268, p. 74.

(4) Seepage

(a) Definition.
The process of water transfer in soils is called seepage. Four particular cases of water seepage are shown below. The symbols used are: A = seepage cross-sectional area (m²), H = hydraulic heat (m), L = seepage length (m), Q = total discharge (m³/s), α = angle of layer (rad), b = length of A (m), d = width of A (m), h = depth (m), i = hydraulic gradient (dimensionless), k = permeability (m/s), v = velocity (m/s).

(b) Homogeneous layer system, horizontal seepage (Fig. 5.07–6)

$$v_x = k_x i_x = k_x H/L = Q_x/A_x$$

$$Q_x = v_x A_x = k_x i_x A_x = k_x H dh/L$$

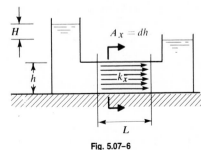

Fig. 5.07–6

(c) Homogeneous layer system, inclined seepage (Fig. 5.07–7)

$$v_\alpha = k_\alpha i_\alpha = k_\alpha H/L = k_\alpha \sin \alpha = Q_\alpha/dh$$

$$Q_\alpha = v_\alpha A_\alpha = k_\alpha i_\alpha A_\alpha = k_\alpha (\sin \alpha) A_\alpha = k_\alpha H dh/L$$

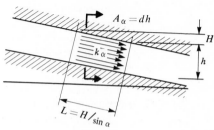

Fig. 5.07–7

(d) Multilayer system, horizontal seepage (Fig. 5.07–8)

$$k_x = \frac{h_1 k_{1x} + h_2 k_{2x} + \cdots + h_n k_{2x}}{h} = \frac{\lambda_x}{h}$$

$$v_x = k_x i_x = H\lambda_x/hL = Q_x/A_x = Q_x/bh$$

$$Q_x = v_x A_x = k_x i_x A_x = Hd\lambda_x/L$$

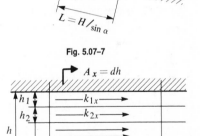

Fig. 5.07–8

(e) Multilayer system, vertical seepage (Fig. 5.07–9)

$$k_y = \frac{h}{h_1/k_{1y} + h_2/k_{2y} + \cdots + h_n/k_{ny}} = \frac{h}{\lambda_y}$$

$$v_y = k_y i_y = H/\lambda_y = Q_y/A_y = Q_y/bd$$

$$Q_y = v_y A_y = k_y i_y A_y = Hbd/\lambda_y$$

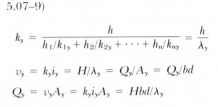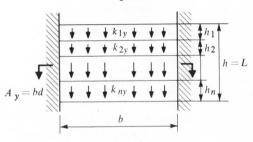

Fig. 5.07–9

(5) Consolidation

(a) Definition and notation. The gradual reduction of thickness of a given layer of soil under the prescribed pressure is called its consolidation. The process of consolidation can in general be represented by the time-compression diagram. The symbols used in consolidation analysis are:

a_v = coefficient of compressibility (m²/kgf)
c_v = coefficient of consolidation (m²/s)
e = void ratio at p (1)
e_0 = void ratio at p_0 (1)
k = coefficient of permeability (m/s)
m_v = coefficient of volume compressibility (m²/s)
p = pressure at $t = t$ (kgf/m²)
p_0 = pressure at $t = 0$ (kgf/m²)
t_c = period of consolidation (min)
w_L = water content (%)
γ_W = weight density of water (kgf/m³)
Δe = change in void ratio (1)
Δp = change in pressure (1)
C_v = compression index (1)
H = length of drainage path (m)
T_v = time factor corresponding to U (1)
U = average consolidation ratio (%)
ΔH = change in H (m)

where (1) designates the dimensionless quantity.

(b) Time-compression diagrams for sand (Fig. 5.07–10) and clay (Fig. 5.07–11) are shown below. The units of time are minutes or days.

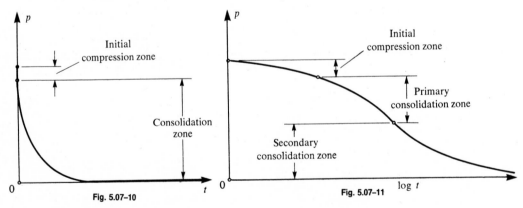

Fig. 5.07–10   Fig. 5.07–11

(c) Parameters of consolidation are

$$a_v = \frac{e_0 - e}{p_0 - p} = -\frac{\Delta e}{\Delta p} \qquad c_v = \frac{k(1 + e_0)}{a_v \gamma_W} \qquad m_v = \frac{a_v}{1 + e_0}$$

$$C_v = \frac{\Delta e}{\log (p/p_0)} \qquad \Delta H = \frac{H \, \Delta e}{1 + e_0} \qquad t_c = \frac{T_v H^2}{c_v}$$

(d) Time factor T_v corresponding to a specific average consolidation ratio U is given for six particular cases of boundary conditions and pore-pressure distribution in Appendix A.34.

5.08 STRESS ANALYSIS OF SOILS

(1) Vertical Loads

(a) **Assumptions.** For analysis, the loads are assumed to be applied normal to the soil surface or normal to a plane parallel to this surface within the soil system which is semi-infinite, weightless, homogeneous, and linearly elastic, with Poisson's ratio $\nu \cong 0.5$.

(b) **Concentrated load** of magnitude P applied at the surface point A (Fig. 5.08–1) develops at the point B (given by the coordinates x, y, z in the soil system) the vertical stress s and the vertical displacement δ expressed as

$$s = \frac{P}{z^2} K_1 \qquad \delta = \frac{P}{ER} L_1 \qquad \begin{cases} P \text{ in kgf} \\ s, E \text{ in kgf/m}^2 \\ x, y, z, R, \delta \text{ in m} \end{cases}$$

where

$$R = \sqrt{x^2 + y^2 + z^2} \qquad K_1 = \frac{3}{2\pi} \left(\frac{z}{R}\right)^5 \qquad L_1 = \frac{3(R^2 + z^2)}{4\pi R^2}$$

and E = modulus of elasticity given in Appendix A.28.

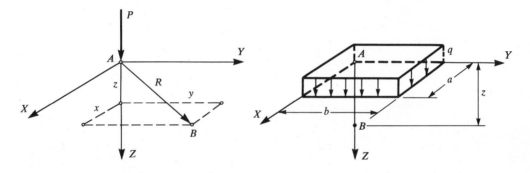

Fig. 5.08–1   **Fig. 5.08–2**

(c) **Rectangular load of constant intensity** q and of sides a, b applied along the axes X, Y, respectively (Fig. 5.08–2), develops under its corner A at the depth z at point B

$$s = qK_2 \qquad \delta = \frac{qz}{E} L_2 \qquad \{q \text{ in kgf/m}^2\}$$

where in terms of

$$m = a/z \qquad m = b/z \qquad t = \sqrt{1 + m^2 + n^2} \qquad \phi = mn/t$$

the equivalents are

$$K_2 = \frac{1}{2\pi} \left[\left(\frac{1}{1 + m^2} + \frac{1}{1 + n^2} \right) \phi + \tan^{-1} \phi \right]$$

$$L_2 = \frac{3}{8\pi} \left(m \ln \frac{t + n}{t - n} + n \ln \frac{t + m}{t - m} \right)$$

and E is the same as in (b).

(2) Static Effect of Soil Water

(a) **Total pressure** p (kgf/m^2) at depth h (m) in a given soil system consists of the following components:

Surface load pressure s (kgf/m^2) produced by the loads applied at the soil surface as shown in Sec. 5.08–1

Intergranular pressure $\bar{p}$ (kgf/m^2), also known as the effective pressure, produced by contacts of mineral particles

Hydrostatic pressure u_w (kgf/m^2), also called the neutral or porewater pressure, produced by hydrostatic stresses in voids

Capillary moisture load pressure u_c (kgf/m^2) produced by the weight of the suspended capillary column of water

(b) **Dry-submerged soil system.** If the soil system consists of a dry layer of thickness h_2 and a saturated layer of arbitrary thickness as shown in Fig. 5.08–3, the pressures developed by the surface load and the soil-water system are:

$$p_1 = h_1 \gamma_{dry} + s_1 \qquad 0 \le h_1 \le h_2$$
$$p_3 = h_2 \gamma_{dry} + (h_3 - h_2) \gamma_{sat} + s_3 \qquad h_2 \le h_3$$

where γ_{dry} = dry weight density (kgf/m^3), γ_w = water weight density, γ_{sat} = saturated weight density, and h_1, h_2, h_3 = depths below the surface.

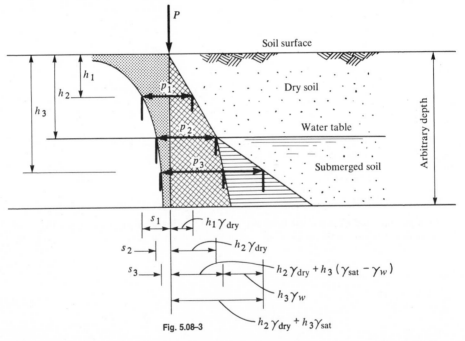

Fig. 5.08–3

(c) **Capillary fringe–submerged soil system.** If the upper layer of soil in Fig. 5.08–3 is also saturated but the water table remains at the level indicated in the same figure, so that a capillary fringe is formed above this water table, the pressures developed at the respective levels are:

$$p_1 = h_1 \gamma_{dry} + h_2 \gamma_w n + s_1 \qquad 0 \le h_1 \le h_2$$
$$p_3 = h_3 \gamma_{sat} + s_3 \qquad h_2 \le h_3$$

where n = porosity in percent, defined in Table 5.07–1.

(3) Shearing Strength of Soils

(a) **Shearing stress** developed in the soil mass at failure is called the shearing strength of that particular soil. This stress, designated by s_t, is frequently called the tangential stress. Because of the complexity of the system, the numerical value of s_t can be determined independently only by laboratory tests. The results of these tests for selected types of soils are given below.

(b) **Strength of cohesionless soils** composed of rigid particles (gravels, sands, and/or silts) is

$$s_t = s_n \tan \phi = p \tan \phi \qquad \left\{ \begin{array}{l} s_t, s_n, p \text{ in kgf/m}^2 \\ \phi \text{ in degrees} \end{array} \right\}$$

where s_n = normal stress, identical to the pressure p defined in Sec. 5.08–2, and ϕ = angle of internal friction at a given p:

$$\phi = \phi_0 + \frac{p - p_0}{p_0} \qquad \{\phi_0 \text{ in degrees}\}$$

In this relation, ϕ_0 is the angle at $p_0 = 500 \text{ kgf/m}^2$. Typical values of ϕ_0 are given below.

Material	ϕ_0, degrees	
	Loose	Dense
Silty sand	30–26	34–30
Uniform fine to medium sand	30–26	38–32
Well-graded sand	34–30	47–40
Sand and gravels	36–33	50–47

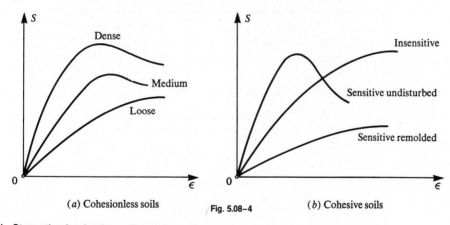

(a) Cohesionless soils   **Fig. 5.08–4**   (b) Cohesive soils

(c) **Strength of cohesive soils** such as clays of low and medium sensitivity is

$$s_t = s_n \tan \phi = p \tan \phi$$

where s_n, p have the same meaning as in (b) above and

$$\phi = 40 - I_p/4 \qquad 0 < I_p \le 50$$
$$\phi = 40 - I_p/6 \qquad 50 < I_p \le 100$$

The plasticity index I_p is defined in Sec. 5.07–1h.

(d) **Stress-strain curves of soils** bear some similarity to the stress-strain curves of other engineering materials as shown in Fig. 5.08–4.

(4) Bearing Capacity of Soils

(a) **Definition.** The ultimate bearing capacity q_u (kgf/m^2) is the load per unit area which causes the collapse of the foundation due to plastic shear failure. For dense sands and insensitive clays this value is well defined; for loose sands and sensitive clays the failure is progressive and q_u is not well defined.

(b) **Presumptive bearing capacity** is based on past experiences with particular types of soils and is included in state and municipal building codes. Typical values of maximum allowable bearing pressures for rocks and soils are given in Appendix A.28.

(c) **Prandtl's formula** gives the approximate value of ultimate bearing capacity q_u of soil under a long footing of length L (m) and width b (m) as

$$q_u = (c \cot \phi + \tfrac{1}{2}\gamma_{dry} b \sqrt{N_\phi})(N_\phi e^{\pi \tan \phi} - 1)$$

where c = cohesion (kgf/m^2), ϕ = angle of internal friction (degrees), γ_{dry} = dry weight density (kgf/m^3), and

$$N_\phi = \tan^2 (45° + \phi/2)$$

The numerical values of ϕ for particular soils are listed in Sec. 5.08–3. The values of cohesion c are given below.

Type of soil	c, kgf/m^2	Type of soil	c, kgf/m^2
Almost liquid clay	500	Soft clay	2,000
Very soft clay	1,000	Medium-soft clay	4,000
Damp, muddy sand	2,000	Medium-hard clay	6,000

(d) **Terzaghi–Meyerhof formula** for circular, square, and rectangular footings includes the effect of soil weight, surcharge, and the standard parameters of soil and is in terms of $\alpha = 45° + \phi/2$.

$$q_u = \tfrac{1}{2}F_1 \gamma b \tan^5 \alpha + 2F_2 c(\tan \alpha + \tan^3 \alpha) + q \tan^4 \alpha$$

where γ = weight density of soil under natural conditions (kgf/m^3), q = pressure of surcharge (kgf/m^2), and F_1, F_2 are the shape parameters given below.

Shape of footing	F_1	F_2
Circle, diameter = b	1.20	0.70
Square, $L = b$	1.25	0.85
Rectangle, $L = 2b$	1.12	0.90
Rectangle, $L = 5b$	1.05	0.70

The angle of internal friction ϕ has the greatest influence. If the angle ϕ is very small, only the cohesion influences the bearing capacity.

6

MECHANICS OF
HEAT AND GASES

6.01 TEMPERATURE AND HEAT

(1) Basic Terms

(a) **Heat** is the total internal energy of matter associated with the motion and configuration of its molecules and atoms (molecular energy) and can be expressed in the units of energy (work) defined in Sec. 3.05–1.

(b) **Temperature** is the measure of the magnitude of the molecular energy per molecule and as such is a quantity related to a selected datum (fixed point of selected scale, Sec. 6.01–2).

(c) **Thermometer** is a mechanical, electrical, or optical device which through a marked scale indicates its own temperature.

(d) **Temperature scale** is defined by two fixed points (two distinct temperatures) whose fixed temperature interval is divided into an arbitrarily selected number of equal parts called *degrees of temperature.* Three such points are the temperatures defined below.

(e) **Lower fixed point** (ice point, melting point) is the temperature T_l of a mixture of pure water and ice at a pressure of 1 atm.

(f) **Upper fixed point** (steam point, boiling point) is the temperature T_u of a mixture of pure water and steam at a pressure of 1 atm.

(g) **Absolute zero temperature** is the temperature T_a of a body whose total molecular energy is zero (average molecular velocity is zero, Sec. 6.02–5f).

(2) Temperature Scales

(a) **Celsius scale:** 1 degree Celsius (after Anders Celsius) = 1°C.

$$T_l = 0°C \qquad \text{fixed temperature interval} = 100°C \qquad T_u = 100°C$$

also called the centigrade scale (Fig. 6.01–1).

(b) **Fahrenheit scale:** 1 degree Fahrenheit (after Daniel Fahrenheit) = 1°F.

$$T_l = 32°F \qquad \text{fixed temperature interval} = 180°F \qquad T_u = 212°F$$

(c) **Rankine scale:** 1 degree Rankine (after William John Rankine) = 1°R.

$$T_l = 492°R \qquad \text{fixed temperature interval} = 180°R \qquad T_u = 672°R$$

(d) **Réaumur scale:** 1 degree Reaumur (after René Antoine Réaumur) = 1°Re.

$$T_l = 0°Re \qquad \text{fixed temperature interval} = 80°Re \qquad T_u = 80°Re$$

(e) **Kelvin scale:** 1 degree Kelvin (after William Thomson, Lord Kelvin) = 1 K.

$$T_a = 0\,K \qquad T_l = 273\,K \qquad \text{fixed temperature interval} = 100\,K \qquad T_u = 373\,K$$

(f) **SI unit of temperature** is the kelvin (K), defined as

$$1/273.16 \cong 1/273$$

of the thermodynamic temperature of the triple point of water, i.e., the temperature at which a mixture of pure ice and water is in equilibrium with water vapor when exposed to the air at standard atmospheric pressure.

(g) Comparison of the fixed points listed above is shown below:

	°C	°F	°R	°Re	K
Absolute zero temperature	−273	−460	0	−218	0
Lower fixed point (freezing of water)	0	+32	+492	0	+273
Upper fixed point (boiling of water)	+100	+212	+672	+80	+373

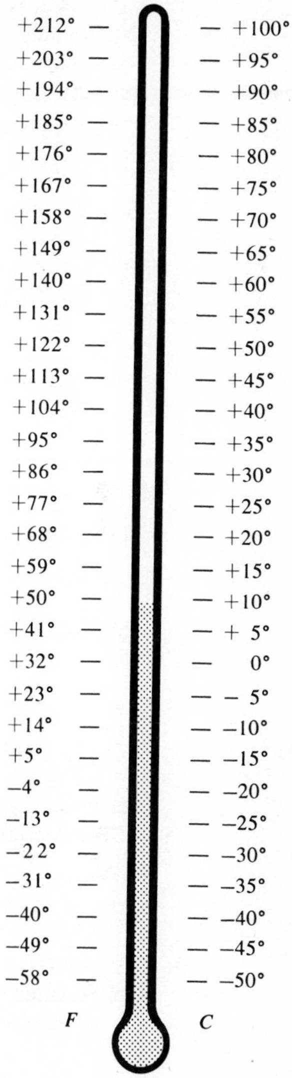

(3) Temperature Relations

(a) Temperature difference ΔT is the difference of two temperatures given in the same scale.

(b) Conversion relations for ΔT are

$$1°C = \tfrac{9}{5}°F = \tfrac{9}{5}°R = \tfrac{4}{5}°Re = 1 \text{ K}$$

$$1°F = \tfrac{5}{9}°C = 1°R = \tfrac{4}{9}°Re = \tfrac{5}{9} \text{ K}$$

$$1°R = 1°F = \tfrac{5}{9}°C = \tfrac{4}{9}°Re = \tfrac{5}{9} \text{ K}$$

$$1°Re = \tfrac{9}{4}°F = \tfrac{9}{4}°R = \tfrac{5}{4}°C = \tfrac{5}{4} \text{ K}$$

$$1 \text{ K} = \tfrac{9}{5}°F = \tfrac{9}{5}°R = \tfrac{4}{5}°Re = 1°C$$

example:

If the temperature of a furnace is increased from 80 to 1200°F, then

$$\Delta T = 1120°F = \tfrac{5}{9} \times 1120°C = 622.2°C = 1 \times 1120°R = 1120°R$$
$$= \tfrac{4}{9} \times 1120°Re = 498°Re = \tfrac{5}{9} \times 1120 \text{ K} = 622.2 \text{ K}$$

(c) Conversion relations for relative temperature T are

$$F = \tfrac{9}{5}C + 32 = R - 460 = \tfrac{9}{4}Re + 32 = \tfrac{9}{5}(K - 255)$$

$$C = \tfrac{5}{9}(F - 32) = \tfrac{5}{9}(R - 492) = \tfrac{5}{4}Re = K - 273$$

$$R = \tfrac{9}{5}C + 492 = F + 460 = \tfrac{9}{4}Re + 492 = \tfrac{9}{5}K$$

$$Re = \tfrac{4}{5}C = \tfrac{4}{9}(F - 32) = \tfrac{4}{9}(R - 492) = \tfrac{4}{5}(K - 273)$$

$$K = C + 273 = \tfrac{5}{9}(F + 460) = \tfrac{5}{9}R = \tfrac{5}{4}Re + 273$$

where C = degrees Celsius, F = degrees Fahrenheit, R = degrees Rankine, Re = degrees Réaumur, and K = kelvins (see also Appendix B.27).

example:

If the standard room temperature is 20°C, then the same temperature on other scales is

$$(\tfrac{9}{5} \times 20° + 32°)F = 68°F \qquad (\tfrac{4}{5} \times 20°)Re = 16°Re$$

$$(\tfrac{9}{5} \times 20° + 492°)R = 528°R \qquad (20° + 273°)K = 293 \text{ K}$$

F		C
+212°	—	+100°
+203°	—	+95°
+194°	—	+90°
+185°	—	+85°
+176°	—	+80°
+167°	—	+75°
+158°	—	+70°
+149°	—	+65°
+140°	—	+60°
+131°	—	+55°
+122°	—	+50°
+113°	—	+45°
+104°	—	+40°
+95°	—	+35°
+86°	—	+30°
+77°	—	+25°
+68°	—	+20°
+59°	—	+15°
+50°	—	+10°
+41°	—	+5°
+32°	—	0°
+23°	—	−5°
+14°	—	−10°
+5°	—	−15°
−4°	—	−20°
−13°	—	−25°
−22°	—	−30°
−31°	—	−35°
−40°	—	−40°
−49°	—	−45°
−58°	—	−50°

Fig. 6.01–1

(4) Measurement of Heat

(a) **Quantity of heat** Q is the amount of molecular energy stored in a body. To heat or to cool a body means to convey (give) energy to its molecules or to withdraw energy from them.

(b) **MKS unit of heat** is 1 kilocalorie (kcal), defined as the quantity of heat required to raise the temperature of 1 kilogram of water from 14.5 to 15.5°C.

$$1 \text{ kilocalorie} = 1 \text{ kcal} = 1,000 \text{ calories} = 1,000 \text{ cal}$$

(c) **FPS unit of heat** is 1 British thermal unit (Btu), defined as the quantity of heat required to raise the temperature of 1 pound of water from 58.5 to 59.5°F.

(d) **Mechanical equivalents.** Since heat is a form of energy, all units of heat can be expressed in units of work.

$$1 \text{ kcal} = 4,184 \text{ J} = 426.65 \text{ m-kgf} \qquad 1 \text{ kcal} = 3.967 \text{ Btu} \qquad 1 \text{ Btu} = 0.2520 \text{ kcal}$$
$$1 \text{ Btu} = 1,055 \text{ J} = 778.13 \text{ ft-lbf} \qquad 1 \text{ kWh} = 860.0 \text{ kcal} \qquad 1 \text{ kWh} = 3,413 \text{ Btu}$$

where kWh = kilowatt-hour.

(5) Heat Capacity

(a) **Specific heat capacity** c of a substance is the quantity of heat required to raise the temperature of its unit mass by 1 degree. The specific heat capacity of the more common substances are listed in Appendix A.

(b) **Units of specific heat capacity** c are

$$\frac{1 \text{ kilocalorie}}{1 \text{ kilogram} \times 1° \text{ Celsius}} = \frac{1 \text{ calorie}}{1 \text{ gram} \times 1° \text{ Celsius}} = \frac{1 \text{ Btu}}{1 \text{ pound} \times 1° \text{ Fahrenheit}}$$

which shows that the magnitude of c is the same number in all these units, and since 1° Celsius = 1 kelvin, the magnitude of c is the same number in the Kelvin scale.

(c) **Heat capacity** q of a body of mass m is the quantity of heat required to raise the temperature of the body by 1 degree.

$$q = cm \qquad \left\{ \begin{array}{l} q \text{ in kcal} \\ c \text{ in kcal/kg-°C} \\ m \text{ in kg} \end{array} \right\}$$

where c = specific heat capacity defined in (a).

(d) **Quantity of heat** Q required to raise the temperature of the body of mass m from the temperature T_1 to T_2 is

$$Q = cm(T_2 - T_1) \qquad \left\{ \begin{array}{l} Q \text{ in kcal} \\ t_1, t_2 \text{ in °C} \end{array} \right\}$$

example:

If the initial temperature $T_1 = 60°F$ of a block of aluminum of $m = 100$ lb and $c = 0.22$ Btu/lb-°F is raised to $T_2 = 150°F$, the heat quantity required for this change is

$$Q = 0.22 \times 100 \times (150 - 60) = 1,980 \text{ Btu}$$

(6) Change in Phase

(a) **Fusion (melting)** is the change of phase (state) of a substance from a solid to a liquid, and the temperature at which this change occurs is called the *melting point* of that substance.

(b) **Latent heat of fusion** L_f of a substance is the heat required per unit of its mass to change the solid to a liquid at its melting temperature (which remains constant during this change of phase).

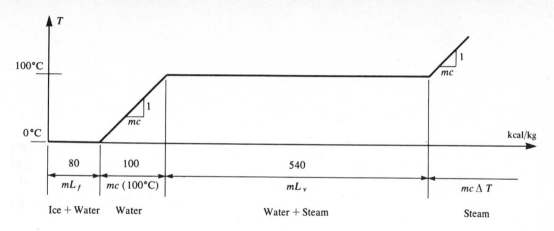

Fig. 6.01-2

(c) Quantity of latent heat of fusion Q_f of a substance of mass m is

$$Q_f = mL_f \qquad \left\{\begin{array}{l} Q_f \text{ in kcal} \\ m \text{ in kg} \\ L_f \text{ in kcal/kg} \end{array}\right\}$$

where $1 \text{ kcal/kg} = \frac{9}{5} \text{ Btu/lb}$.

example:

For an ice cube of $m = 1,000 \text{ kg}$ at a pressure of 1 atm and a temperature of $0°C = 32°F$,

$L_f = 80 \text{ kcal/kg} = 144 \text{ Btu/lb}$ (Appendix A)

and

$Q_f = 1,000 \times 80 = 80,000 \text{ kcal} = 1,000 \times 2.205 \times 144 = 317,520 \text{ Btu}$

(d) Vaporization (boiling) is the change of phase (state) of a substance from a liquid to a vapor, and the temperature at which this change occurs is called the *boiling point* of that substance.

(e) Latent heat of vaporization L_v of a substance is the heat required per unit of its mass to change the liquid to a vapor at its boiling temperature (which remains constant during this change of phase).

(f) Quantity of latent heat of vaporization Q_v of a substance of mass m is

$$Q_v = mL_v \qquad \left\{\begin{array}{l} Q_v \text{ in kcal} \\ m \text{ in kg} \\ L_v \text{ in kcal/kg} \end{array}\right\}$$

where $1 \text{ kcal/kg} = \frac{9}{5} \text{ Btu/lb}$.

example:

For a volume of water of mass $m = 1,000 \text{ kg}$ at a pressure of 1 atm and a temperature of $100°C = 212°F$,

$L_v = 540 \text{ kcal/kg} = 972 \text{ Btu/lb}$ (Appendix A)

and

$Q_v = 1,000 \times 540 = 540,000 \text{ kcal} = 1,000 \times 2.205 \times 972 = 2,143,260 \text{ Btu}$

The variation of temperature and the related change in thermal energy are shown graphically in Fig. 6.01–2.

6.02 IDEAL GASES

(1) Molecular Mass

(a) **Dalton's hypothesis** states that each basic element is composed of atoms each of which has a definite fixed mass and that each compound (combination of basic elements) is made up of molecules each of which contains a definite number of atoms of each constituent element.

(b) **Nuclide** is an atom of a particular mass and of a particular element. Isotopes are nuclides of different masses belonging to the same element.

(c) **Atomic mass unit (amu)** is defined as one-twelfth of the atomic mass of the predominant carbon isotope C^{12}.

$$1 \text{ amu} = 1 \text{ u} = (1.6605 \pm 0.0001) \times 10^{-27} \text{ kg}$$

(d) **Atomic mass** A of an element is the ratio of the average of the mass of isotopes (forming the element) to one-twelfth the mass of C^{12}. Since atomic mass is a ratio, it is a dimensionless constant.

(e) **Molecular mass** M of a gas is the sum of atomic masses of all the isotopes included in the molecule of that gas and is also dimensionless.

example:

A molecule of oxygen (O_2) contains two atoms, each of atomic mass 16. Thus, $M = 16 \times 2 = 32$. A molecule of carbon dioxide (CO_2) contains one atom of C of atomic mass 12 and two atoms of O. Thus

$$M = 12 \times 1 + 16 \times 2 = 44$$

where the value of each particular atomic mass is taken from Appendix A.22.

(f) **One mole (mol)** is the molecular mass M of a substance in grams and one kilomole (kmol) is the molecular mass M of the same substance in kilograms.

example:

The values of A and M of the substances used in this example are given in Appendix A.22.

(α) Hydrogen (H_2) has two atoms, each of atomic mass $A_H = 1$. Its molecular mass $M = 2 \times 1 = 2$, its 1 mole $= M \times g = 2$ g, and the absolute mass of one of its molecules $= 2 \times 12 \times 1.66 \times 10^{-24}$ g $= 3.98 \times 10^{-23}$ g, where according to (c) and (d), $12 \times 1.66 \times 10^{-24}$ g is the atomic mass of carbon C^{12}.

(β) Water (H_2O) has two atoms of hydrogen, each of atomic mass $A_H = 1$, and one atom of oxygen of atomic mass $A_O = 16$. Its molecular mass $M = 18 \times 12 \times 1.66 \times 10^{-24}$ g $= 3.59 \times 10^{-22}$ g.

(g) **Avogadro's number** N_A is the ratio of the number of molecules N to the number of moles n of a gas, and it is the same constant in all cases.

$$N_A = \frac{N}{n} = 6.02205 \times 10^{23}/\text{mol} = 6.02205 \times 10^{26}/\text{kmol}$$

example:

In the mass of 100 g of CO_2 the number of moles is

$$n = \frac{m}{M} = \frac{100}{44} = 2.273$$

where M was computed in (e), and the number of molecules in this mass is

$$N = n \times N_A = 2.273 \times 6.02205 \times 10^{23} = 1.37 \times 10^{24} \text{ molecules}$$

(2) Gas Laws

(a) **State of confined gas** is specified by three quantities: volume V, pressure p, and temperature T. At low pressures and high temperatures all gases obey three simple laws [(b) to (d) below] and are called *ideal* (perfect) gases.

(b) **Boyle's law.** When the temperature T of a gas sample is kept constant, the product of pressure p and volume V is constant.

$$p_1 V_1 = p_2 V_2 = \cdots = p_r V_r = \text{constant} \qquad \begin{Bmatrix} p \text{ in kgf/m}^2 \\ V \text{ in m}^3 \end{Bmatrix}$$

(c) **Charles' law.** When the pressure p of a gas sample is kept constant, the volume of the sample V is directly proportional to its absolute temperature T.

$$\frac{V_1}{T_1} = \frac{V_2}{T_2} = \cdots = \frac{V_r}{T_r} = \text{constant} \qquad \begin{Bmatrix} V \text{ in m}^3 \\ T \text{ in K} \end{Bmatrix}$$

where $T_1, T_2, \ldots, T_r$ = absolute temperatures in the Kelvin scale corresponding to states $1, 2, \ldots, r$.

example:

If a balloon filled with air has $V_1 = 300$ m^3 at 0°C, its volume V_2 at 67°C under the same pressure is

$$V_2 = \frac{300}{273} \times 340 = 373.6 \text{ m}^3$$

where $T_1 = 0°C = 273$ K and $T_2 = 67°C = (273 + 67)$ K.

(d) **General gas law.** For any gas sample the product of pressure p and volume V divided by its absolute temperature T is constant.

$$\frac{p_1 V_1}{T_1} = \frac{p_2 V_2}{T_2} = \cdots = \frac{p_r V_r}{T_r} = nR \qquad \begin{Bmatrix} p \text{ in kgf/m}^2 \\ V \text{ in m}^3 \\ T \text{ in K} \\ R \text{ in m-kgf/kmol-K} \\ n \text{ in kmol} \end{Bmatrix}$$

where $T_1, T_2, \ldots, T_r$ = absolute temperature in the Kelvin scale, R = constant defined in Sec. 6.02–3, and n = number of kilomoles (Sec. 6.02–1f).

(e) **Dalton's law of partial pressures** states that the total pressure p of a gas mixture in a given container of volume V at a temperature T equals the sum of the pressures that each component gas would exert if it alone occupied that container.

$$p = p_1 + p_2 + \cdots + p_r = \frac{(n_1 + n_2 + \cdots + n_r)RT}{V}$$

where p = total pressure, $p_1, p_2, \ldots, p_r$ = pressures of respective component gases, $n_1, n_2, \ldots, n_r$ = number of kilomoles of respective component gases, and R, T, V are defined in (d).

(3) Gas Constants

(a) **One kilomole** (Sec. 6.02–1f) of any gas under a particular pressure and temperature occupies the same volume and specifically at 273 K and 1 atm:

Volume of 1 kilomole = 22.414 m^3
Volume of 1 mole = 22.414 liter = 22,414 cm^3

(b) Relation of mass m and molecular mass M is

$$m \text{ (in kg)} = n \text{ (in kmol)} \times M \text{ (in kg/kmol)}$$

where n = number of kilomoles in mass m and M = molecular mass in kilograms per kilomole [see example in (d) below].

(c) Relation of volume V_0 and molecular volume V_m is

$$V_0 \text{ (in m}^3\text{)} = n \text{ (in kmol)} \times 22.414 \text{ m}^3/\text{kmol}$$

where V_0 = volume of gas at $T_0 = 273$ K and $p_0 = 1$ atm.

(d) Universal gas constant R is

$$R = \frac{1 \text{ atm} \times 22.414 \text{ m}^3}{273 \text{ K} \times 1 \text{ kmol}} = 0.0821 \text{ atm-m}^3/\text{kmol-K}$$

$$= \frac{(1.0133 \times 10^5 \text{ N/m}^2)(22.414 \text{ m}^3)}{273 \text{ K} \times 1 \text{ kmol}} = 8{,}315 \text{ J/kmol-K}$$

$$= \frac{(1.0133 \times 10^5 \text{ N/m}^2)(22.414 \text{ m}^3)}{9.807 \times 273 \text{ K} \times 1 \text{ kmol}} = 848 \text{ m-kgf/kmol-K}$$

and since $1 \text{ J} = 2.39 \times 10^{-4} \text{ kcal}$,

$$R = \frac{8{,}315 \times 2.39 \times 10^{-4} \text{ kcal}}{1 \text{ kmol} \times 1 \text{ K}} = 1.987 \text{ kcal/kmol-K}$$

example:

If at 20°C and 1 atm a volume 3.2 m³ of an ideal gas weighs 2.1 kg, then by the formula of Sec. 6.02–2d, the number of kilomoles in this volume is

$$n = \frac{pV}{RT} = \frac{1 \times 3.2}{0.0821 \times (273 + 20)} = 0.133 \text{ kmol}$$

and the mass per mol is

$$M = \frac{m}{n} = \frac{2.1 \text{ kg}}{0.133 \text{ kmol}} = 15.8 \text{ g/mol}$$

(e) Mass density ρ of a gas at a temperature T and a pressure p is

$$\rho = \frac{m}{V} = \frac{pMkg}{RT} \qquad \left\{ \begin{array}{l} \rho \text{ in kg/m}^3 \\ p \text{ in kgf/m}^2 \\ Mkg \text{ in kg/kmol} \\ R \text{ in m-kgf/kmol-K} \\ T \text{ in K} \end{array} \right\}$$

where p, Mkg, R, T are those of Sec. 6.02–2d.

(f) Mass density ρ_0 of a gas at 273 K and 1 atm is

$$\rho_0 = \frac{m}{V_0} = \frac{mTp_0}{VT_0p} = \frac{Mkg}{22.414}$$

where V_0 = volume at $p_0 = 1$ atm and $T_0 = 273$ K and V = volume at a pressure p and a temperature T.

example:

The mass density of the gas in (d) at $p = 1$ atm and $T = 20°C = 293$ K is

$$\rho = \frac{m}{V} = \frac{2.1 \text{ kg}}{3.2 \text{ m}^3} = 0.656 \text{ kg/m}^3$$

and at $p_0 = 1$ atm and $T_0 = 0°C = 273$ K it is $\quad \rho_0 = \frac{mT}{VT_0} = \frac{2.1 \times 293}{3.2 \times 273} = 0.704 \text{ kg/m}^3$

which must check $\quad \rho_0 = \frac{Mkg}{\text{volume of 1 kmol}} = \frac{15.8}{22.414} = 0.704 \text{ kg/m}^3$

where Mkg is given in (d) and the volume of 1 kmol is given in (a).

(4) Specific Heat and Heat Capacity

(a) Specific heat capacity c of gases up to 200°C may be considered to be independent of temperature. At higher temperatures the average specific heat is

$$\bar{c} = \frac{\bar{c}_2 T_2 - \bar{c}_1 T_1}{T_2 - T_1} \qquad \left\{ \begin{array}{l} \bar{c} \text{ in kcal/kg-°K} \\ T \text{ in °K} \end{array} \right\}$$

where $\bar{c}_2 = (c_2 - c_0)/2$, $\bar{c}_1 = (c_1 - c_0)/2$, $c_0 =$ specific heat capacity at $T_0 = 273$ K, $c_1 =$ specific heat capacity at T_1, and $c_2 =$ specific heat capacity at T_2.

(b) Average heat capacity $\bar{q}$ of a gas at higher temperature is

$$\bar{q} = m\bar{c} \qquad \left\{ \begin{array}{l} \bar{q} \text{ in kcal/K} \\ \bar{c} \text{ in kcal/kg-K} \end{array} \right\}$$

where $\bar{c}$ is either $\bar{c}_v =$ average specific heat at constant volume or $\bar{c}_p =$ average specific heat at constant pressure, both defined below.

(c) Specific heat capacity c_v at constant volume is the amount of heat that must be added to a unit mass of gas to increase its temperature by 1 degree when the volume is kept constant.

$$c_v = \frac{C_v}{Mkg} \qquad \left\{ \begin{array}{l} c_v \text{ in kcal/kg-K} \\ C_v \text{ in kcal/kmol-K} \\ Mkg \text{ in kg/kmol} \end{array} \right\}$$

where $C_v =$ specific heat capacity per kilomole at constant volume (for numerical values see below) and $Mkg =$ molecular mass in kilograms (Secs. 6.02–1e, f).

(d) Specific heat capacity c_p at constant pressure is the amount of heat that must be added to a unit mass of gas to increase its temperature by 1 degree when the pressure is kept constant.

$$c_p = \frac{C_p}{Mkg} \qquad \left\{ \begin{array}{l} c_p \text{ in kcal/kg-K} \\ C_p \text{ in kcal/kmol-K} \\ Mkg \text{ in kg/kmol} \end{array} \right\}$$

where $C_v =$ specific heat capacity per kilomole at constant pressure (for numerical values see below) and $Mkg =$ molecular mass in kilograms (Secs. 6.02–1e, f).

(e) Difference of C_p **and** C_v equals the universal gas constant R (Sec. 6.02–3d).

$$C_p - C_v = R = 1.987 \text{ kcal/kmol-K} \qquad \text{or} \qquad c_p - c_v = \frac{R}{Mkg}$$

where Mkg is specific for a given gas and for

Monatomic gas:	$C_v = 3R/2$	$C_p = 5R/2$
Diatomic gas:	$C_v = 5R/2$	$C_p = 7R/2$
Polyatomic gas:	$C_v = 6R/2$	$C_p = 8R/2$

For numerical values of Mkg, c_v, and c_p, refer to Appendix A.

(5) Kinetic Theory of Gases

(a) Pressure p exerted by gases on the walls of the containing vessel is

$$p = \frac{1}{3} N_1 \mu \bar{v}^2 \qquad \begin{Bmatrix} p \text{ in kg/m-s}^2 & \mu \text{ in kg} \\ N_1 \text{ in } 1/\text{m}^3 & \bar{v} \text{ in m/s} \end{Bmatrix}$$

where N_1 = number of molecules in unit volume, μ = mass of one molecule, and $\bar{v}$ = average velocity of molecules in the unit volume.

(b) Volume pressure P defined as $P = pV$ is

$$P = \frac{1}{3} N_1 \mu \bar{v}^2 V = \frac{1}{3} \mu \bar{v}^2 n N_A \qquad \begin{Bmatrix} P \text{ in kg-m}^2/\text{s}^2 \\ n \text{ in kmol} \\ N_A \text{ in } 1/\text{kmol} \end{Bmatrix}$$

where according to Avogadro's law the number of molecules in the volume V is

$$N = N_1 V = n N_A$$

and N_A = Avogadro's constant (Sec. 6.02–1g).

(c) General gas law equation (Sec. 6.02–2d) expressed in terms of (b) is

$$\frac{1}{3} \mu \bar{v}^2 n N_A = nRT$$

where n = number of kilomoles in V, R = universal gas constant, and T = temperature in K.

(d) Kinetic energy E_k of one molecule of gas in (c) is then

$$E_k = \frac{1}{2} \mu \bar{v}^2 = \frac{3}{2} \frac{R}{N_A} T = \frac{3}{2} kT$$

where k = Boltzmann's constant (see below).

(e) Boltzmann's constant k is

$$k = \frac{R}{N_A} = \frac{8{,}315 \text{ J/kmol-K}}{6.02205 \times 10^{26}/\text{kmol}} = 1.381 \times 10^{-23} \text{ J/K}$$

where J/K = joules per kelvin.

(f) Root-mean-square velocity $\bar{v}$ is

$$\bar{v} = \sqrt{\frac{3kT}{\mu}} \qquad \{\bar{v} \text{ in m/s}\}$$

where k, T, and μ are defined in (e), (c), and (a), respectively.

example:

The root-mean-square velocity of H_2 of $\mu = 3.35 \times 10^{-27}$ kg at $T = 360$ K is

$$\bar{v} = \sqrt{\frac{3 \times 1.38 \times 10^{-23} \times 360}{3.35 \times 10^{-27}}} = \sqrt{4.449 \times 10^6} = 2{,}109 \text{ m/s}$$

and at $p_0 = 1$ atm and $T = 360$ K, the number of molecules in 1 m^3 of H_2 is

$$N = \frac{3 p_0}{\mu \bar{v}^2} = \frac{3 \times 1.033 \times 10^4}{3.35 \times 10^{-27} \times 2{,}109^2} = 2.08 \times 10^{24}$$

Since k and μ are constant, the velocity $\bar{v}$ increases with the square root of temperature and is zero at 0 K. The number of molecules per 1 m^3 increases with pressure p and decreases with the square of the average velocity, $\bar{v}^2$.

6.03 TEMPERATURE VOLUME CHANGE

(1) Thermal Expansion of Solids

(a) Volume change. When the temperature of a solid rises from T_1 to T_2, the solid expands and the change in volume is called the *temperature expansion,* which is proportional to the dimensions of the solid. Inversely, if the temperature drops from T_2 to T_1, the solid shrinks and the change in volume is called the *temperature contraction.*

(b) Linear expansion Δl is the change in length of a solid (Fig. 6.03–1) due to the change in temperature $\Delta T = T_2 - T_1$.

$$\Delta l = l_2 - l_1 = \alpha l_1 (T_2 - T_1) \qquad \begin{cases} \Delta l, l_1, l_2 \text{ in m} \\ T_1, T_2 \text{ in K} \\ \alpha \text{ in 1/K} \end{cases}$$

from which $l_2 = l_1[1 + \alpha(T_2 - T_1)]$

where l_1 = length at T_1, l_2 = length at T_2, and α = coefficient of linear expansion (see below).

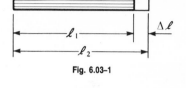

Fig. 6.03–1

(c) Coefficient of linear expansion α of a solid is the change in its unit length per 1 degree change in temperature.

$$\alpha = \frac{N_s}{\text{K}} = \frac{N_s}{°\text{C}} = \frac{\tfrac{5}{9} N_s}{°\text{F}}$$

where N_s = number depending on the substance and practically independent of temperature (see Appendices A.67 and A.68).

(d) Area expansion ΔA is the change in area of a solid (Fig. 6.03–2) due to the change in temperature $\Delta T = T_2 - T_1$.

$$\Delta A = A_2 - A_1 = 2\alpha A_1 (T_2 - T_1) \qquad \begin{cases} \Delta A, A_1, A_2 \text{ in m}^2 \\ T_1, T_2 \text{ in K} \\ \alpha \text{ in 1/K} \end{cases}$$

from which $A_2 = A_1[1 + 2\alpha(T_2 - T_1)]$

where A_1 = area at T_1, A_2 = area at T_2, and α = same coefficient as in (c).

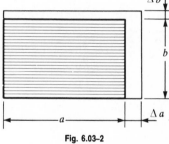

Fig. 6.03–2

(e) Volume expansion ΔV is the change in volume of a solid (Fig. 6.03–3) due to the change in temperature $\Delta T = T_2 - T_1$.

$$\Delta V = V_2 - V_1 = 3\alpha V_1 (T_2 - T_1) \qquad \begin{cases} \Delta V, V_1, V_2 \text{ in m}^3 \\ T_1, T_2 \text{ in K} \\ \alpha \text{ in 1/K} \end{cases}$$

from which $V_2 = V_1[1 + 3\alpha(T_2 - T_1)]$

where V_1 = volume at T_1, V_2 = volume at T_2, and α = same coefficient as in (c).

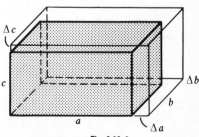
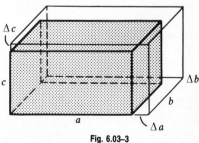

Fig. 6.03–3

(2) Thermal Expansion of Liquids

(a) **Volume change** ΔV of a liquid due to the change of temperature from T_1 to T_2 is similar to that of solids:

$$\Delta V = V_2 - V_1 = \beta V_1 (T_2 - T_1) \qquad \left\{ \begin{array}{l} \Delta V, V_1, V_2 \text{ in } m^3 \\ T_1, T_2 \text{ in K} \\ \beta \text{ in } 1/K \end{array} \right\}$$

from which

$$V_2 = V_1 [1 + \beta (T_2 - T_1)]$$

where V_1 = volume at T_1, V_2 = volume at T_2, and β = coefficient of volume expansion of liquid (see below).

(b) **Coefficient of volume expansion** β of a liquid is the change in its unit volume per degree change in temperature:

$$\beta = \frac{N_f}{K} = \frac{N_f}{^\circ C} = \frac{\frac{5}{9} N_f}{^\circ F}$$

where N_f = number depending on the substance and is independent of temperature for a large range of temperature (see Appendix A.70).

(c) **Anomalous expansion** is a characteristic peculiar to water, which does not expand linearly with increasing temperature. Between 0 and 4°C, water contracts; above 4°C it expands. This behavior can be seen clearly when one plots a curve using Celsius temperature as the horizontal coordinate and the reciprocal of the mass density of water at the respective temperature as the vertical coordinate. (Appendix A.57 gives values for the mass density of water at various temperatures on the Celsius scale.)

(3) Thermal Expansion of Gases

(a) **Volume change** of a gas under a constant pressure $p_1 = p_2$ due to the change in temperature from T_1 to T_2 is

$$\Delta V = V_2 - V_1 = \frac{V_1}{T_1} (T_2 - T_1) \qquad \left\{ \begin{array}{l} \Delta V, V_1, V_2 \text{ in } m^3 \\ T_1, T_2 \text{ in K} \end{array} \right\}$$

from which

$$V_2 = V_1 \left(1 + \frac{T_2 - T_1}{T_1} \right)$$

where V_1 = volume at T_1 and V_2 = volume at T_2.

(b) **Mass density change** of a gas under a constant pressure $p_1 = p_2$ due to the change in temperature from T_1 to T_2 is

$$\Delta \rho = \rho_2 - \rho_1 = \rho_1 \left(\frac{T_1}{T_2} - 1 \right) = \frac{pMkg}{RT_1} \left(\frac{T_1}{T_2} - 1 \right)$$

from which

$$\rho_2 = \rho_1 \frac{T_1}{T_2} = \frac{pMkg}{RT_2} \qquad \left\{ \begin{array}{l} \rho \text{ in } kg/m^3 \\ T_1, T_2 \text{ in K} \end{array} \right\}$$

where ρ_1 = mass density at T_1, ρ_2 = mass density at T_2, and p, Mkg, R are those of Sec. 6.02–3e.

6.04 HEAT TRANSFER

(1) Modes of Heat Transfer

(a) **Three modes** by which thermal energy is transferred from one point to another in a given medium (when there exist temperature differences) are conduction, convection, and radiation.

(b) **Conduction** is the process of heat transfer within a medium by which the thermal energy is transmitted from one molecule to the adjacent molecules in a purely thermal motion during which the mass of the medium is at rest (static heat carrier).

example:

The most typical example of conduction is the performance of a kitchen stove, where the heat energy of the stove is conducted through the walls of the pot to the food being cooked.

(c) **Convection** is the process of heat transfer during which the thermal energy is transferred from one place to another by the motion of the mass of the medium (kinetic heat carrier).

example:

The circulation of hot air for heating and of cool air for air conditioning of homes are typical examples of convection.

(d) **Radiation** is the process of heat transfer by means of electromagnetic waves during which the mass of the medium plays no role in the transfer of heat.

example:

The transfer of heat energy from the sun to the earth is a typical radiation process during which the space between them plays no role in this process.

(2) Heat Conduction through a Homogeneous Medium

(a) **Quantity of heat** Q that flows in the normal direction through a homogeneous medium of thickness d (Fig. 6.04–1) in the time interval t is

$$Q = kiAt = \frac{k(T_a - T_b)At}{d} \qquad \left\{ \begin{array}{l} Q \text{ in kcal} \\ k \text{ in kcal/m-s-°C} \\ T_a, T_b \text{ in °C} \\ A \text{ in m}^2 \\ t \text{ in s} \\ d \text{ in m} \end{array} \right\}$$

where k = thermal conductivity (see below), T_a, T_b = temperatures at the near face a and the far face b, respectively, and A = transfer area.

(b) **Temperature gradient** i is the drop of temperature per unit thickness of the medium:

$$i = \frac{T_a - T_b}{d} \qquad \{i \text{ in °C/m}\}$$

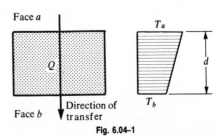

Fig. 6.04–1

(c) **Thermal conductivity** k of a particular substance is the time rate of heat conduction per unit area and per unit temperature gradient (Appendixes A.67 and A.68). Materials with high k are called *good conductors* (such as silver, copper, gold, etc.), and materials with low k are called *good insulators* (such as glass, wool, cork, asbestos, etc.).

(d) **Average thermal conductivity** $\bar{k}$. Usually, k is not a constant but varies with temperature and can be represented by the average value

$$\bar{k} = \frac{k_a + k_b}{2}$$

where k_a, k_b = thermal conductivities at T_a and T_b, respectively.

(e) **Units of** k are

$$\frac{\text{kilocalorie}}{\text{meter} \times \text{second} \times \text{degree Celsius}} = \text{kcal/m-s-}°\text{C}$$

$$\frac{\text{British thermal unit} \times \text{inch}}{\text{foot}^2 \times \text{hour} \times \text{degree Fahrenheit}} = \text{Btu-in./ft}^2\text{-h-}°\text{F}$$

$$\frac{\text{watt}}{\text{meter} \times \text{kelvin}} = \text{W/m-K}$$

and their conversion relations are

$$1 \text{ kcal/m-s-}°\text{C} = 29{,}000 \text{ Btu-in./ft}^2\text{-h-}°\text{F} = 4{,}184 \text{ W/m-K}$$

where W = 1 watt = 1 joule/second = 1 newton × meter/second.

(f) **In pipes,** the quantity of heat Q that flows radially through the thickness $(R_b - R_a)$ of a pipe of length l in the time interval t is

$$Q = \frac{2\pi k l (T_a - T_b) t}{\ln (R_b / R_a)} \qquad \begin{cases} k \text{ in kcal/m-s-}°\text{C} \\ l, R_a, R_b \text{ in m} \\ T_a, T_b \text{ in }°\text{C} \end{cases}$$

where k = thermal conductivity defined in (c), T_a, T_b = inside and outside temperatures, and R_a, R_b = inside and outside radii.

(3) Heat Conduction through a Multilayer Medium

(a) **Layers in series.** The quantity of heat Q that flows in the normal direction through the multilayer medium of Fig. 6.04–2 (in which each layer is homogeneous and has a different thermal conductivity) in the time interval t is

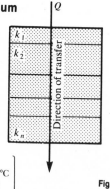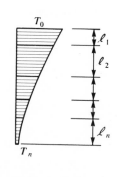

Fig. 6.04–2

$$Q = \frac{A(T_0 - T_n)t}{\dfrac{l_1}{k_1} + \dfrac{l_2}{k_2} + \cdots + \dfrac{l_n}{k_n}} \qquad \begin{cases} Q \text{ in kcal} \\ k \text{ in kcal/m-s-}°\text{C} \\ T_0, T_n \text{ in }°\text{C} \\ A \text{ in m}^2 \\ t \text{ in s} \\ l \text{ in m} \end{cases}$$

where T_0, T_n = temperatures of the near and far faces, respectively, and A, t, l, k have similar meaning to those in Sec. 6.04–2a.

(b) Layers in parallel. The quantity of heat Q that flows in the normal direction through the multistrip medium of Fig. 6.04–3 (in which each strip is homogeneous and has a different thermal conductivity) in the time interval t is

$$Q = \frac{(k_1 A_1 + k_2 A_2 + \cdots + k_n A_n)(T_a - T_b)t}{l}$$

where T_a, T_b = temperatures on the near and far faces, respectively, and A, t, l, k have similar meaning to those of Sec. 6.04–2a.

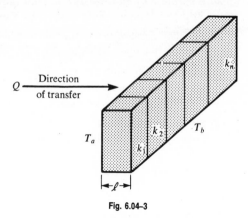

Direction of transfer

Fig. 6.04–3

(c) In multilayer pipes the quantity of heat Q that flows radially through the thickness $(R_n - R_0)$ of a pipe of length l in the time interval t is

$$Q = \frac{2\pi l (T_0 - T_n)t}{\dfrac{1}{k_1}\ln\dfrac{R_1}{R_0} + \dfrac{1}{k_2}\ln\dfrac{R_2}{R_1} + \cdots + \dfrac{1}{k_n}\ln\dfrac{R_n}{R_{n-1}}}$$

where T_0, T_n = inside and outside temperatures, R_0, R_1, ..., R_n = radii of the respective layers, and k_1, k_2, ..., k_n = thermal conductivities of the same layers.

(4) Heat Convection

(a) Newton's law of cooling states that the heat transfer from a stationary or moving fluid of temperature T_f to the surrounding medium of temperature T_m during the interval t is

$$Q = hA(T_f - T_m)t$$

where h = convection coefficient (see below) and A = area of transfer.

(b) Convection coefficient h of a particular fluid and medium of transfer is the time rate of heat convection per unit area and per degree of temperature (Table 6.04–1).

(c) Units of h are

$$\frac{\text{kilocalorie}}{\text{meter}^2 \times \text{second} \times \text{degree Celsius}} = \text{kcal/m}^2\text{-s-}^\circ\text{C}$$

$$\frac{\text{British thermal unit}}{\text{foot}^2 \times \text{hour} \times \text{degree Fahrenheit}} = \text{Btu/ft}^2\text{-h-}^\circ\text{F}$$

$$\frac{\text{watt}}{\text{meter}^2 \times \text{kelvin}} = \text{W/m}^2\text{-K}$$

and their conversion relations are

$$1\ \text{kcal/m}^2\text{-s-}^\circ\text{C} = 738\ \text{Btu/ft}^2\text{-h-}^\circ\text{F} = 4,184\ \text{W/m}^2\text{-K}$$

where $W = 1$ watt (Sec. 6.04–2e).

TABLE 6.04–1

Fluid	h, kcal/m^2-s-$^\circ$C
Static air Air flow	0.001–0.003 0.012–0.020
Static water Water flow	0.120–0.240 0.300–0.500
Static steam Steam flow	2.000–2.500 2.700–3.400

example:

If a flat vertical wall of temperature $T_m = 120^\circ$ and area $A = 100\ \text{m}^2$ is exposed to an air flow at temperature $T_f = 20^\circ$ for 8 hours, then the heat loss is

$$Q = 0.015 \times 100 \times (20 - 120) \times 8 \times 3{,}600 = 4.32 \times 10^6\ \text{kcal}$$

(5) Thermal Radiation

(a) **Ability** of a given substance to emit radiation when heated is proportional to its ability to absorb radiation. Substances which are good radiators are equally good absorbers.

(b) **Kirchhoff's principle** of radiation states that the ratio of rates of radiation of any two surfaces of equal temperature is equal to the ratio of the absorption of these two surfaces.

(c) **Ideal absorber**, which is also an ideal radiator, is the perfectly blackbody. Although such a body does not exist, it is a useful concept used as the standard for comparing the emissivity and absorptance.

(d) **Stefan–Boltzmann law** states that the time rate of radiation R of a perfect radiator is proportional to the fourth power of its absolute temperature.

$$R = \sigma T^4 \qquad \begin{cases} R \text{ in kcal/m}^2\text{-h} \\ \sigma \text{ in kcal/m}^2\text{-h-K}^4 \\ T \text{ in K} \end{cases}$$

where σ = Stefan–Boltzmann constant (see below) and T = absolute temperature in K (in the SI system) and in °R (in the FPS system).

(e) **Stefan–Boltzmann constant** σ is

$$\sigma = \frac{4.88 \times 10^{-8} \text{ kcal}}{\text{m}^2 \times \text{h} \times \text{K}^4} = \frac{5.67 \times 10^{-8} \text{ W}}{\text{m}^2 \times \text{K}^4} = \frac{0.173 \times 10^{-8} \text{ Btu}}{\text{ft}^2 \times \text{h} \times °\text{R}^4}$$

where °R = degree Rankine.

(f) **Quantity of heat** Q transferred by radiation between two parallel equal areas $A_1 = A_2 = A$ in the time t is

$$Q = \bar{e}A\left[\left(\frac{T_1}{100}\right)^4 - \left(\frac{T_2}{100}\right)^4\right]10^8 t \qquad \begin{cases} \bar{e} \text{ in kcal/m}^2\text{-h-K}^4 \\ A \text{ in m}^2 \\ T_1, T_2 \text{ in K} \end{cases}$$

where $\bar{e}$ = radiation constant, T_1, T_2 = absolute temperatures of the areas, and t = time in hours if $\bar{e}$ is expressed in terms of kcal or Btu and in seconds if $\bar{e}$ is expressed in watts (see below).

(g) **Radiation constant** $\bar{e}$ is

$$\bar{e} = \frac{\sigma}{\dfrac{1}{e_1} + \dfrac{1}{e_2} - 1} \qquad \begin{cases} \sigma \text{ in kcal/m}^2\text{-h-K}^4 \\ e_1, e_2 \text{ dimensionless} \end{cases}$$

where e_1, e_2 = emissivities of the respective areas (Table 6.04–2) and σ = Stefan–Boltzmann constant defined in (e).

(h) **Parallel areas of different size** transfer by radiation a quantity of heat

$$Q = \bar{f}A_1\left[\left(\frac{T_1}{100}\right)^4 - \left(\frac{T_2}{100}\right)^4\right]10^8 t \qquad \{Q \text{ in kcal}\}$$

where

$$\bar{f} = \frac{\sigma}{\dfrac{1}{e_1} + \dfrac{A_1}{A_2}\left(\dfrac{1}{e_2} - 1\right)} \qquad \{\bar{f} \text{ dimensionless}\}$$

which for very large A_2 yields $\bar{f} = \sigma e_1$, where σ is given in (e).

TABLE 6.04–2

Material	T, K	e, dimensionless
Aluminum, polished	373–773	0.053
Aluminum, rough polished	373–773	0.082
Aluminum, commercial, new	373–773	0.350
Aluminum, commercial, oxidized	373–773	0.520
Brass, polished, new	373–573	0.100
Brass, polished, oxidized	373–573	0.600
Copper, polished, new	293	0.200
Copper, polished, oxidized	293	0.660
Copper, black	293	0.790
Gold, polished	293	0.015
Iron, polished, new	293	0.050
Iron, polished, oxidized	293	0.250
Iron, rough	293	0.550
Iron, gray	293	0.820
Ice	273	0.615
Water	293	0.666
Lumber	293	0.900
Glass	293	0.922
Brick, red	293	0.930
Perfectly black body	293	1.000

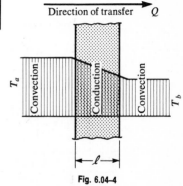

Direction of transfer Q

Fig. 6.04–4

(6) Overall Heat Transfer

(a) **Wall transfer.** The quantity of heat Q transferred in the interval t through three stages depicted in Fig. 6.04–4 is

$$Q = \frac{A(T_a - T_b)t}{\dfrac{1}{h_1} + \dfrac{l}{k} + \dfrac{1}{h_2}}$$

where A = transfer area, T_a, T_b = temperatures of the profile shown in Fig. 6.04–4, h_1, h_2 = convection coefficients (Sec. 6.04–4b) of the near and far side fluids, respectively, and k = thermal conductivity (Sec. 6.04–2c).

(b) **Pipe transfer.** Similarly, the quantity of heat Q transferred in the interval t through the three stages depicted in Fig. 6.04–5 is

$$Q = \frac{2\pi l(T_a - T_b)t}{\dfrac{1}{h_a R_a} + \dfrac{\ln(R_b/R_a)}{k} + \dfrac{1}{h_b R_b}}$$

where l = length of pipe, T_a, T_b = temperatures of the profile shown in Fig. 6.04–5, R_a, R_b = inside and outside radii, and h_a, k, h_b = respective transfer constants of Secs. 6.04–4b and 6.04–2c.

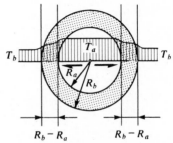

Fig. 6.04–5

6.05 THERMODYNAMICS

(1) Classification and Definitions

(a) **Thermodynamics** is concerned with the transformation of heat energy into mechanical energy, and vice versa. The principles of thermodynamics can be stated in terms of observable (macroscopic) quantities such as pressure, volume, temperature, work, and energy.

(b) **Thermodynamic processes** such as work and transformation of energy are classified as reversible or irreversible (below) and as isochoric, isobaric, isothermic, adiabatic, or polytropic (Sec. 6.05–2).

(c) **Reversible process** meets two criteria:

(α) The process after completion may be induced to occur in an exactly reversed order so that the whole system returns to its initial state.

(β) The energy transformed to a new form can be returned to the initial form (without loss).

If one or both of these criteria are not satisfied, the process is termed *irreversible*.

(d) **Mechanical work** U_m done by a gas expanding from the initial volume V_1 to a new volume V_2 against a variable pressure p is

$$U_m = \int_{V_1}^{V_2} p \, dV \qquad \left\{ \begin{array}{l} U_m \text{ in m-kgf} \\ p \text{ in kgf/m}^2 \\ V \text{ in m}^3 \end{array} \right\}$$

and equals the shaded area of the pV diagram of Fig. 6.05–1a and b.

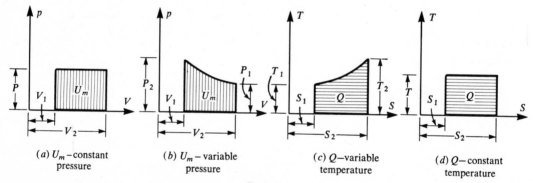

(a) U_m–constant pressure   (b) U_m–variable pressure   (c) Q–variable temperature   (d) Q–constant temperature

Fig. 6.05–1

(e) **Thermal energy stored** in a mass m of gas (of an average specific heat capacity $\bar{c}$) by the change in its temperature from T_1 to T_2 is

$$E_T = Q = \bar{c}m(T_2 - T_1) \qquad \left\{ \begin{array}{l} E_t \text{ in kcal} \\ \bar{c} \text{ in kcal/kg-K} \\ m \text{ in kg} \\ T \text{ in K} \end{array} \right\}$$

where $\bar{c} = \bar{c}_v$ if V = constant, and $\bar{c} = \bar{c}_p$ if p = constant ($\bar{c}_v$, $\bar{c}_p$ are defined in Secs. 6.02–4c and d).

(f) **Enthalpy** H is the sum of E_{Tv} under constant volume V and pV and equals E_{Tp} under constant pressure p.

$$H = \bar{c}_v m(T_2 - T_1) + pV = \bar{c}_p m(T_2 - T_1)$$

where $\bar{c}_v$ and $\bar{c}_p$ = specific heat capacities of Secs. 6.02–4c and d, respectively.

(g) Entropy S is a state function defined by the differential relation $dS = dQ/T$, and if a gas passes from state 1 to state 2 in a reversible process (Fig. 6.05–1c and d), the change in entropy is

$$\Delta S = S_2 - S_1 = \int_1^2 \frac{dQ}{T} \qquad \text{or} \qquad Q = \int_1^2 T\, dS \qquad \left\{ \begin{array}{l} Q \text{ in kcal} \\ S \text{ in kcal/K} \\ T \text{ in K} \end{array} \right\}$$

where Q = quantity of heat and T = absolute temperature.

(2) First Law of Thermodynamics

(a) Total amount of internal energy in a system is generally indeterminate, and only changes in this energy are measurable. These changes are determined by the first law of thermodynamics stated below.

(b) First law of thermodynamics. Work U_m that is performed on a mechanical system by external forces equals the heat Q absorbed by the system less the increase in internal energy ΔU_i.

$$U_m = Q - \Delta U_i \qquad \text{or} \qquad Q = U_m + \Delta U_i$$

where the units of work, heat, and energy are interchangeable according to the following identity:

$$1 \text{ kcal} = 4{,}184 \text{ J} = 427 \text{ m-kgf} = 1/860 \text{ kWh} = 1/633 \text{ hp-h (MKS)} = 3.967 \text{ Btu}$$

(c) Particular cases of processes governed by this principle are introduced in (d) to (g) below. The nomenclature and constants used below are those of Secs. 6.02–3 and 6.02–4.

(d) Isochoric process (V = constant) is one in which the volume V remains constant and the temperature T and pressure p change (Fig. 6.05–2). The governing equations of this process are

$$\frac{p_1}{T_1} = \frac{p_2}{T_2} \qquad V_1 = V_2 = \frac{mRT_1}{Mkgp_1}$$

$$Q = \bar{c}_v m (T_2 - T_1) = \Delta U_i$$

$$U_m = 0 \qquad \Delta S = \bar{c}_v m \ln \frac{T_2}{T_1}$$

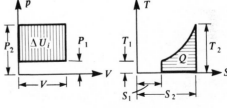

Fig. 6.05–2

example:

A fixed quantity of nitrogen of $m = 10$ kg is under pressure $p_1 = 20{,}000$ kgf/m² at $T_1 = 300$ K in a closed cylinder. If the pressure is increased to $p_2 = 30{,}000$ kgf/m², then the temperature increases to

$$T_2 = \frac{p_2 T_1}{p_1} = \frac{30{,}000 \times 300}{20{,}000} = 450 \text{ K}$$

and the generated heat is

$$Q = \bar{c}_v m (T_2 - T_1) = 0.177 \times 10 \times 150 = 265.5 \text{ kcal}$$

where $\bar{c}_v = 0.177$ kcal/kg-K is given in Appendix A.
The volume of the cylinder is

$$V = \frac{m}{Mkg} \frac{RT_1}{p_1} = \frac{10 \times 848 \times 300}{28 \times 20{,}000} = 4.54 \text{ m}^3$$

where $Mkg = 28$ kg/kmol, also given in Appendix A. Note for nitrogen N_2 that $A_N = 14$ and $Mkg = (2 \times 14)$ kg/kmol.

(e) Isobaric process (p = constant) is one in which the pressure p is constant and the volume V and temperature T change (Fig. 6.05–3). The governing equations of this process are

$$\frac{V_1}{V_2} = \frac{T_1}{T_2} \qquad p_1 = p_2 = \frac{mRT_1}{MkgV_1}$$

$$Q = U_m + \Delta U_i$$

$$Q = \bar{c}_p m (T_2 - T_1)$$

$$U_m = p(V_2 - V_1)$$

$$\Delta U_i = \bar{c}_v m (T_2 - T_1)$$

$$\Delta S = \bar{c}_p m \ln \frac{T_2}{T_1} = \bar{c}_p m \ln \frac{V_2}{V_1}$$

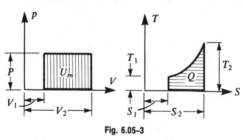

Fig. 6.05-3

(f) Isothermal process (T = constant) is one in which the temperature T is constant and the volume V and pressure p change (Fig. 6.05–4). The governing equations of this process are

$$p_1 V_1 = p_2 V_2$$

$$T_1 = T_2 = \frac{Mkgp_1 V_1}{mR}$$

$$Q = U_m \qquad \Delta U_i = 0$$

$$U_m = p_1 V_1 \ln \frac{p_1}{p_2} = p_1 V_1 \ln \frac{V_2}{V_1}$$

$$= \frac{mRT_1}{Mkg} \ln \frac{p_1}{p_2} = \frac{mRT_1}{Mkg} \ln \frac{V_2}{V_1}$$

$$\Delta S = \frac{mR}{Mkg} \ln \frac{p_1}{p_2} = \frac{mR}{Mkg} \ln \frac{V_2}{V_1}$$

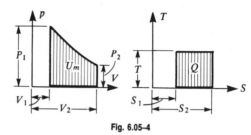

Fig. 6.05-4

(g) Adiabatic process (Q = 0) is one in which no heat is added to or removed from the system and the product of p by V^α is constant (Fig. 6.05–5). The governing equations of this process in terms of $\alpha = \bar{c}_p / \bar{c}_v$ are

$$\frac{p_1}{p_2} = \left(\frac{V_2}{V_1}\right)^\alpha \qquad \frac{V_1}{V_2} = \sqrt[\alpha]{\frac{p_2}{p_1}}$$

$$\frac{T_1}{T_2} = \left(\frac{V_2}{V_1}\right)^{\alpha-1} = \sqrt[\alpha]{\left(\frac{p_1}{p_2}\right)^{\alpha-1}}$$

$$U_m = -\Delta U_i = \bar{c}_v m (T_1 - T_2)$$

$$= \frac{mR}{(\alpha - 1)Mkg} (T_1 - T_2)$$

$$= \frac{p_1 V_1 - p_2 V_2}{(\alpha - 1)}$$

$$Q = 0 \qquad \Delta S = 0$$

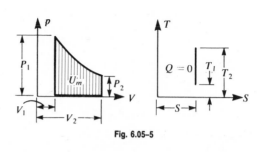

Fig. 6.05-5

example:

A mass $m = 10\,\text{kg}$ of nitrogen has a volume $V_1 = 4.54\,\text{m}^3$ at $p_1 = 20{,}000\,\text{kgf/m}^2$ and $T_1 = 300\,\text{K}$. If it expands adiabatically to $V_2 = 9.08\,\text{m}^3$, the new pressure is

$$p_2 = p_1 \left(\frac{V_1}{V_2}\right)^\alpha = 20{,}000 \left(\frac{4.54}{9.08}\right)^{1.407} = 7{,}500\,\text{kgf/m}^2$$

where $\alpha = \bar{c}_p/\bar{c}_v = \dfrac{0.249}{0.177} = 1.407$ and the new temperature is

$$T_2 = \frac{p_2 V_2 T_1}{p_1 V_1} = \frac{7{,}500 \times 9.08 \times 300}{20{,}000 \times 4.54} = 225\,\text{K} = -48°\text{C}$$

The mechanical work

$$U_m = \bar{c}_v m (T_1 - T_2) = 0.177 \times 10 \times (300 - 225) = 133\,\text{kcal} = -\Delta U_i$$

and $Q = U_m + \Delta U_i = 0$ since no heat was added to the system.

(h) Polytropic process is the general case of a thermodynamic process. The governing equations of this process in terms of

$$\alpha = \frac{\bar{c}_p}{\bar{c}_v} \qquad \beta = \frac{\bar{c} - \bar{c}_p}{\bar{c} - \bar{c}_v} = \frac{\log p_1 - \log p_2}{\log V_2 - \log V_1} \qquad \text{where } \bar{c}_p - \bar{c}_v = \frac{R}{Mkg}$$

are

$$\frac{p_1}{p_2} = \left(\frac{V_2}{V_1}\right)^\beta = \left(\frac{T_1}{T_2}\right)^{\beta/(\beta-1)} \qquad \frac{V_1}{V_2} = \left(\frac{p_2}{p_1}\right)^{1/\beta} = \left(\frac{T_2}{T_1}\right)^{1/(\beta-1)} \qquad \frac{T_1}{T_2} = \left(\frac{p_1}{p_2}\right)^{(\beta-1)/\beta} = \left(\frac{V_2}{V_1}\right)^{\beta-1}$$

$$Q = \frac{\beta - \alpha}{1 - \alpha}\, U_m = \frac{\beta - \alpha}{\beta - 1}\,\Delta U_i$$

$$= \bar{c} m (T_2 - T_1) = \frac{\beta - \alpha}{\beta - 1}\, \bar{c}_v m (T_2 - T_1) = \frac{\beta - \alpha}{(\beta - 1)\alpha}\, \bar{c}_p m (T_2 - T_1)$$

$$= \frac{\alpha - \beta}{(1 - \alpha)(1 - \beta)}\,(p_2 V_2 - p_1 V_1) = \frac{(\beta - \alpha)mR}{(1 - \alpha)(1 - \beta)Mkg}\,(T_2 - T_1)$$

$$U_m = \frac{1 - \alpha}{\beta - \alpha}\, Q = \frac{1 - \alpha}{\beta - 1}\,\Delta U_i$$

$$= \frac{1 - \alpha}{\beta - \alpha}\, \bar{c} m (T_2 - T_1) = \frac{1 - \alpha}{\beta - 1}\, \bar{c}_v m (T_2 - T_1) = \frac{1 - \alpha}{(\beta - 1)\alpha}\, \bar{c}_p m (T_2 - T_1)$$

$$= \frac{1}{\beta - 1}\,(p_2 V_2 - p_1 V_1) = \frac{mR}{(1 - \beta)Mkg}\,(T_2 - T_1)$$

$$\Delta U_i = \frac{\beta - 1}{\beta - \alpha}\, Q = \frac{\beta - 1}{1 - \alpha}\, U_m$$

$$= \frac{\beta - 1}{\beta - \alpha}\, \bar{c} m (T_2 - T_1) = \bar{c}_v m (T_2 - T_1) = \frac{1}{\alpha}\, \bar{c}_p m (T_2 - T_1)$$

$$= \frac{1}{1 - \alpha}\,(p_2 V_2 - p_1 V_1) = \frac{mR}{(1 - \alpha)Mkg}\,(T_2 - T_1)$$

$$\Delta S = m \left[\bar{c}_v \ln \frac{T_2}{T_1} + (\bar{c}_p - \bar{c}_v) \ln \frac{V_2}{V_1}\right]$$

$$= m \left[\bar{c}_p \ln \frac{T_2}{T_1} - (\bar{c}_p - \bar{c}_v) \ln \frac{p_2}{p_1}\right]$$

$$= m \left[\bar{c}_v \ln \frac{p_2}{p_1} + \bar{c}_p \ln \frac{V_2}{V_1}\right]$$

(3) Second Law of Thermodynamics

(a) Ideal cycle is a closed reversible process, in which there is no dissipation of energy, and the total potential and kinetic energy of which remains constant.

(b) Perfectly reversible system, such as that described above, is of course not possible since the conversion of heat into mechanical work always produces energy losses and the total energy of the system cannot remain constant.

(c) Verbal formulation of the second law of thermodynamics is based on this experience and occurs in two major forms:

    (α) *Clausius' formulation.* Heat cannot flow from a body of some temperature to a body of another temperature without simultaneously producing some other effects in its surrounding.

    (β) *Kelvin–Planck formulation.* It is impossible to construct an engine which, when operating in a cycle, will produce no other effects than the extraction of heat from a reservoir and the performance of an equivalent amount of work.

(d) Mathematical formulation of the second law of thermodynamics is a statement of inequality; i.e., there is always some loss of energy which is not recoverable by the system.

(e) Heat engine is a thermodynamic device which converts heat energy into work, and its operation is described in three steps (Fig. 6.05–6):

    (α) A quantity of heat Q_1 is supplied to the engine from a reservoir at a high temperature T_1 (heat input).

    (β) A mechanical work U_m is performed by the engine by the use of a portion and only a portion of the heat input.

    (γ) A remaining portion of heat Q_2 is released to a reservoir at a lower temperature T_2.

(f) Efficiency of a heat engine is the ratio of the work output to the heat input.

$$\text{Eff.} = \frac{\text{work output}}{\text{heat input}} = \frac{Q_1 - Q_2}{Q_1} = \frac{T_1 - T_2}{T_1}$$

where Q_1 = supplied heat, Q_2 = rejected heat, T_1 = temperature of Q_1, and T_2 = temperature of Q_2.

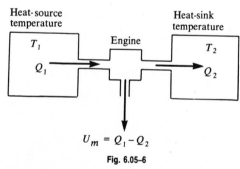

Fig. 6.05–6

(4) Carnot Cycle

(a) Classical ideal cycle is the four-phase cycle of the Carnot engine of Fig. 6.05–7, which consists of a cylinder filled with gas compressed by a movable piston.

(b) Phase 1-2 (Fig. 6.05–7a). First, the cylinder is placed in contact with a heat source of temperature T_1. The inflow of heat Q_1 causes the isothermal expansion of gas from volume V_1 and pressure p_1 to volume V_2 and pressure p_2, which produces external work $U_{m,1\text{-}2}$ (curve 1-2, Fig. 6.05–8).

$$U_{m,1\text{-}2} = Q_1 = \frac{m}{Mkg}\, RT_1 \ln \frac{V_2}{V_1} \qquad V_2 > V_1, \quad p_2 < p_1, \quad T_{1,1} = T_{1,2} = T_1$$

which is the process of Sec. 6.05–2f.

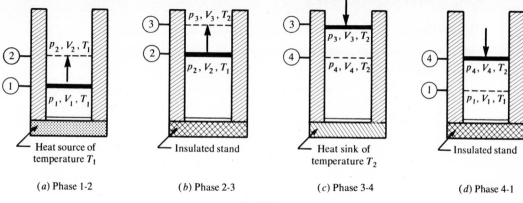

(a) Phase 1-2   (b) Phase 2-3   (c) Phase 3-4   (d) Phase 4-1

Fig. 6.05–7

(c) Phase 2-3 (Fig. 6.05–7b). Then, the cylinder is placed on an insulated stand, where the gas continues to expand adiabatically to volume V_3 as the pressure drops to the lowest level p_3 and the temperature drops from T_1 to T_2, producing additional external work $U_{m,2\text{-}3}$ (curve 2-3, Fig. 6.05–8).

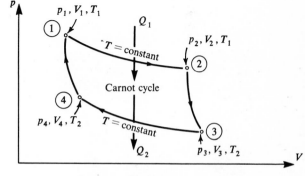

Fig. 6.05–8

$$U_{m,2\text{-}3} = m\bar{c}_v(T_2 - T_1)$$

$$V_3 > V_2, \quad p_3 < p_2, \quad T_2 < T_1$$

which is the process of Sec. 6.05–2g.

(d) Phase 3-4 (Fig. 6.05–7c). Next, the cylinder is removed from the insulated stand and placed on a heat sink of temperature T_2. The outflow of heat Q_2 causes the isothermal compression of gas from volume V_3 and pressure p_3 to volume V_4 and pressure p_4 and produces external work $U_{m,3\text{-}4}$ (curve 3-4, Fig. 6.05–8).

$$U_{m,3\text{-}4} = Q_2 = \frac{m}{Mkg}\,RT_2\ln\frac{V_4}{V_3} \qquad V_4 < V_3, \quad p_4 > p_3, \quad T_{3,2} = T_{4,2} = T_2$$

which is the process of Sec. 6.05–2f.

(e) Phase 4-1 (Fig. 6.05–7d). Finally, the cylinder is again placed on the insulated stand, where the gas continues to contract adiabatically to the initial volume V_1 as the pressure increases to the initial pressure p_1 and the temperature rises to the initial temperature T_1, producing external work $U_{m,4\text{-}1}$ (Fig. 6.05–8).

$$U_{m,4\text{-}1} = m\bar{c}_v(T_1 - T_2) \qquad V_1 < V_4, \quad p_1 > p_4, \quad T_1 > T_2$$

which is the process of Sec. 6.05–2g.

(f) Total work of one cycle is then

$$U_m = U_{m,1\text{-}2} + U_{m,2\text{-}3} + U_{m,3\text{-}4} + U_{m,4\text{-}1} = \frac{mR}{Mkg}(T_1 - T_2)\ln\frac{V_2}{V_1}$$

which is essentially the difference of two isothermic works (processes of Sec. 6.05–2f).

6.06 THERMAL MACHINES

(1) Otto Cycle

(a) Internal-combustion engine is the most widely used gasoline engine. The performance of this engine can be defined in the following four steps called *strokes*:

(1-2) *Intake stroke* (Fig. 6.06–1a), with the piston moving downward to increase the volume and with the intake valve open to admit the gas-air mixture.

(2-3) *Compression stroke* (Fig. 6.06–1b), with the intake and exhaust valve closed and the piston moving upward, causing a rise of pressure.

(3-4) *Power stroke* (Fig. 6.06–1c) takes place as the piston reaches the top position at which the mixture is ignited, causing a sharp increase in pressure and temperature when the expanded mixture forces the piston downward thus producing external work.

(4-1) *Exhaust stroke* (Fig. 6.06–1d) pushes the burned mixture out of the cylinder through the exhaust valve.

This cycle repeats as long as the supply of fuel continues.

Intake valve

Exhaust valve

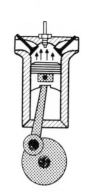

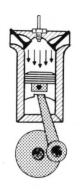

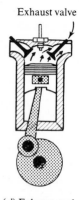

(*a*) Intake stroke　　　(*b*) Compression stroke　　　(*c*) Power stroke　　　(*d*) Exhaust stroke

Fig. 6.06–1

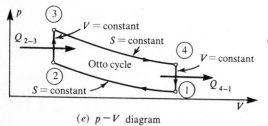

(*e*) p–V diagram

(b) Otto cycle is the ideal representation of the real cycle of the gasoline engine (Fig. 6.06–1e). The phases of this cycle are

(1-2) Adiabatic compression: $\qquad Q_{1\text{-}2} = 0$

(2-3) Constant-volume heat addition: $\quad Q_{2\text{-}3} = m\bar{c}_v(T_3 - T_2)$

(3-4) Adiabatic expansion: $\qquad Q_{3\text{-}4} = 0$

(4-1) Constant-volume heat rejection: $\quad Q_{4\text{-}1} = m\bar{c}_v(T_4 - T_1)$

where T_1, T_2, T_3, T_4 = absolute temperature.

(c) Efficiency of the Otto cycle is

$$\text{Eff.} = \frac{\text{net work}}{\text{heat addition}} = \frac{Q_{2\text{-}3} - Q_{4\text{-}1}}{Q_{2\text{-}3}} = 1 - \frac{T_1}{T_2} = 1 - \left(\frac{V_2}{V_1}\right)^{\alpha-1}$$

where V_1, V_2 = maximum and minimum volumes and $\alpha = \bar{c}_p/\bar{c}_v$ (Secs. 6.02–4c and d).

(2) Diesel Cycle

(a) **Diesel engine** first compresses air to a high temperature, then injects fuel, causing spontaneous combustion which pushes the piston downward.

(b) **Diesel cycle** is the ideal representation of the real cycle of the diesel engine (Fig. 6.06–2), and it consists also of four phases:

(1-2) Adiabatic compression: $Q_{1\text{-}2} = 0.$
(2-3) Constant-pressure heat addition: $Q_{2\text{-}3} = m\bar{c}_p(T_3 - T_2)$
(3-4) Adiabatic expansion: $Q_{3\text{-}4} = 0$
(4-1) Constant-volume heat rejection: $Q_{4\text{-}1} = m\bar{c}_v(T_4 - T_1)$

where the symbols are the same as in Sec. 6.06–1.

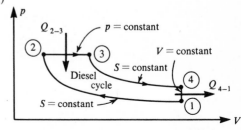

(c) **Efficiency of the diesel cycle** is

$$\text{Eff.} = 1 - \frac{\bar{c}_v\,(T_4 - T_1)}{\bar{c}_p\,(T_3 - T_2)} = 1 - \frac{T_4 - T_1}{\alpha\,(T_3 - T_2)}$$

where $\alpha = \bar{c}_p/\bar{c}_v$.

Fig. 6.06–2

(3) Brayton Cycle

(a) **Turbojet engine** and its variations have vastly affected military and civilian aircraft transportation; their thermodynamic processes can be idealized by the Brayton cycle.

(b) **Brayton cycle** also consists of four phases defined below (Fig. 6.06–3):

(1-2) Adiabatic compression: $Q_{1\text{-}2} = 0$
(2-3) Constant-pressure heat addition: $Q_{2\text{-}3} = m\bar{c}_p(T_3 - T_2)$
(3-4) Adiabatic expansion: $Q_{3\text{-}4} = 0$
(4-1) Constant-pressure heat rejection: $Q_{4\text{-}1} = m\bar{c}_p(T_4 - T_1)$

where the symbols are the same as in Sec. 6.06–1.

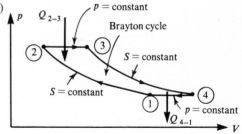

(c) **Efficiency of the Brayton cycle** is

$$\text{Eff.} = 1 - \frac{T_4 - T_1}{T_3 - T_2} = 1 - \left(\frac{p_1}{p_2}\right)^{\alpha/(1-\alpha)}$$

where $p_1, p_2 = $ pressures and $\alpha = \bar{c}_p/\bar{c}_v$.

Fig. 6.06–3

(4) Rankine Cycle

(a) **Steam engine** is the oldest thermal machine, the operation of which is represented by the diagram of Fig. 6.06–4.

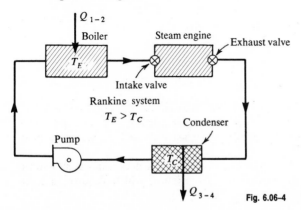

Fig. 6.06–4

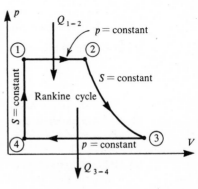

Fig. 6.06–5

(b) Rankine cycle is the ideal representation of the real cycle of the steam engine (Fig. 6.06–5), and it consists of four phases defined below:

(1-2) Constant-pressure heat addition: $Q_{1\text{-}2} = m\bar{c}_p T_E$
(2-3) Isothermal expansion: $Q_{2\text{-}3} = 0$
(3-4) Constant-pressure heat rejection: $Q_{3\text{-}4} = m\bar{c}_p T_C$
(4-1) Isothermal compression: $Q_{4\text{-}1} = 0$

(c) Efficiency of the Rankine cycle is

$$\text{Eff.} = \frac{T_E - T_C}{T_C}$$

where T_E = temperature of evaporation and T_C = temperature of condensation.

(5) Reversed Rankine Cycle

(a) Refrigerator acts like a reversed heat engine. A schematic diagram of a refrigerator is shown in Fig. 6.06–6.

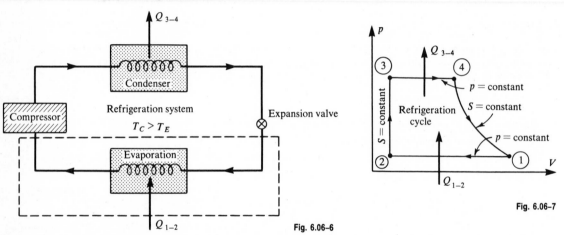

Fig. 6.06–7

Fig. 6.06–6

(b) Reversed Rankine cycle is the ideal representation of the real cycle of the refrigerator (Fig. 6.06–7), and it consists of four phases defined below:

(1-2) Isothermal heat addition: $Q_{1\text{-}2} = m\bar{c}_p T_E$
(2-3) Adiabatic compression: $Q_{2\text{-}3} = 0$
(3-4) Isothermal heat rejection: $Q_{3\text{-}4} = m\bar{c}_p T_C$
(4-1) Adiabatic expansion: $Q_{4\text{-}1} = 0$

(c) Efficiency of the reversed Rankine cycle is

$$\text{Eff.} = \frac{T_E}{T_C - T_E}$$

where T_E = temperature of evaporation and T_C = temperature of condensation.

(d) MKS refrigeration ton is the amount of heat required to freeze 1,000 kg of water at 0°C in 24 hours. 1 MKS refr. ton = (80 kcal/kg) × 1,000 kg = 80,000 kcal, which requires a heat outflow of 55.55 kcal/min for 24 hours.

(e) FPS refrigeration ton is the amount of heat required to freeze 2,000 lb of water at 32°F in 24 hours. 1 FPS refr. ton = (144 Btu/lb) × 2,000 lb = 288,000 Btu, which requires a heat outflow of 200 Btu/min for 24 hours.

7

ELECTROSTATICS AND ELECTRIC CURRENT

7.01 ELECTRIC FORCE

(1) Electrons, Protons, and Neutrons

(a) Molecules of matter are composed of atoms, which in turn are composed primarily of electrons surrounding a central core called the *nucleus*.

(b) Nucleus of an atom consists of a number of protons (p^+) and (except for hydrogen) one or more neutrons (n^0).

(c) Electron (e^-) is a negatively charged particle of mass

$$m_e = 9.10953 \times 10^{-31} \, \text{kg}$$

(d) Proton is a positively charged particle of mass

$$m_p = 1.67265 \times 10^{-27} \, \text{kg}$$

(e) Neutron is a particle of zero charge and of mass

$$m_n = 1.67482 \times 10^{-27} \, \text{kg}$$

(f) Atom of matter under normal conditions contains equal numbers of electrons and protons and is consequently in a neutral state.

example:

The neon atom, for example, consists of 10 electrons, 10 protons, and 10 neutrons, as shown in Fig. 7.01–1.

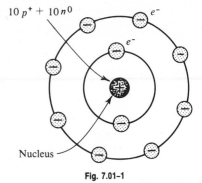

$10 \, p^+ + 10 \, n^0$

e^-

e^-

Nucleus —

Fig. 7.01–1

(2) Electric Charge

(a) Positive ion is an atom from which one or more electrons have been removed and thus is positively charged.

(b) Negative ion is an atom to which one or more electrons have been added and thus is negatively charged.

(c) Ionization (electrification) is the process of adding or subtracting electrons to or from the atoms of matter.   This can be accomplished by physical or chemical processes.

example:

If a hard-rubber rod is rubbed with a cat's fur, the electrons are transferred from the fur to the rod, by which the rod becomes negatively charged, leaving the fur positively charged.

(d) Electrostatic charge Q is defined as the excess number of electrons (negative electric charge) or of protons (positive electric charge) in a body.

(e) Electronic unit charge e is the charge of one proton.   The charge of one electron is then $-e$.

$$1e = 1.6022 \times 10^{-19} \, \text{C}$$

where C = coulomb, defined below.

(f) SI unit of electrostatic charge is 1 coulomb (after Charles Augustin de Coulomb), designated by C, and defined as the electrostatic charge which when placed at a distance of 1 meter from an equal charge of the same sign produces a repulsive force of 9×10^9 newtons (see also Secs. 7.03–1c and d).

$$1 \text{ coulomb} = 1 \text{ C} = 3 \times 10^4 (\sqrt{10 \text{ N}}) \text{m}$$

$$1 \text{ statcoulomb} = 1 \text{ sC} = (\sqrt{\text{dyne}}) \text{cm}$$

where N = newton (Sec. 3.02–4a).

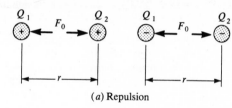

(g) Fractions of 1 C are

$$1 \text{ microcoulomb} = 1 \, \mu\text{C} = 10^{-6} \text{ C}$$
$$1 \text{ nanocoulomb} = 1 \text{ nC} = 10^{-9} \text{ C}$$
$$1 \text{ picocoulomb} = 1 \text{ pC} = 10^{-12} \text{ C}$$

(a) Repulsion

(h) Conversion relations are

$$1 \text{ C} = 6.24 \times 10^{18} e = 3 \times 10^9 \text{ sC}$$
$$1 \text{ sC} = 2.08 \times 10^9 e = 1/3 \times 10^{-9} \text{ C}$$

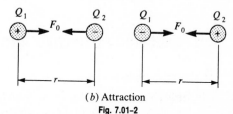

(b) Attraction

Fig. 7.01-2

(3) Coulomb's Law

(a) Electric force F_0 between two point charges Q_1 and Q_2 in a vacuum is directly proportional to the product of these two charges and is inversely proportional to the square of the distance r between them.

$$F_0 = k_0 \frac{Q_1 Q_2}{r^2} \qquad \left\{ \begin{array}{l} F \text{ in N} \\ Q \text{ in C} \\ r \text{ in m} \end{array} \right\}$$

where F_0 = repulsion or attraction according to whether Q_1 and Q_2 have equal or opposite signs (Fig. 7.01–2) and k_0 = proportionality constant, defined below.

(b) Proportionality constant k_0 is

$$k_0 = \frac{1}{4\pi\epsilon_0} = 8.98755 \times 10^9 \text{ N-m}^2/\text{C}^2 = 8.98755 \times 10^9 \text{ kg-m}^3/\text{s}^2\text{-C}^2$$

where $\epsilon_0 = 8.8542 \times 10^{-12}$ C²/N-m² is called the *permittivity of free space* (of vacuum).

(c) Permittivity of medium. When the medium between Q_1 and Q_2 is not a vacuum, the electric force in (a) becomes

$$F = k \frac{Q_1 Q_2}{r^2} \qquad \{k \text{ in N-m}^2/\text{C}^2\}$$

where

$$k = \frac{1}{4\pi\epsilon} = \frac{1}{4\pi c \epsilon_0} = \frac{k_0}{c}$$

and c = dimensionless dielectric constant, specific for the given medium (for dry air, $c \cong 1$, $\epsilon \cong \epsilon_0$; for other materials, see Appendix A).

example:

If two electrons of charges $Q_1 = Q_2 = 1.6 \times 10^{-19}$ C are placed 2 cm apart, their repulsive force in a medium of c = 1.8 is

$$F = \frac{8.99 \times 10^9 \times (1.6 \times 10^{-19})^2}{1.8 \times (2 \times 10^{-2})^2} = 3.2 \times 10^{-25} \text{ N} = 3.2 \times 10^{-25} \text{ kg-m/s}^2$$

(d) Charge Q required in each suspended pith ball of Fig. 7.01–3 is

$$Q = \sqrt{\frac{mgr^3}{2hk}} \qquad \begin{cases} m \text{ in kg} \\ g \text{ in m/s}^2 \\ r, h \text{ in m} \\ k \text{ in kg-m}^3/\text{s}^2 - \text{C}^2 \end{cases}$$

where m = mass of one ball, g = 9.81 m/s², r = distance of their mass centers, h = elevation, and k = constant of proportionality defined in (c).

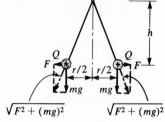

Fig. 7.01–3

7.02 ELECTRIC POTENTIAL AND CAPACITANCE

(1) Electric Field

(a) Definition. An electric field is said to exist in a region of space in which an electric charge produces an electric force (Fig. 7.02–1).

(b) Intensity E of an electric field at a given point is the force exerted on a unit electric charge at that point and is a vector quantity (Fig. 7.02–2).

$$E = \frac{kQ}{r^2} \qquad \begin{cases} E \text{ in N/C} \\ k \text{ in N-m}^2/\text{C}^2 \\ Q \text{ in C} \\ r \text{ in m} \end{cases}$$

where Q = electric charge and k and r are defined in Sec. 7.01–3.

(c) SI unit of intensity of electric field is

$$1 \text{ newton/coulomb} = 1 \text{ N/C}$$

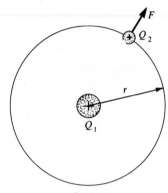

Fig. 7.02–1

(d) Direction of the intensity E at a given point of the electric field is the direction in which a positive charge would move if it were placed at that point.

(e) Electric force F acting on an electric charge Q' placed in an electric field at a point of intensity E is

$$F = EQ' \qquad \{Q' \text{ in C}\}$$

example:

In an electric field generated by an electric charge $Q = -1.2 \times 10^{-7}$ C, the intensity at the distance $r = 0$ m (Fig. 7.02–3) is

$$E = \frac{k_0 Q}{cr^2} = \frac{-8.99 \times 10^9 \times 1.2 \times 10^{-7}}{1.2 \times 10^2} = -8.99 \text{ N/C}$$

where $c = 1.2$ = dielectric constant of the given medium (Sec. 7.01–3c) and k_0 = proportionality constant of vacuum (Sec. 7.01–3b).

If a charge $Q' = -0.3 \times 10^{-3}$ C is placed at $r = 10$ m, the repulsive force becomes

$$F = 8.99 \times 0.3 \times 10^{-3} = 0.270 \times 10^{-3} \text{ N}$$

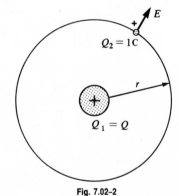

Fig. 7.02–2

(f) Electric intensity E_i at the point i of an electric field generated by a number of electric charges $Q_1, Q_2, \ldots, Q_n$ is the vector sum of their intensities at the point i.

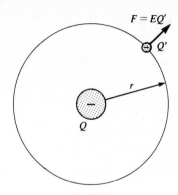

Fig. 7.02–3

example:

If two electric charges $Q_1 = 1 \times 10^{-6}\,\mathrm{C}$ and $Q_2 = -2 \times 10^{-6}\,\mathrm{C}$ are located in a vacuum in the XY plane as shown in Fig. 7.02–4a, the intensity of this field at i is given by

$$E_{i1} = k_0 \frac{Q_1}{r_{i1}^2} = 8.99 \times 10^9 \frac{1 \times 10^{-6}}{5^2 + 3^2} = +2.6 \times 10^2\,\mathrm{N/C}$$

$$E_{i2} = k_0 \frac{Q_2}{r_{i2}^2} = 8.99 \times 10^9 \frac{-2 \times 10^{-6}}{13^2 + 5^2} = -0.9 \times 10^2\,\mathrm{N/C}$$

where $(+)$ indicates repulsion and $(-)$ indicates attraction.

The components of E_i are computed from the geometry of Fig. 7.02–4a as

$$E_{ix} = (-2.6 \times 10^2)\frac{5}{\sqrt{34}} + (0.9 \times 10^2)\frac{13}{\sqrt{194}} = -1.39 \times 10^2\,\mathrm{N/C}$$

$$E_{iy} = (-2.6 \times 10^2)\frac{3}{\sqrt{34}} + (-0.9 \times 10^2)\frac{5}{\sqrt{194}} = -1.66 \times 10^2\,\mathrm{N/C}$$

The magnitude is

$$E_i = \sqrt{(-1.39 \times 10^2)^2 + (-1.66 \times 10^2)^2} = 2.16 \times 10^2\,\mathrm{N/C}$$

and the direction cosines are

$$\cos \alpha = -\frac{1.39}{2.16} = -0.64 \qquad \cos \beta = -\frac{1.66}{2.16} = -0.77$$

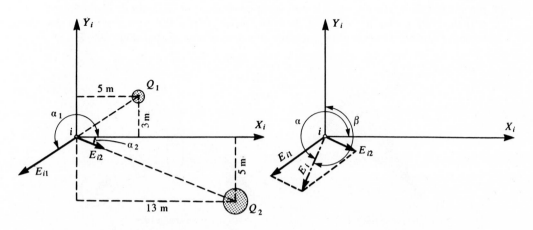

(a) Component intensities E_{i1}, E_{i2} (b) Resultant intensity E_i

Fig. 7.02–4

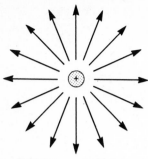

(a) Field lines, one positive charge

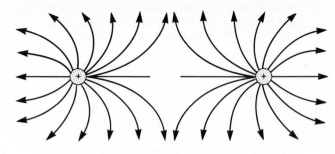

(c) Field lines, two equal charges of like signs

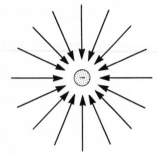

(b) Field lines, one negative charge

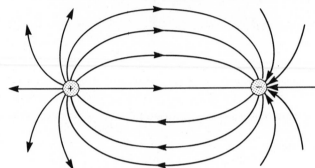

(d) Field lines, two equal charges of unlike signs

Fig. 7.02–5

(2) Electric Field Lines, Tubes, and Flux

- (a) **Electric field line** is an imaginary line in the electric field whose tangent at each point is the direction of the intensity of the field at that point.
- (b) **Sense** of the electric field line is such that it points from the positive charge in the direction of the negative charge (Fig. 7.02–5).
- (c) **Electric tube** is a region of the electric field bounded by electric field lines (Fig. 7.02–6).
- (d) **Electric flux** Φ through an electric tube is the product of the magnitude of the intensity E and the cross-sectional area A of the tube, provided that the area is small so that the intensity can be assumed to be constant and normal to the area (Fig. 7.02–7).

$$\Phi = EA = \frac{QA}{4\pi\epsilon r^2} \qquad \{\Phi \text{ in N-m}^2/\text{C}\}$$

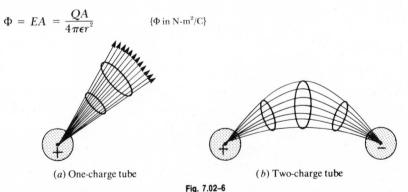

(a) One-charge tube (b) Two-charge tube

Fig. 7.02–6

where Q, ϵ, r are introduced in Sec. 7.01–3.

(e) Gauss' law states that the total electric flux Φ passing through a closed surface equals the product of $1/\epsilon$ by the total charge Q inside the closed surface.

$$\Phi = \frac{Q}{\epsilon} = \frac{Q}{c\epsilon_0}$$

where c, ϵ_0 are those of Sec: 7.01–3.

example:

Electric flux of a sphere is

$$\Phi = \frac{QA}{4\pi\epsilon r^2} = \frac{Q4\pi r^2}{4\pi r^2\epsilon} = \frac{Q}{\epsilon}$$

where the surface of the sphere $A = 4\pi r^2$ and Φ is independent of r.

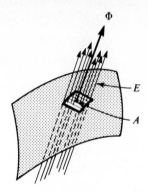

Fig. 7.02-7

(3) Electric Potential

(a) Definition. The electric potential V of an electric field at a point is the work required to move a unit positive charge from infinity to that point against the electric charge of that field.

$$V = \frac{kQ}{r} \qquad \{V \text{ in N-m/C}\}$$

where Q = electric charge generating the field, r = distance of the point of V to the location of Q, and k = proportionality constant of Sec. 7.01–3c.

(b) Difference ΔV of electric potential between two points 1 and 2 of a given electric field is the work required to move a unit positive charge from point 1 to point 2 against the electric forces of that field.

$$\Delta V = V_1 - V_2 = kQ\left(\frac{1}{r_1} - \frac{1}{r_2}\right) = E_1 r_1 - E_2 r_2 \qquad \left\{\begin{matrix} E \text{ in N/C} \\ r \text{ in m} \end{matrix}\right\}$$

where r_1, r_2 = distance of points 1, 2 from the location of Q, respectively, and E_1, E_2 = intensities of the field at 1, 2, respectively.

(c) SI unit of electric potential is 1 volt (after Alessandro Volta), designated by V and defined as

$$1 \text{ volt} = 1 \text{ V} = 1\frac{\text{joule}}{\text{coulomb}} = 1 \text{ J/C} = 1\frac{\text{newton} \times \text{meter}}{\text{coulomb}} = 1 \text{ N-m/C}$$

and

$$1\frac{\text{volt}}{\text{meter}} = 1 \text{ V/m} = 1\frac{\text{newton}}{\text{coulomb}} = 1 \text{ N/C}$$

(d) Equipotential surface connecting all points of the field of equal potential is always normal to the electric field lines.

(e) Potential energy of the charge Q' with respect to the charge Q placed at distance r is

$$E_p = -\frac{kQQ'}{r} = -Fr \qquad \{E_p \text{ in N-m}\}$$

which can be interpreted as the product of the force F (attraction or repulsion) of the charges by their distance r.

(f) Electronvolt (eV) is the unit of energy defined as the work of a charge e of one electron moved through a difference of 1 volt.

$$1 \text{ eV} = 1.6022 \times 10^{-19} \text{ C} \times \text{V} = 1.6022 \times 10^{-19} \text{ J} \quad \text{or} \quad 1 \text{ J} = 6.2414 \times 10^{18} \text{ eV}$$

where e is defined in Sec. 7.01–2e.

(4) Electric Capacitance

(a) Conductor is a medium through which an electric charge may be easily transferred (conducted). Most metals are good conductors.

(b) Insulator is a medium which resists the transfer of an electric charge. Rubber, porcelain, phenol-formaldehyde resins, and air are good insulators.

(c) Capacitor (condenser) consists of two closely spaced conductors of equal and opposite charges separated by an insulator. Its function, called *capacitance*, is the storage of electric charge.

(d) Capacitance C is the ratio of the magnitude of the electric charge Q of either conductor to the potential difference ΔV between the conductors.

$$C = \frac{Q}{\Delta V} \qquad \left\{ \begin{array}{l} Q \text{ in C} \\ \Delta V \text{ in V} \end{array} \right\}$$

(e) SI unit of capacitance is 1 farad (after Michael Faraday), designated by the symbol F, and defined as the capacitance of a capacitor between whose plates there appears a difference of potential of 1 volt when it is charged by an electric charge of 1 coulomb.

$$1 \text{ farad} = 1 \text{ F} = 1 \frac{\text{coulomb}}{\text{volt}} = 1 \text{ C/V}$$

and

$$1 \text{ F/m} = 1 \text{ C/V-m} = 1 \text{ C}^2/\text{N-m}^2$$

(f) Fractions and multiples of 1 F are

$$1 \text{ kilofarad} = 1 \text{ kF} = 10^3 \text{ F}$$
$$1 \text{ microfarad} = 1 \ \mu\text{F} = 10^{-6} \text{ F}$$
$$1 \text{ nanofarad} = 1 \text{ nF} = 10^{-9} \text{ F}$$
$$1 \text{ picofarad} = 1 \text{ pF} = 10^{-12} \text{ F}$$

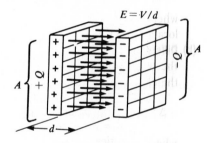

Fig. 7.02–8

(5) Parallel-Plate Capacitors

(a) Capacitance. If two parallel plates, each of area A and separated by the distance d, are charged with equal and opposite charges $\pm Q$, then the system is called a *parallel-plate capacitor* (Fig. 7.02–8), and its capacitance is

$$C = \frac{\epsilon A}{d} \qquad \left\{ \begin{array}{l} C \text{ in F} \\ A \text{ in m}^2 \\ d \text{ in m} \end{array} \right\}$$

where ϵ = permittivity of the insulator defined in Sec. 7.01–3c (medium between plates).

(b) Dielectric constant c (Sec. 7.01–3c) for a particular medium is defined as the ratio of the electric capacitance C of the medium between two plates (Fig. 7.02–8) to the capacitance C_0 in a vacuum.

$$c = \frac{C}{C_0} = \frac{\epsilon}{\epsilon_0} \qquad \left\{ \begin{array}{l} c \text{ dimensionless} \\ C \text{ in F} \\ \epsilon \text{ in F/m or in C}^2/\text{N-m}^2 \end{array} \right\}$$

where $\epsilon_0 = 8.8542 \times 10^{-12}$ F/m.

(c) **Density** σ of the charge Q is defined as the intensity of the charge per unit area:

$$\sigma = \epsilon \frac{V}{d} = \frac{Q}{A} \qquad \{\sigma \text{ in C/m}^2\}$$

and is related to the field intensity E as

$$\sigma = \epsilon E$$

from which

$$E = \frac{Q}{\epsilon A} \qquad \{E \text{ in N/C}\}$$

where N = newton.

(d) **Stress between plates** is

$$s = \frac{1}{2}\frac{\sigma^2}{\epsilon} = \frac{\epsilon V^2}{2d^2} = \frac{1}{2\epsilon}\left(\frac{Q}{A}\right)^2 \qquad \{s \text{ in N/m}^2\}$$

and the total force of attraction is $F = Q^2/2\epsilon A$.

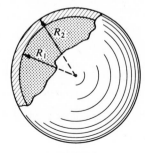

Fig. 7.02-9

(6) Special Capacitors

(a) **Spherical capacitor** of Fig. 7.02–9 has a capacitance

$$C = \frac{4\pi\epsilon R_1 R_2}{R_2 - R_1} \cong \frac{cR_1 R_2}{(9 \times 10^9)(R_2 - R_1)}$$

where c = dielectric constant (Appendix A).

(b) **Cylindrical capacitor** of Fig. 7.02–10 has a capacitance

$$C = \frac{4\pi\epsilon l}{2\ln(R_2/R_1)} \cong \frac{cl}{(18 \times 10^9)\ln(R_2/R_1)}$$

(c) **Multiplate capacitor** of Fig. 7.02–11, consisting of n parallel plates of the same area A spaced d distance apart, has a capacity

$$C = \frac{\epsilon A (n-1)}{d} \cong \frac{cA(n-1)}{1.13 \times 10^{11} d}$$

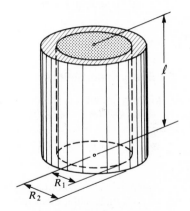

Fig. 7.02-10

(d) **Capacitors connected in series** (Fig. 7.02–12) have the same charge Q in each plate so that

$$Q = Q_1 = Q_2 = \cdots = Q_n$$

their total potential V equals the sum of the individual potentials so that

$$V = V_1 + V_2 + \cdots + V_n$$

and their total capacitance is

$$C = \frac{1}{\dfrac{1}{C_1} + \dfrac{1}{C_2} + \cdots + \dfrac{1}{C_n}}$$

where $C_1, C_2, \ldots, C_n$ = capacitances of the respective units.

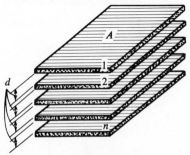

Fig. 7.02-11

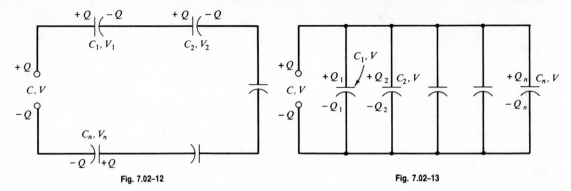

Fig. 7.02–12   Fig. 7.02–13

(e) Capacitors connected in parallel (Fig. 7.02–13) have the same potential across each unit so that

$$V = V_1 = V_2 = \cdots = V_n$$

their total charge Q equals the sum of the individual charges so that

$$Q = Q_1 + Q_2 + \cdots + Q_n$$

and their total capacitance is

$$C = C_1 + C_2 + \cdots + C_n$$

where again $C_1, C_2, \ldots, C_n$ = capacitances of the respective units.

(7) Energy and Strength

(a) Energy stored in the capacitor, defined as the product of the average potential difference and the total charge, is

$$E_p = \frac{1}{2}QV = \frac{1}{2}CV^2 = \frac{1}{2}\frac{Q^2}{C} \qquad \left\{ \begin{matrix} E_p \text{ in J} \\ Q \text{ in C} \\ V \text{ in V} \\ C \text{ in F} \end{matrix} \right\}$$

where V = potential difference, C = capacitance, and E_p is called the *electrostatic* potential of the capacitor.

(b) Dielectric strength of a given material is the maximum electric field intensity beyond which the material ceases to be an insulator and becomes a conductor.

(c) Conversion relations between

$$\text{J} = \text{joule} \qquad \text{C} = \text{coulomb} \qquad \text{V} = \text{volt} \qquad \text{F} = \text{farad}$$

are

$$\text{J} = \text{V} \times \text{C} = \frac{\text{C}^2}{\text{F}} = \text{V}^2 \times \text{F}$$

$$\text{C} = \text{V} \times \text{F} = \frac{\text{J}}{\text{V}} = \sqrt{\text{J} \times \text{F}}$$

$$\text{V} = \frac{\text{C}}{\text{F}} = \frac{\text{J}}{\text{C}} = \sqrt{\frac{\text{J}}{\text{F}}}$$

$$\text{F} = \frac{\text{C}}{\text{V}} = \frac{\text{J}}{\text{V}^2} = \frac{\text{C}^2}{\text{J}}$$

7.03 DIRECT ELECTRIC CURRENT

(1) Motion of Electric Charge

(a) Direct electric current (abbreviated dc) is a one-direction transfer of the electric charge Q through a conductor from one point to another point of this conductor.

(b) Intensity of current I is the electric charge transferred through the conductor in a unit of time.

$$I = \frac{Q}{t} \qquad \begin{cases} I \text{ in A} \\ Q \text{ in C} \\ t \text{ in s} \end{cases}$$

where t = time.

(c) SI unit of electric current is 1 ampere (after André Marie Ampère), designated by A and defined as the constant current which, if maintained in two straight parallel conductors of infinite length and negligible circular cross sections placed 1 meter apart in a vacuum, will produce between them a force equal to 2×10^{-7} newton per meter length.

$$1 \text{ ampere} = 1 \text{ A} = 6.2414 \times 10^{18} e/s = 3 \times 10^4 (\sqrt{10 \text{ N}}) \text{ m/s} = 1 \text{ coulomb/second} = 1 \text{ C/s}$$

where e = electronic unit charge (Sec. 7.01–2e), C = coulomb (Sec. 7.01–2f), and N = newton.

(d) SI unit of quantity of electricity is 1 coulomb, designated by C, and defined as the quantity of electric charge transported in 1 second by the current of 1 ampere.

$$1 \text{ coulomb} = 1 \text{ C} = 6.2414 \times 10^{18} e = 3 \times 10^4 (\sqrt{10 \text{ N}}) \text{ m} = 1 \text{ ampere} \times \text{second} = 1 \text{ A-s}$$

where A, N, and e are the same as in (c).

(2) Voltage and Resistance

(a) Voltage (potential difference) V between two points in a conductor is the work required to transfer a unit charge from one point to the other.

$$V = \frac{U}{Q} \qquad \begin{cases} U \text{ in J} \\ Q \text{ in C} \end{cases}$$

(b) SI unit of voltage is 1 volt, designated by V and defined as the difference of electric potential between two points of a conducting wire carrying a constant current of 1 ampere when the power dissipated between these points is equal to 1 watt (see also Sec. 7.02–3c).

$$1 \text{ volt} = 1 \text{ V} = 1 \text{ J/C} = 1 \text{ W/A}$$

where J = joule, C = coulomb, W = watt, and A = ampere.

(c) Ohm's law states that for a given conductor at a given temperature the intensity I of an electric current is linearly proportional to the potential difference V (voltage) between the ends of the conductor.

$$I = \frac{V}{R} = GV \qquad \begin{cases} I \text{ in A} & R \text{ in } \Omega \\ V \text{ in V} & G \text{ in S} \end{cases}$$

where R = constant of proportionality called *resistance*, depending upon the temperature, material, and shape of the conductor (Sec. 7.03–4) and $G = 1/R$ = *conductance*.

(d) SI unit of resistance is 1 ohm, designated by Ω (after Georg Simon Ohm), and defined as the electric resistance between two points of a conductor when a constant difference of potential of 1 volt, applied between these two points, produces in this conductor a current of 1 ampere.

$$1 \text{ ohm} = 1 \, \Omega = 1 \, \text{V/A} = 1 \, \text{W/A}^2$$

(e) SI unit of conductance is 1 siemens, designated by S (after Ernst Werner von Siemens), and defined as the electric conductance of a conductor in which a current of 1 ampere is produced by an electric potential difference of 1 volt.

$$1 \text{ siemens} = 1 \, \text{S} = 1/\Omega = 1 \, \text{A/V}$$

(3) Electromotive Force, Power, and Energy

(a) Transient electric current occurs in a conductor for a short period of time while a capacitor is being discharged.

(b) Constant electric current occurs in a conductor when a capacitor is continuously and uniformly supplied with electric charge so that the potential difference remains constant. The voltage thus supplied is called the *electromotive force* (EMF):

$$\text{EMF} = V \qquad \{\text{EMF in V}\}$$

 **(c) Power** P of a source of supply (battery or generator) is defined as

$$P = \frac{\text{energy}}{\text{time}} = \frac{QV}{t} = IV = I^2R = \frac{V^2}{R} \qquad \begin{cases} P \text{ in W} & V \text{ in V} \\ Q \text{ in C} & R \text{ in } \Omega \\ I \text{ in A} & t \text{ in s} \end{cases}$$

where Q = charge, V = potential difference, I = intensity, and R = resistance.

(d) SI unit of electric power is 1 watt, designated by W and defined as the work of 1 joule per second.

$$1 \text{ watt} = 1 \, \text{W} = 1 \, \text{joule/second} = 1 \, \text{J/s} = 1 \, \text{N-m/s}$$

where N = newton (see also Appendix B).

(e) Electromotive work U_E done by transferring a charge Q between the ends of a conductor of potential difference V and resistance R is

$$U_E = QV = IVt = I^2Rt = \frac{V^2t}{R} \qquad \begin{cases} U_E \text{ in J} & I \text{ in A} \\ Q \text{ in C} & R \text{ in } \Omega \\ V \text{ in V} & t \text{ in s} \end{cases}$$

where t = time and R = resistance.

(f) Heat U_T generated in the conductor by the transfer defined above is equal to the electromotive work.

$$U_T = U_E \qquad \{U_T \text{ in J}\}$$

example:

If a current of $I = 20 \, \text{A}$ flows through a conductor of $R = 200 \, \Omega$ for $t = 1$ hour, then

$$P = I^2R = 20^2 \times 200 = 80{,}000 \, \text{W}$$

$$U_E = I^2Rt = 80{,}000 \times 3{,}600 = 2.88 \times 10^8 \, \text{J}$$

$$U_T = 288 \times 10^6 \times 0.239 \times 10^{-3} = 6.88 \times 10^4 \, \text{kcal}$$

(4) Resistivity

(a) Resistance R of a homogeneous conductor of constant cross section A and length l is

$$R = \frac{\rho l}{A} \qquad \begin{cases} R \text{ in } \Omega \\ l \text{ in m} \\ A \text{ in m}^2 \\ \rho \text{ in } \Omega\text{-m} \end{cases}$$

where ρ = resistivity or specific resistance defined below.

(b) Resistivity ρ is a specific constant whose magnitude depends on material and temperature. Numerical values of ρ for the most common technical materials are given in Appendix A. The reciprocal of resistivity is called *conductivity*.

example:

The resistance R of a hardened copper wire of resistivity $\rho = 1.75 \times 10^{-8} \Omega$-m, length $l = 10$ m, and diameter $d = 8$ mm is

$$R = \frac{1.75 \times 10^{-8} \times 10}{\frac{\pi}{4} \times (8 \times 10^{-3})^2} = 3.48 \times 10^{-3} \Omega$$

(c) Temperature increase produces an increase in the resistivity of most metals, and the resistivity of all metals approaches zero as the temperature approaches $0 \text{ K} = -273°C$.

(d) Rise in resistance due to rise in temperature can be expressed as

$$R_T = R_0(1 + \alpha_0 T) \qquad \begin{cases} R_T, R_0 \text{ in } \Omega \\ \alpha_0 \text{ in } 1/°C \\ T \text{ in } °C \end{cases}$$

where R_T = resistance at temperature T, R_0 = resistance at $0°C$, and α_0 = temperature coefficient of resistance at $0°C$ (Appendix A).

(e) Alternative formula for the rise in resistance is

$$\frac{R_2}{R_1} = \frac{M + T_2}{M + T_1} \qquad \begin{cases} R_1, R_2 \text{ in } \Omega \\ T_1, T_2, M \text{ in } °C \end{cases}$$

where R_1 = resistance at temperature T_1, R_2 = resistance at temperature T_2, $M = 1/\alpha_0$, and α_0 is defined in (d).

example:

The resistance of an aluminum wire at $0°C$ is $R_0 = 0.2 \Omega$. If $\alpha_0 = 0.0039/°C$, the resistance at $T = 60°C$ is

$$R_T = 0.2(1 + 0.0039 \times 60) = 0.246 \Omega$$

By (e),

$$R_2 = \frac{0.2(256.4 + 60)}{256.4} = 0.246 \Omega$$

(f) Resistance of nonmetals is enormous compared with that of metals (greater by a factor of about 10^{20}), and for this reason they are used as insulation (Appendix A). However, one nonmetal, carbon, is a fairly good conductor, $\rho = 4000 \times 10^{-8} \Omega$-m at $20°C$ (Appendix A).

7.04 FUNDAMENTALS OF DC CIRCUITS

(1) Classification and Definitions

(a) **Simple dc circuit**, where dc designates direct current, is an electric system consisting of a single source of electromotive force (EMF) connected to a single resistor as shown in Fig. 7.04–1. Its governing equation is

$$EMF = V = IR \qquad \begin{Bmatrix} V \text{ in V} \\ I \text{ in A} \\ R \text{ in } \Omega \end{Bmatrix}$$

where V = voltage, I = intensity of current, and R = resistance.

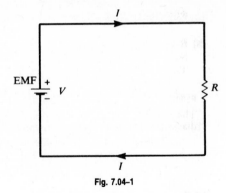

Fig. 7.04–1

(b) **Equivalent simple dc circuit** is the representation of a circuit consisting of a single source of EMF and a system of resistors, whose resistance is expressed by a single total resistance (see below).

(c) **Resistors in series.** If n resistors of resistance $R_1, R_2,..., R_n$ are connected in series (Fig. 7.04–2), their total resistance is

$$R = R_1 + R_2 + \cdots + R_n$$

the intensity of current through each resistor is the same,

$$I = I_1 = I_2 = \cdots = I_n$$

the voltages of the respective parts are

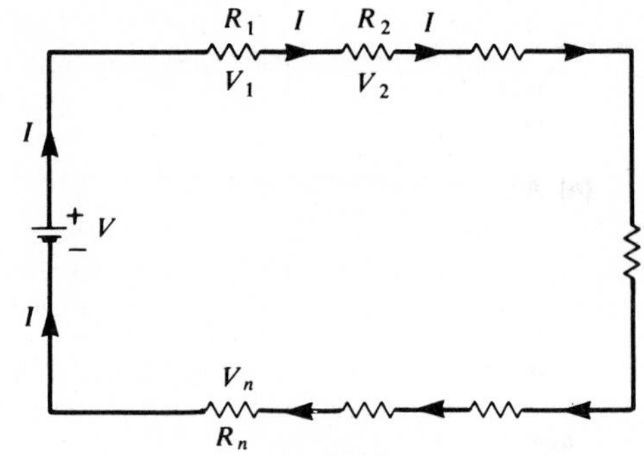

$$V_1 = IR_1 \qquad V_2 = IR_2 \qquad \cdots \qquad V_n = IR_n$$

and the total external voltage is

$$V = V_1 + V_2 + \cdots + V_n = I(R_1 + R_2 + \cdots + R_n) = IR$$

Fig. 7.04–2

example:

The series dc circuit of Fig. 7.04–3 consists of four resistances $R_1 = 2\,\Omega$, $R_2 = 4\,\Omega$, $R_3 = 10\,\Omega$, $R_4 = 14\,\Omega$ and is supplied by a battery of EMF = 150 V. The total resistance is

$$R = 2 + 4 + 10 + 14 = 30\,\Omega$$

The intensity of the current is

$$I = 150/30 = 5\,A$$

and the particular voltages are

$$V_1 = 5 \times 2 = 10\,V \qquad V_2 = 5 \times 4 = 20\,V \qquad V_3 = 5 \times 10 = 50\,V \qquad V_4 = 5 \times 14 = 70\,V$$

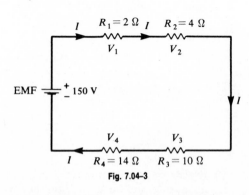

Fig. 7.04–3

(d) Resistors in parallel. If n resistors of resistance $R_1, R_2, \ldots, R_n$ are connected in parallel (Fig. 7.04–4), their total resistance is

$$R = \cfrac{1}{\cfrac{1}{R_1} + \cfrac{1}{R_2} + \cdots + \cfrac{1}{R_n}}$$

the total intensity of the current is

$$I = I_1 + I_2 + \cdots + I_n$$

and the voltages of the respective parts are

$$V = V_1 = V_2 = \cdots = V_n$$

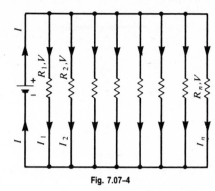

example:

The parallel dc circuit of Fig. 7.04–5 consists of four resistances $R_1 = 2\,\Omega$, $R_2 = 4\,\Omega$, $R_3 = 10\,\Omega$, $R_4 = 14\,\Omega$ and is supplied by a battery of EMF = 200 V. The total resistance is

$$R = \frac{1}{\frac{1}{2} + \frac{1}{4} + \frac{1}{10} + \frac{1}{14}} = 1.085\,\Omega$$

The total intensity of current is

$$I = \frac{200}{1.085} = 184.3\,\text{A}$$

The voltages are $V = V_1 = V_2 = V_3 = V_4 = 200$ V, and the intensities of current in the respective branches are

Fig. 7.07–4

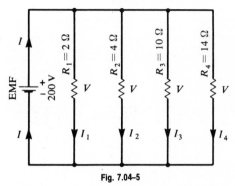

Fig. 7.04–5

$$I_1 = \frac{200}{2} = 100\,\text{A} \qquad I_2 = \frac{200}{4} = 50\,\text{A} \qquad I_3 = \frac{200}{10} = 20\,\text{A} \qquad I_4 = \frac{200}{14} = 14.3\,\text{A}$$

which must check $I = I_1 + I_2 + I_3 + I_4 = 184.3\,\text{A}$.

(e) Internal resistance. When the resistance of the source of EMF is considered, the governing equation of the circuit becomes

$$V = I(R_s + R_c) \qquad \begin{cases} V \text{ in V} \\ I \text{ in A} \\ R \text{ in } \Omega \end{cases}$$

where R_s = resistance of source and R_c = resistance of circuit.

example:

The series dc circuit of Fig. 7.04–6 with resistances $R_1 = 10\,\Omega$ and $R_2 = 20\,\Omega$ is supplied by a 62-V battery of $R_s = 1\,\Omega$. The intensity of the current is

$$I = \frac{62}{1 + 10 + 20} = 2\,\text{A}$$

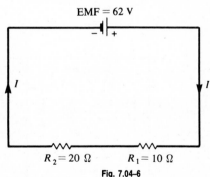

Fig. 7.04–6

The voltages are

$$V_1 = IR_1 = 2 \times 10 = 20\,\text{V} \qquad V_2 = IR_2 = 2 \times 20 = 40\,\text{V} \qquad V_s = IR_s = 2 \times 1 = 2\,\text{V}$$

which must check $V = V_1 + V_2 + V_3 = 62\,\text{V}$.

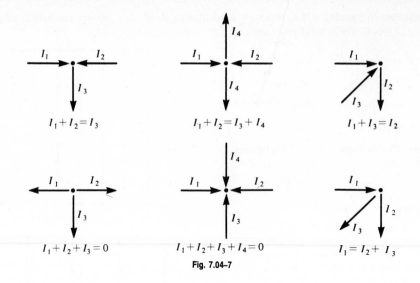

$$I_1 + I_2 = I_3 \qquad I_1 + I_2 = I_3 + I_4 \qquad I_1 + I_3 = I_2$$

$$I_1 + I_2 + I_3 = 0 \qquad I_1 + I_2 + I_3 + I_4 = 0 \qquad I_1 = I_2 + I_3$$

Fig. 7.04–7

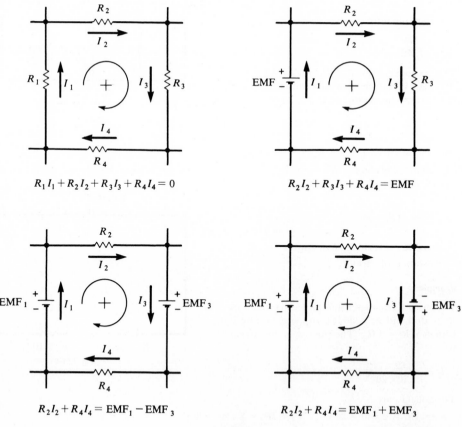

$$R_1 I_1 + R_2 I_2 + R_3 I_3 + R_4 I_4 = 0 \qquad R_2 I_2 + R_3 I_3 + R_4 I_4 = \text{EMF}$$

$$R_2 I_2 + R_4 I_4 = \text{EMF}_1 - \text{EMF}_3 \qquad R_2 I_2 + R_4 I_4 = \text{EMF}_1 + \text{EMF}_3$$

Fig. 7.04–8

(2) Kirchhoff's Laws

(a) **Complex dc circuit** is an electrical system consisting of several resistances and several generators arranged in several loops forming a network.

(b) **Analysis of complex circuits** is based on two laws called *Kirchhoff's laws* (presented below).

(c) **Kirchhoff's first law (nodal law).** The sum of currents arriving at the junction point (nodal point) of a network equals the sum of currents leaving that junction point.

$$\Sigma I \text{ (arriving)} = \Sigma I \text{ (leaving)}$$

(d) **Typical nodal equations** obtained by applying the nodal law at the nodes of Fig. 7.04–7 are shown below each node in that figure.

(e) **Kirchhoff's second law (mesh law).** The sum of voltage about a closed loop (mesh of a network) equals the sum of electromotive forces of this loop.

$$\Sigma V \text{ (in one loop)} = \Sigma EMF \text{ (in one loop)}$$

or ΣIR (in one loop) $= \Sigma EMF$ (in one loop).

(f) **Typical mesh equations** obtained by applying the mesh law to the loops of Fig. 7.04–8 are shown below each loop in that figure.

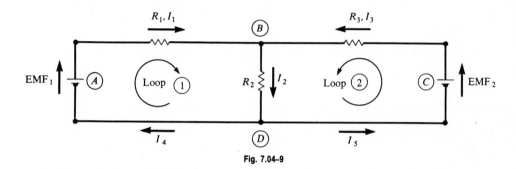

Fig. 7.04–9

example:

The network of Fig. 7.04–9 consists of three resistors of $R_1 = 5\,\Omega$, $R_2 = 10\,\Omega$, $R_3 = 20\,\Omega$ and two generators of $EMF_1 = 20$ V and $EMF_2 = 30$ V. To find the intensity and direction of the current, the direction is first assumed for each loop as shown in Fig. 7.04–9.

Then according to (c), at

Node A: $I_4 = I_1$ Node C: $I_5 = I_3$

Node B: $I_2 = I_1 + I_3$ Node D: $I_2 = I_4 + I_5$

where I_1 and I_3 are selected as the unknowns.

Next, according to (d), in

Loop 1: $EMF_1 = I_1 R_1 + (I_1 + I_3) R_2$ Loop 2: $EMF_2 = I_3 R_3 + (I_1 + I_3) R_2$

$\quad\quad\quad\quad 20 = 5I_1 + 10(I_1 + I_3)$ $\quad\quad\quad\quad\quad 30 = 20I_3 + 10(I_1 + I_3)$

from which $I_1 = \dfrac{6}{7}$ A, $I_3 = \dfrac{5}{7}$ A, and the positive sign of I_1 and I_3 indicates that the direction was assumed correctly. A minus sign would indicate the opposite.

The remaining intensities are obtained from the nodal equations as follows:

$$I_2 = \frac{6}{7} + \frac{5}{7} = \frac{11}{7}\,A \qquad I_4 = \frac{6}{7}\,A \qquad I_5 = \frac{5}{7}\,A$$

7.05 DC CIRCUIT CALCULATIONS

(1) Equivalent of Parallel Resistances

(a) Two parallel resistances. The equivalent resistance of the network of Fig. 7.05–1 is

$$R = \frac{R_1 R_2}{R_1 + R_2} = \frac{1}{G_1 + G_2}$$

where

$$G_1 = \frac{1}{R_1} \qquad \text{and} \qquad G_2 = \frac{1}{R_2}$$

The particular intensities are

$$I_1 = I \frac{G_1}{G_1 + G_2} \qquad I_2 = I \frac{G_2}{G_1 + G_2}$$

and

$$I = V/R = I_1 + I_2$$

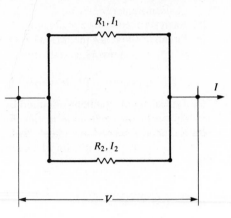

Fig. 7.05–1

example:

If $R_1 = 10\,\Omega$, $R_2 = 20\,\Omega$, and $V = 100\,\text{V}$, then

$$G_1 = \frac{1}{10} = \frac{0.1}{\Omega} \qquad G_2 = \frac{1}{20} = \frac{0.05}{\Omega} \qquad R = \frac{1}{0.1 + 0.05} = 6.7\,\Omega$$

$$I = \frac{V}{R} = \frac{100}{6.7} = 15\,\text{A} \qquad I_1 = 15 \times \frac{2}{3} = 10\,\text{A} \qquad I_2 = 15 \times \frac{1}{3} = 5\,\text{A}$$

which must check $I = I_1 + I_2 = 15\,\text{A}$.

(b) Three parallel resistances. The equivalent resistance of the network of Fig. 7.05–2 is

$$R = \frac{R_1 R_2 R_3}{R_1 R_2 + R_2 R_3 + R_3 R_1} = \frac{1}{G_1 + G_2 + G_3}$$

where G has the same meaning as in (a). The particular intensities are

$$I_1 = I \frac{G_1}{G_1 + G_2 + G_3} \qquad I_2 = I \frac{G_2}{G_1 + G_2 + G_3}$$

$$I_3 = I \frac{G_3}{G_1 + G_2 + G_3}$$

and $I = V/R = I_1 + I_2 + I_3$.

(c) _n_ parallel resistances. In a general case, the equivalent resistance is

$$R = \frac{1}{G_1 + G_2 + \cdots + G_n} = \frac{1}{\Sigma G}$$

$$I_1 = I \frac{G_1}{\Sigma G} \qquad I_2 = I \frac{G_2}{\Sigma G} \quad \cdots \quad I_n = I \frac{G_n}{\Sigma G}$$

and $I = V/R = I_1 + I_2 + \cdots + I_n$.

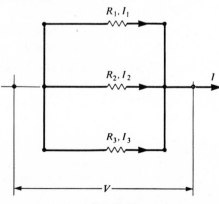

Fig. 7.05–2

(2) Equivalent of Parallel and Series Resistances

(a) Single loop of Fig. 7.05-3 has

$$R = R_1 + \cfrac{1}{\cfrac{1}{R_2} + \cfrac{1}{R_3}} + R_4 = \frac{1}{G_1} + \frac{1}{G_2 + G_3} + \frac{1}{G_4}$$

$$I_1 = I \qquad I_2 = I\frac{G_2}{G_2 + G_3} \qquad I_3 = I\frac{G_3}{G_2 + G_3} \qquad I_4 = I$$

and $I = V/R = I_1 = I_2 + I_3 = I_4$.

example:

If $R_1 = R_2 = R_3 = R_4 = 10\,\Omega$ and $V = 100\,V$, then $G_1 = G_2 = G_3 = G_4 = 1/10 = 0.1/\Omega$,

$$R = \frac{1}{0.1} + \frac{1}{0.1 + 0.1} + \frac{1}{0.1} = 25\,\Omega \qquad I = \frac{V}{R} = \frac{100}{25} = 4\,A$$

$$I_1 = 4\,A \qquad I_2 = 4 \times 1/2 = 2\,A \qquad I_3 = 4 \times 1/2 = 2\,A \qquad I_4 = 4\,A$$

which must check $I = I_1 = I_2 + I_3 = I_4 = 4\,A$.

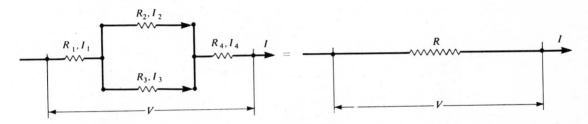

Fig. 7.05-3

(b) Double loop of Fig. 7.05-4 has

$$R = R_1 + \cfrac{1}{\cfrac{1}{R_2} + \cfrac{1}{R_3} + \cfrac{1}{R_4}} + R_5 = \frac{1}{G_1} + \frac{1}{G_2 + G_3 + G_4} + \frac{1}{G_5}$$

$$I_1 = I \qquad I_2 = I\frac{G_2}{G_2 + G_3 + G_4} \qquad I_3 = I\frac{G_3}{G_2 + G_3 + G_4} \qquad I_4 = I\frac{G_4}{G_2 + G_3 + G_4} \qquad I_5 = I$$

and

$$I = V/R = I_1 = I_2 + I_3 + I_4 = I_5$$

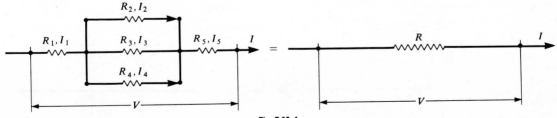

Fig. 7.05-4

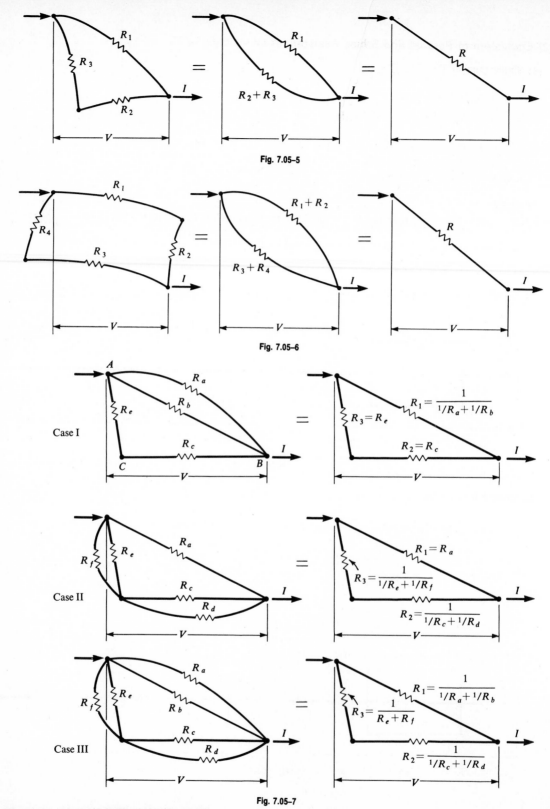

Fig. 7.05-5

Fig. 7.05-6

Case I

Case II

Case III

Fig. 7.05-7

(c) Simple triangle of Fig. 7.05–5 has

$$R = \frac{1}{\dfrac{1}{R_1} + \dfrac{1}{R_2 + R_3}} = \frac{1}{G_1 + H_{23}}$$

$$I_1 = IG_1R \qquad I_2 = I_3 = IH_{23}R$$

and

$$I = V/R = I_1 + I_2 = I_1 + I_3$$

(d) Simple rectangle of Fig. 7.05–6 has

$$R = \frac{1}{\dfrac{1}{R_1 + R_2} + \dfrac{1}{R_3 + R_4}} = \frac{1}{H_{12} + H_{34}} = \frac{1}{\Sigma H}$$

$$I_1 = I_2 = I\frac{H_{12}}{\Sigma H} \qquad I_3 = I_4 = I\frac{H_{34}}{\Sigma H}$$

and

$$I = V/R = I_1 + I_4 = I_2 + I_3$$

example:

If $R_1 = 10\,\Omega$, $R_2 = 20\,\Omega$, $R_3 = 30\,\Omega$, and $V = 100\,\text{V}$, then

$$R = \frac{1}{\dfrac{1}{10} + \dfrac{1}{20 + 30}} = 8.3\,\Omega$$

$$I_1 = IRG_1 = VG_1 = 100 \times 0.1 = 10\,\text{A} \qquad I_2 = I_3 = IRH_{23} = VH_{23} = 100 \times 0.02 = 2\,\text{A}$$

$$I = 10 + 2 = 12\,\text{A}$$

which must check $V = IR = 12 \times 8.3 = 100\,\text{V}$.

(e) Triangles with double sides of Fig. 7.05–7 must be transformed to triangles with simple sides and solved by (*c*).

example:

If in case II of Fig. 7.05–7, $R_a = R_c = R_d = R_e = R_f = 10\,\Omega$ and $V = 100\,\text{V}$, then

$$R_1 = 10\,\Omega \qquad R_2 = \frac{1}{\frac{1}{10} + \frac{1}{10}} = 5\,\Omega \qquad R_3 = \frac{1}{\frac{1}{10} + \frac{1}{10}} = 5\,\Omega$$

and by (*c*),

$$R = \frac{1}{\dfrac{1}{10} + \dfrac{1}{5 + 5}} = 5\,\Omega$$

$$I_1 = IRG_1 = VG_1 = 100 \times 0.1 = 10\,\text{A} \qquad I_2 = I_3 = IRH_{23} = VH_{23} = 100 \times 0.1 = 10\,\text{A}$$

and

$$I = I_1 + I_2 = 20\,\text{A} \qquad \text{or} \qquad I = \frac{V}{R} = \frac{100}{5} = 20\,\text{A}$$

(f) Simple rectangle with diagonal of Fig. 7.05–8 has

$$R = \frac{1}{\dfrac{1}{R_1 + R_2} + \dfrac{1}{R_3} + \dfrac{1}{R_4 + R_5}} = \frac{1}{H_{12} + G_3 + H_{45}}$$

$$I_1 = I_2 = IH_{12}R \qquad I_3 = IG_3R \qquad I_4 = I_5 = IH_{45}R$$

and

$$I = V/R = I_1 + I_3 + I_5 = I_2 + I_3 + I_4$$

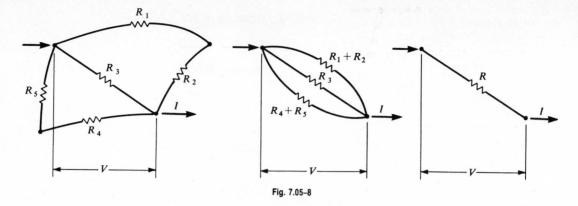

Fig. 7.05-8

(3) Star and Delta Connections

(a) Relations. When delta branches (Fig. 7.05–9a) or star branches (Fig. 7.05–9b) are present in a network, it is always possible to convert the delta to a star, and vice versa.*

(b) Delta-to-star conversion yields

$$R_{A0} = \frac{R_{AB}R_{CA}}{\Sigma R} \qquad R_{B0} = \frac{R_{AB}R_{BC}}{\Sigma R} \qquad R_{C0} = \frac{R_{CA}B_{BC}}{\Sigma R}$$

where R_{A0}, R_{B0}, R_{C0} = resistances of the star system, R_{AB}, R_{BC}, R_{CA} = resistances of the delta system, and $\Sigma R = R_{AB} + R_{BC} + R_{CA}$.

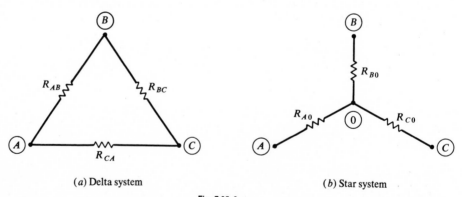

$\quad$ (a) Delta system $\qquad\qquad\qquad\qquad\qquad\qquad$ (b) Star system

Fig. 7.05-9

(c) Star-to-delta conversion yields

$$R_{AB} = \frac{\Sigma RR}{R_{C0}} \qquad R_{BC} = \frac{\Sigma RR}{R_{A0}} \qquad R_{CA} = \frac{\Sigma RR}{R_{B0}}$$

where R_{A0}, R_{B0}, R_{C0}, R_{AB}, R_{BC}, R_{CA} = same as in (b) and

$$\Sigma RR = R_{A0}R_{B0} + R_{B0}R_{C0} + R_{C0}R_{A0}$$

*The conformation discussed here as "star" is also called "wye."

(d) Application of delta-to-star conversion. The system of Fig. 7.05–10a is converted into the simple rectangular system of Fig. 7.05–10b, where R_{A0}, R_{B0}, R_{C0} are those of (b). This system can be further reduced to the series system of Fig. 7.05–10c, where R_{A0} is now known and the resistance of the parallel system $0D$ is

$$R_{0D} = \frac{1}{\dfrac{1}{R_{B0} + R_1} + \dfrac{1}{R_{C0} + R_2}}$$

Finally, the resistance of AD is

$$R_{AD} = R_{A0} + R_{0D}$$

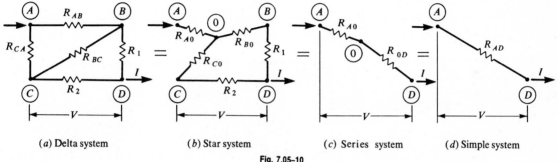

| (a) Delta system | (b) Star system | (c) Series system | (d) Simple system |

Fig. 7.05–10

(e) Applications of star-to-delta conversion. The system of Fig. 7.05–11a is converted into the triangular system of Fig. 7.05–11b, where R_{AD}, R_{BC}, R_{CA} are those of (c). This system can be further reduced to the parallel system of Fig. 7.05–11c, where R_{CA} is now known and

$$R_5 = \frac{1}{\dfrac{1}{R_{BC}} + \dfrac{1}{R_3}} \qquad R_6 = \frac{1}{\dfrac{1}{R_{AB}} + \dfrac{1}{R_4}}$$

Finally, the resistance of AC is

$$R_7 = \frac{1}{\dfrac{1}{R_{CA}} + \dfrac{1}{R_5 + R_6}}$$

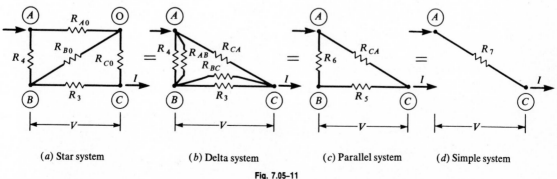

| (a) Star system | (b) Delta system | (c) Parallel system | (d) Simple system |

Fig. 7.05–11

(4) Superposition Method

(a) Superposition theorem. If there is more than one source of EMF, then the system can be solved for each source separately and the results of the separate solutions summed in the final solution.

(b) Application of superposition to the system of Fig. 7.05–12a requires the introduction of two separate systems, shown in Fig. 7.05–12b and c, in which

$$I_1 = \frac{\text{EMF}_1}{R} \qquad \text{and} \qquad I_2 = \frac{\text{EMF}_2}{R}$$

where

$$R = \frac{1}{\dfrac{1}{R_{AB}} + \dfrac{1}{R_{CD}}}$$

and the superposition yields

$$I = I_1 - I_2 = \frac{\text{EMF}_1 - \text{EMF}_2}{\dfrac{1}{R_{AB}} + \dfrac{1}{R_{CD}}}$$

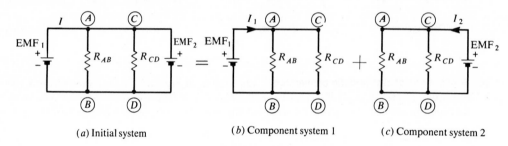

(*a*) Initial system   (*b*) Component system 1   (*c*) Component system 2

Fig. 7.05–12

8

MAGNETISM AND ELECTRODYNAMICS

8.01 MAGNETIC FORCES AND FIELDS

(1) Magnetic Forces

(a) **Magnet** is a body which possesses the property of attracting iron, and if a magnet is suspended as shown in Fig. 8.01–1, its principal axis will assume a position not far off the geographic north-south direction.

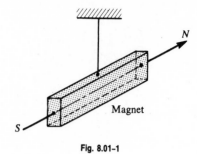

Fig. 8.01–1

(b) **Simplest forms of magnet** are the straight bar magnet and the horseshoe magnet.

(c) **Ends of magnet** are designated as the north-seeking pole and the south-seeking pole according to the orientation defined in (a).

(d) **Magnetic force** is generated by the motion of electrons (by electric current) and can exist between two electric charges if and only if both charges are moving relative to a fixed datum.

(e) **Positive and negative forces** in a magnet always occur in pairs and are of equal intensity.

(f) **Law of magnetic forces states** that like poles repel each other and unlike poles attract each other.

(2) Coulomb's Law

(a) **Force F_0 of attraction or repulsion** between the poles of two magnets in vacuum is directly proportional to the product of the pole strengths $(p_1 \cdot p_2)$ and inversely proportional to the square of the distance (r^2) between them (Fig. 8.01–2).

$$F_0 = k_0 \frac{p_1 p_2}{r^2} \qquad \begin{cases} F_0 \text{ in N} \\ p_1, p_2 \text{ in A-m} \\ k_0 \text{ in N/A}^2 \\ r \text{ in m} \end{cases}$$

where F_0 = attraction or repulsion according to whether p_1 and p_2 are of unlike or like pole strengths, k_0 = proportionality constant defined in (c), and A = ampere defined below.

(b) **SI unit of magnetic pole strength** is

1 ampere × meter = 1 A-m = $3 \times 10^4 (\sqrt{10 \text{ N}}) \text{m}^2/\text{s}$

where 1 ampere = 1 coulomb/second (Sec. 7.03–1c).

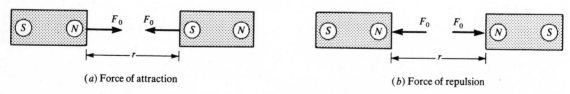

(a) Force of attraction   (b) Force of repulsion

Fig. 8.01–2

(c) Proportionality constant k_0 is

$$k_0 = \frac{\mu_0}{4\pi} = 10^{-7}\,\text{N/A}^2$$

where

$$\mu_0 = 4\pi \times 10^{-7}\,\text{N/A}^2$$

is an experimental constant called the *permeability of free space* (in vacuum).

(d) Permeability of medium. When the medium between p_1 and p_2 is not a vacuum, the magnetic force in (a) becomes

$$F = k\frac{p_1 p_2}{r^2}$$

where

$$k = \frac{\mu}{4\pi} = \frac{c\mu_0}{4\pi} = 10^{-7}\,c$$

and $c = \mu/\mu_0$ is the ratio of the permeability of the given medium and of the free space defined in (c).

(e) Unit pole strength, designated by ups, is the strength of a magnetic pole which exerts in vacuum a force of 1 dyne upon another magnetic pole of the same strength placed 1 centimeter away.

$$1\,\text{ups} = (\sqrt{\text{dyne}})\text{cm} = 10^{-5}\,(\sqrt{10\,\text{N}})\text{m} = \frac{\text{coulomb}}{3 \times 10^9} = 1\,\text{statcoulomb}$$

where 1 coulomb and 1 statcoulomb are defined in Sec. 7.01–2f.

(3) Magnetic Field

(a) Definition. Since one pole of a magnet exerts forces on the other pole of the same magnet, there is a force field surrounding each magnet called its *magnetic field*.

(b) Flux density B of a magnetic field is a vector defining the magnitude and direction of the field.

$$B = k\frac{p}{r^2} = \frac{\mu p}{4\pi r^2} \cong 10^{-7}\frac{p}{r^2} \qquad \left\{\begin{array}{l} B \text{ in T} \\ k \text{ in N/A}^2 \\ p \text{ in A-m} \\ r \text{ in m} \end{array}\right\}$$

where k, μ, p, and r are defined in Sec. 8.01–1.

(c) SI unit of flux density B is 1 tesla (after Nikola Tesla), designated by T and defined as 1 newton per ampere-meter.

$$1\,\text{tesla} = 1\,\text{T} = 1\frac{\text{N}}{\text{A-m}} = 1\frac{\text{kg}}{\text{C-s}} = 1\frac{\text{kg}}{\text{A-s}^2}$$

where C = coulomb, A = ampere, and N = newton.

(d) Direction of the flux density B at a given point in the magnetic field is the direction in which a positive unit pole strength would move if it were placed at that point.

(e) Field intensity H is proportional to B.

$$H = \frac{B}{\mu} = \frac{p}{4\pi r^2} \qquad \left\{\begin{array}{l} H \text{ in A/m} \\ \mu \text{ in N/A}^2 \\ p \text{ in A-m} \\ B \text{ in T} \end{array}\right\}$$

where μ, p, r are those of (b).

(f) **Magnetic force** F acting on a pole of strength p' placed in a magnetic field at a point of flux density B is

$$F = Bp' = \mu H p' \qquad \left\{ \begin{array}{l} F \text{ in N} \\ p' \text{ in A-m} \end{array} \right\}$$

(g) **Magnetic intensity** H_i at the point i of a magnetic field generated by a number of magnetic poles is the vector sum of their intensities at that point.

(4) Magnetic Field Lines, Moments, and Flux

(a) **Magnetic field line** is an imaginary line in the magnetic field whose tangent at each point is the direction of the intensity of the field at that point.

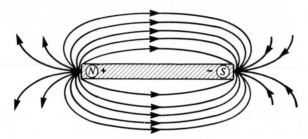

(a) Dipole magnetic field

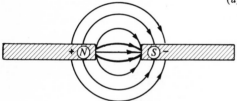

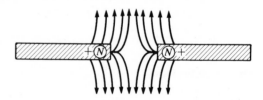

(b) Two unlike-pole magnetic fields   (c) Two like-pole magnetic fields

Fig. 8.01–3

(b) **Sense of the magnetic field line** is such that it points from the positive (north) pole in the direction of the negative (south) pole (Fig. 8.01–3).

(c) **Magnetic momemt** of a small magnet of length l and of pole strength p is $M_p = pl$. If this magnet is placed in a homogeneous magnetic field of intensity H, the torque of the field on this magnet is

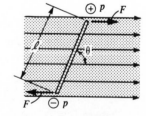

$$M_f = \underbrace{- \mu p H}_{\text{force}} \ \underbrace{l \sin \theta}_{\text{arm}} = - \mu M_p H \sin \theta \qquad \left\{ \begin{array}{l} M \text{ in N-m} \\ M_p \text{ in A-m}^2 \\ H \text{ in T} \\ \theta \text{ in rad} \end{array} \right\}$$

Fig. 8.01–4

where θ = position angle of the segment l (Fig. 8.01–4).

(d) **Magnetic flux** Φ of a magnetic field of density B through an area A normal to the field lines is the product of B and A (provided that the area is small so that the intensity can be assumed to be constant).

$$\Phi = BA = \frac{\mu p A}{4\pi r^2} \qquad \left\{ \begin{array}{l} \Phi \text{ in Wb} \\ B \text{ in T} \\ p \text{ in A-m} \\ r \text{ in m} \end{array} \right\}$$

where $B = \mu H$ and μ, r, p are defined in Sec. 8.01–2.

(e) SI unit of magnetic flux Φ is 1 weber (after Wilhelm Eduard Weber), designated by Wb and defined as the magnetic flux through an area of $1\,m^2$ normal to the magnetic field whose flux density is 1 tesla.

$$1\,\text{weber} = 1\,\text{Wb} = 1\,T \times m^2 = 1\frac{N \times m}{A} = 1\frac{kg \times m^2}{C \times s} = 10^8\,\text{maxwells}$$

where T = tesla, A = ampere, C = coulomb, and N = newton.

(f) SI unit of magnetic flux density B is

$$1\,\text{weber/meter}^2 = 1\,\text{Wb/m}^2 = 1\,T = 10^4\,\text{gauss}$$

(g) Permeability in vacuum μ_0 (Sec. 8.01–2c) in terms of webers and teslas is

$$\mu_0 = 4\pi \times 10^{-7}\,\text{Wb/A-m} = 4\pi \times 10^{-7}\,\text{T-m/A}$$

8.02 ELECTROMAGNETIC FIELDS

(1) Long Straight Wire

(a) Electric current of constant intensity I flowing through a long straight wire generates in the surrounding medium a magnetic field, whose field lines are concentric circles in planes normal to the axis of the wire (Fig. 8.02–1).

(b) Direction of the magnetic field is determined by the right-hand rule as shown in Fig. 8.02–1, where the thumb of the right hand points in the direction of the current and the fingers holding the wire indicate the direction of the magnetic field lines.

(c) Magnetic flux density B of this field is

$$B = \frac{\mu I}{2\pi r} = \frac{2 \times 10^{-7} I}{r} \qquad \left\{ \begin{array}{l} B \text{ in T} \\ \mu \text{ in N/A}^2 \\ I \text{ in A} \\ r \text{ in m} \end{array} \right\}$$

where r = distance of B from the axis of the wire and μ is defined in Sec. 8.01–2d.

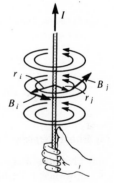

Fig. 8.02–1

example:

The flux density of the magnetic field of an electric current $I = 20\,A$, at the distance $r = 10\,cm$ from the wire in a vacuum, is

$$B = \frac{4\pi \times 10^{-7} \times 20}{2\pi \times 10/100} = 4 \times 10^{-5}\,T = 4 \times 10^{-5}\,\text{Wb/m}^2$$

(2) Two Long Straight Wires

(a) Force per unit length between two parallel wires carrying a current of intensity I is directly proportional to the square of I and inversely proportional to the distance between the wires (Fig. 8.02–2).

$$F = \frac{\mu I^2}{2\pi d} = \frac{2 \times 10^{-7} c I^2}{d} \qquad \left\{ \begin{array}{l} F \text{ in N/m} \\ \mu \text{ in N/A}^2 \\ I \text{ in A} \\ d \text{ in m} \end{array} \right\}$$

where μ is defined in Sec. 8.01–2d. The force is an attraction if the currents have the same direction and is a repulsion if they are of opposite directions.

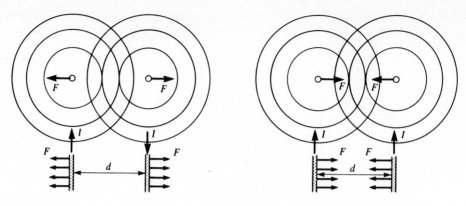

Fig. 8.02–2

(b) Ampere, introduced as the SI unit of current in Sec. 7.03–1c and defined as

1 ampere = 1 coulomb/second

is defined here by the process of (a) as the constant current which, if maintained in two straight parallel conductors of infinite length, of negligible circular sections, and placed 1 meter apart in vacuum, will produce between these conductors a force equal to 2×10^{-7} newton per meter length.

(c) General case of (a) consists of two wires of currents I_1 and I_2. Then the force per unit length is

$$F = \frac{\mu}{2\pi} \frac{I_1 I_2}{d} = \frac{2 \times 10^{-7} c I_1 I_2}{d} \qquad \{F \text{ in N/m}\}$$

where μ is the same as in (a).

(3) Short Straight Wire

(a) Elemental magnetic flux density dB at a distance r from an element dl of a short wire carrying a current I (Fig. 8.02–3) is

$$dB = \frac{\mu I \, dl \, \sin\theta}{4\pi r^2} \qquad \{dB \text{ in T}\}$$

where θ = angle between r and the current at dl.

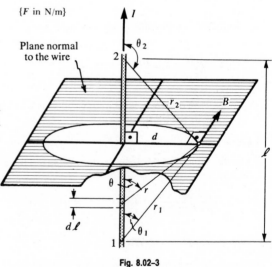

Fig. 8.02–3

(b) Total magnetic flux density B at the same point (Fig. 8.02–3) is

$$B = \int_{\theta_1}^{\theta_2} dB = \frac{\mu I (\cos\theta_1 - \cos\theta_2)}{4\pi d} \qquad \{B \text{ in T}\}$$

where θ_1, θ_2 = positive angles and d = normal distance.

(4) Loops, Coils, Solenoids, and Toroids

(a) **Flat circular loop** of radius r, carrying a constant current I (Fig. 8.02–4), produces a magnetic field whose flux density is

$$B = \frac{\mu I}{2r} = \frac{6.28 \times 10^{-7}\, cI}{r} \qquad \{B \text{ in T}\}$$

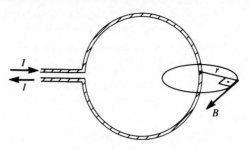

Fig. 8.02–4

(b) **Closely wound flat circular coil** of n turns and radius r, carrying a constant current I, produces a magnetic field whose flux density at the center is

$$B = \frac{\mu n I}{2r} = \frac{6.28n \times 10^{-7}\, cI}{r} \qquad \{B \text{ in T}\}$$

example:

A circular coil of wire of $n = 100$ and $r = 10\,\text{cm}$ has a central magnetic flux density $B = 15 \times 10^{-5}\,\text{T}$. The current required to produce this flux density in vacuum is

$$I = \frac{2rB}{\mu n} = \frac{2 \times 10^{-1} \times 15 \times 10^{-5}}{4\pi \times 10^{-7} \times 100} = 0.24\,\text{A}$$

(c) **Solenoid** consisting of n circular turns of wire per length l, forming a cylindrical helix (Fig. 8.02–5) with radius r, carrying a constant current I, produces a magnetic field whose flux density at any interior point is

$$B = \frac{\mu n I}{l} = \frac{12.567 \times 10^{-7}\, cnI}{l} \qquad \{B \text{ in T}\}$$

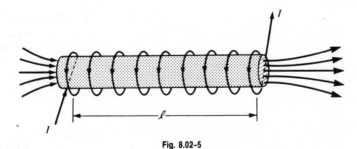

Fig. 8.02–5

example:

The flux density at the center of the hollow core of a solenoid of length $l = 20\,\text{cm}$, having 8 turns per cm, and carrying a current of 3 A, is

$$B = 4\pi \times 10^{-7} \times 8 \times 300 = 3 \times 10^{-3}\,\text{T}$$

and if the radius of the core is $r = 10\,\text{cm}$, the magnetic flux in this core is

$$\Phi = BA = 3 \times 10^{-3} \times \pi \times 0.1^2 = 9.42 \times 10^{-5}\,\text{Wb}$$

(d) Toroid is a solenoid bent into a closed ring (Fig. 8.02–6). Its magnetic field is entirely within the toroid, and its density is

$$B = \frac{\mu n I}{\pi d} = \frac{4 \times 10^{-7} c n I}{l} \qquad \{B \text{ in T}\}$$

where n = total number of turns and $l = \pi d$ = average circumference.

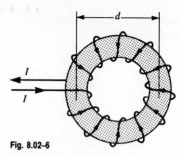

Fig. 8.02–6

(5) Systems of Wires

(a) Four straight wires. The force F between two circuits each composed of two wires of length l located in parallel planes at distance c and carrying currents I_1 and I_2, as shown in Fig. 8.02–7, is

$$F = \frac{4 abcl\mu I_1 I_2}{\pi[(a+b)^2 + c^2][(a-b)^2 + c^2)]} \qquad \begin{cases} F \text{ in N} \\ I \text{ in A} \\ \mu \text{ in N/A}^2 \\ a, b, c, l \text{ in m} \end{cases}$$

where F is an attraction for the direction of currents shown and is a repulsion if the current is reversed in one of the circuits.

(b) Two parallel circular loops of equal size. The force F between two parallel coaxial circular loops of radius r located a distance d apart and carrying currents I_1 and I_2, as shown in Fig. 8.02–8, is

$$F = \frac{\mu r I_1 I_2}{d} \qquad \begin{cases} F \text{ in N} \\ I \text{ in A} \\ \mu \text{ in N/A}^2 \\ r, d \text{ in m} \end{cases}$$

where the sign of F follows the rule of (a) and $d < r$.

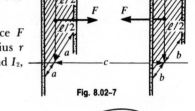

Fig. 8.02–7

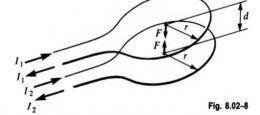

Fig. 8.02–8

(c) Two parallel circular loops of unequal size. The force F between two parallel coaxial circular loops of radius r_1 and r_2 located a distance d apart and carrying currents I_1 and I_2 is

$$F = \frac{\mu r_1 d I_1 I_2}{(r_1 - r_2)^2 + d^2} \qquad \begin{cases} F \text{ in N} \\ I \text{ in A} \\ \mu \text{ in N/A}^2 \\ d, r_1, r_2 \text{ in m} \end{cases}$$

where the sign of F follows the rule of (a) and $r_1 > r_2 > d$.

(d) Two parallel rectangular loops of equal size. The force F between two parallel coaxial rectangular loops of sides a and b located a distance d apart and carrying currents I_1 and I_2 is

$$F = \frac{\mu(a+b)I_1 I_2}{2\pi d} \qquad \begin{cases} F \text{ in N} \\ I \text{ in A} \\ \mu \text{ in N/A}^2 \\ a, b, d \text{ in m} \end{cases}$$

where the sign of F follows the rule of (a) and $a \lesssim b$.

(6) Forces and Moments in Magnetic Field

(a) Parallel wire. The force F on a straight wire of length l carrying a current I parallel to the field lines of the magnetic field of constant density B (Fig. 8.02–9) is zero.

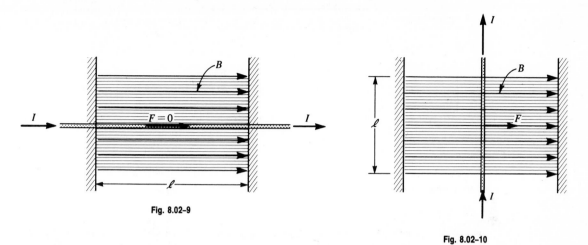

Fig. 8.02–9

Fig. 8.02–10

(b) Transverse wire. The force F on a straight wire carrying a current I normal to the field lines of the magnetic field of constant density B (Fig. 8.02–10) is

$$F = BlI \qquad \left\{ \begin{array}{l} F \text{ in N} \\ B \text{ in N/A-m} \\ l \text{ in m} \\ I \text{ in A} \end{array} \right\}$$

(c) Inclined wire. The force F on a straight wire of length l carrying a current I in a magnetic field of constant density B (Fig. 8.02–11) is given by

$$F_x = BlI \cos \theta$$

$$F_y = BlI \sin \theta$$

where F_x and F_y are the longitudinal and transverse components of

$$F = BlI \qquad \{F_x, F_y, F \text{ in N}\}$$

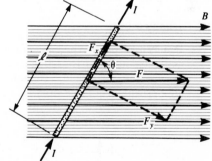

Fig. 8.02–11

(d) Torque M on a loop (Fig. 8.02–12) carrying a current I in a field of magnetic density B is

$$M = BAI \cos \alpha \qquad \left\{ \begin{array}{l} M \text{ in N-m} \\ B \text{ in N/A-m} \\ A \text{ in m}^2 \\ I \text{ in A} \end{array} \right\}$$

where A = area enclosed by the loop, which in this case is a rectangle, although it may have any other shape, such as a circle, an ellipse, etc. For $\alpha = 0°$, M is a maximum; for $\alpha = 90°$, M is zero.

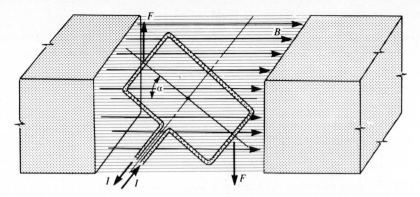

Fig. 8.02-12

(e) Torque M on a coil closely wound in a plane having n turns of wire and carrying a current I in a field of magnetic density B is

$$M = nBAI \cos \alpha$$

where A = area enclosed by the exterior boundary of the coil and α = angle of the plane of the coil with the direction of B.

example:

A circular coil of $r = 10\,\text{cm}$ and consisting of 60 turns carries a current of 30 A in a magnetic field of density $B = 1.2 \times 10^{-2}\,\text{T}$.
When $\alpha = 30°$,

$$M = 60 \times 1.2 \times 10^{-2} \times \pi \times 0.1^2 \times 30 \times 0.5 = 0.33\,\text{N-m}$$

8.03 ELECTROMAGNETIC INDUCTION

(1) Faraday's Law

(a) Duality principle. When an electric current flows through a conductor, a magnetic field is generated in the surrounding medium. In turn, when a conductor moves across the field lines of a magnetic field, a change in voltage is induced in the conductor which gives rise to an electromotive force (to a current) as shown in Fig. 8.03–1.

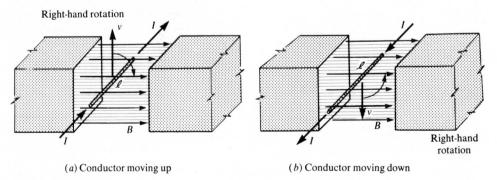

(a) Conductor moving up (b) Conductor moving down

Fig. 8.03–1

(b) Difference in potential induced in a conductor moving with velocity v normal to the density vector B of the magnetic field of width l is

$$V = Bvl \qquad \left\{ \begin{array}{l} V \text{ in V} \\ B \text{ in T} \\ v \text{ in m/s} \\ l \text{ in m} \end{array} \right\}$$

and equals the electromotive force (EMF).

(c) Direction of current produced by the motion described above is defined by the right-hand rotation of the velocity vector v toward the density vector B (Fig. 8.03–2).

(d) SI unit of potential is 1 volt, designated by V and defined as the difference of electric potential between two points of a conducting wire carrying a constant current of 1 ampere when the power dissipated between these points is equal to 1 watt.

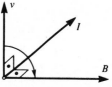

$$1 \text{ volt} = 1 \text{ V} = 1 \text{ T} \times \text{m}^2/\text{s} = 1 \text{ Wb/s} = 1\,\Omega \times \text{A}$$
$$= 1 \text{ N} \times \text{m/A} \times \text{s} = 1 \text{ J/C} = 1 \text{ W/A}$$

where T = tesla, Wb = weber, Ω = ohm, N = newton, J = joule, and W = watt.

Fig. 8.03–2

example:

If a wire of length $l = 1$ m moves normal to the field lines of a magnetic field of density $B = 1$ T and generates an EMF = 2 V, then its velocity must be

$$v = \frac{V}{Bl} = \frac{2}{1 \times 1} = 2 \text{ m/s}$$

(e) Change in magnetic flux with respect to time also induces a change in potential.

$$V = \frac{d\Phi}{dt} = \text{EMF} \qquad \left\{ \begin{array}{l} V \text{ in V} \\ d\Phi/dt \text{ in Wb/s} \end{array} \right\}$$

where $\Phi = BA$ is the magnetic flux (Sec. 8.01–4d) and

$$\frac{d\Phi}{dt} = \frac{dB}{dt} A \qquad \text{or} \qquad \frac{d\Phi}{dt} = B \frac{dA}{dt}$$

where the first change is the time change in density and the second one is the time change in area of the cross section of the field.

(f) Moving coil of n turns with velocity v in a magnetic field of density B develops

$$\text{EMF} = V = nBvl \sin \alpha$$

where l = total length of the wire in the field and α = angle between B and v.

(g) Stationary coil of n turns in a magnetic field varying with time develops a voltage

$$V = n \frac{dB}{dt} A \sin \alpha$$

or

$$V = nB \frac{dA}{dt} \sin \alpha$$

where dB/dt, dA/dt = those of (e) and α = angle of B with the normal of the plane of the coil.

(h) Rotating coil of n turns about the axis normal to the field lines of a magnetic field of density B develops

$$\text{EMF} = V = nBA\omega \sin \alpha$$

where A = area of the coil, ω = angular velocity of rotation, and $\alpha = \omega t =$ instantaneous position angle.

example:

The rectangular coil of Fig. 8.03–3 consists of $n = 50$ and $A = 10 \times 20$ cm; it rotates with $\omega = 100$ rad/s in a magnetic field of density $B = 2 \times 10^{-2}$ Wb/m^2.

The induced voltage is

$$V = 50 \times 2 \times 10^{-2} \times 0.1 \times 0.2 \times 100 \times \sin 100t$$

$$= 2 \sin 100t$$

where $\alpha = 100t$.

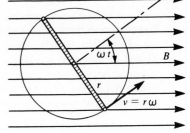

Fig. 8.03-3

For $\alpha = 0°$: $t = 0$	$\sin 0° = 0$	and	$V = 0$
For $\alpha = 30°$: $t = \dfrac{\pi}{600}$	$\sin 30° = 0.5$	and	$V = 1$ V
For $\alpha = 90°$: $t = \dfrac{\pi}{200}$	$\sin 90° = 1$	and	$V = 2$ V

(2) Self-inductance

(a) Change in current I in a circuit induces a change in the flux density of the surrounding medium, which in turn induces changes in the voltage of the circuit. The voltage V at a given time t is

$$V = \frac{IR}{1 - e^{-\alpha t}} \qquad \left\{ \begin{array}{ll} V \text{ in V} & L \text{ in H} \\ I \text{ in A} & \alpha \text{ in 1/s} \\ R \text{ in } \Omega & t \text{ in s} \end{array} \right\}$$

where $\alpha = R/L$, L = constant of self-inductance, R = resistance, and $e = 2.71828$.

(b) SI unit of electric inductance is 1 henry (after Joseph Henry), designated by H and defined as the inductance of a closed circuit in which an electromotive force of 1 volt is produced when the electric current in the circuit varies uniformly at a rate of 1 ampere per second.

$$1 \text{ henry} = 1 \text{ H} = 1 \text{ V} \times \text{s/A} = 1 \text{ T} \times \text{m}^2/\text{A} = 1 \text{ Wb/A} = 1 \text{ N} \times \text{m/A}^2 = 1 \text{ J/A}^2 = 1 \Omega \times \text{s}$$

where V = volt, T = tesla, Wb = weber, Ω = ohm, N = newton, J = joule and W = watt.

(c) Coil of n **turns.** Self-inductance L of a coil of n turns is

$$L = \frac{n\Phi}{I} \qquad \left\{ \begin{array}{l} L \text{ in H} \\ \Phi \text{ in Wb} \\ I \text{ in A} \end{array} \right\}$$

where Φ = magnetic flux produced by I.

(d) Single-layer solenoid. Self-inductance L of a cylindrical solenoid of n turns with core permeability μ is

$$L = \frac{n^2 \mu A}{l} \qquad \left\{ \begin{array}{ll} L \text{ in H} & I \text{ in A} \\ \mu \text{ in N/A}^2 & n \text{ dimensionless} \\ A \text{ in m}^2 & l \text{ in m} \end{array} \right\}$$

where μ is defined in Sec. 8.01–2d and $l =$ length of the solenoid.

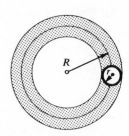

Fig. 8.03-4

(e) Circular toroid. Self-inductance L of a circular toroid of n turns (Fig. 8.03–4) with core permeability μ is

$$L = \mu n^2 (R - \sqrt{R^2 - r^2})$$

where $l = 2\pi R$.

(f) Rectangular toroid. Self-inductance L of a rectangular toroid of n turns (Fig. 8.03–5) with core permeability μ is

$$L = \frac{\mu n^2 b \ln (R/r)}{2\pi}$$

where $l = \pi (R + r)$ and $\ln (R/r) =$ natural logarithm of R/r.

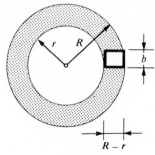

Fig. 8.03–5

(3) Mutual inductance

(a) Change in current I_1 of circuit 1 induces a change in the neighboring circuit 2, and vice versa. Thus

$$\Delta V_1 = M_{12} \frac{dI_2}{dt} \qquad \text{and} \qquad \Delta V_2 = M_{21} \frac{dI_1}{dt} \qquad \left\{ \begin{array}{l} \Delta V \text{ in V} \\ dI/dt \text{ in A/s} \\ M \text{ in H} \end{array} \right\}$$

where $M_{12} = M_{21} =$ constant of mutual inductance.

(b) Two equal-sized coils. Mutual inductance in two coils in which a current I_1 in the first one establishes a flux Φ_2 through n_2 turns in the other one is

$$M_{12} = \frac{n_2 \Phi_2}{I_1} \qquad \left\{ \begin{array}{l} M \text{ in H} \\ n_2 \text{ dimensionless} \\ \Phi \text{ in Wb} \\ I \text{ in A} \end{array} \right\}$$

and vice versa.

(c) Two equal-sized parallel turns. Mutual inductance of two equal-sized circular coaxial turns of radius r placed in parallel planes at a distance d apart is

$$M_{12} = \mu r \alpha^2 [(\alpha^2 + 12) \ln \alpha - 2(\alpha^2 - 2)] \qquad \left\{ \begin{array}{l} M \text{ in H} \\ n \text{ dimensionless} \\ \mu \text{ in N/A}^2 \\ d, r \text{ in m} \end{array} \right\}$$

where $\alpha = 8r/d$ and $\mu =$ permeability of medium (Sec. 8.01–2d).

(d) Two equal-length solenoids. Mutual inductance in two concentric solenoids of length l and of number of turns n_1, n_2 with internal cross-sectional area A is

$$M_{12} = \frac{\mu n_1 n_2 A}{l} \qquad \left\{ \begin{array}{l} M \text{ in H} \\ \mu \text{ in N/A}^2 \\ n \text{ dimensionless} \\ A \text{ in m}^2 \\ l \text{ in m} \end{array} \right\}$$

where $\mu =$ permeability of medium (Sec. 8.01–2d).

example:

If two coils of turns $n_1 = 600$ and $n_2 = 400$ are placed near each other and the current in the first coil $I_1 = 4$ A induces flux $\Phi_1 = 4 \times 10^{-4}$ Wb and $\Phi_2 = 2 \times 10^{-4}$ Wb, the self-inductance in the first one is

$$L_1 = \frac{n_1 \Phi_1}{I_1} = \frac{600 \times 4 \times 10^{-4}}{4} = 0.06 \text{ H}$$

and the mutual inductance is

$$M_{21} = \frac{n_2 \Phi_2}{I_1} = \frac{400 \times 2 \times 10^{-4}}{4} = 0.02 \text{ H}$$

If the current in the first one is turned off after 0.2 second,

$$\text{EMF}_2 = \Delta V_2 = \frac{n_2 \Delta \Phi_2}{\Delta t} = \frac{400 \times (2 \times 10^{-4} - 0)}{0.2} = 0.4 \text{ V}$$

or

$$\text{EMF}_2 = \Delta V_2 = M_{21} \frac{\Delta I_1}{\Delta t} = 0.02 \times \frac{4 - 0}{0.2} = 0.4 \text{ V}$$

(4) Energy of Magnetic Field

(a) Change in current. When the current in an electric circuit increases from 0 to I, the total energy input is

$$U_I = \int_0^I LI \, dI \qquad \left\{ \begin{array}{l} U_I \text{ in J} \\ L \text{ in H} \\ I \text{ in A} \end{array} \right\}$$

where L = inductance.

(b) Constant permeability and inductance. If μ and L are constant in (a), the total energy input is

$$U_I = \frac{1}{2} LI^2 = \frac{1}{2} \frac{\mu A I^2}{l} \qquad \left\{ \begin{array}{ll} U_I \text{ in J} & \mu \text{ in N/A}^2 \\ L \text{ in H} & l \text{ in m} \\ I \text{ in A} & A \text{ in m}^2 \end{array} \right\}$$

where A = area of coil, l = axial length of coil, and μ = permeability of the medium.

(c) Solenoids and toroids. If μ and L are constant, the energy of the magnetic field of a solenoid or toroid of n turns is

$$U_I = \frac{1}{2} \frac{\mu n^2 A I^2}{l}$$

where μ, n, l, and A have the same meaning as in (b).

(5) Magnetic Properties of Matter

(a) Basic classification. With respect to their magnetic conductivity, materials are classified as paramagnetic, diamagnetic, and ferromagnetic.

(b) Paramagnetic material conducts magnetic lines only slightly better than a vacuum; that is, its relative permeability $c = \mu/\mu_0$ is slightly greater than 1.

(c) Diamagnetic material conducts magnetic lines less readily than a vacuum; that is, its relative permeability $c = \mu/\mu_0$ is slightly less than 1.

(d) Ferromagnetic material conducts magnetic lines several hundred times better than a vacuum; that is, its relative permeability $c = \mu/\mu_0$ is many times greater than 1.

8.04 ALTERNATING ELECTRIC CURRENT

(1) Classification and Definitions

(a) Alternating electric current, abbreviated ac, is one which is produced by an alternating potential, and consequently its magnitude and direction vary accordingly (they alternate).

(b) Simplest alternating current is the simple sinusoidal current defined by

$$\text{EMF} = V = V_{max}\sin\omega t$$

$$I = I_{max}\sin\omega t$$

$$\left\{\begin{matrix}\text{EMF, } V \text{ in V} & \omega \text{ in rad/s}\\ I \text{ in A} & t \text{ in s}\end{matrix}\right\}$$

where V = potential (voltage), I = intensity of current, V_{max} = maximum potential, I_{max} = maximum intensity of current, ω = angular frequency (see below), and t = time.

(c) Graphical representation of alternating current is shown by the two sine curves of **Fig. 8.04–1**, which indicate that the magnitudes of V and I vary with time and their sign (sense) changes with time.

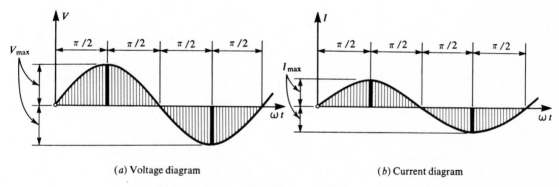

(*a*) Voltage diagram   (*b*) Current diagram

Fig. 8.04–1

(d) Period τ is the time required for the current to go through one complete cycle of variation.

1 cycle = 360 electrical degrees = 2π electrical radians

(e) Frequency f is the number of cycles per unit of time and is the reciprocal of period τ.

$$f = \frac{1}{\tau} \qquad \left\{\begin{matrix}f \text{ in Hz}\\ \tau \text{ in s}\end{matrix}\right\}$$

(f) Angular frequency ω is the number of revolutions in radians per unit of time.

$$\omega = 2\pi f$$

(g) SI unit of frequency is 1 hertz (after Heinrich Hertz), designated by Hz and defined as 1 cycle per second.

1 hertz = 1 Hz = 1 cycle/second = 1 s^{-1}

(h) Relations of τ, f, and ω are

$$\tau = \frac{1}{f} = \frac{2\pi}{\omega} \qquad f = \frac{1}{\tau} = \frac{\omega}{2\pi} \qquad \omega = \frac{2\pi}{\tau} = 2\pi f$$

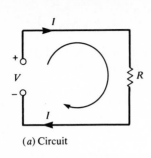

(a) Circuit

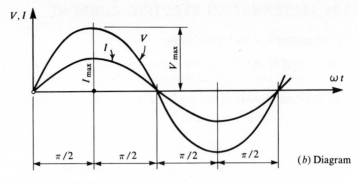

(b) Diagram

Fig. 8.04–2

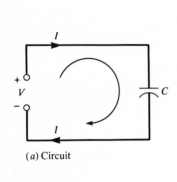

(a) Circuit

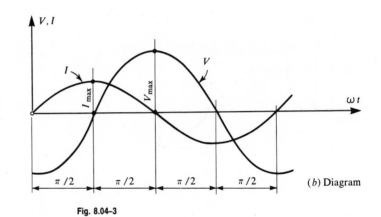

(b) Diagram

Fig. 8.04–3

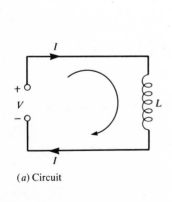

(a) Circuit

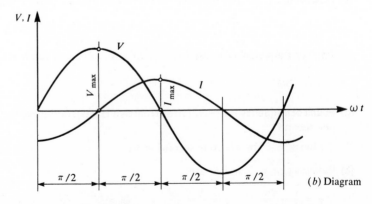

(b) Diagram

Fig. 8.04–4

278   Magnetism and Electrodynamics

(2) Extreme and Effective Values

(a) Extreme values of V and I occur when $\omega t = \pi/2, 3\pi/2$ and are $\pm V_{max}, \pm I_{max}$, as indicated in Fig. 8.04–1.

(b) Effective values of V and I are the corresponding values of direct electric current which produce the same heating effect.

$$V_{eff} = \frac{V_{max}}{\sqrt{2}} = 0.707\,V_{max} \qquad I_{eff} = \frac{I_{max}}{\sqrt{2}} = 0.707\,I_{max}$$

(c) Measuring instruments are calibrated in effective values.

example:

Household voltage of 120 V (V_{eff}) is actually ±170 V and the corresponding current of 10 A (I_{eff}) is ±14.4 A.

The parameters of household current are

$$f = 60 \text{ Hz} \qquad \omega = 120\,\pi/\text{s} \qquad \tau = (1/60)\,\text{s}$$

(3) Resistance, Capacitance, and Inductance

(a) Constant resistance. When the resistance R is constant and the capacitance C and the inductance L are negligible, the ac circuit of Fig. 8.04–2 is governed by

$$V = RI_{max} \sin(2\pi ft)$$
$$I = I_{max} \sin(2\pi ft)$$

$$\left\{ \begin{array}{ll} V \text{ in V} & I \text{ in A} \\ R \text{ in } \Omega & 2\pi ft \text{ in rad} \end{array} \right\}$$

which shows that the voltage and current reach their peak and zero values at the same time; that is, they are in phase. The average power in this case is $P = 1/2\,RI_{max}^2$.

(b) Constant capacitance. When the capacitance C is constant and the resistance R and the inductance L are negligible, the ac circuit of Fig. 8.04–3 is governed by

$$V = X_C I_{max} \sin\left(2\pi ft - \frac{\pi}{2}\right)$$
$$I = I_{max} \sin(2\pi ft)$$

$$\left\{ \begin{array}{ll} V \text{ in V} & X_C \text{ in } \Omega \\ I \text{ in A} & C \text{ in F} \\ 2\pi ft \text{ in rad} & \end{array} \right\}$$

where the capacitive reactance is $X_C = \dfrac{1}{2\pi fC}$

and C = capacitance. The graphs of Fig. 8.04–3 show that voltage lags current by the phase angle $\phi = \pi/2$. The average power delivered to the capacitor in this case is zero.

(c) Constant inductance. When the inductance L is constant and the resistance R and the capacitance C are negligible, the ac circuit of Fig. 8.04–4 is governed by

$$V = X_L I_{max} \sin(2\pi ft)$$
$$I = I_{max} \sin\left(2\pi ft - \frac{\pi}{2}\right)$$

$$\left\{ \begin{array}{ll} V \text{ in V} & X_L \text{ in } \Omega \\ I \text{ in A} & L \text{ in H} \\ 2\pi ft \text{ in rad} & \end{array} \right\}$$

where the inductive reactance is $X_L = 2\pi fL$ and L = inductance. The graphs of Fig. 8.04–4 show that voltage leads current by the phase angle $\phi = \pi/2$. The average power delivered to the inductance in this case is zero.

(4) *RC* Circuit—Steady State

(a) *RC* circuit. When a sinusoidal voltage V of a given frequency f is applied to a circuit of constant resistance R, constant capacitance C, and negligible inductance ($L \cong 0$), the circuit (Fig. 8.04–5) is called an *RC* circuit.

(b) Governing equations of the *RC* circuit are

$$I = \frac{V_{max}}{Z_{RC}} \sin\left(2\pi ft + \phi_{RC}\right)$$

$$V = Z_{RC} I_{max} \sin\left(2\pi ft\right)$$

$$\left\{\begin{array}{l} I \text{ in A} \\ V \text{ in V} \\ Z \text{ in } \Omega \\ 2\pi ft, \phi \text{ in rad} \end{array}\right\}$$

where Z_{RC} = impedance, ϕ_{RC} = phase angle, and the maximum current is $I_{max} = V_{max}/Z_{RC}$.

(c) Impedance Z_{RC} computed from Fig. 8.04–6 is

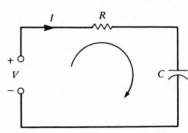

Fig. 8.04–5

$$Z_{RC} = \sqrt{R^2 + X_C^2} = \sqrt{R^2 + \left(\frac{1}{2\pi fC}\right)^2}$$

(d) Phase angle ϕ_{RC} computed from Fig. 8.04–6 is given by

$$\tan \phi_{RC} = \frac{X_C}{R} = \frac{1}{2\pi fRC} \qquad \left\{\begin{array}{l} C \text{ in F} \\ R \text{ in } \Omega \end{array}\right\}$$

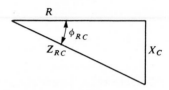

Fig. 8.04–6

example:

In Fig. 8.04–5 if $V_{max} = 120$ V, $f = 60$ Hz, $C = 8 \times 10^{-5}$ F, and $R = 40\,\Omega$, then

$$X_C = \frac{1}{2\pi \times 60 \times 8 \times 10^{-5}} = 33\,\Omega \qquad Z_{RC} = \sqrt{40^2 + 33^2} \cong 52\,\Omega$$

$$\tan \phi_{RC} = 33/52 = 0.63 \qquad \phi_{RC} = 0.56\,\text{rad} \qquad I_{max} = 120/52 = 2.3\,\text{A}$$

(5) *RL* Circuit—Steady State

(a) *RL* circuit. When a sinusoidal voltage V of a given frequency f is applied to a circuit of constant resistance R, constant inductance L, and negligible capacitance ($C \cong 0$), the circuit (Fig. 8.04–7) is called an *RL* circuit.

(b) Governing equations of the *RL* circuit are

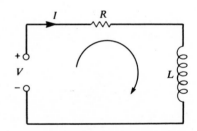

Fig. 8.04–7

$$I = \frac{V_{max}}{Z_{RL}} \sin\left(2\pi ft - \phi_{RL}\right)$$

$$V = Z_{RL} I_{max} \sin\left(2\pi ft\right)$$

$$\left\{\begin{array}{l} I \text{ in A} \\ V \text{ in V} \\ Z \text{ in } \Omega \\ 2\pi ft \text{ in rad} \\ \phi \text{ in rad} \end{array}\right\}$$

where Z_{RL} = impedance, ϕ_{RL} = phase angle, and the maximum current is $I_{max} = V_{max}/Z_{RL}$.

(c) Impedance Z_{RL} computed from Fig. 8.04–8 is

$$Z_{RL} = \sqrt{R^2 + X_L^2} = \sqrt{R^2 + (2\pi fL)^2}$$

(d) Phase angle ϕ_{RL} computed from Fig. 8.04–8 is given by

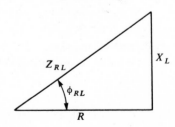

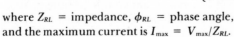

Fig. 8.04–8

$$\tan \phi_{RL} = \frac{X_L}{R} = \frac{2\pi fL}{R} \qquad \left\{\begin{array}{l} L \text{ in H} \\ R \text{ in } \Omega \end{array}\right\}$$

example:

In Fig. 8.04–7 if $V_{max} = 120$ V, $f = 60$ Hz, $L = 0.1$ H, and $R = 40\,\Omega$, then

$$X_C = 2\pi \times 60 \times 0.1 = 38\,\Omega \qquad Z_{RC} = \sqrt{40^2 + 38^2} \cong 55\,\Omega$$

$$\tan \phi_{RL} = 38/55 = 0.69 \qquad \phi_{RL} = 0.60 \text{ rad} \qquad I_{max} = 120/55 = 2.2 \text{ A}$$

(6) RLC Circuit—Steady State

(a) RLC **circuit.** When the sinusoidal voltage V of a given frequency f is applied to a circuit of constant resistance R, constant capacitance C, and constant inductance L, the circuit (Fig. 8.04–9) is called an RLC circuit.

(b) **Governing equations** of the RLC circuit are

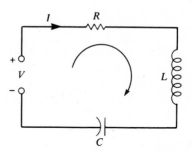

$$I = \frac{V_{max}}{Z} \sin (2\pi ft - \phi)$$

$$V = ZI_{max} \sin (2\pi ft)$$

where $Z =$ impedance, $\phi =$ phase angle, and the maximum current is $I_{max} = V_{max}/Z$.

Fig. 8.04–9

(c) **Impedance** Z computed from Fig. 8.04–10 is

$$Z = \sqrt{R^2 + (X_L - X_C)^2} = \sqrt{R^2 + \left(2\pi fL - \frac{1}{2\pi fC}\right)^2}$$

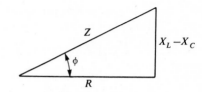

(d) **Phase angle** ϕ computed from Fig. 8.04–10 is given by

Fig. 8.04–10

$$\tan \phi = \frac{X_L - X_C}{R} = \frac{2\pi fL - \dfrac{1}{2\pi fC}}{R} \qquad \begin{cases} L \text{ in H} \\ C \text{ in F} \\ R \text{ in } \Omega \end{cases}$$

(e) **Resonance** in an RLC circuit occurs when $X_L = X_C$. From this condition, the resonant frequency of the current is

$$f = \frac{1}{2\pi \sqrt{CL}} \qquad \begin{cases} f \text{ in Hz} & V \text{ in V} \\ C \text{ in F} & R \text{ in } \Omega \\ L \text{ in H} & I \text{ in A} \end{cases}$$

and

$$V = RI_{max} \sin (t/\sqrt{CL}) \qquad I = I_{max} \sin (t/\sqrt{CL}) \qquad \{t/\sqrt{CL} \text{ in rad}\}$$

which shows that the current is in phase.

(f) **Power** used in an RLC circuit is classified as apparent power P_a, active power P_p, and reactive power P_q.

$$P_a = VI \qquad P_p = VI \cos \phi \qquad P_q = VI \sin \phi \qquad \begin{cases} P_a \text{ in V-A} \\ P_p \text{ in W} \\ P_q \text{ in Var} \end{cases}$$

where in terms of $X_L = 2\pi fL$ and $X_c = \dfrac{1}{2\pi fC}$,

$$\cos \phi = \frac{R}{Z} = \frac{R}{\sqrt{R^2 + (X_L - X_C)^2}} \qquad \sin \phi = \frac{X_L - X_C}{Z} = \frac{X_L - X_C}{\sqrt{R^2 + (X_L - X_C)^2}}$$

and Var $=$ reactive V-A $=$ reactive volt-ampere.

8.05 COMPLEX ALGEBRA OF AC SYSTEMS

(1) Forms of Representation

(a) Complex number

$$N = a + bj \qquad \text{(where } j = \sqrt{-1}\text{)}$$

consists of the real part a and the imaginary part bj and can be represented as a vector in the complex plane of Fig. 8.05–1.

(b) Complex plane vector, called a *planar*, consists of the real component a plotted along the real axis in real units and the imaginary component b in units of j (imaginary units).

(c) Trigonometric form of N based on the geometry of Fig. 8.05–1 is

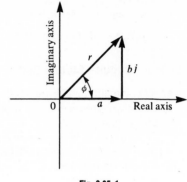

Fig. 8.05–1

$$\hat{N} = a + bj = r\,(\cos\phi + j\sin\phi) = r\underline{/\phi}$$

where $r = \sqrt{a^2 + b^2}$ is the magnitude of N and $\sin\phi = b/\sqrt{a^2+b^2}$, $\cos\phi = a/\sqrt{a^2+b^2}$ are its direction functions.

(d) Exponential form of N based on the series expansion of its trigonometric form is

$$\hat{N} = a + bj = \sqrt{a^2 + b^2}\,e^{j\phi} = re^{j\phi} = r\underline{/\phi}$$

where $e = 2.71828\ldots$ (base of natural logarithms) and ϕ = angle between the real axis and N in Fig. 8.05–1.

example:

If $\hat{N} = 3 + 4j$, then $r = \sqrt{3^2 + 4^2} = 5$, $\tan\phi = 4/3 = 1.33$, $\cos\phi = 3/5 = 0.6$, $\sin\phi = 4/5 = 0.8$, $\phi = 0.927\,\text{rad} = 53°06'$, and $\hat{N} = 5(\cos 0.927 + j\sin 0.927) = 5e^{0.927j} = 5\underline{/0.927}\,\text{rad}$.

(e) Addition and subtraction of complex numbers can best be done in the orthogonal form as

$$(a + bj) + (c + dj) = (a + c) + (b + d)j \qquad (a + bj) + (a - bj) = 2a$$

$$(a + bj) - (c + dj) = (a - c) + (b - d)j \qquad (a + bj) - (a - bj) = 2bj$$

(f) Multiplication and division of complex numbers can be best done in the exponential form as

$$(a + bj)(a + dj) = r_1 e^{j\phi} \cdot r_2 e^{j\phi_2} = r_1 r_2 e^{j(\phi_1 + \phi_2)}$$

$$(a + bj) : (c + dj) = \frac{r_1 e^{j\phi_1}}{r_2 e^{j\phi_2}} = \frac{r_1}{r_2}\,e^{j(\phi_1 - \phi_2)}$$

where $r_1 = \sqrt{a^2 + b^2}$, $r_2 = \sqrt{c^2 + d^2}$, and ϕ_1, ϕ_2 are computed from $\tan\phi_1 = b/a$, $\tan\phi_2 = d/c$.

example:

If $\hat{N}_1 = 3 + 4j$ and $\hat{N}_2 = 12 + 9j$, then $r_1 = 5$, $r_2 = 15$, $\tan\phi_1 = 4/3 = 1.33$, $\phi_1 = 0.927$, $\tan\phi_2 = 9/12 = 0.75$, $\phi_2 = 0.643$, and

$$N_1 N_2 = 5 \times 15 \times e^{j(0.927 + 0.643)} = 75e^{1.570j} = 75\underline{/1.570}\,\text{rad}$$

$$N_1 : N_2 = \frac{5}{15}\,e^{j(0.927 - 0.643)} = 0.33e^{0.284j} = 0.33\underline{/0.284}\,\text{rad}$$

(2) Impedance and Admittance

(a) Impedance triangles in Figs. 8.04–6, 8.04–8, and 8.04–10 indicate the possibility of complex plane representation of Z as shown below.

$$\hat{Z}_{RC} = R + jX_C = (\sqrt{R^2 + X_C^2})\underline{/\phi_{RC}} = Z_{RC}\underline{/\phi_{RC}} \qquad \text{(Fig. 8.04–6)}$$

$$\hat{Z}_{RL} = R + jX_L = (\sqrt{R^2 + X_L^2})\underline{/\phi_{RL}} = Z_{RL}\underline{/\phi_{RL}} \qquad \text{(Fig. 8.04–8)}$$

$$\hat{Z} = R + j(X_L - X_C) = [\sqrt{R^2 + (X_L - X_C)^2}]\underline{/\phi} = Z\underline{/\phi} \qquad \text{(Fig. 8.04–10)}$$

where $X_C = 1/\omega C$, $X_L = \omega L$, $\omega = 2\pi f$, and $\underline{/\phi}$ are the symbols of Secs. 8.04–4 and 8.04–5.

(b) Conductance $\hat{G}$ of an ac circuit of resistance R, reactance X, and impedance Z is defined as

$$\hat{G} = \frac{R}{R^2 + X^2} + j(0) = \frac{R}{Z^2} + j(0) = G\underline{/0} \qquad \left\{ \begin{array}{l} G \text{ in } 1/\Omega \\ R, X \text{ in } \Omega \end{array} \right\}$$

where $X = X_L - X_C$.

(c) Susceptance $\hat{B}$ of an ac circuit of resistance R, reactance X, and impedance Z is defined as

$$\hat{B} = j\frac{X}{R^2 + X^2} = j\frac{X}{Z^2} = B\underline{/\pi/2} \qquad \{B \text{ in } 1/\Omega\}$$

(d) Admittance $\hat{Y}$ is the reciprocal of impedance Z, defined by

$$\hat{Y} = G - jB = \frac{R}{Z^2} - j\frac{X}{Z^2} = Y\underline{/\phi} \qquad \{Y \text{ in } 1/\Omega\}$$

where $Y = \sqrt{G^2 + B^2} = 1/Z$ and $\tan \phi = -B/G = -X/R$.

(3) Series AC Circuit

(a) Vector voltage $\hat{V}$. When two or more vector impedances $\hat{Z}_1, \hat{Z}_2, \ldots, \hat{Z}_n$ are connected in series and the same current I flows through each, then

$$\hat{V} = \hat{Z}_1\hat{I} + \hat{Z}_2\hat{I} + \cdots + \hat{Z}_n\hat{I} = \hat{I}[(R_1 + R_2 + \cdots + R_n) + j(X_1 + X_2 + \cdots + X_n)]$$
$$= \hat{I}(R + jX) = \hat{I}\hat{Z}$$

where $\hat{Z} = $ total impedance vector.

(b) Current $\hat{I}$ flowing in the direction of $\hat{V} = a + jb$ in a circuit of impedance $\hat{Z} = R + jX$ is

$$\hat{I} = \frac{a + jb}{R + jX} = \frac{aR + bX}{R^2 + X^2} + j\frac{bR - aX}{R^2 + X^2} = \frac{V}{Z}\underline{/(\phi_V - \phi_Z)}$$

where $V = \sqrt{a^2 + b^2}$, $Z = \sqrt{R^2 + X^2}$, and the angles ϕ_V, ϕ_Z are given by

$$\tan \phi_V = \frac{b}{a} \qquad \tan \phi_Z = \frac{X}{R}$$

example:

The circuit of Fig. 8.05–2a is supplied with $\text{EMF}_{\text{eff}} = 250$ V, its current is 5 A, and the resistances of the circuit are $R_1 = 10\,\Omega$ and $R_2 = 20\,\Omega$.

The impedance triangle of Fig. 8.05–2b yields

$$IX_L = I\sqrt{Z^2 - (R_1 + R_2)^2} = \sqrt{I^2 Z^2 - I^2(R_1 + R_2)^2}$$
$$= \sqrt{V^2 - I^2(R_1 + R_2)^2} = \sqrt{250^2 - 5^2(10 + 20)^2} = 200 \text{ V}$$

and $X_L = 200/5 = 40\,\Omega$, $\hat{Z} = 30 + 40j = (50\underline{/0.927 \text{ rad}})\,\Omega$, where

$$\tan \phi_Z = 40/30 = 1.33 \qquad \phi_Z = 0.927 \text{ rad}$$

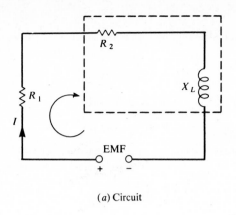

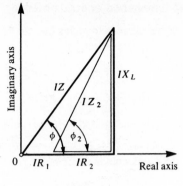

| (a) Circuit | (b) Impedance triangle |

Fig. 8.05-2

(4) Parallel AC Circuit

(a) Vector current $\hat{I}$. When two or more vector impedances $\hat{Z}_1, \hat{Z}_2, \ldots, \hat{Z}_n$ are connected in parallel and the voltage $\hat{V}$ is the same for each, then

$$\hat{I} = \frac{\hat{V}}{\hat{Z}_1} + \frac{\hat{V}}{\hat{Z}_2} + \cdots + \frac{\hat{V}}{\hat{Z}_n} = \hat{V}\left(\frac{1}{\hat{Z}_1} + \frac{1}{\hat{Z}_2} + \cdots + \frac{1}{\hat{Z}_n}\right) = \hat{V}(\hat{Y}_1 + \hat{Y}_2 + \cdots + \hat{Y}_n)$$

$$= \hat{V}[(G_1 + G_2 + \cdots + G_n) - j(B_1 + B_2 + \cdots + B_n)] = \hat{V}(G - jB) = \hat{V}\hat{Y}$$

where $\hat{Y}_1, \hat{Y}_2, \ldots, \hat{Y}_n$ = respective vector admittances (Sec. 8.05-2d), $G_1, G_2, \ldots, G_n$ = respective scalar conductances, and $B_1, B_2, \ldots, B_n$ = respective scalar susceptances.

(b) Vector voltage $\hat{V}$ producing a vector current $\hat{I} = c + jd$ in the circuit of admittance $\hat{Y} = G - jB$ is

$$\hat{V} = \frac{c + jd}{G - jB} = \frac{cG + dB}{G^2 + B^2} + j\frac{dG - cB}{G^2 + B^2} = \frac{I}{Y} \underline{/(\phi_I - \phi_Y)}$$

where $I = \sqrt{c^2 + d^2}$, $Y = \sqrt{G^2 + B^2}$, and the angles ϕ_I and ϕ_Y are given by

$$\tan \phi_I = \frac{d}{c} \qquad \tan \phi_Y = -\frac{B}{G}$$

example:

If the circuit of Fig. 8.05-3a is supplied with EMF = 300 V and its impedance vectors are $\hat{Z}_1 = (20 + 20j)\Omega$ and $\hat{Z}_2 = (10 + 30j)\Omega$, then

$$\hat{Y}_1 = \frac{20}{20^2 + 20^2} - j\frac{20}{20^2 + 20^2} = \frac{0.025 - 0.025j}{\Omega}$$

$$\hat{Y}_2 = \frac{10}{10^2 + 30^2} - j\frac{30}{10^2 + 30^2} = \frac{0.01 - 0.03j}{\Omega}$$

$$\hat{Y} = \hat{Y}_1 + \hat{Y}_2 = \frac{0.035 - 0.055j}{\Omega}$$

$$\hat{I} = \hat{V}\hat{Y} = 300 \times 0.035 - 300 \times 0.055j = (10.5 - 16.5j) \text{ A}$$

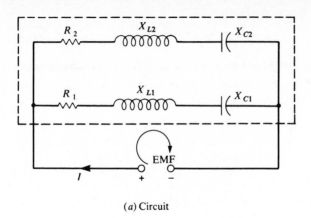

| (a) Circuit | (b) Admittance triangle |

Fig. 8.05–3

(5) Power of the AC Circuit

(a) Active power P_p delivered to or from an ac circuit of $\hat{I} = c + dj$ and $\hat{V} = a + bj$ is

$$P_p = (ac + bd) = \sqrt{c^2 + d^2}\sqrt{a^2 + b^2}\cos(\phi_V - \phi_I)$$

where ϕ_V and ϕ_I are given by

$$\tan\phi_V = \frac{b}{a} \qquad \tan\phi_I = \frac{d}{c}$$

example:

If $\hat{I} = (10\underline{/45°})$ A and $\hat{V} = (120\underline{/60°})$ V, then

$$\hat{I} = 10(\cos 45° + j\sin 45°) = (7.07 + 7.07j)\ \text{A}$$

$$\hat{V} = 120(\cos 60° + j\sin 60°) = (60 + 104j)\ \text{V}$$

and

$$P_p = 424 + 735 = 1{,}159\ \text{W} \qquad \text{or} \qquad P_p = 10 \times 120 \times \cos(60° - 45°) = 1{,}159\ \text{W}$$

(b) Reactive power P_q represents energy supplied to the system during the time when the inductance is being discharged. In terms of (a),

$$P_q = (ad - bc) = \sqrt{a^2 + b^2}\sqrt{c^2 + d^2}\sin(\phi_V - \phi_I)$$

where ϕ_V and ϕ_I are the same as in (a).

example:

For the current of example (a),

$$P_q = 735 - 424 = 311\ \text{Var} \qquad \text{or} \qquad P_q = 10 \times 120 \times \sin(60° - 45°) = 311\ \text{Var}$$

where Var = reactive volt-ampere.

Magnetism and Electrodynamics **285**

8.06 CATALOG OF BASIC AC CIRCUITS

(1) Notation

(a) **Basic elements** and their simple combinations of the ac circuit are shown in Sec. 8.06.   Each case is represented by a symbolic diagram, a complex plane diagram, and the analytical expressions for $\hat{Z}$, Z, $\hat{Y}$, Y, $\tan \phi_Z$, and $\tan \phi_Y$.

(b) **Symbols** used in this section are

$j = \sqrt{-1}$ (dimensionless) $\qquad\phi_Y$ = admittance phase angle (rad)
f = frequency (Hz) $\qquad\phi_Z$ = impedance phase angle (rad)
C = capacitance (F) $\qquad\pi$ = 3.14159 (dimensionless)
L = inductance (H) $\qquad\omega = 2\pi f$ = angular frequency (rad/s)
R = resistance (Ω) $\qquad\hat{R}$ = resistance vector (Ω)
V = voltage (V) $\qquad\hat{V}$ = voltage vector (V)
I = current (A) $\qquad\hat{I}$ = current vector (A)
Y = admittance ($1/\Omega$) $\qquad\hat{Y}$ = admittance vector ($1/\Omega$)
Z = impedance (Ω) $\qquad\hat{Z}$ = impedance vector (Ω)

(c) **Scalar relations:**

$$X_L = 2\pi fL \qquad\qquad X_C = 1/2\pi fC \qquad\qquad X = X_L - X_C$$

$$Z = \sqrt{R^2 + X^2} \qquad\qquad G = R/Z^2 \qquad\qquad B = X/Z^2$$

$$Y = \sqrt{G^2 + B^2} \qquad \tan \phi_Z = X/R \qquad \tan \phi_Y = -X/R$$

$$V = IZ \qquad\qquad I = VY$$

(d) **Vector relations:**

$$\hat{Z} = R + jX = Ze^{j\phi_Z} = Z\underline{/\phi_Z} \qquad \hat{Y} = G - jB = Ye^{j\phi_Y} = Y\underline{/\phi_Y}$$
$$\hat{V} = \hat{I}\hat{Z} \qquad\qquad\qquad\qquad \hat{I} = \hat{V}\hat{Y}$$

(2) Series Combinations

(a) *RL* **system** (Fig. 8.06–1).

$$\hat{Z} = R + j\omega L$$

$$Z = \sqrt{R^2 + (\omega L)^2}$$

$$\hat{Y} = \frac{R}{R^2 + (\omega L)^2} - j\frac{\omega L}{R^2 + (\omega L)^2}$$

$$Y = \sqrt{\left(\frac{R}{R^2 + (\omega L)^2}\right)^2 + \left(\frac{\omega L}{R^2 + (\omega L)^2}\right)^2}$$

$$\tan \phi_Z = \omega L/R = -\tan \phi_Y$$

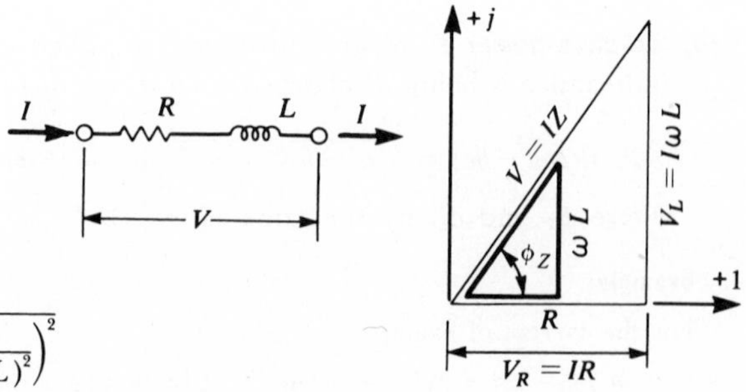

Fig. 8.06–1

(b) *RC* **system** (Fig. 8.06–2).

$$\hat{Z} = R - j(1/\omega C)$$

$$Z = \sqrt{R^2 + (1/\omega C)^2}$$

$$\hat{Y} = \frac{R}{R^2 + (1/\omega C)^2} + j\frac{1/\omega C}{R^2 + (1/\omega C)^2}$$

$$Y = \sqrt{\left(\frac{R}{R^2 + (1/\omega C)^2}\right)^2 + \left(\frac{1/\omega C}{R^2 + (1/\omega C)^2}\right)^2}$$

$$\tan \phi_Z = -1/\omega CR = -\tan \phi_Y$$

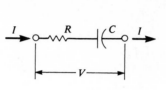
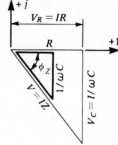

Fig. 8.06-2

(c) *LC* **system** (Fig. 8.06–3).

$$\hat{Z} = j(\omega L - 1/\omega C)$$

$$Z = \omega L - 1/\omega C$$

$$\hat{Y} = -j\frac{1}{\omega L - 1/\omega C}$$

$$Y = \frac{1}{\omega L - 1/\omega C}$$

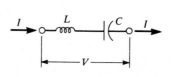
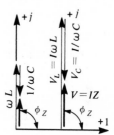

Fig. 8.06-3

For $(\omega L - 1/\omega C) > 0$: $\tan \phi_Z = +\infty$ $\tan \phi_Y = -\infty$

For $(\omega L - 1/\omega C) < 0$: $\tan \phi_Z = -\infty$ $\tan \phi_Y = +\infty$

(d) *RLC* **system** (Fig. 8.06–4).

$$\hat{Z} = R + j(\omega L - 1/\omega C)$$

$$Z = \sqrt{R^2 + (\omega L - 1/\omega C)^2}$$

$$\hat{Y} = \frac{R}{Z^2} - j\frac{(\omega L - 1/\omega C)}{Z^2}$$

$$Y = \frac{1}{\sqrt{R^2 + (\omega L - 1/\omega C)^2}}$$

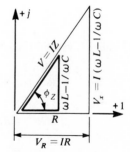

Fig. 8.06-4

$$\tan \phi_Z = \frac{\omega L - 1/\omega C}{R} \qquad \tan \phi_Y = \frac{1/\omega C - \omega L}{R}$$

(3) Parallel Combinations

(a) *RR* **system** (Fig. 8.06–5).

$$\hat{Z} = \frac{R_1 R_2}{R_1 + R_2} + j(0)$$

$$Z = \frac{R_1 R_2}{R_1 + R_2}$$

$$\hat{Y} = \frac{R_1 + R_2}{R_1 R_2} + j(0)$$

$$Y = \frac{R_1 + R_2}{R_1 R_2} \qquad \tan \phi_Z = 0 = \tan \phi_Y$$

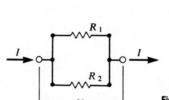
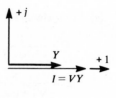

Fig. 8.06-5

(b) *LL* **system** (Fig. 8.06–6).

$$\hat{Z} = j\frac{\omega L_1 L_2}{L_1 + L_2}$$

$$Z = \frac{\omega L_1 L_2}{L_1 + L_2}$$

$$\hat{Y} = -j\frac{L_1 + L_2}{\omega L_1 L_2}$$

$$Y = \frac{L_1 + L_2}{\omega L_1 L_2} \qquad \tan \phi_Z = \infty = -\tan \phi_Y$$

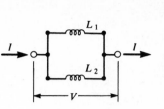

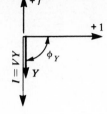

Fig. 8.06–6

(c) *CC* **system** (Fig. 8.06–7)

$$\hat{Z} = -j\frac{1}{\omega(C_1 + C_2)}$$

$$Z = \frac{1}{\omega(C_1 + C_2)}$$

$$\hat{Y} = j\omega(C_1 + C_2)$$

$$Y = \omega(C_1 + C_2) \qquad \tan \phi_Z = -\infty = -\tan \phi_Y$$

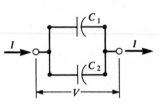

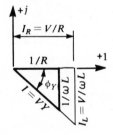

Fig. 8.06–7

(d) *RL* **system** (Fig. 8.06–8).

$$\hat{Z} = \frac{R(\omega L)^2}{R^2 + (\omega L)^2} + j\frac{R^2 \omega L}{R^2 + (\omega L)^2}$$

$$Z = \frac{\omega L R}{\sqrt{R^2 + (\omega L)^2}}$$

$$\hat{Y} = \frac{1}{R} - j\frac{1}{\omega L}$$

$$Y = \frac{\sqrt{R^2 + (\omega L)^2}}{\omega L R} \qquad \tan \phi_Z = \frac{R}{\omega L} = -\tan \phi_Y$$

Fig. 8.06–8

(e) *RC* **system** (Fig. 8.06–9).

$$\hat{Z} = \frac{R}{1 + (\omega C R)^2} - j\frac{\omega C R^2}{1 + (\omega C R)^2}$$

$$Z = \frac{R}{\sqrt{1 + (\omega C R)^2}}$$

$$\hat{Y} = \frac{1}{R} + j\omega C$$

$$Y = \sqrt{(1/R)^2 + (\omega C)^2} \qquad \tan \phi_z = -\omega C R = -\tan \phi_Y$$

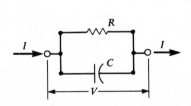

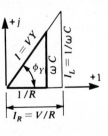

Fig. 8.06–9

(f) *LC* **system** (Fig. 8.06–10).

$$\hat{Z} = j\frac{1}{1/\omega L - \omega C}$$

$$Z = \frac{1}{1/\omega L - \omega C}$$

$$\hat{Y} = -j(1/\omega L - \omega C) \qquad \text{If } (1/\omega L - \omega C) > 0: \quad \tan \phi_Z = +\infty = -\tan \phi_Y$$

$$Y = 1/\omega L - \omega C \qquad \text{If } (1/\omega L - \omega C) < 0: \quad \tan \phi_Z = -\infty = -\tan \phi_Y$$

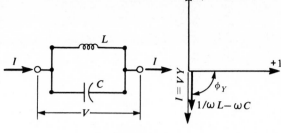

Fig. 8.06–10

(g) *RLC* **system** (Fig. 8.06–11).

$$\hat{Z} = \frac{1/R}{N} + j\frac{1/\omega L - \omega C}{N}$$

$$N = (1/R)^2 + (1/\omega L - \omega C)^2$$

$$Z = 1/\sqrt{N}$$

$$\hat{Y} = 1/R - j(1/\omega L - \omega C)$$

$$Y = \sqrt{N} \qquad \tan \phi_Z = \frac{1/\omega L - \omega C}{1/R} = -\tan \phi_Y$$

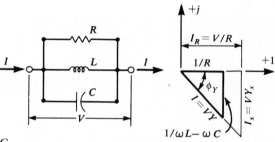

Fig. 8.06–11

(4) Parallel-Series Combinations

(a) (*RR* **parallel** + *R*) **system** (Fig. 8.06–12).

$$\hat{Z}_{12} = \frac{R_1 R_2}{N}$$

$$\hat{Z}_3 = R_3 \qquad \hat{Z} = \hat{Z}_{12} + \hat{Z}_3$$

$$N = R_1 + R_2$$

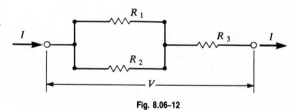

Fig. 8.06–12

(b) (*RR* **parallel** + *L*) **system** (Fig. 8.06–13).

$$\hat{Z}_{12} = \frac{R_1 R_2}{N}$$

$$\hat{Z}_3 = j\omega L_3 \qquad \hat{Z} = \hat{Z}_{12} + \hat{Z}_3$$

$$N = R_1 + R_2$$

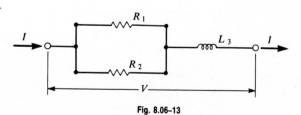

Fig. 8.06–13

(c) (RR parallel $+ C$) system (Fig. 8.06–14).

$$\hat{Z}_{12} = \frac{R_1 R_2}{N}$$

$$\hat{Z}_3 = \frac{-j}{\omega C_3} \qquad \hat{Z} = \hat{Z}_{12} + \hat{Z}_3$$

$$N = R_1 + R_2$$

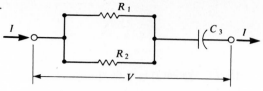
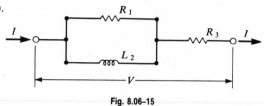

Fig. 8.06–14

(d) (RL parallel $+ R$) system (Fig. 8.06–15).

$$\hat{Z}_{12} = \frac{R_1(\omega L_2)^2}{N} + j\frac{R_1^2 \omega L_2}{N}$$

$$\hat{Z}_3 = R_3 \qquad \hat{Z} = \hat{Z}_{12} + \hat{Z}_3$$

$$N = R_1^2 + (\omega L_2)^2$$

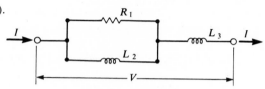

Fig. 8.06–15

(e) (RL parallel $+ L$) system (Fig. 8.06–16).

$$\hat{Z}_{12} = \frac{R_1(\omega L_2)^2}{N} + j\frac{R_1^2 \omega L_2}{N}$$

$$\hat{Z}_3 = j\omega L_3 \qquad \hat{Z} = \hat{Z}_{12} + \hat{Z}_3$$

$$N = R_1^2 + (\omega L_2)^2$$

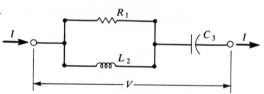

Fig. 8.06–16

(f) (RL parallel $+ C$) system (Fig. 8.06–17).

$$\hat{Z}_{12} = \frac{R_1(\omega L_2)^2}{N} + j\frac{R_1^2 \omega L_2}{N}$$

$$\hat{Z}_3 = \frac{-j}{\omega C_3} \qquad \hat{Z} = \hat{Z}_{12} + \hat{Z}_3$$

$$N = R_1^2 + (\omega L_2)^2$$

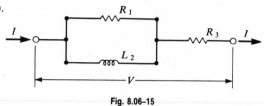

Fig. 8.06–17

(g) (RC parallel $+ R$) system (Fig. 8.06–18).

$$\hat{Z}_{12} = \frac{R_1}{N} - j\frac{R_1^2 \omega C_2}{N}$$

$$\hat{Z}_3 = R_3 \qquad \hat{Z} = \hat{Z}_{12} + \hat{Z}_3$$

$$N = 1 + (R_1 \omega C_2)^2$$

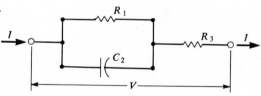

Fig. 8.06–18

(h) (RC parallel $+ L$) system (Fig. 8.06–19).

$$\hat{Z}_{12} = \frac{R_1}{N} - j\frac{R_1^2 \omega C_2}{N}$$

$$\hat{Z}_3 = \omega L_3 \qquad \hat{Z} = \hat{Z}_{12} + \hat{Z}_3$$

$$N = 1 + (R_1 \omega C_2)^2$$

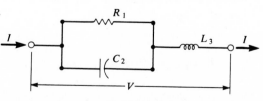

Fig. 8.06–19

(i) $(RC$ **parallel** $+ C)$ **system** (Fig. 8.06–20).

$$\hat{Z}_{12} = \frac{R_1}{N} - j\frac{R_1^2\omega C_2}{N}$$

$$\hat{Z}_3 = \frac{-j}{\omega C_3} \qquad \hat{Z} = \hat{Z}_{12} + \hat{Z}_3$$

$$N = 1 + (R_1\omega C_2)^2$$

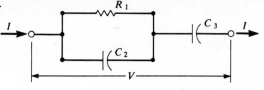

Fig. 8.06–20

(j) $(LC$ **parallel** $+ R)$ **system** (Fig. 8.06–21).

$$\hat{Z}_{12} = j\frac{\omega L_1}{N}$$

$$\hat{Z}_3 = R_3 \qquad \hat{Z} = \hat{Z}_{12} + \hat{Z}_3$$

$$N = 1 - \omega^2 L_1 C_2$$

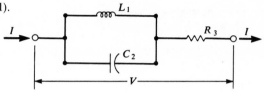

Fig. 8.06–21

(k) $(LC$ **parallel** $+ L)$ **system** (Fig. 8.06–22).

$$\hat{Z}_{12} = j\frac{\omega L_1}{N}$$

$$\hat{Z}_3 = j\omega L_3 \qquad \hat{Z} = \hat{Z}_{12} + \hat{Z}_3$$

$$N = 1 - \omega^2 L_1 C_2$$

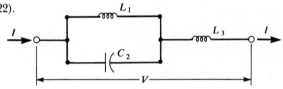

Fig. 8.06–22

(l) $(LC$ **parallel** $+ C)$ **system** (Fig. 8.06–23).

$$\hat{Z}_{12} = j\frac{\omega L_1}{N}$$

$$\hat{Z}_3 = \frac{-j}{\omega C_3} \qquad \hat{Z} = \hat{Z}_{12} + \hat{Z}_3$$

$$N = 1 - \omega^2 L_1 C_2$$

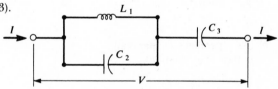

Fig. 8.06–23

(5) Parallel-Parallel Combinations

(a) $(RR$ **parallel** $+ R$ **parallel)** **system** (Fig. 8.06–24).

$$\hat{Y}_{12} = \frac{R_1 + R_2}{N}$$

$$\hat{Y}_3 = \frac{1}{R_3} \qquad \hat{Y} = \hat{Y}_{12} + \hat{Y}_3$$

$$N = R_1 R_2$$

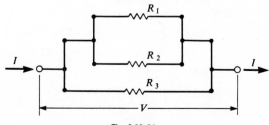

Fig. 8.06–24

(b) $(RR$ **parallel** $+ L$ **parallel)**
system (Fig. 8.06–25).

$$\hat{Y}_{12} = \frac{R_1 + R_2}{N}$$

$$\hat{Y}_3 = \frac{-j}{\omega L_3} \qquad \hat{Y} = \hat{Y}_{12} + \hat{Y}_3$$

$$N = R_1 R_2$$

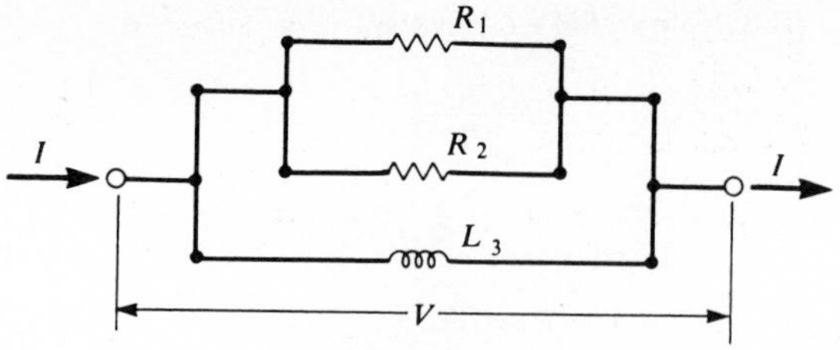

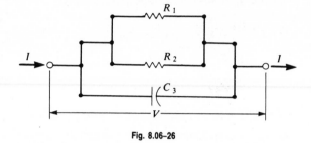

Fig. 8.06–25

(c) $(RR$ **parallel** $+ C$ **parallel)**
system (Fig. 8.06–26).

$$\hat{Y}_{12} = \frac{R_1 + R_2}{N}$$

$$\hat{Y}_3 = j\omega C_3 \qquad \hat{Y} = \hat{Y}_{12} + \hat{Y}_3$$

$$N = R_1 R_2$$

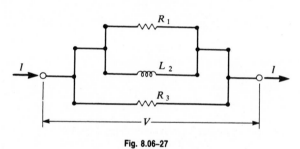

Fig. 8.06–26

(d) $(RL$ **parallel** $+ R$ **parallel)**
system (Fig. 8.06–27).

$$\hat{Y}_{12} = \frac{\omega L_2}{N} - j\frac{R_1}{N}$$

$$\hat{Y}_3 = \frac{1}{R_3} \qquad \hat{Y} = \hat{Y}_{12} + \hat{Y}_3$$

$$N = \omega R_1 L_2$$

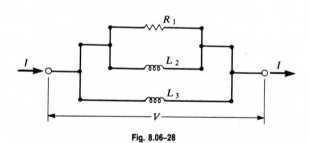

Fig. 8.06–27

(e) $(RL$ **parallel** $+ L$ **parallel)**
system (Fig. 8.06–28).

$$\hat{Y}_{12} = \frac{\omega L_2}{N} - j\frac{R_1}{N}$$

$$\hat{Y}_3 = \frac{-j}{\omega L_3} \qquad \hat{Y} = \hat{Y}_{12} + \hat{Y}_3$$

$$N = \omega R_1 L_2$$

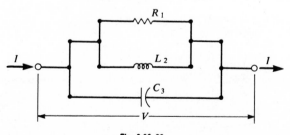

Fig. 8.06–28

(f) $(RL$ **system** $+ C$ **parallel)**
system (Fig. 8.06–29).

$$\hat{Y}_{12} = \frac{\omega L_2}{N} - j\frac{R_1}{N}$$

$$\hat{Y}_3 = j\omega C_3 \qquad \hat{Y} = \hat{Y}_{12} + \hat{Y}_3$$

$$N = \omega R_1 L_2$$

Fig. 8.06–29

(g) (RC parallel $+ R$ parallel)
system (Fig. 8.06–30).

$$\hat{Y}_{12} = \frac{1}{N} + j\frac{\omega R_1 C_2}{N}$$

$$\hat{Y}_3 = \frac{1}{R_3} \qquad \hat{Y} = \hat{Y}_{12} + \hat{Y}_3$$

$$N = R_1$$

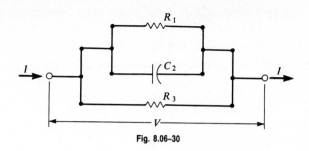

Fig. 8.06–30

(h) (RC parallel $+ L$ parallel)
system (Fig. 8.06–31).

$$\hat{Y}_{12} = \frac{1}{N} + j\frac{\omega R_1 C_2}{N}$$

$$\hat{Y}_3 = \frac{-j}{\omega L_3} \qquad \hat{Y} = \hat{Y}_{12} + \hat{Y}_3$$

$$N = R_1$$

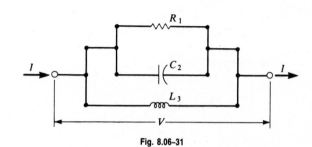

Fig. 8.06–31

(i) (RC parallel $+ C$ parallel)
system (Fig. 8.06–32).

$$\hat{Y}_{12} = \frac{1}{N} + j\frac{\omega R_1 C_2}{N}$$

$$\hat{Y}_3 = j\omega C_3 \qquad \hat{Y} = \hat{Y}_{12} + \hat{Y}_3$$

$$N = R_1$$

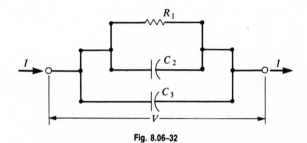

Fig. 8.06–32

(6) Quality Factor

(a) Definition. The quality factor of coils, capacitors, and circuits is defined as

$$\bar{Q} = 2\pi \frac{\text{maximum stored energy}}{\text{energy dissipated per cycle}}$$

(b) Particular cases of this factor are:

For combination of Sec. 8.06–2a $\bar{Q} = \omega L/R$
For combination of Sec. 8.06–2b $\bar{Q} = 1/\omega CR$
For combination of Sec. 8.06–2d $\bar{Q} = \sqrt{LC}/CR$
where the last case corresponds to the *resonant state*.

8.07 TRANSIENTS IN ELECTRIC CURRENTS

(1) Classification and Definitions

(a) **Transient state** of all electric current is the state of transition between the change of EMF and/or change in impedance condition and the establishment of a steady state (when I, V, and Z stay constant).

(b) **Electric circuit elements** are classified into two categories: active and passive. The active circuit element is the source of electromotive force. The passive circuit elements are resistors, capacitors, and inductors.

(c) **Resistor voltage change** is

$$\Delta V_R = V_{R1} - V_{R2} = RI \qquad \begin{cases} \Delta V, \text{ V in V} \\ R \text{ in } \Omega \\ I \text{ in A} \end{cases}$$

where the current flows from the point of higher voltage V_1 to the point of lower voltage V_2.

(d) **Uncharged capacitor** becomes charged when connected to two points of different voltage, and its change in voltage is

$$\Delta V_C = V_{C2} - V_{C1} = \frac{1}{C} \int_{t_1}^{t_2} I \, dt \qquad \begin{cases} \Delta V, \text{ V in V} \quad I \text{ in A} \\ C \text{ in F} \qquad t \text{ in s} \end{cases}$$

where V_{C1} = voltage at t_1 and V_{C2} = voltage at t_2.

(e) **Inductor voltage change** is

$$\Delta V_L = V_{L2} - V_{L1} = L \frac{dI}{dt} \qquad \begin{cases} \Delta V, \text{ V in V} \quad I \text{ in A} \\ L \text{ in H} \qquad t \text{ in s} \end{cases}$$

where V_{L1} = voltage at t_1 and V_{L2} = voltage at t_2.

(f) **Superposition** of (c), (d), and (e) yields the governing equation of the transient state.

$$\Delta V = RI + L \frac{dI}{dt} + \frac{1}{C} \int I \, dt$$

is the change in voltage in the transient state.

(g) **Three basic cases** of transient state, RC transient state, RL transient state, and RLC transient state, are analyzed in this section.

(2) *RL* System, DC Circuit

(a) **System.** When a constant $EMF = V$ is applied at $t = 0$ to a circuit of constant resistance R and constant inductance L (Fig. 8.07–1a) in which an initial current I_0 is already flowing, the expression for current at any subsequent time t given by the solution of the equation of Sec. 8.07–1f is

$$I = \frac{V}{R} - \left(\frac{V}{R} - I_0 \right) e^{-\alpha t} \qquad \begin{cases} I \text{ in A} \qquad L \text{ in H} \\ V \text{ in V} \qquad \alpha \text{ in 1/s} \\ R \text{ in } \Omega \qquad t \text{ in s} \end{cases}$$

where $\alpha = R/L$, $e = 2.71828$, I = current at t, and I_0 = current at $t = 0$.

(b) **Time constant** $t_m = 1/\alpha$ in seconds is the time required for the current I to build up to 63.2 percent of its ultimate value V/R.

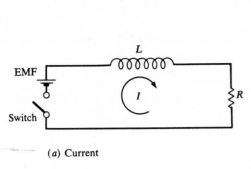

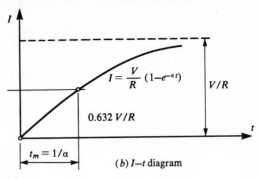

(a) Current

(b) I–t diagram

Fig. 8.07–1

(c) Initial current. If the initial current $I_0 = 0$, equation (a) reduces to

$$I = \frac{V}{R}(1 - e^{-\alpha t})$$

which is shown graphically in Fig. 8.07–1b. Theoretically, after an infinite time the current I reaches the maximum value V/R. In practical cases, it is a matter of seconds.

(d). Short-circuit current is obtained by putting $V = 0$ in (a); that is,

$$I = I_0 e^{-\alpha t}$$

from which the current at $t = 0$ is I_0, at $t = 1/\alpha$ is $0.368 I_0$, and at $t = \infty$ is zero.

(3) *RC* System, DC Circuit

(a) System. When a constant EMF $= V$ is applied at $t = 0$ to a circuit of constant resistance R and of constant capacitance C with initial charge Q_0 (Fig. 8.07–2), the current at any subsequent time t given by the solution of the equation of Sec. 8.07–1f is

$$I = \left(\frac{V}{R} - \frac{V_0}{R}\right) e^{-\beta t} \qquad \begin{cases} I \text{ in A} & C \text{ in F} \\ V \text{ in V} & t \text{ in s} \\ R \text{ in } \Omega & \beta \text{ in 1/s} \end{cases}$$

where $\beta = 1/RC$, $e = 2.71828$, $V_0 = $ voltage in capacitor at $t = 0$, and $I = $ current at t.

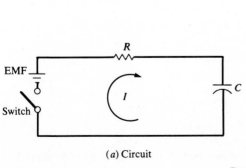

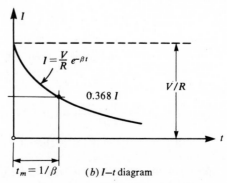

(a) Circuit

(b) I–t diagram

Fig. 8.07–2

(b) Capacitor charge Q is the time integral of the current; that is,

$$Q = \int_0^t I \, dt = CV - (CV - Q_0)e^{-\beta t} \qquad \{Q \text{ in C}\}$$

where Q_0 = initial charge at $t = 0$.

(c) Initial charge. If the initial charge $Q_0 = 0$, equation (a) reduces to

$$I = \frac{V}{R} e^{-\beta t}$$

which is shown graphically in Fig. 8.07–2b. Theoretically, after an infinite time the current I becomes zero. In practical cases, it is a matter of seconds. Similarly, if $Q_0 = 0$, from (b),

$$Q = CV(1 - e^{-\beta t})$$

which is shown graphically in Fig. 8.07–3.

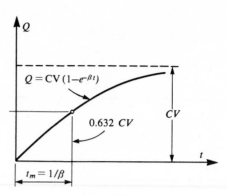

Fig. 8.07-3

(d) Time constant $t_m = 1/\beta$ in seconds is the time required for the charge Q to build up to 63.2 percent of its ultimate value CV.

(e) Short-circuit current is obtained by putting $V = 0$ in (a); that is,

$$I = \frac{V_0}{R} e^{-\beta t} = I_0 e^{-\beta t}$$

from which the current at $t = 0$ is I_0, at $t = 1/\beta$ is $0.368 I_0$, and at $t = \infty$ is zero.

(4) *RLC* System, DC Circuit

(a) System. When a constant EMF $= V$ is applied at $t = 0$ to the circuit of constant resistance R, constant inductance L, and constant capacitance C with initial values I_0 and V_0 at $t = 0$, the expression for current at any subsequent time t given by the solution of the equation of Sec. 8.07–1f is

$$I = \frac{V - V_0 - (\alpha - \beta)I_0}{2L\beta} e^{-(\alpha - \beta)t} - \frac{V - V_0 - (\alpha + \beta)I_0}{2L\beta} e^{-(\alpha + \beta)t} \qquad \left\{ \begin{array}{ll} I \text{ in A} & \alpha, \beta \text{ in 1/s} \\ Q \text{ in C} & t \text{ in s} \\ R \text{ in } \Omega & V \text{ in V} \\ L \text{ in H} & C \text{ in F} \end{array} \right\}$$

where $\alpha = R/2L$ and $\beta = \sqrt{R^2/4L^2 - 1/LC}$.

(b) Case I. If $\beta = 0$,

$$I = \left[I_0 + \frac{2(V - V_0) - RI_0}{2L} t \right] e^{-\alpha t}$$

and if $I_0 = 0$ and $V_0 = 0$, then $I = (V/L)te^{-\alpha t}$.

(c) Case II. If $R^2/4L^2 < 1/LC$, $\lambda = \sqrt{1/LC - R^2/4L^2}$,

$$I = \left[\frac{2(V - V_0) - RI_0}{2L\beta} \sin \lambda t + I_0 \cos \lambda t \right] e^{-\alpha t}$$

and if $I_0 = 0$ and $V_0 = 0$, then $I = (V/L\beta)e^{-\alpha t} \sin \lambda t$.

(d) Case III. If $R^2/4L^2 > 1/LC$, I is given by (a) and if $I_0 = 0$ and $V_0 = 0$, then $I = (V/L\beta)e^{-\alpha t} \sinh \beta t$.

(e) **Time constant** defined in each case as the time required for the current to build up to its maximum value is

Case I: $\quad t_m = \dfrac{2L}{R}$ $\qquad$ (in seconds)

Case II: $\quad t_m = \dfrac{1}{\lambda}\tan^{-1}\dfrac{\lambda}{\alpha}$ $\qquad$ (in seconds)

Case III: $\quad t_m = \dfrac{1}{\beta}\tanh^{-1}\dfrac{\beta}{\alpha}$ $\qquad$ (in seconds)

(5) *RL* System, AC Circuit

(a) **System.** When an alternating voltage of angular frequency ω given as $V_{max}\sin(\omega t + \theta)$ is applied at $t = 0$ to an RL series circuit of constant resistance R and constant inductance L, the current I at any subsequent time t given as the solution of the general equation in Sec. 8.07–1f is

$$I = I_t + I_s \qquad \left\{ \begin{array}{ll} I \text{ in A} & \alpha, \omega \text{ in rad/s} \\ V \text{ in V} & t \text{ in s} \\ R \text{ in } \Omega & \theta, \phi \text{ in rad} \\ L \text{ in H} & Z \text{ in } \Omega \end{array} \right\}$$

where the *transient current*

$$I_t = -\frac{V_{max}}{Z} e^{-\alpha t} \sin(\theta - \phi)$$

and the *steady-state current*

$$I_s = \frac{V_{max}}{Z} \sin(\omega t + \theta - \phi)$$

in which

$$\alpha = \frac{R}{L} \qquad Z = \sqrt{R^2 + \omega^2 L^2} \qquad \phi = \tan^{-1}\frac{\omega L}{R}$$

(b) **Transient current** I_t is governed by the decay factor $e^{-\alpha t}$, which becomes zero in a relatively short time, and the constant, $\sin(\theta - \phi)$, which depends upon the time in the cycle at which the voltage has been applied. For $\theta - \phi = 0, 2\pi, 4\pi, \ldots, 2k\pi, \ldots$ this constant is zero and the current goes directly into the steady state. For $\theta - \phi = \pi/2, 3\pi/2,$ $\ldots, (2k-1)\pi/2, \ldots$ this constant equals 1 and the current has the maximum possible amplitude. The *steady-state current* I_s lags the application of voltage by the phase angle ϕ.

(6) *RC* System, AC Circuit

(a) **System.** When the same voltage is applied to an RC series circuit of constant resistance R and constant capacitance C with the initial charge Q_0, the current I at any subsequent time t given as the solution of the general equation in Sec. 8.07–1f is

$$I = I_t + I_s \qquad \left\{ \begin{array}{ll} I \text{ in A} & \beta, \omega \text{ in rad/s} \\ V \text{ in V} & t \text{ in s} \\ R \text{ in } \Omega & \theta, \phi \text{ in rad} \\ C \text{ in F} & Z \text{ in } \Omega \\ Q \text{ in C} \end{array} \right\}$$

where *the transient current*

$$I_t = \left[\frac{V_{max}}{R}\sin\theta - \frac{V_{max}}{Z}\sin(\theta + \phi) + \beta Q_0 \right] e^{-\beta t}$$

and the *steady-state current*

$$I_s = \frac{V_{max}}{R}\sin(\omega\tau + \theta + \phi)$$

in which $\beta = \dfrac{1}{RC}$ $\qquad \phi = \tan^{-1}\dfrac{1}{\omega RC}$ $\qquad Z = \sqrt{R^2 + \left(\dfrac{1}{\omega C}\right)^2}$

(b) **Transient current** I_t is again governed by the decay factor $e^{-\beta t}$ and the constant in the brackets. The *steady-state current* I_s leads the applied voltage by the phase angle ϕ.

(7) *RLC* System, AC Circuit

(a) **System.** Finally, when the same voltage is applied to an *RLC* series circuit of constant R, L, C with initial charge Q_0, the current I at any subsequent time t given by the general solution of the equation in Sec. 8.07–1f is

$$I = I_t + I_s \qquad \left\{ \begin{array}{ll} I \text{ in A} & Q \text{ in C} \\ V \text{ in V} & \lambda,\ \alpha,\ \beta,\ \omega \text{ in rad/s} \\ R \text{ in } \Omega & t \text{ in s} \\ L \text{ in H} & \theta,\ \phi \text{ in rad} \\ C \text{ in F} & Z \text{ in } \Omega \end{array} \right\}$$

where the *transient current* I_t takes on one of the forms given in (b), (c), (d) below and where the *steady-state current*

$$I_s = \frac{V_{\max}}{Z} \sin (\omega t + \theta - \phi)$$

in which
$$\alpha = \frac{R}{2L} \qquad\qquad \phi = \tan^{-1} \frac{\omega L - \dfrac{1}{\omega C}}{R}$$

$$Z = \sqrt{R^2 + \left(\omega L - \frac{1}{\omega C} \right)^2} \qquad K_1 = -\frac{V_{\max}}{Z} \sin (\theta - \phi)$$

(b) **Case I.** If $(R/2L)^2 = 1/LC$, then

$$I_t = (K + K_2 t)e^{-\alpha t}$$

where α, K_1 are given in (a) above and

$$K_2 = \frac{V_{\max}}{L} \left[\sin \theta - \frac{R}{2Z} \sin (\theta - \phi) - \frac{\omega L}{Z} \cos (\theta - \phi) \right] - \frac{Q_0}{LC}$$

(c) **Case II.** If $(R/2L)^2 < 1/LC$ and $\lambda = \sqrt{(1/LC) - (R/2L)^2}$, then

$$I_t = (K_1 \cos \lambda t + K_2 \sin \lambda t)e^{-\alpha t}$$

where α, K_1 are given in (a) above and

$$K_2 = \frac{V_{\max}}{\lambda L} \left[\sin \theta - \frac{R}{2Z} \sin (\theta - \phi) - \frac{\omega L}{Z} \cos (\theta - \phi) \right] - \frac{Q_0}{\lambda LC}$$

(d) **Case III.** If $(R/2L)^2 > 1/LC$ and $\beta = \sqrt{(R/2L)^2 - (1/LC)}$, then

$$I_t = (K_1 \cosh \beta t + K_2 \sinh \beta t)e^{-\alpha t}$$

where α, K_1 are given in (a) above and

$$K_2 = \frac{V_{\max}}{L} \left[\sin \theta + \frac{R}{2Z} \sin (\theta - \phi) - \frac{\omega L}{Z} \cos (\theta - \phi) \right] - \frac{Q_0}{\beta LC}$$

(e) **Classification.** Case I represents the critically damped transient, case II represents the underdamped (oscillatory) transient, and case III represents the overdamped transient. The graphs of these transients are formally identical to the graphs of free linear vibrations with damping depicted in Sec. 9.01–5.

9

VIBRATION AND ACOUSTICS

9.01 MECHANICAL VIBRATION

(1) Classification and Definitions

(a) **Vibration** (oscillation) of a mass particle is defined as its motion about its position of equilibrium and can be harmonic or nonharmonic.

(b) **Harmonic vibration** is a repetitive motion whose time-displacement diagram is a sine curve (or cosine curve) of constant amplitude and period (Sec. 9.01–2).

(c) **Nonharmonic vibration** is a nonrepetitive motion whose time-displacement curve is nonuniform with variable amplitude and/or period.

(d) **Longitudinal vibration** is a vibration of a mass particle parallel to the direction of energy transfer (propagation).

(e) **Transverse vibration** is a vibration of the mass particle normal to the direction of energy propagation.

(f) **Mechanical vibration** is produced by mechanical causes (forces, displacements, velocities).

(2) Mechanical Models

(a) **Concept.** The vibrations introduced above can be represented by two classes of mechanical models, called the *hookean models* and the *Kelvin models* (Figs. 9.01–1 and 9.01–2).

(b) **Three basic elements** are used for the construction of these models:

(α) *Lumped mass element,* assumed to be perfectly rigid and represented by a solid ball

(β) *Elastic element,* assumed to be massless, elastic, and represented by a spring

(γ) *Viscoelastic element,* assumed to be massless, viscoelastic, and represented by a dashpot (damper)

(c) **Hookean model** of Fig. 9.01–1 is the simplest mechanical model; it consists of a lumped mass ball attached by an elastic spring to a rigid foundation. By definition, this model has one degree of freedom, allowing linear vibration of the ball along the X axis. Since no retarding forces are acting on the ball, its motion is called *free vibration.*

(d) **Kelvin model** of Fig. 9.01–2 is the next most frequently used vibration model; it consists of a lumped mass ball attached by an elastic spring and a dashpot to a rigid foundation. By definition, this model also has one degree of freedom, allowing linear vibration of the ball along the X axis. Since the dashpot produces a retarding (damping) force on the ball, its motion is called *damped vibration.*

(3) Free Linear Vibration

(a) **Equations of motion.** When the motion of the hookean model is produced by the initial displacement x_0 and/or by the initial velocity $\dot{x}_0$ of the ball, the motion is called *free linear vibration,* defined by

$$x = \underbrace{x_0 \cos \omega t}_{\text{curve }(a)} + \underbrace{\frac{\dot{x}_0}{\omega} \sin \omega t}_{\text{curve }(b)}$$

$$\dot{x} = -\omega x_0 \sin \omega t + \dot{x}_0 \cos \omega t$$

$$\ddot{x} = -\omega^2 x_0 \cos \omega t - \omega \dot{x}_0 \sin \omega t$$

$$\left\{\begin{array}{ll} x \text{ in m} & k \text{ in N/m} \\ \dot{x} \text{ in m/s} & m \text{ in kg} \\ \ddot{x} \text{ in m/s}^2 & \omega \text{ in rad/s} \\ t \text{ in s} \end{array}\right\}$$

where x = displacement from the equilibrium position at time t, $\dot{x} = dx/dt$ = velocity at t, $\ddot{x} = d^2x/dt^2$ = acceleration at t, $\omega = \sqrt{k/m}$ = natural circular frequency, k = linear spring constant, and m = mass of the ball.

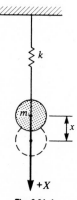

Fig. 9.01-1

(b) Graphical representation of this motion is the time-displacement curve constructed by superposing curves (a) and (b), as shown in Fig. 9.01–3a and b, or by means of the modified equations given below.

$$x = A\cos(\omega t - \alpha) \qquad \text{or} \qquad x = A\sin(\omega t + \beta) \qquad \left\{\begin{array}{l} x \text{ in m} \\ \omega \text{ in rad/s} \\ \alpha,\ \beta \text{ in rad} \end{array}\right\}$$

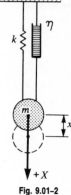

where A = natural amplitude and $\alpha,\ \beta$ = phase angles.

(c) Natural amplitude A, defined as the greatest displacement of the lumped mass m from its equilibrium position, is

$$A = \sqrt{x_0^2 + \left(\frac{\dot{x}_0}{\omega}\right)^2} \qquad \{A \text{ in m}\}$$

where $x_0,\ \dot{x}_0$ = initial conditions defined in (a) and ω = natural circular frequency defined in (a).

Fig. 9.01–2

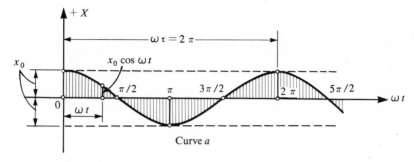

Curve a

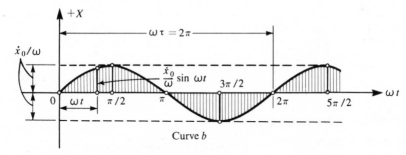

Curve b

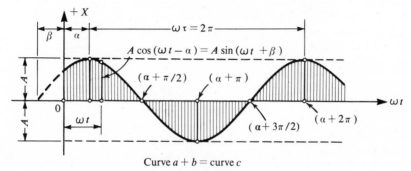

Curve $a + b =$ curve c

Fig. 9.01–3

(d) Phase angles α and β, defining the shift of the harmonic curve on the ωt axis in Fig. 9.01–3c, are given by

$$\tan \alpha = \frac{\dot{x}_0}{\omega x_0} \qquad \text{and} \qquad \tan \beta = \frac{\omega x_0}{\dot{x}_0}$$

and are related as shown in Fig. 9.01–4; $\alpha + \beta = \pi/2$.

(e) Natural period τ, defined as the time required for one complete cycle (after which a repetition of the same motion takes place during the same time τ), is

$$\tau = 2\pi/\omega = 2\pi \sqrt{m/k} \qquad \{\tau \text{ in s}\}$$

(f) Natural frequency f, defined as the number of cycles per unit of time, is

$$f = \frac{1}{\tau} = \frac{\omega}{2\pi} = \frac{\sqrt{k/m}}{2\pi} \qquad \{f \text{ in Hz}\}$$

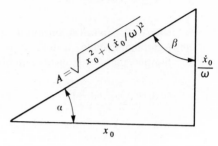

Fig. 9.01–4

example:

In Fig. 9.01–1 if $m = 1$ kg, $k = 10^6$ N/m, $x_0 = 1$ cm, and $\dot{x}_0 = 10^3$ cm/s, then

$$\omega = \sqrt{\frac{k}{m}} = 10^3 \text{ rad/s} \qquad A = \sqrt{(10^{-2})^2 + \left(\frac{10}{10^3}\right)^2} = 1.4 \times 10^{-2} \text{ m}$$

$$\tau = \frac{2\pi}{\omega} = 6.28 \times 10^{-3} \text{ s} \qquad f = \frac{\omega}{2\pi} = 159 \text{ Hz}$$

$$\alpha = \tan^{-1} \frac{\dot{x}_0}{\omega x_0} = \frac{\pi}{4} \qquad \beta = \tan^{-1} \frac{\omega x_0}{\dot{x}_0} = \frac{\pi}{4}$$

Note: $\alpha + \beta = \pi/2$ is valid in all cases.

(4) Forced Linear Vibration

(a) Time-dependent force. When a time-dependent force $F(t)$ is applied on the lumped mass of the hookean model at $t = 0$, as shown in Fig. 9.01–5, and the initial conditions are the same as in Sec. 9.01–3a, the motion of the lumped mass m is called *forced linear vibration*, defined by

$$x = x_0 \cos \omega t + \frac{\dot{x}_0}{\omega} \sin t + \Phi$$

$$\dot{x} = -\omega x_0 \sin \omega t + \dot{x}_0 \cos \omega t + \dot{\Phi}$$

$$\ddot{x} = -\omega^2 x_0 \cos \omega t - \omega \dot{x}_0 \sin \omega t + \ddot{\Phi}$$

$$\left\{ \begin{array}{ll} x, \Phi \text{ in m} & k \text{ in N/m} \\ \dot{x} \text{ in m/s} & m \text{ in kg} \\ \ddot{x} \text{ in m/s}^2 & \omega \text{ in rad/s} \\ t \text{ in s} & \end{array} \right\}$$

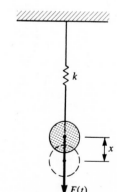

Fig. 9.01–5

where the symbols are those of Sec. 9.01–3a and Φ = displacement due to $F(t)$ at t when $x_0 = 0$, $\dot{x}_0 = 0$; $\dot{\Phi} = d\Phi/dt$ = velocity due to $F(t)$ at t when $x_0 = 0$, $\dot{x}_0 = 0$; and $\ddot{\Phi} = d^2\Phi/dt^2$ = acceleration due to $F(t)$ at t when $x_0 = 0$, $\dot{x}_0 = 0$.

(b) Rectangular pulse. If $F(t) = P$ between $t = 0$ and $t = t_1$ (Fig. 9.01–6),

$$\Phi = \frac{P}{k}(1 - \cos \omega t)$$

If P is removed at $t = t_1$,

$$\Phi = \frac{P}{k}[\cos(\omega t - \omega t_1) - \cos \omega t]$$

where in both cases ω = natural angular frequency and k = linear spring constant.

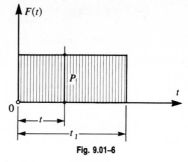

Fig. 9.01–6

(c) Increasing triangular pulse. If $F(t) = Pt/t_1$ between $t = 0$ and $t = t_1$ (Fig. 9.01–7),

$$\Phi = \frac{P}{kt_1}\left(t - \frac{\sin \omega t}{\omega}\right)$$

If P is removed at $t = t_1$,

$$\Phi = \frac{P}{kt_1}\left[t_1 \cos(\omega t - \omega t_1) + \frac{\sin(\omega t - \omega t_1)}{\omega} - \frac{\sin \omega t}{\omega}\right]$$

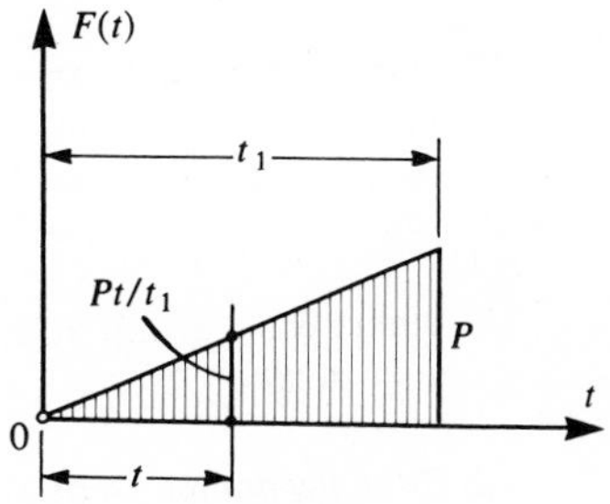

(d) Decreasing triangular pulse. If $F(t) = P(t_1 - t)/t_1$ between $t = 0$ and $t = t_1$ (Fig. 9.01–8),

$$\Phi = \frac{P}{k}\left(1 - \cos \omega t - \frac{t}{t_1} + \frac{\sin \omega t}{\omega t_1}\right)$$

For $t > t_1$,

$$\Phi = \frac{P}{k}\left[-\cos \omega t - \frac{\sin(\omega t - \omega t_1)}{\omega t_1} + \frac{\sin \omega t}{\omega t_1}\right]$$

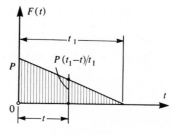

(e) Sine pulse. If $F(t) = P \sin \lambda t$ for all values of t (Fig. 9.01–9), then

$$\Phi = \frac{P\omega^2}{k(\omega^2 - \lambda^2)}\left(\sin \lambda t - \frac{\lambda}{\omega}\sin \omega t\right)$$

where $\lambda = 2\pi/t_1$.

(f) Cosine pulse. If $F(t) = P \cos \lambda t$ for all values of t (Fig. 9.01–10), then

$$\Phi = \frac{P\omega^2}{k(\omega^2 - \lambda^2)}(\cos \lambda t - \cos \omega t)$$

where again $\lambda = 2\pi/t_1$.

Fig. 9.01–7

Fig. 9.01–8

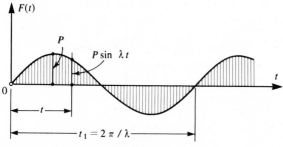

Fig. 9.01–9

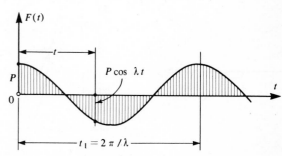

Fig. 9.01–10

(5) Free Linear Vibration with Damping

(a) Types of motion.   When the motion of the Kelvin model (Fig. 9.01–2) is produced by the initial displacement x_0 and/or by the initial velocity $\dot{x}_0$ of the ball, the motion is called *free linear vibration with damping*.   According to the relationship of the parameters of the model, the motion is classified as *underdamped motion, critically damped motion,* or *overdamped motion*.

(b) Parameters of motion are

$$\gamma = \frac{\eta}{2m} \qquad \omega = \sqrt{\frac{k}{m}} \qquad \begin{Bmatrix} \eta \text{ in kg/s} & m \text{ in kg} \\ k \text{ in N/m} & \gamma, \omega \text{ in rad/s} \end{Bmatrix}$$

where η = linear damping constant, k = linear spring constant, and m = lumped mass.

(c) Underdamped motion.   When $\gamma^2 < \omega^2$ and $\lambda = \sqrt{\omega^2 - \gamma^2}$, the motion of the Kelvin model of Fig. 9.01–2 is called underdamped free linear vibration, defined by

$$x = e^{-\gamma t}\left(x_0 \cos \lambda t + \frac{\dot{x}_0 + \gamma x_0}{\lambda} \sin \lambda t\right) \qquad \begin{Bmatrix} x \text{ in m} & t \text{ in s} \\ \dot{x} \text{ in m/s} & \lambda, \gamma \text{ in rad/s} \end{Bmatrix}$$

where $x_0, \dot{x}_0$ = initial conditions, γ, ω = parameters introduced in (b), e = 2.71828, and $\dot{x} = dx/dt$.

(d) Critically damped motion.   When $\gamma^2 = \omega^2$, the motion of the Kelvin model of Fig. 9.01–2 is called critically damped free linear vibration, defined by

$$x = e^{-\gamma t}[x_0(1 + \gamma t) + \dot{x}_0 t] \qquad \begin{Bmatrix} x \text{ in m} & \gamma \text{ in rad/s} \\ \dot{x} \text{ in m/s} & t \text{ in s} \end{Bmatrix}$$

where $x_0, \dot{x}_0, \gamma,$ and e have the same meaning as in (c) and $\dot{x} = dx/dt$.

(e) Overdamped motion.   When $\gamma^2 > \omega^2$ and $\lambda = \sqrt{\gamma^2 - \omega^2}$, the motion of the Kelvin model of Fig. 9.01–2 is called overdamped free linear vibration, defined by

$$x = e^{-\gamma t}\left(x_0 \cosh \lambda t + \frac{\dot{x}_0 + \gamma x_0}{\lambda} \sinh \lambda t\right) \qquad \begin{Bmatrix} x \text{ in m} & t \text{ in s} \\ \dot{x} \text{ in m/s} & \gamma, \lambda \text{ in rad/s} \end{Bmatrix}$$

where $x_0, \dot{x}_0, \gamma, \omega,$ and e have the same meaning as in (c) and $\dot{x} = dx/dt$.

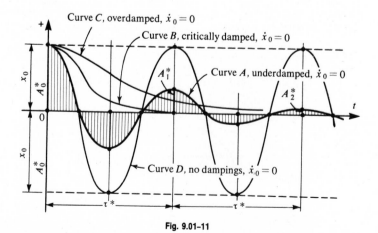

Fig. 9.01–11

(f) Graphical representation. The three graphs of Fig. 9.01–11 represent the time-displacement curves of free linear vibration with damping, each corresponding to one of the cases mentioned above. Curve A, representing underdamped motion, shows that this motion is oscillatory with diminishing amplitudes but constant periods. Curves B and C, representing critically damped and overdamped motions, respectively, show that the magnitude of damping is so large that the motion is no longer oscillatory and decreases exponentially with time (aperiodic motions). Curve D, representing free vibration without damping, is shown for comparison.

(6) Free Linear Underdamped Vibration

(a) Displacement equations. As in Sec. 9.01–3b, the displacement equation of free linear underdamped vibration of Sec. 9.01–5c may be expressed in simpler forms as

$$x = A \cos(\lambda t - \alpha^*) \qquad \text{or} \qquad x = A \sin(\lambda t + \beta^*) \qquad \left\{ \begin{array}{l} x, A \text{ in m} \\ \alpha^*, \beta^* \text{ in rad} \\ \lambda \text{ in rad/s} \\ t \text{ in s} \end{array} \right\}$$

where $A = e^{-\gamma t}\sqrt{x_0^2 + (\dot{x}_0 + \gamma x_0)^2/\lambda^2}$ and $\alpha^*, \beta^* = $ phase angles.

(b) Diminishing amplitude A_n^*, defined as the greatest distance of the lumped mass from its equilibrium position at the nth multiple of the period τ^* (Fig. 9.01–11), is

$$A_n^* = e^{-\gamma n \tau^*}\sqrt{x_0^2 + \frac{(\dot{x}_0 + \gamma x_0)^2}{\lambda^2}} = A_0^* e^{-\gamma n \tau^*}$$

where $x_0, \dot{x}_0 = $ initial conditions defined in Sec. 9.01–5a, $\gamma, \lambda = $ parameters of motion defined in Sec. 9.01–5b, $n = 0, 1, 2, \ldots$, and $e = 2.71828$.

(c) Phase angles α^* and β^*, representing the shift of the decaying harmonic curve on the λt axis, are given by

$$\tan \alpha^* = \frac{\dot{x}_0 + \gamma x_0}{\lambda x_0} \qquad \text{and} \qquad \tan \beta^* = \frac{\lambda x_0}{\dot{x}_0 + \gamma x_0}$$

and are related as shown in Fig. 9.01–12.

(d) Period τ^*, defined as the constant time interval between two successive amplitudes of the same sign, is

$$\tau^* = \frac{2\pi}{\lambda} = \frac{2\pi}{\sqrt{\omega^2 - \gamma^2}} \qquad \{\tau^* \text{ in s}\}$$

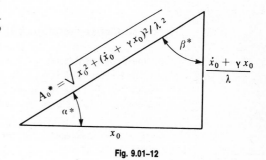

Fig. 9.01–12

(e) Frequency f^* is the reciprocal of τ^* given by

$$f^* = \frac{1}{\tau^*} = \frac{\lambda}{2\pi} = \frac{\sqrt{\omega^2 - \gamma^2}}{2\pi} \qquad \{f^* \text{ in Hz}\}$$

example:

A machine of mass $m = 100$ kg is mounted on four springs of total stiffness $k = 10,000$ N/m and on four dampers of total damping constant $\eta = 1,200$ kg/s, as shown in Fig. 9.01–13. If the system is initially at rest and the conditions $x_0 = 0.01$ m and $\dot{x}_0 = 0.02$ m/s are imparted to the mass at $t = 0$, the resulting vibration has

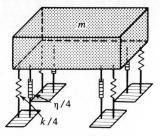

$$\omega = \sqrt{k/m} = \sqrt{10,000/100} = 10 \text{ rad/s}$$

$$\gamma = \eta/2m = 1,200/2 \times 100 = 6 \text{ rad/s}$$

$$\lambda = \sqrt{\omega^2 - \gamma^2} = \sqrt{10^2 - 6^2} = 8 \text{ rad/s}$$

Fig. 9.01–13

and its time-displacement equation is

$$x = e^{-6t}\left(0.01 \cos 8t + \frac{0.02 + 6 \times 0.01}{8} \sin 8t\right) = 0.01 e^{-6t}(\cos 8t + \sin 8t)$$

or in terms of $A = e^{-6t}\sqrt{(0.01)^2 + (0.01)^2} = 0.014 e^{-6t}$ and $\alpha^* = \tan^{-1} 0.01/0.01 = \pi/4$,

$$x = 0.014 e^{-6t} \cos(8t - \pi/4)$$

The period and frequency of this underdamped vibration are

$$\tau^* = 2\pi/\lambda = 2\pi/8 = 0.785 \text{ s} \qquad f^* = 1/\tau^* = 1.27 \text{ Hz}$$

The ratio of two successive amplitudes of like sign is

$$\frac{e^{-\gamma(n+1)\tau^*}}{e^{-\gamma n \tau^*}} = e^{-\gamma \tau^*} = e^{-2\pi\gamma/\lambda} = e^{-4.71} = 0.009$$

which shows a very low rate of damping governed by the ratio γ/λ.

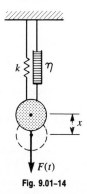

Fig. 9.01–14

(7) Forced Linear Vibration with Damping

(a) Time-dependent force. When a time-dependent force $F(t)$ is applied on the lumped mass of the Kelvin model at $t = 0$, as shown in Fig. 9.01–14, and the initial conditions are the same as in Sec. 9.01–5a, the motion of the lumped mass m is called *forced linear vibration with damping*, defined by

$$x = \begin{cases} e^{-\gamma t}\left(x_0 \cos \lambda t + \dfrac{\dot{x}_0 + \gamma x_0}{\lambda} \sin \lambda t\right) + \Phi_1 & \text{if } \gamma^2 < \omega^2, \ \lambda = \sqrt{\omega^2 - \gamma^2} \\[2mm] e^{-\gamma t}[x_0(1 + \gamma t) + \dot{x}_0 t] + \Phi_2 & \text{if } \gamma^2 = \omega^2, \ \lambda = 0 \\[2mm] e^{-\gamma t}\left(x_0 \cosh \lambda t + \dfrac{\dot{x}_0 + \gamma x_0}{\lambda} \sinh \lambda t\right) + \Phi_3 & \text{if } \gamma^2 > \omega^2, \ \lambda = \sqrt{\gamma^2 - \omega^2} \end{cases}$$

where the first part of each equation corresponds to the respective cases of Secs. 9.01–5c, d, and e and the second part, designated by Φ_1, Φ_2, and Φ_3, respectively, is the displacement of the respective vibrations due to $F(t)$ when $x_0 = 0$, $\dot{x}_0 = 0$.

(b) Evaluation of Φ_1, Φ_2, and Φ_3 is very involved and only a few special cases have been solved in a closed form. The simplest case, $F(t) = P$, represented by the diagram of Fig. 9.01–15, has

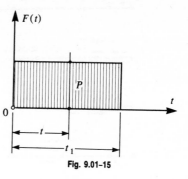

Fig. 9.01–15

$$\Phi_1 = \frac{P}{k\lambda}[\lambda - e^{-\gamma t}(\gamma \sin \lambda t + \lambda \cos \lambda t)]$$

$$\Phi_2 = \frac{P}{k}[1 - e^{-\gamma t}(1 + \gamma)]$$

$$\Phi_3 = \frac{P}{k\lambda}[\lambda - e^{-\gamma t}(\gamma \sinh \lambda t + \lambda \cosh \lambda t)]$$

If P is removed at $t = t_1$,

$$\Phi_1 = \frac{P}{k\lambda}\{e^{-\gamma(t-t_1)}[\gamma \sin \lambda(t - t_1) + \lambda \cos \lambda(t - t_1)] - e^{-\gamma t}(\gamma \sin \lambda t + \lambda \cos \lambda t)\}$$

$$\Phi_2 = \frac{P}{k}[e^{-\gamma(t-t_1)}(1 + \gamma) - e^{-\gamma t}(1 + \gamma)]$$

$$\Phi_3 = \frac{P}{k\lambda}\{e^{-\gamma(t-t_1)}[\gamma \sinh \lambda(t - t_1) + \lambda \cosh \lambda(t - t_1)] - e^{-\gamma t}(\gamma \sinh \lambda t + \lambda \cosh \lambda t)\}$$

where in Φ_1, $\lambda = \sqrt{\omega^2 - \gamma^2}$; in Φ_3, $\lambda = \sqrt{\gamma^2 - \omega^2}$; and ω, γ = same as in Sec. 9.01–5b.

9.02 WAVE MECHANICS

(1) Classification and Definitions

(a) **Wave motion** of a mass particle of matter is defined as its vibration about its position of equilibrium during which its wave energy is transferred (propagation) from one point to another without physical transfer of matter itself between these points.

(b) **Waves are classified** according to the source (cause) as mechanical or electromagnetic, longitudinal or transverse, and as harmonic or nonharmonic.

(c) **Mechanical waves**, such as vibration of elastic springs, waves of liquid surfaces, and sound waves in air, are typical examples of mechanical waves produced by a mechanical source.

(d) **Electromagnetic waves**, such as light and radio signals, are motions by electrical and magnetic disturbances.

(e) **Longitudinal wave** (Fig. 9.02–1) is one in which the vibration of the mass particle is in the direction of energy propagation.

(f) **Transverse wave** (Fig. 9.02–2) is one in which the vibration of the mass particle is perpendicular to the direction of energy propagation.

(2) Mechanical Waves

(a) **Source and medium.** In order for mechanical wave motion to develop, it is necessary to have a disturbance (source causing the motion) and a medium through which the disturbance can be transmitted.

(b) **Speed** with which a wave moves in a medium is dependent upon the elasticity of the medium and the inertia of its particles.

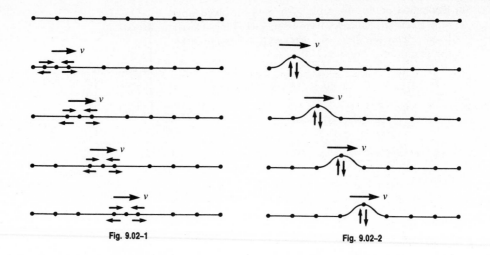

Fig. 9.02–1   Fig. 9.02–2

(c) Speed v **of a longitudinal wave** depends on the elasticity and density of the medium.

In thin rods: $v = \sqrt{E/\rho}$

In extended solids: $v = \sqrt{(K + 4G/3)/\rho}$

In liquids: $v = \sqrt{K/\rho}$

In gases: $v = \sqrt{\alpha p/\rho}$

$$\left\{ \begin{array}{l} v \text{ in m/s} \\ E,\ G,\ K,\ p \text{ in N/m}^2 \\ \rho \text{ in kg/m}^3 \\ \alpha \text{ dimensionless} \end{array} \right\}$$

where E = modulus of elasticity, G = modulus of rigidity, K = bulk modulus, ρ = mass density, p = pressure, $\alpha = \bar{c}_p/\bar{c}_v$, $\bar{c}_p$ = average specific heat capacity at constant pressure, and $\bar{c}_v$ = average specific heat capacity at constant volume.

(d) Speed v **of a transverse wave** in a stretched elastic string (wire) is

$$v = \sqrt{\frac{Sl}{m}} \qquad \left\{ \begin{array}{l} S \text{ in N} \\ l \text{ in m} \\ m \text{ in kg} \end{array} \right\}$$

where S = tension in string, l = length of string, and m = mass of string.

(3) Sinusoidal Wave Motion

(a) Single-pulse motion. The wave motions shown in Figs. 9.02–1 and 9.02–2 are called single-pulse motions since they were produced by a single disturbance (one pulse).

(b) Harmonic-pulse motion. If the disturbances are repeated harmonically, the wave motion forms a continuous wave train in which the displacement x of each mass particle is a simple harmonic function of time.

$$x = A_0 \sin (2\pi ft)$$

where A_0 = local amplitude, f = frequency, and t = time.

example:

If the end of a string is attached to a weight hanging on an elastic spring, as shown in Fig. 9.02–3, and if the weight is pulled downward a small distance and released, the weight will be in a state of free vibration, which in turn generates in the string a simple harmonic wave train.

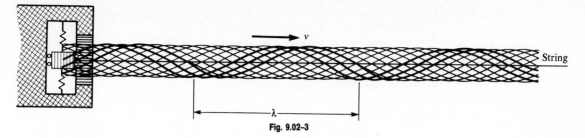

Fig. 9.02–3

(c) **Two mass particles** of a wave train are said to be *in phase* if at a given time they have the same displacement and if they move in the same direction.

(d) **Wavelength** λ of a wave train is the distance between any two particles which are in phase. In the case of harmonic wave motion, the speed v of wave propagation is

$$v = f\lambda = \lambda/\tau \qquad \begin{cases} f \text{ in Hz} \\ \lambda \text{ in m} \\ \tau \text{ in s} \end{cases}$$

where f = frequency and τ = period of the source producing the wave motion.

(e) **Damped wave motion.** If a wave motion is propagating around a center of disturbance, the amplitude diminishes with distance from the center and the wave motion has the characteristics of one of the damped vibrations discussed in Sec. 9.01–5.

(4) Geometry of Wave Motion

(a) **Independence principle.** If two or more wave trains exist simultaneously in a given medium, each travels through the medium as if the others did not exist; that is, one is independent of the others.

(b) **Superposition principle.** If two or more wave trains exist simultaneously in a given medium, the resultant displacement of any mass particle at a given time is the vector sum of the respective displacements of that particle caused by these wave trains at that time.

(c) **Fourier series** is the consequence of the principle of superposition; that is, general wave motion may be represented as a superposition of several (or infinitely many) simple harmonic wave motions.

(d) **Interference.** The joint effect of two or more wave trains is called *interference*, and it can be constructive or destructive interference.

example:

Two waves reinforce each other to the maximum if they have no path difference (Fig. 9.02–4a), and they weaken each other if they have a path difference of one-half wavelength (Fig. 9.02–4b).

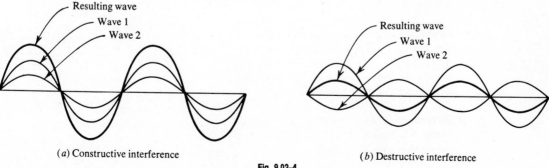

(a) Constructive interference   (b) Destructive interference

Fig. 9.02–4

(e) Reflection of waves. When the wave train passing through a medium encounters an interface with another medium which it cannot enter, then it is entirely reflected from that interface; during reflection the angle of incidence equals the angle of reflectance (Fig. 9.02–5).

(f) Refraction of waves. When the wave train passes obliquely from one medium into another, then the relation of the angle of incidence α_1 and the angle of refraction α_2 is related to the velocity of incidence v_1 and the velocity of refractance v_2 (Fig. 9.02–6) as

$$\frac{\sin \alpha_1}{\sin \alpha_2} = \frac{v_1}{v_2}$$

If $v_1 < v_2$, then $\alpha_1 < \alpha_2$ and the wave train refracts from the normal. If $v_1 > v_2$, then $\alpha_1 > \alpha_2$ and the wave train refracts toward the normal.

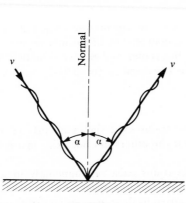

Fig. 9.02–5

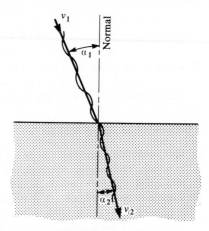

Fig. 9.02–6

(5) Standing Waves

(a) Concept. When two sinusoidal wave trains of equal amplitudes and identical frequencies travel in opposite directions in a confined medium (string, bar, pipe, etc.), the resulting wave pattern is called a *standing wave* (Fig. 9.02–7).

(b) Nodes of a standing wave are the points of the medium which remain at rest during the motion. The points of maximum amplitude occur midway between the nodes and are called *antinodes*.

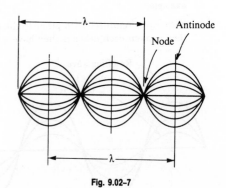

Fig. 9.02–7

(c) Wavelength λ of a standing wave is the distance between alternate nodes or between alternate antinodes.

(d) Elastic string hinged at its ends can develop any number of transverse standing-wave patterns called *normal modes* (modal shapes) as shown in Fig. 9.02–8. Their wavelength is

$$\lambda_n = \frac{2l}{n} \qquad n = 1, 2, 3, \ldots$$

and the frequency of their vibration is

$$f_n = \frac{nv}{2l} = \frac{n}{2l}\sqrt{\frac{Sl}{m}} \qquad \begin{cases} f_n \text{ in Hz} \\ v \text{ in m/s} \\ l \text{ in m} \\ S \text{ in N} \\ m \text{ in kg} \end{cases}$$

where l = length of string, v = speed of waves, S = tension in string, and m = mass of string.

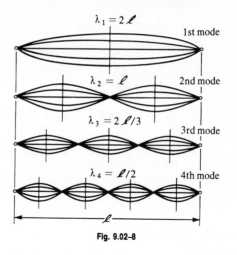

Fig. 9.02–8

TABLE 9.02–1 Modal Shapes of Elastic Bars

	Simple beam	Cantilever beam	Fixed end beam	Propped end beam
Initial shape	Hinge ——— Roller	Fixed ——— Free	Fixed ——— Fixed	Fixed ——— Roller
1st mode	$C = 1.57$	$C = 0.56$	$C = 3.56$	$C = 2.45$
2nd mode	$C = 6.28$ $\quad l/2 \quad l/2$	$C = 3.51$ $\quad .23l$	$C = 9.82$ $\quad l/2 \quad l/2$	$C = 7.95$ $\quad .56\,l \quad .44l$
3rd mode	$C = 14.13$ $\quad l/3 \;\; l/3 \;\; l/3$	$C = 9.82$ $\quad .13l$ $\quad .51\,l \quad .49\,l$	$C = 19.24$ $\quad .36\,l \;\; .28l \;\; .36\,l$	$C = 16.59$ $\quad .38l \;\; .31\,l \;\; .31\,l$
4th mode	$C = 25.13$ $\quad l/4 \; l/4 \; l/4 \; l/4$	$C = 19.24$ $\quad .09\,l$ $\quad .35\,l \;\; .29\,l \;\; .36\,l$	$C = 31.81$ $\quad .28\,l \; .22\,l \; .22\,l \; .28\,l$	$C = 28.37$ $\quad .28\,l \; .24l \; .24l \; .24l$

example:

If a steel wire of length $l = 2$ m and mass $m = 2 \times 10^{-2}$ kg is under tension $S = 900$ N, then the first frequency of its free vibration is

$$f_1 = \frac{1}{2 \times 2} \sqrt{\frac{900 \times 2}{2 \times 10^{-2}}} = 75 \text{ Hz}$$

and

$$f_2 = 2f_1 = 150 \text{ Hz} \qquad f_3 = 3f_1 = 225 \text{ Hz} \qquad \dots \qquad f_n = nf_1$$

(e) Elastic straight bar of constant cross section similarly can develop any number of normal modes whose shapes depend on the end conditions of the bar. The frequency of their transverse vibration is

$$f_n = C\sqrt{\frac{EI}{ml^3}} \qquad \left\{ \begin{array}{l} f_n \text{ in Hz} \\ C \text{ dimensionless} \\ E \text{ in N/m}^2 \\ I \text{ in m}^4 \\ m \text{ in kg} \\ l \text{ in m} \end{array} \right\}$$

where C = frequency coefficient (Table 9.02–1), E = modulus of elasticity, I = moment of inertia of the cross section, m = mass of bar, and l = length of bar.

9.03 GENERAL ACOUSTICS

(1) Nature of Sound

(a) Sound is a longitudinal mechanical wave motion in an elastic medium and is classified according to its frequency as audible by the human ear (frequency from 20 to 20,000 Hz), as *infrasonic* or below the response of the human ear (frequency below 20 Hz), and as *ultrasonic*, above the response of the human ear (frequency above 20,000 Hz).

(b) Acoustics is the systematic investigation of the nature, origin, and propagation of sound.

(c) Production of sound is dependent upon the source of mechanical vibration and the elastic medium through which the disturbance can travel (propagate). All vibrating solids, liquids, and gases produce sound. No sound can propagate in a vacuum.

(2) Speed of Sound

(a) In wires and rods, the speed v_L of a longitudinal sound wave is

$$v_L = \sqrt{E/\rho} \qquad \left\{ \begin{array}{l} v_L \text{ in m/s} \\ E \text{ in N/m}^2 \\ \rho \text{ in kg/m}^3 \end{array} \right\}$$

where E = modulus of elasticity and ρ = mass density of wire.

example:

The speed of sound in a steel wire of $E = 2.1 \times 10^6$ kgf/cm^2 = $2.1 \times 10^6 \times 9.81 \times 10^4$ N/m^2 = 2.06×10^{11} N/m^2 and $\rho = 7.8$ g/cm^3 = 7.8×10^3 kg/m^3 is

$$v_L = \sqrt{\frac{2.06 \times 10^{11}}{7.8 \times 10^3}} = 5,140 \text{ m/s}$$

(b) In extended solids, the speed v_L of a longitudinal sound wave is

$$v_L = \sqrt{\frac{K + 4G/3}{\rho}} \qquad \begin{Bmatrix} v_L \text{ in m/s} \\ K, G \text{ in N/m}^2 \\ \rho \text{ in kg/m}^3 \end{Bmatrix}$$

where K = bulk modulus, G = modulus of rigidity, and ρ = mass density of solid.

example:

The speed of sound in a steel bar of $E = 2.06 \times 10^{11}$ N/m², $\nu = 0.3$, and $\rho = 7.8 \times 10^3$ kg/m³ is

$$v_L = \sqrt{\frac{1.72 \times 10^{11} + 1.33 \times 0.79 \times 10^{11}}{7.8 \times 10^3}} = 5{,}960 \text{ m/s}$$

where according to Sec. 4.03–3e,

$$K = \frac{E}{3(1 - 2\nu)} = \frac{2.06 \times 10^{11}}{3 \times (1 - 2 \times 0.3)} = 1.72 \times 10^{11} \text{ N/m}^2$$

$$G = \frac{E}{2(1 + \nu)} = \frac{2.06 \times 10^{11}}{2 \times (1 + 0.3)} = 0.79 \times 10^{11} \text{ N/m}^2$$

(c) In liquids, the speed v_L of a longitudinal sound wave is

$$v_L = \sqrt{\frac{K}{\rho}} \qquad \begin{Bmatrix} v_L \text{ in m/s} \\ K \text{ in N/m}^2 \\ \rho \text{ in kg/m}^3 \end{Bmatrix}$$

where K = bulk modulus and ρ = mass density of liquid.

example:

The speed of sound in water of $K = 2.1 \times 10^9$ N/m² and $\rho = 1{,}000$ kg/m³ is

$$v_L = \sqrt{\frac{2.1 \times 10^9}{10^3}} = 1{,}450 \text{ m/s}$$

(d) In gases, the speed v_L of a longitudinal sound wave is

$$v_L = \sqrt{\frac{\alpha p}{\rho}} = \sqrt{\frac{\alpha RT}{Mkg}} \qquad \begin{Bmatrix} v_L \text{ in m/s} \\ \alpha \text{ dimensionless} \\ p \text{ in N/m}^2 \\ \rho \text{ in kg/m}^3 \\ R \text{ in m-N/kmol-K} \\ T \text{ in K} \\ Mkg \text{ in kg/kmol} \end{Bmatrix}$$

where $\alpha = \bar{c}_p/\bar{c}_v$, $\bar{c}_p$ = average specific heat capacity of gas at constant pressure, $\bar{c}_v$ = average specific heat capacity at constant volume, p = pressure, ρ = mass density, R = universal gas constant (Sec. 6.02–3d), Mkg = molecular mass of gas in kilograms (Sec. 6.02–1f), and T = temperature in degrees Kelvin. For monatomic gases $\alpha = 1.67$, and for most diatomic gases $\alpha = 1.40$.

example:

The speed of sound in helium gas of $\alpha = 1.66$, $T = 700°C$, and $Mkg = 4$ kg/kmol is

$$v_L = \sqrt{\frac{1.66 \times 8{,}315 \times (273 + 700)}{4}} = 1.83 \times 10^3 \text{ m/s}$$

where $R = 8{,}315$ J/kmol K as given in Sec. 6.02–3d.

(e) In air, the speed v_L of a longitudinal sound wave is

$$v_L = 331 \sqrt{\frac{T}{273}} \qquad \left\{ \begin{array}{l} v_L \text{ in m/s} \\ T \text{ in K} \end{array} \right\}$$

or approximately

$$v_L \cong 331 + 0.6(T - 273)$$

where T is absolute temperature. The speed is independent of barometric pressure, frequency, and wavelength.

example:

The speed of sound at 40°C is

$$v_L = 331 \sqrt{\frac{273 + 40}{273}} = 354.4 \text{ m/s}$$

or, by the approximate formula,

$$v_L \cong 331 + 0.6(313 - 273) = 355 \text{ m/s}$$

(3) Vibrating Air Column

(a) Closed pipe—first harmonic. When a longitudinal wave is set up in a closed pipe (Fig. 9.03–1*a*), the fundamental mode of oscillation of the air column has the displacement node at the closed end, and the sound is called the *first harmonic of the closed pipe* (fundamental tone). The basic wavelength λ_1 and the fundamental frequency f_1 are

$$\lambda_1 = 4l \qquad f_1 = \frac{v_L}{4l} \qquad \left\{ \begin{array}{l} \lambda, l \text{ in m} \\ v_L \text{ in m/s} \\ f \text{ in Hz} \end{array} \right\}$$

where l = length of pipe and v_L = speed of sound in air (Sec. 9.03–2*e*).

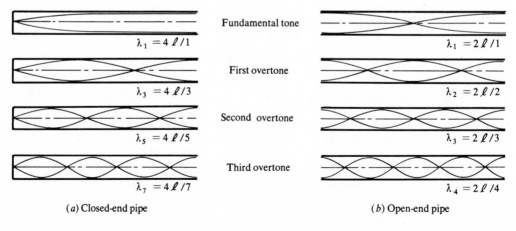

$\lambda_1 = 4\ell/1$	Fundamental tone $\quad \lambda_1 = 2\ell/1$
$\lambda_3 = 4\ell/3$	First overtone $\quad \lambda_2 = 2\ell/2$
$\lambda_5 = 4\ell/5$	Second overtone $\quad \lambda_3 = 2\ell/3$
$\lambda_7 = 4\ell/7$	Third overtone $\quad \lambda_4 = 2\ell/4$

(*a*) Closed-end pipe   (*b*) Open-end pipe

Fig. 9.03–1

(b) Closed pipe—higher harmonics. The higher modes of this oscillation have shapes as shown in Fig. 9.03–1a in which the last displacement node is again at the closed end. Their sounds are called *higher harmonics* (overtones). Their wavelengths and frequencies are

$$\lambda_n = \frac{4l}{n} \qquad f_n = \frac{n v_L}{4l} \qquad n = 3, 5, 7, \ldots$$

which shows that only odd harmonics can develop in a closed pipe.

example:

The frequencies of a closed pipe of $l = 3$ m at 20°C are

$$f_1 = \frac{v_L}{4l} = \frac{331 + 0.6 \times 20}{4 \times 3} = 28.6 \text{ Hz}$$

$$f_3 = 3f_1 = 85.8 \text{ Hz} \qquad f_5 = 143 \text{ Hz} \qquad \cdots$$

(c) Open pipe—first harmonic. When a longitudinal wave is set up in an open pipe (Fig. 9.03–1b), the fundamental mode of oscillation of the air column has the displacement antinodes at the open ends of the pipe, and the sound produced by this vibration is again called the *first harmonic of the open pipe* (fundamental tone). The fundamental wavelength λ_1 and the fundamental frequency f_1 are

$$\lambda_1 = 2l \qquad f_1 = \frac{v_L}{2l} \qquad \begin{cases} \lambda, l \text{ in m} \\ v_L \text{ in m/s} \\ f \text{ in Hz} \end{cases}$$

where l = length of pipe and v_L = speed of sound in air (Sec. 9.03–2e).

(d) Open pipe—higher harmonics. The higher modes of this vibration have shapes as shown in Fig. 9.03–1b in which the first and last displacement antinodes are again at the open ends. The sounds produced by these vibrations are called *higher harmonics* (overtones). Their wavelengths and frequencies are

$$\lambda_n = \frac{2l}{n} \qquad f_n = \frac{n v_L}{2l} \qquad n = 2, 3, 4, \ldots$$

which shows that the open pipe has all possible harmonics and is therefore used in many musical instruments.

example:

The frequencies of an open pipe of $l = 3$ m at 20°C are

$$f_1 = \frac{v_1}{2l} = \frac{331 + 0.6 \times 20}{2 \times 3} = 57.2 \text{ Hz} \qquad f_2 = 2f_1 = 114.4 \text{ Hz} \qquad f_3 = 3f_1 = 171.6 \text{ Hz} \qquad \cdots$$

9.04 AUDIBLE-SOUND ACOUSTICS

(1) Intensity of Audible Sounds

(a) Sensory effects of sounds denoted by physiologists as loudness, pitch, and quality are correlated with the measurable parameters of sound denoted by physicists as intensity, frequency, and wave shape.

Loudness	corresponds to	*intensity*
Pitch	corresponds to	*frequency*
Quality	corresponds to	*wave shape*

(b) Speed v_p of sound at the center of wavelength λ is

$$v_P = A\omega = 2\pi f A \qquad \begin{cases} v_P \text{ in m/s} \\ A \text{ in m} \\ f \text{ in Hz} \end{cases}$$

where A = amplitude, f = frequency, and $\omega = 2\pi f$ = angular frequency.

(c) Pressure p of sound is

$$p = \rho v_L v_P = 2\pi f A \rho v_L \qquad \begin{cases} p \text{ in N/m}^2 \\ \rho \text{ in kg/m}^3 \\ v_L, v_P \text{ in m/s} \end{cases}$$

where ρ = density of medium and v_L = speed of sound in air.

(d) Intensity of sound (of sound wave) is the amount of wave energy transmitted per unit time per unit area normal to the direction of sound propagation; that is, intensity of sound is the power transmitted per unit area.

$$I = \frac{p v_P}{2} = \frac{\rho v_P^2 v_L}{2} = 2\pi^2 f^2 A^2 \rho v_L \qquad \{I \text{ in W/m}^2\}$$

where the symbols are those of (b) and (c).

(2) Classification of Audible Sounds

(a) Hearing threshold I_0 is the minimum intensity of audible sound (zero sound intensity), below which the normal human ear is incapable of hearing.

$$I_0 = 10^{-12} \text{ W/m}^2$$

(b) Pain threshold I_p is the maximum intensity of audible sound that the normal human ear can hear without pain.

$$I_p = 1 \text{ W/m}^2$$

(c) Two thresholds defined above give the two fixed points of the normal hearing range of the human ear (Table 9.04–1).

TABLE 9.04–1

Sound	Intensity, W/m^2	Intensity level, dB
Hearing threshold	1×10^{-12}	0
Rustling leaves	1×10^{-11}	10
Whisper	1×10^{-10}	20
Home radio	1×10^{-8}	40
Home conversation	3×10^{-6}	65
Busy street traffic	3×10^{-5}	75
Train in open	1×10^{-4}	90
Train in tunnel	1×10^{-2}	100
Pain threshold	1	120

(d) Difference in intensity level of two sound waves of intensities I_1 and I_2 is

$$\Delta L = \log \frac{I_1}{I_2} = N \text{ B} \qquad \begin{cases} I_1, I_2 \text{ in W/m}^2 \\ \Delta L \text{ in B} \end{cases}$$

where N = pure number and B = bel (defined below).

(e) One bel, designated by the symbol B (after Alexander Graham Bell), is ΔL of $I_1/I_2 = 10$. Since 1 bel is too large for practical applications, a more useful unit,

$$1 \text{ decibel} = 1 \text{ dB} = 1/10 \text{ B}$$

is used, which is the smallest difference in loudness the normal human ear can detect.

example:

If two sounds have intensities of $I_1 = 2 \times 10^{-4}$ W/m^2 and $I_2 = 1 \times 10^{-10}$ W/m^2, then

$$\Delta L = \log \frac{I_1}{I_2} = \log (2 \times 10^6) = 6.3 \text{ B} = 63 \text{ dB}$$

If $I_1 = 2 \times 10^{-4}$ W/m^2 and $I_2 = 1 \times 10^{-12}$ W/m$^2 = I_0$, then

$$\Delta L = \log \frac{I_1}{I_0} = \log (2 \times 10^8) = 8.3 \text{ B} = 83 \text{ dB}$$

If $I_1 = 1$ W/m^2 and $I_2 = 1 \times 10^{-12}$ W/m$^2 = I_0$ then

$$\Delta L = \log \frac{I_1}{I_0} = \log (1 \times 10^{12}) = 12 \text{ B} = 120 \text{ dB}$$

(3) Sound Effects

(a) Loss of sound intensity due to sound propagation through a wall of thickness d (Fig. 9.04–1) is

$$\Delta I = I_a - I_b = I_a (1 - 10^{-\delta})$$

where I_a = approach intensity, I_b = exit intensity, and δ = coefficient given in Table 9.04–2.

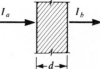

Fig. 9.04–1

TABLE 9.04–2

Material	d, cm	δ	Material	d, cm	δ
Brick wall	5	3	Wooden partition	4	3.5
Brick wall	10	4	Glass wall	2	2
Brick wall	20	5	Glass wall	3	2.5
Concrete wall	10	4.5	Wooden door		2
Concrete wall	20	5	Wooden door + storm door		2.5
Concrete wall	30	5.5	Glass window		1
Wooden partition	2	3	Glass window + storm window		2

example:

If the busy street traffic noise of $I_a = 3 \times 10^{-5}$ W/m^2 propagates through a brick wall of $d = 10$ cm, the intensity of the noise inside the house is

$$I_b = I_a 10^{-4} = 3 \times 10^{-5} \times 10^{-4} = 3 \times 10^{-9} \text{ W/m}^2$$

which is below the intensity of a home radio.

(b) Doppler effect is the apparent change in frequency of a sound source when there is relative motion between the source and the observer. Several cases of this effect are given below. The symbols used are

f_O = frequency of sound to observer (in Hz)
f_S = frequency of sound from source (in Hz)
v_O = speed of observer (in m/s)
v_S = speed of source (in m/s)
v_L = speed of sound (in m/s)

(c) Source stationary, observer in motion. If $v_S = 0$ and the observer moves toward the source,

$$f_O = f_S \frac{v_L + v_O}{v_L}$$

If the observer moves away from the source,

$$f_O = f_S \frac{v_L - v_O}{v_L}$$

(d) Source in motion, observer stationary. If $v_O = 0$ and the source moves toward the observer,

$$f_O = f_S \frac{v_L}{v_L - v_S}$$

If the source moves away from the observer,

$$f_O = f_S \frac{v_L}{v_L + v_S}$$

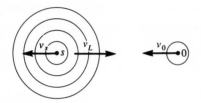

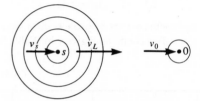

(a) Motion to the left $\qquad\qquad\qquad\qquad$ (b) Motion to the right

Fig. 9.04–2

(e) Source and observer in motion. If the source and the observer move in the same direction to the left (Fig. 9.04–2a),

$$f_O = f_S \frac{v_L + v_O}{v_L + v_S}$$

If the source and the observer move in the same direction to the right (Fig. 9.04–2b),

$$f_O = f_S \frac{v_L - v_O}{v_L - v_S}$$

If the source and the observer move toward each other,

$$f_O = f_S \frac{v_L + v_O}{v_L - v_S}$$

If the source and the observer move away from each other,

$$f_O = f_S \frac{v_L - v_O}{v_L + v_S}$$

10
GEOMETRICAL AND WAVE OPTICS

10.01 WAVE OPTICS

(1) Nature of Light

(a) **Light** is electromagnetic radiation that is capable of affecting the sensor (retina) of the human eye, and it is different from other forms of radiation emitted by hot bodies by its energy, which is of the order of 2.4×10^{-19} to 4.8×10^{-19} J.

(b) **Radiation of electromagnetic energy** can occur at any frequency but it always travels at the speed of light, which in vacuum is

$$c = 2.997\,925 \times 10^8 \text{ m/s} \cong 3 \times 10^8 \text{ m/s}$$
$$\cong 1.86 \times 10^5 \text{ mi/sec}$$

(c) **Wavelength** λ of electromagnetic radiation is related to its frequency f as

$$c = f\lambda \qquad \begin{cases} c \text{ in m/s} \\ f \text{ in Hz} \\ \lambda \text{ in m} \end{cases}$$

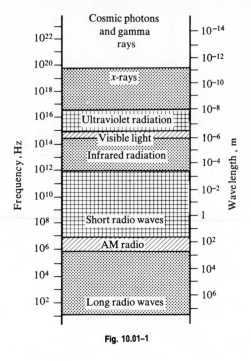

Fig. 10.01–1

(d) **Electromagnetic spectrum** spreads over a very large range of frequencies, as shown in Fig. 10.01–1. The visible region lies in the range of

$$\lambda = 3.8 \times 10^{-7} \text{ to } 7.8 \times 10^{-7} \text{ m}$$
$$f = 3.8 \times 10^{14} \text{ to } 7.9 \times 10^{14} \text{ Hz}$$

(e) **Units of wavelengths** are

$$1 \text{ micron} = 1\,\mu = 10^{-6} \text{ m} = 1 \text{ micrometer} = 1\,\mu\text{m}$$
$$1 \text{ millimicron} = 1\,m\mu = 10^{-9} \text{ m} = 1 \text{ nanometer} = 1 \text{ nm}$$
$$1 \text{ angstrom} = 1\,\text{Å} = 10^{-10} \text{ m}$$

where the nanometer (millimicron) is the most commonly used unit.

example:

Wavelengths and frequencies of spectral colors are

Violet light:	$\lambda = 450 \text{ nm} = 4.5 \times 10^{-7} \text{ m}$	$f = 6.67 \times 10^{14} \text{ Hz}$
Blue light:	$\lambda = 480 \text{ nm} = 4.8 \times 10^{-7} \text{ m}$	$f = 6.25 \times 10^{14} \text{ Hz}$
Green light:	$\lambda = 520 \text{ nm} = 5.2 \times 10^{-7} \text{ m}$	$f = 5.77 \times 10^{14} \text{ Hz}$
Yellow light:	$\lambda = 580 \text{ nm} = 5.8 \times 10^{-7} \text{ m}$	$f = 5.17 \times 10^{14} \text{ Hz}$
Orange light:	$\lambda = 600 \text{ nm} = 6.0 \times 10^{-7} \text{ m}$	$f = 5.00 \times 10^{14} \text{ Hz}$
Red light:	$\lambda = 640 \text{ nm} = 6.4 \times 10^{-7} \text{ m}$	$f = 4.69 \times 10^{14} \text{ Hz}$

The sensitivity of the human eye to lights of various wavelengths is shown in Fig. 10.01–2.

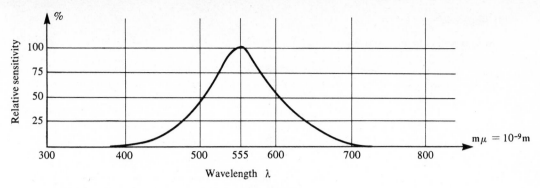

Fig. 10.01-2

(2) Radiometry

(a) Transfer of radiant energy (of light) from a source involves the flow of radiation, called the *flux* F_R, measured in watts (W).

(b) Radiant intensity I_R of a light source is its radiant flux per unit solid angle.

$$I_R = \frac{F_R}{\Omega} \qquad \left\{ \begin{array}{l} I_R \text{ in W/sr} \\ F_R \text{ in W} \\ \Omega \text{ in sr} \end{array} \right\}$$

(c) Solid angle Ω is defined as the ratio of the area S on the surface of a sphere to its radius squared (R^2) (Fig. 10.01–3).

$$\Omega = \frac{S}{R^2} \qquad \left\{ \begin{array}{l} S \text{ in m}^2 \\ R \text{ in m} \end{array} \right\}$$

(d) SI unit of solid angle is 1 steradian, designated by the symbol sr, and defined as the solid angle subtended at the vertex by a spherical sector whose spherical part of the surface is equal to the square of the radius of the sphere.

example:

The solid angle subtended by the whole sphere is

$$\Omega = \frac{4\pi R^2}{R^2} = 4\pi \text{ sr}$$

The solid angle subtended by $1/4\pi = 0.07957$ of the surface of the sphere is

$$\Omega = \frac{4\pi R^2/4\pi}{R^2} = 1 \text{ steradian}$$

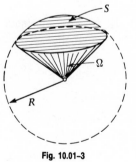

Fig. 10.01–3

(e) Irradiance E_R is the radiant flux per unit area of surface.

$$E_R = \frac{F_R}{S} \qquad \left\{ \begin{array}{l} E_R \text{ in W/m}^2 \\ F_R \text{ in W} \\ S \text{ in m}^2 \end{array} \right\}$$

(3) Photometry

(a) **Luminous flux** F_L is that part of the total radiant flux F_R emitted from a light source which is capable of affecting the sensors of the retina of the human eye.

(b) **Conversion chart.** The luminous flux F_L is a subjective phenomenon (the human eye is not equally sensitive to all colors), and it is possible that waves outside the visible spectrum radiate energy without emitting any luminous flux at all. Since the radiant flux F_R is energy per unit time and can be measured, the radiant flux must be converted into the luminous flux by means of the luminosity curve of Fig. 10.01–4.

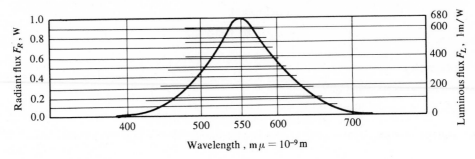

Fig. 10.01–4

(c) **Luminous intensity** I_L of a light source is its luminous flux per unit solid angle.

$$I_L = \frac{F_L}{\Omega} \qquad \begin{cases} I_L \text{ in cd} \\ F_L \text{ in lm} \\ \Omega \text{ in sr} \end{cases}$$

(d) **SI unit of luminous intensity** is 1 candela, designated by the symbol cd, and defined as such a value that the luminous intensity of a full radiator at the solidification temperature of platinum is 60 candelas per centimeter squared.

(e) **SI unit of luminous flux** is 1 lumen, designated by the symbol lm, and defined as the luminous flux emitted in a solid angle of 1 steradian by a uniform point source of intensity of 1 candela.

1 lumen = 1 lm = 1 candela × steradian = 1 cd-sr

example:

If a 1-candela source radiates normally upon a spherical surface of area 1 m² at a distance of 1 m from the source or upon a surface of area 1 ft² at a distance of 1 ft, the luminous flux thus emitted is 1 lumen.

(f) **Alternative definition.** 1 lumen is 1/680 watt of yellow-green light of wavelength 550 nm (which according to the conversion chart of Fig. 10.01–4 has 100 percent intensity).

example:

If F_R of this light is 5 W, the corresponding $F_L = 680 \times 5 = 3,400$ lm.

If F_R of a light of $\lambda = 500$ nm is 5 W, the corresponding $F_L = 3,400 \times 0.5 = 1,700$ lm, where 0.5 is the coefficient from the chart in Fig. 10.01–4.

(g) Isotropic luminous flux of a point source is

$$F_L = 4\pi I_L \qquad \left\{\begin{array}{l} F_L \text{ in lm} \\ \pi \text{ in sr} \\ I_L \text{ in cd} \end{array}\right\}$$

where 4π = solid angle of a sphere and I_L = luminous intensity defined in (c). Consequently a one-candela source radiates a total luminous flux of 4π lumens.

(h) Luminous intensity I_B of a beam of light emitted by a source of intensity I_L on an area S at a distance d from the source is

$$I_B = \frac{4\pi I_L}{\Omega} = \frac{4\pi d^2 I_L}{S} \qquad \left\{\begin{array}{l} I_B,\ I_L \text{ in cd} \\ \pi,\ \Omega \text{ in sr} \\ d \text{ in m} \\ S \text{ in m}^2 \end{array}\right\}$$

example:

A reflector equipped with a bulb of $I_L = 100$ cd is directed on a vertical wall located 50 m away and illuminates on it an area $S = 300$ m^2. The total luminous flux emitted by the bulb is $F_L = 4\pi I_L = 4 \times 3.14159 \times 100 = 1{,}256$ lm. The solid angle of the area S is

$$\Omega = \frac{S}{R^2} = \frac{300}{50^2} = 0.120 \text{ sr}$$

and the intensity of the beam of light emitted on this area is

$$I_B = \frac{F_L}{\Omega} = \frac{1256}{0.120} = 10{,}467 \text{ lm/sr} = 10{,}467 \text{ cd}$$

(i) Luminous efficiency is the ratio of the luminous flux F_L to the radiant flux F_R.

$$\text{Eff.} = \frac{F_L}{F_R} \qquad \left\{\begin{array}{l} \text{Eff. in lm/W} \\ F_L \text{ in lm} \\ F_R \text{ in W} \end{array}\right\}$$

(4) Illumination

(a) Illuminance E_L of a surface S is the luminous flux F_L per unit of that area (Fig. 10.01–5).

$$E_L = \frac{F_L}{S} = \frac{F_L}{\Omega R^2} = \frac{I_L}{R^2} \qquad \left\{\begin{array}{ll} E_L \text{ in lm/m}^2 & R \text{ in m} \\ F_L \text{ in lm} & I_L \text{ in cd} \\ S \text{ in m}^2 & \Omega \text{ in sr} \end{array}\right\}$$

where R = distance between the source and the surface, I_L = intensity of luminous flux, and Ω = solid angle.

(b) SI unit of illuminance is 1 lux, designated by the symbol lx, and defined as the illuminance of 1 lumen per meter squared.

$$1 \text{ lux} = 1 \text{ lx} = 1 \text{ lm/m}^2$$

(c) Projected surface. If the flux F_L makes an angle θ with the normal to the surface as shown in Fig. 10.01–6, then

$$E_L = \frac{I_L \cos \theta}{R^2}$$

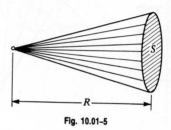

Fig. 10.01–5

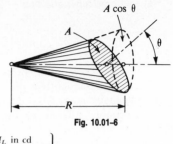

Fig. 10.01–6

(d) **Principle of photometry** states that the luminous intensities I_{L1} and I_{L2} of two sources producing equal illuminance E_L on the same surface are directly proportional to the squares of their distances R_1 and R_2 from that surface; that is, if

$$E_L = \frac{I_{L1}}{R_1^2} = \frac{I_{L2}}{R_2^2} \qquad \text{then} \qquad \frac{I_{L1}}{I_{L2}} = \left(\frac{R_1}{R_2}\right)^2 \qquad \begin{cases} I_L \text{ in cd} \\ E_L \text{ in lm/m}^2 \\ R \text{ in m} \end{cases}$$

example:

If two sources of $I_{L1} = 100$ cd and $I_{L2} = 400$ cd are located 20 m apart, then there are two points on the straight line connecting them which have the same illuminance E_L from each source. From Fig. 10.01–7,

$$\frac{I_{L1}}{I_{L2}} = \left(\frac{R_1}{R_2}\right)^2$$

$$\frac{100}{400} = \left(\frac{20 - x}{-x}\right)^2$$

and $x = 13.3$ m or 40 m.

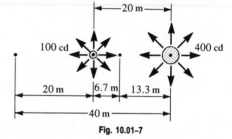

Fig. 10.01–7

10.02 GEOMETRY OF A LIGHT RAY

(1) Geometric Representation

(a) **Geometrical optics** investigates the geometry of reflection (Sec. 10.02–2) and refraction (Sec. 10.02–3) of light and is based on the application of Huygens' principles.

(b) **Huygens' first principle** states that every point on an advancing wave front of light can be considered a source of secondary waves, which in an isotropic medium move forward as spherical wavelets (Fig. 10.02–1).

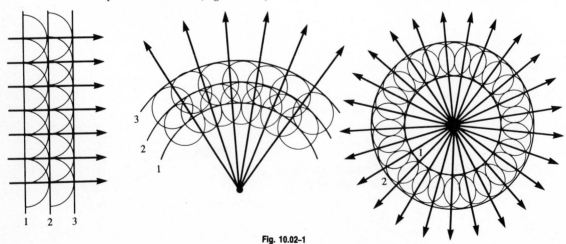

Fig. 10.02–1

(c) **Huygens' second principle.** The new position of the wave front is the envelope of wavelets emitted from all points of the wave front in the previous position.

(d) **Ray of light** is the geometric representation of the wave motion of light by a straight line perpendicular to the wave front.

(2) Reflection of Light

(a) **Reflection.** When a light beam strikes the boundary between two media (such as air and water), a part of that light is reflected. This is the reason for the visibility of nonluminous bodies (Fig. 10.02–2). Reflection of light occurs in accordance with two empirical laws defined below.

(b) **First law of reflection** states that the incident ray, the reflected ray, and the normal to the surface of reflection lie in the same plane.

(c) **Second law of reflection** states that the angle of incidence θ_i equals the angle of reflection θ_r.

$$\theta_i = \theta_r$$

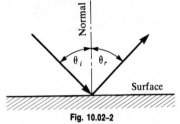

Fig. 10.02–2

(d) **Reflection coefficient** μ_r of a light ray at normal incidence ($\theta_i = 0$) from the plane surface is

$$\mu_r = \frac{I_{Lr}}{I_{Li}}$$

where I_{Li} and I_{Lr} = intensities of the incident and reflected light, respectively. Numerical values of μ_r for the most commonly used metals are given in Appendix A. For light of $\lambda = 500$ nm, μ_r of mirror alloy = 0.833, μ_r of steel = 0.550, μ_r of gold = 0.415, and μ_r of silver = 0.950.

(e) **Intensity of reflected light** I_{Lr} increases with the angle of incidence θ_i.

$$I_{Lr} = I_{Li} - I_{Li}(1 - \mu_r) \cos \theta_i$$

where μ_r = reflection coefficient given in (d).

example:

If $I_{Li} = 100$ cd, $\theta_i = 60°$, and $\mu_r = 0.5$, then

$$I_{Lr} = 100 - 100(1 - 0.5) \times 0.5 = 75 \text{ cd}$$

For $\theta_i = 0°$, $I_{Lr} = 50$ cd, and for $\theta_i = 90°$, $I_{Lr} = 100$ cd.

(3) Refraction of Light

(a) **Refraction.** When a light beam strikes the boundary between two transparent media, a certain part of it is reflected, but in general a much larger part passes into the second medium with a sudden change in direction, called *refraction* (Fig. 10.02–3). This phenomenon occurs in accordance with three empirical laws defined below.

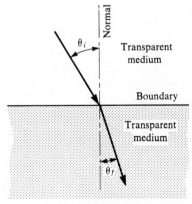

Fig. 10.02–3

(b) **First law of refraction** states that the incident ray, the refracted ray, and the normal to the plane surface of refraction lie in the same plane.

(c) **Second law of refraction** states that the ratio of the sine of the angle of incidence θ_i measured from the normal to the surface of refraction to the sine of the angle of refraction θ_f measured from the same normal is a constant μ_f, specific for the two media, and equals the ratio of the speed of incidence c_i to the speed of refraction c_f.

$$\frac{\sin \theta_i}{\sin \theta_f} = \frac{c_i}{c_f} = \mu_f \qquad \left\{ \begin{array}{l} \theta_i,\ \theta_f \text{ in rad} \\ c_i,\ c_f \text{ in m/s} \\ \mu_f \text{ dimensionless} \end{array} \right\}$$

(d) **Third law of refraction** states that the path of incidence and refraction of the light ray at the interface of two media is reversible (Fig. 10.02–4).

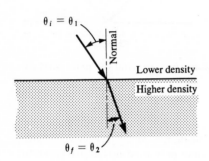

Fig. 10.02–4

(4) Index of Refraction and Wavelength

(a) **Index of refraction** μ_m is a particular case of μ_f introduced in (c) and is defined as the ratio of the speed of light c in vacuum to the speed of light c_m in a particular medium.

$$\mu_m = \frac{c}{c_m} = \frac{\sin \theta}{\sin \theta_m}$$

where θ = incidence angle and θ_m = refraction angle. For air, $\mu_m = 1$; for water, $\mu_m = 1.33$; for commercial plate glass, $\mu_m = 1.5$ (Appendix A).

example:

If the light ray passes from air into water with $\theta = 30° = 0.524$ rad, then

$$\sin \theta_m = \frac{\sin (\pi/6)}{\mu_m} = \frac{0.5}{1.33} = 0.375 \qquad \theta_m = 22°03' = 0.385 \text{ rad}$$

and

$$c_m = \frac{c}{\mu_m} = \frac{3 \times 10^8}{1.33} = 2.26 \times 10^8 \text{ m/s}$$

(b) Relation between μ_f **and** μ_m **is given as**

$$\mu_f = \frac{\mu_{m2}}{\mu_{m1}} = \frac{c/c_{m2}}{c/c_{m1}} = \frac{c_{m1}}{c_{m2}}$$

where μ_{m1} = index of refraction of the incidence medium, μ_{m2} = index of refraction of the refraction medium, c_{m1} = speed of light in the incidence medium, and c_{m2} = speed of light in the refraction medium. If μ_{m1} = 1 (air), then $\mu_f = \mu_{m2}$.

example:

If for water μ_{m1} = 1.33 and for glass μ_{m2} = 1.5, then

$$\mu_f = 1.50/1.33 = 1.128$$

and since $c_{m1} = 2.26 \times 10^8$ m/s given in (a),

$$c_{m2} = c_{m1}/\mu_f = 2.26/1.128 = 2 \times 10^8 \text{ m/s}$$

(c) Wavelength λ_m **of light in a material medium is**

$$\lambda_m = \lambda/\mu_m = \lambda c_m/c \qquad \{\lambda \text{ in m}\}$$

and the number N_m of waves in a medium of thickness d_m is

$$N_m = d_m/\lambda_m = \mu_m d_m/\lambda \qquad \begin{Bmatrix} d_m \text{ in m} \\ N_m \text{ dimensionless} \end{Bmatrix}$$

where λ = wavelength of light in vacuum (Fig. 10.02–5).

(d) Frequency f_m **of the same light in the same material medium is the same as in vacuum:**

$$f_m = f \qquad \{f \text{ in Hz}\}$$

but the speed in terms of λ_m is different:

$$c_m = c/\mu_m \qquad \{c \text{ in m/s}\}$$

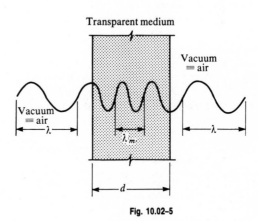

Fig. 10.02-5

example:

The wavelength of green light in vacuum is $\lambda = 5.2 \times 10^{-7}$ m/s and its frequency $f = 5.77 \times 10^{14}$ Hz.

If it passes through a plate glass of $d_m = 3$ cm and $\mu_m = 1.5$,

$$\lambda_m = \frac{5.2 \times 10^{-7}}{1.5} = 3.47 \times 10^{-7} \text{ m/s} \qquad f_m = 5.77 \times 10^{14} \text{ Hz}$$

$$c_m = \frac{3 \times 10^8}{1.5} = 2 \times 10^8 \text{ m/s},$$

$$N_m = \frac{3 \times 10^{-2}}{3.47 \times 10^{-7}} = 8.65 \times 10^4$$

(5) Internal Reflection and Refraction

(a) **Total internal reflection.** When light tends to pass obliquely from a medium of high density to another medium of lower density, under certain conditions total reflection in the first medium may occur (Fig. 10.02–6).

(b) **Critical angle** θ_c is the limiting angle of incidence in a medium which produces an angle of refraction $\theta_f = 90° = \pi/2$ rad. Thus $\sin \theta_f = 1$ and

$$\sin \theta_c = \frac{\mu_{m2}}{\mu_{m1}}$$

where μ_{m1} = index of refraction of denser medium and μ_{m2} = index of refraction of less dense medium. For water and air, $\theta_c \cong 48°$; for glass and air, $\theta_c \cong 41°$; and for glass and water, $\theta_c \cong 62°$.

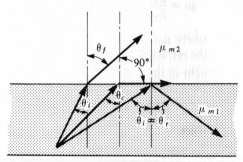

Fig. 10.02–6

(c) **Displacement.** When a light beam passes from one medium through another medium and exits in a third medium which is of the same density as the first medium and if the interfaces of incidence and exit are parallel (Fig. 10.02–7), the displacement of the light beam is

$$d = \frac{h}{\cos \theta_m} \sin(\theta_i - \theta_m) \qquad \begin{cases} d, h \text{ in m} \\ \theta_m, \theta_i \text{ in rad} \end{cases}$$

where θ_m = angle of refraction, θ_i = angle of incidence, and h = thickness of the medium of refraction.

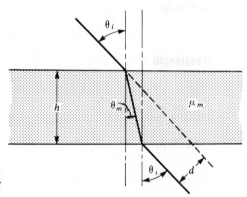

Fig. 10.02–7

example:

If the light beam strikes a glass plate of $h = 0.1$ m under $\theta_i = 45° = \pi/4$ rad, then

$$\sin \theta_m = \frac{\sin \theta_i}{\mu_m} = \frac{0.707}{1.5} = 0.47133 \qquad \theta_m = 28°07' = 0.491 \text{ rad}$$

$$d = \frac{h \sin(\theta_i - \theta_m)}{\cos \theta_m} = \frac{0.1 \times 0.29074}{0.88186} = 0.033 \text{ m} = 3.3 \text{ cm}$$

(d) **Deviation angle** D of a symmetrical triangular prism is the angle between the directions of incidence and of emergence of the light ray, as shown in Fig. 10.02–8a.

(e) Minimum deviation angle D_m of the same prism shown in Fig. 10.02–8b is defined by the conditions

$$\theta_i = \theta_e \qquad \theta_1 = \theta_2 = \alpha/2$$

and is

$$D_m = 2\theta_i - \alpha$$

where α = apex angle of the prism, θ_i is computed from the relation $\sin\theta_i = \mu_m \sin(\alpha/2)$, and μ_m = index of refraction of the prism.

example:

For a glass prism of $\alpha = 60°$ and $\mu_m = 1.5$, $\sin\theta_i = 1.5\sin 30° = 0.75$, $\theta_i = 48°35'$, and

$$D_m = 2 \times (48°35') - 60° = 37°10'$$

(f) Index of refraction μ_m in terms of a given D_m and α is

$$\mu_m = \frac{\sin\dfrac{\alpha + D_m}{2}}{\sin(\alpha/2)}$$

where α and D_m are those of Fig. 10.02–8.

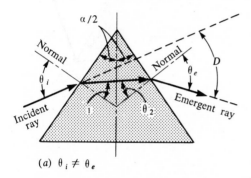

(a) $\theta_i \neq \theta_e$ (b) $\theta_i = \theta_e$

Fig. 10.02–8

(g) Apparent depth. As known from experience, refraction causes a submerged object in a liquid to appear closer to the surface than it actually is. The relation between the true depth a and the apparent depth b is

$$\frac{a}{b} = \sqrt{\frac{\mu_f^2 - \sin^2\theta_i}{\cos^2\theta_i}}$$

where μ_f = index of refraction of the two media (Sec. 10.02–4b) and θ_i = angle shown in Fig. 10.02–9. If $\theta_i \cong 0$,

$$\frac{a}{b} \cong \mu_f = \frac{\mu_{m2}}{\mu_{m1}}$$

where $\mu_{m2} > \mu_{m1}$.

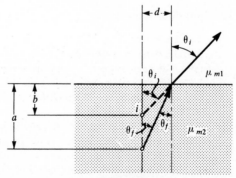

Fig. 10.02–9

If a coin is seen under $\theta_i = 30°$ at $b = 10$ m on the bottom of a pool filled with water, then the depth of the pool measured from the free surface of water is

$$a = 10\sqrt{\frac{1.33^2 - \sin^2 30°}{\cos^2 30°}} = 10 \times 1.42 = 14.2 \text{ m}$$

The approximative formula yields in this case an incorrect value, $a = 10 \times 1.33 = 13.3$ m, since θ_i is too large.

10.03 MIRRORS

(1) Plane Mirrors

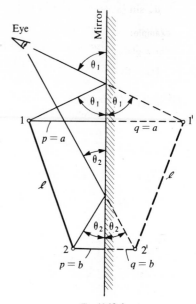

Fig. 10.03–1

(a) **Light reflection** from a smooth surface (such as polished metal) is called *regular* (specular) reflection and from a rough surface (such as concrete) is called *diffused* reflection.

(b) **Plane mirror** is a highly polished plane surface which forms a virtual, upright image whose size equals the real object, with the image distance q equal to the object distance p (Fig. 10.03–1).

(c) **Geometric relations** between the object and its virtual image are defined below.

(α) *Symmetrical properties:*
  Image distance p = object distance q
  Image height h = object height h'
  Angle of reflectance θ' = angle of incidence θ

(β) *Antisymmetrical properties:*
  Image left side = object right side
  Image right side = object left side

example:

The printed page of a book in a plane mirror appears as an image of the same size, but the letters are shown in reversed order and sense.

(2) Spherical Mirrors

(a) **Curved mirror** is a highly polished curved surface which produces a real or virtual image of a real object. Most curved mirrors used in practical applications are spherical, obtained by cutting off a portion of a sphere as shown in Fig. 10.03–2.

(b) **Two kinds of spherical mirrors** are the *concave mirror*, with its reflecting surface on the inside of the spherical surface (Fig. 10.03–3), and the *convex mirror*, with its reflecting surface on the outside of the spherical surface (Fig. 10.03–4).

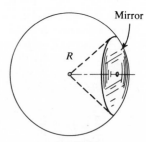

Fig. 10.03–2

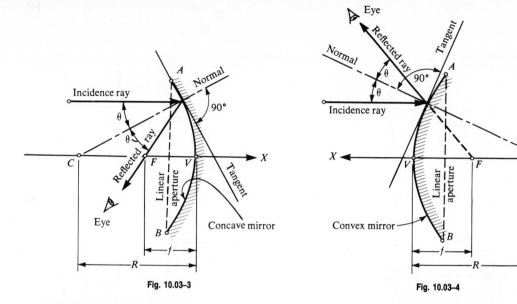

Fig. 10.03-3

Fig. 10.03-4

(c) **Principal focus** F of a spherical mirror lies on the principal axis X of the mirror and is the point of intersection of lines of action of all reflected light rays which strike the mirror parallel to the X axis. The focus is real for a concave mirror (Fig. 10.03–5) and virtual for a convex mirror (Fig. 10.03–6).

(d) **Geometry of spherical mirrors** of Figs. 10.03–3 and 10.03–4 is defined by the center of curvature C, the radius R of the reflecting spherical surface, the position of the principal axis X, the vertex V, the linear aperture $\overline{AB}$, and the focal length

$$\overline{FV} = f = R/2$$

(e) **Mirror equation** is the relationship

$$\frac{1}{p} + \frac{1}{q} = \frac{2}{R} = \frac{1}{f} \qquad \{p, q, R, f \text{ in m}\}$$

from which

$$p = \frac{qf}{q-f} \qquad q = \frac{pf}{p-f} \qquad f = \frac{pq}{p+q}$$

where p = object distance from the mirror, q = image distance from the mirror, R = radius of curvature of the mirror, and $f = R/2$ = focal distance.

(f) **Concave mirror** produces a real image (in front of the mirror) or a virtual image (behind the mirror) of the object according to the ratio $p:f$.

If $p > f$, the image is real and inverted (Fig. 10.03–7a to c).
If $p = f$, no image is formed (Fig. 10.03–7d).
If $p < f$, the image is virtual and upright (Fig. 10.03–7e).

The size l' of the image in terms of the size of the object l is

$$l' = -ql/p = Ml \qquad \left\{ \begin{array}{l} l, l', p, q \text{ in m} \\ M \text{ dimensionless} \end{array} \right\}$$

where p, q, and f are those of (e), and M = magnification factor.

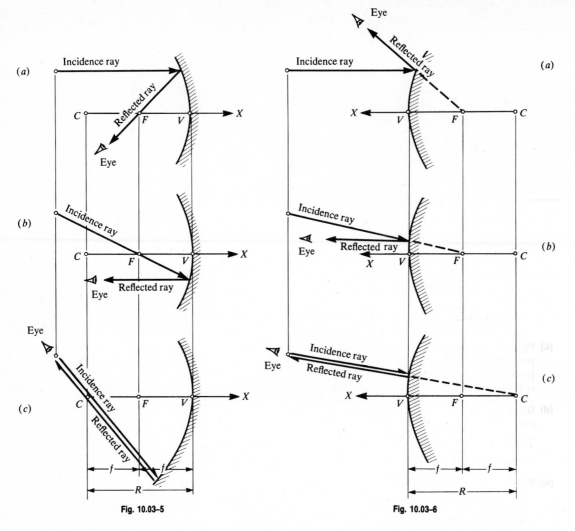

Fig. 10.03–5   Fig. 10.03–6

examples:

The application of the mirror equation (*e*) and the magnification equation (*f*) is shown below, where the mirror is concave, $R = 8$ m, $f = 4$ m, and $l = 3$ m.

The geometric construction in each case follows the number sequence of light rays, and two rays are necessary to locate one image point.

In Fig. 10.03–7*a*, $p = 12$ m,

$$q = \frac{12 \times 4}{12 - 4} = 6 \text{ m} \qquad l' = -\frac{6 \times 3}{12} = -1.5 \text{ m}$$

the image l' is real and inverted, and $|l'/l| = \frac{1}{2}$ (reduction).

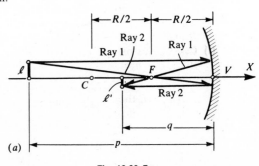

Fig. 10.03–7*a*

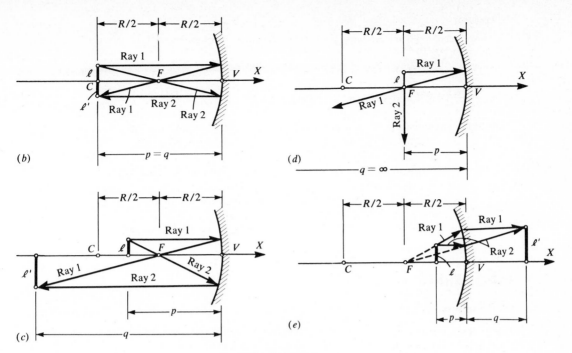

(b)

(d)

(c)

(e)

Fig. 10.03–7 (Continued)

In Fig. 10.03–7b, $p = 8$ m,

$$q = \frac{8 \times 4}{8 - 4} = 8 \text{ m} \qquad l' = -\frac{8 \times 3}{8} = -3 \text{ m}$$

the image l' is real and inverted, and $|l'/l| = 1$ (equal size).

In Fig. 10.03–7c, $p = 6$ m,

$$q = \frac{6 \times 4}{6 - 4} = 12 \text{ m} \qquad l' = -\frac{12 \times 3}{6} = -6 \text{ m}$$

the image l' is real, inverted, and $|l'/l| = 2$ (enlarged).

In Fig. 10.03–7d, $p = 4$ m,

$$q = \frac{4 \times 4}{4 - 4} = \infty$$

and no image is produced.

In Fig. 10.03–7e, $p = 2$ m,

$$q = \frac{2 \times 4}{2 - 4} = -4 \text{ m} \qquad l' = -\frac{(-4) \times 3}{2} = 6 \text{ m}$$

the image is virtual and upright, and $|l'/l| = 2$ (enlarged).

(g) Convex mirror produces only virtual, erect, and smaller images. The size of the image l' in terms of the size of the object l is given by the equation of (f). Note that R and f in convex mirrors are negative quantities (Fig. 10.03–8).

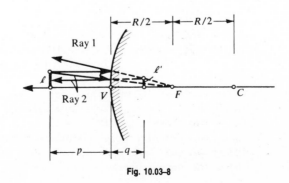

Fig. 10.03–8

example:

The object of $l = 3$ m is located at $p = 12$ m in front of a convex mirror of $R = -8$ m and $f = -4$ m, as shown in Fig. 10.03–8.

The distance of the virtual image is $q = \dfrac{(-12) \times 4}{12 + 4} = -3$ m, and the size of the image is $l' = -\dfrac{(-3) \times 3}{12} = 0.75$ m (reduced).

(h) Sign convention of the parameters p, q, R, f, and M is given below:

p is positive when the object is in front of the mirror.

q is positive when the image is real (in front).

q is negative when the image is virtual (behind).

R, f are positive in a concave mirror.

R, f are negative in a convex mirror.

M is positive when image is upright.

M is negative when image is inverted.

10.04 LENSES

(1) Geometry

(a) Lens is a transparent object that alters the direction of light rays in such a way that they either converge or diverge.

(b) Lenses are classified as double convex, plano-convex, concavo-convex, double concave, plano-concave, and convexo-concave (Fig. 10.04–1).

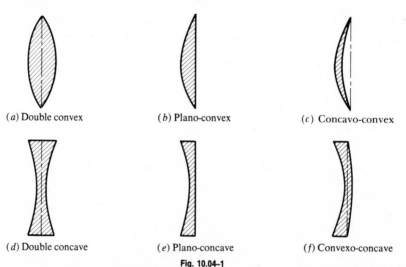

(*a*) Double convex   (*b*) Plano-convex   (*c*) Concavo-convex

(*d*) Double concave   (*e*) Plano-concave   (*f*) Convexo-concave

Fig. 10.04–1

(c) Curved surfaces of lenses may be of any shape, such as spherical, cylindrical, or paraboloidal, but since the spherical surface is easier to produce, most lenses are spherical.

(d) Converging lenses (Fig. 10.04–1*a* to *c*), also called *positive* lenses, are thicker at the center than at the rim and converge a beam of parallel light rays to a real focus (Fig. 10.04–2).

(e) Diverging lenses (Fig. 10.04–1*d* to *f*), also called *negative* lenses, are thinner at the center than at the rim and diverge a beam of parallel light rays from a virtual focus (Fig. 10.04–3).

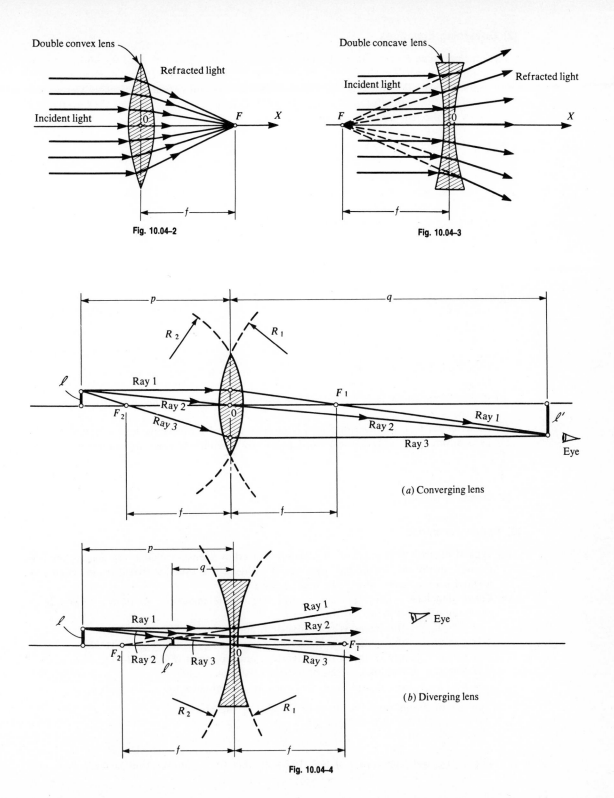

Fig. 10.04–2

Fig. 10.04–3

(a) Converging lens

(b) Diverging lens

Fig. 10.04–4

(2) Governing Equations

(a) Principal focus F of a spherical thin lens lies on the principal axis X of the lens and is the point of intersection of lines of action of reflected light rays which strike the lens parallel to the X axis. The focus is real for converging lenses and virtual for diverging lenses.

(b) Focal length f is the distance of the principal focus F from the center of the lens 0 and is given by the lens-maker's equation as

$$\frac{1}{f} = (\mu_m - 1)\left(\frac{1}{R_1} + \frac{1}{R_2}\right) \qquad \left\{\begin{matrix} f, R_1, R_2 \text{ in m} \\ \mu_m \text{ dimensionless} \end{matrix}\right\}$$

or

$$f = \frac{R_1 R_2}{(\mu_m - 1)(R_1 + R_2)}$$

where R_1, R_2 = radii of curvature of the two spherical surfaces and μ_m = index of refraction (Sec. 10.02–4a), which for optical glass is $\mu_m \cong 1.54$ (Appendix A).

(c) Lens equation for spherical thin lenses (Fig. 10.04–4) is

$$\frac{1}{p} + \frac{1}{q} = (\mu_m - 1)\left(\frac{1}{R_1} + \frac{1}{R_2}\right) = \frac{1}{f}$$

from which

$$p = \frac{qf}{q - f} \qquad q = \frac{pf}{p - f} \qquad f = \frac{pq}{p + q}$$

where p = object distance from the lens, q = image distance from the lens, and f = focal length.

(d) Radius of curvature R is positive if the surface is convex and negative if the surface is concave.

(e) Focal length f of a converging lens is positive and of a diverging lens is negative.

(f) Lens power P is the reciprocal value of the focal length of the lens in meters.

$$P = \frac{1}{f} \qquad \{P \text{ in m}^{-1}\}$$

(g) Diopter is the lens power of a lens of $f = 1$ m.

(3) Image Formation

(a) Type of image. The image of an object produced by a lens is called a *real image* when it is formed on the side of the lens opposite the object; it is called a *virtual image* when it is formed on the same side of the lens as the object.

(b) Converging lens. The type, sense, and size of the image of an object viewed by a converging lens depends on the ratio $p : f$.

If $p > 2f$, the image is real, inverted, and reduced (Fig. 10.04–5a).
If $p = 2f$, the image is real, inverted, and of equal size (Fig. 10.04–5b).
If $f < p < 2f$, the image is real, inverted, and enlarged (Fig. 10.04–5c).
If $p = f$, the image disappears (Fig. 10.04–5d).
If $p < f$, the image is virtual, upright, and enlarged (Fig. 10.04–5e).

The size of the image l' in terms of the size of the object l is

$$l' = -\frac{ql}{p} = Ml \qquad \left\{\begin{matrix} l, l', p, q \text{ in m} \\ M \text{ dimensionless} \end{matrix}\right\}$$

where p, q, and f are those of Sec. 10.04–2c and M = magnification factor.

The application of the lens equation (10.04–2c) and the magnification equation (b) is shown below, where $R_1 = 20\,\text{m}$, $R_2 = 20\,\text{m}$, $\mu_m = 1.5$, $l = 4\,\text{m}$, and

$$f = \frac{20 \times 20}{(1.5 - 1)(20 + 20)} = 20\,\text{m}$$

In Fig. 10.04–5a, $p = 60\,\text{m}$,

$$q = \frac{60 \times 20}{60 - 20} = 30\,\text{m}$$

$$l' = -\frac{30 \times 4}{60} = -2\,\text{m}$$

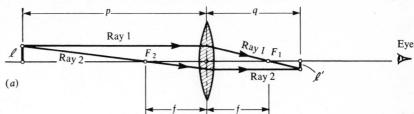

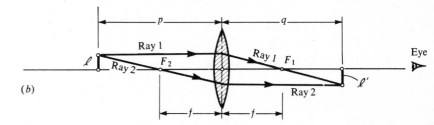

(a)

In Fig. 10.04–5b, $p = 40\,\text{m}$,

$$q = \frac{40 \times 20}{40 - 20} = 40\,\text{m}$$

$$l' = -\frac{40 \times 4}{40} = -4\,\text{m}$$

(b)

In Fig. 10.04–5c, $p = 30\,\text{m}$,

$$q = \frac{30 \times 20}{30 - 20} = 60\,\text{m}$$

$$l' = -\frac{60 \times 4}{30} = -8\,\text{m}$$

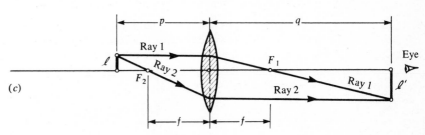

(c)

In Fig. 10.04–5d, $p = 20\,\text{m}$,

$$q = \frac{20 \times 20}{20 - 20} = \infty$$

and no image appears.

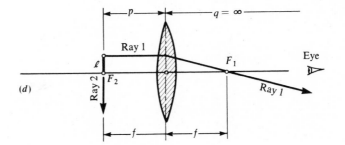

(d)

In Fig. 10.04–5e, $p = 12\,\text{m}$,

$$q = \frac{12 \times 20}{12 - 20} = -30\,\text{m}$$

$$l' = -\frac{(-30) \times 4}{12} = 10\,\text{m}$$

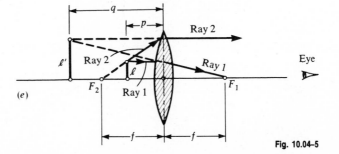

(e)

Fig. 10.04–5

(c) Diverging lens produces only virtual, upright, and reduced images of objects at finite distance p (Fig. 10.04–6). The size of the image in terms of the size l of the object is given by equation (b). The parameters q, f, and R of a diverging lens are negative.

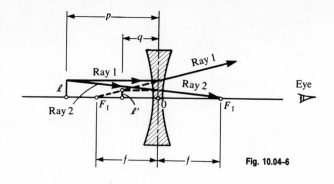

Fig. 10.04–6

example:

The object of $l = 4$ m is viewed by a divergent lens of $R_1 = R_2 = -20$ m, $\mu_m = 1.5$, $f = -20$ m, at $p = 30$ m, as shown in Fig. 10.04–6.

$$q = \frac{30 \times (-20)}{30 + 20} = -12 \text{ m} \quad \text{and} \quad l' = -\frac{(-12) \times 4}{30} = 1.6 \text{ m}$$

(d) Sign convention of the parameters p, q, R, f, and M is given below.

p is positive for a real object.
q is positive for a real image and negative for a virtual image.
R, f are positive for a converging lens and negative for a divergent lens.
M is positive when image is upright and negative when image is inverted.

Appendix A
PHYSICAL TABLES

A.01 GENERAL PHYSICAL CONSTANTS*

Constant	Symbol	Value†	Unit‡	Section
Atomic mass unit	u	1.660 57 (−27)	kg	6.02–1c
Avogadro number	N_A	6.022 05 (+26)	$kmol^{-1}$	6.02–1g
Boltzmann constant	k	1.380 66 (−23)	$J\text{-}K^{-1}$	6.02–5e
Electric field constant	ϵ_0	8.854 20 (−12)	$C^2\text{-}N^{-1}\text{-}m^{-2}$	7.01–3b
Electronvolt	eV	1.602 19 (−19)	J	7.02–3f
Elementary charge	e	1.602 19 (−19)	C	7.01–2e
Gravitation constant	G	6.672 00 (−11)	$N\text{-}m^2\text{-}kg^{-2}$	3.02–2c
Gravitational acceleration	g	9.806 65 (+00)	$m\text{-}s^{-2}$	3.02–2e
Magnetic field constant	μ_0	1.256 64 (−06)	$N\text{-}A^{-2}$	8.01–2c
Mass of electron	m_e	9.109 53 (−31)	kg	7.01–1c
Mass of neutron	m_n	1.674 95 (−27)	kg	7.01–1e
Mass of proton	m_p	1.672 65 (−27)	kg	7.01–1d
Speed of light in vacuum	c	2.997 93 (+08)	$m\text{-}s^{-1}$	10.01–1b
Standard atmosphere	atm	1.013 25 (+05)	$N\text{-}m^{-2}$	5.01–6g
Standard kilomole volume	V_m	2.241 40 (+01)	$m^3\text{-}kmol^{-1}$	6.02–3a
Technical calorie§	cal	4.186 00 (+00)	J	6.01–4d
Universal gas constant	R	8.314 30 (+03)	$J\text{-}K^{-1}\text{-}kmol^{-1}$	6.02–3d

*Definitions and applications of these constants are given in the section of this book indicated in the last column.
†Values of constants are given in the modified scientific notation. For example: 6.022 05 (+ 26) = $6.022\,05 \times 10^{26}$, 1.380 66 (− 23) = $1.380\,66 \times 10^{-23}$.
‡Glossary of symbols of units, **Appendices B.02 through B.05.**
§Thermochemical calorie = 4.184 J, Appendix B.29.

A.02 RECOMMENDED VALUES OF THE FUNDAMENTAL PHYSICAL CONSTANTS

The values of constants given in this table are those recommended by the CODATA Task Group on Fundamental Constants, E. R. Cohen, Chairman.

Quantity	Symbol	Value	Units*	
			SI	CGS
Speed of light in vacuum	c	2.997 924 580	10^8 m · s^{-1}	10^{10} cm · s^{-1}
Fine-structure constant,	α	7.297 350 6	10^{-3}	10^{-3}
$[\mu_0 c^2/4\pi](e^2/\hbar c)$	α^{-1}	137.036 04		
Electron charge	e	1.602 189 2	10^{-19} C	10^{-20} emu
		4.803 242		10^{-10} esu
Planck constant	h	6.626 176	10^{-34} J · s	10^{-27} erg · s
	$\hbar = h/2\pi$	1.054 588 7	10^{-34} J · s	10^{-27} erg · s
Avogadro constant	N	6.022 045	10^{26} kmol^{-1}	10^{23} mol^{-1}
Atomic mass unit	u	1.660 565 5	10^{-27} kg	10^{-24} g
Electron rest mass	m_e	9.109 534	10^{-31} kg	10^{-28} g
$Nm_e = M_e$	M_e	5.485 802 6	10^{-4} u	10^{-4} u
Proton rest mass	m_p	1.672 648 5	10^{-27} kg	10^{-24} g
$Nm_p = M_p$	M_p	1.007 276 471	u	u
Ratio of proton mass to electron mass	m_p/m_e	1836.151 52		
Neutron rest mass	m_n	1.674 954 3	10^{-27} kg	10^{-24} g
$Nm_n = M_n$	M_n	1.008 665 0	u	u
Electron charge to mass ratio	e/m_e	1.758 804 7	10^{11} C · kg^{-1}	10^7 emu · g^{-1}
		5.272 764		10^{17} esu · g^{-1}
Magnetic flux quantum,	Φ_0	2.067 850 6	10^{-15} Wb	10^{-7} G · cm^2
$[c]^{-1}(hc/2e)$	h/e	4.135 701	10^{-15} J · s · C^{-1}	10^{-7} erg · s · emu^{-1}
		1.379 521 5		10^{-17} erg · s · esu^{-1}
Josephson frequency-voltage ratio	$2e/h$	4.835 939	10^{14} Hz · V^{-1}	
Quantum of circulation	$h/2m_e$	3.636 945 5	10^{-4} J · s · kg^{-1}	erg · s · g^{-1}
	h/m_e	7.273 891	10^{-4} J · s · kg^{-1}	erg · s · g^{-1}
Faraday constant	$\mathscr{F}$	9.648 456	10^7 C · kmol^{-1}	10^4 C · mol^{-1}
		2.892 534 2		10^{14} esu · mol^{-1}
Rydberg constant, $[\mu_0 c^2/4\pi]^2(m_e e^4/4\pi\hbar^3 c)$	R_∞	1.097 373 177	10^7 m^{-1}	10^5 cm^{-1}
Bohr radius, $[\mu_0 c^2/4\pi]^{-1}(\hbar^2/m_e e^2) = \alpha/4\pi R_\infty$	a_0	5.291 770 6	10^{-11} m	10^{-9} cm
Classical electron radius, $[\mu_0 c^2/4\pi](e^2/m_e c^2) = \alpha^3/4\pi R_\infty$	r_0	2.817 938 0	10^{-15} m	10^{-13} cm
Free electron g-factor, or electron magnetic moment in Bohr magnetons	$g_{j/2} = \mu_e/\mu_B$	1.001 159 656 7		
Free muon g-factor, or muon magnetic moment in units of $[c](e\hbar/2m_\mu c)$	$g_{\mu/2}$	1.001 166 16		

Quantity	Symbol	Value	Units* SI	Units* CGS
Bohr magneton, $[c]$ $(e\hbar/2m_e c)$	μ_B	9.274 078	10^{-24} J $\cdot$ T^{-1}	10^{-21} erg $\cdot$ G^{-1}
Electron magnetic moment	μ_e	9.284 832	10^{-24} J $\cdot$ T^{-1}	10^{-21} erg $\cdot$ G^{-1}
Gyromagnetic ratio of	γ_p'	2.675 130 1	10^{8} rad $\cdot$ s^{-1} $\cdot$ T^{-1}	10^{4} rad $\cdot$ s^{-1} $\cdot$ G^{-1}
protons in H_2O	$\gamma_p'/2\pi$	4.257 602	10^{7} Hz $\cdot$ T^{-1}	10^{3} Hz $\cdot$ G^{-1}
γ_p' corrected for	γ_p	2.675 198 7	10^{8} rad $\cdot$ s^{-1} $\cdot$ T^{-1}	10^{4} rad $\cdot$ s^{-1} $\cdot$ G^{-1}
diamagnetism of H_2O	$\gamma_p/2\pi$	4.257 711	10^{7} Hz $\cdot$ T^{-1}	10^{3} Hz $\cdot$ G^{-1}
Magnetic moment of protons in H_2O in Bohr magnetons	μ_p'/μ_B	1.520 993 22	10^{-3}	10^{-3}
Proton magnetic moment in Bohr magnetons	μ_p/μ_B	1.521 032 209	10^{-3}	10^{-3}
Ratio of electron and proton magnetic moments	μ_e/μ_p	658.210 688 0		
Proton magnetic moment	μ_p	1.410 617 1	10^{-26} J $\cdot$ T^{-1}	10^{-23} erg $\cdot$ G^{-1}
Magnetic moment of protons in H_2O in nuclear magnetons	μ_p'/μ_N	2.792 774 0		
μ_p'/μ_N corrected for diamagnetism of H_2O	μ_p/μ_N	2.792 845 6		
Nuclear magneton, $[c]$ $(e\hbar/2m_p c)$	μ_N	5.050 824	10^{-27} J $\cdot$ T^{-1}	10^{-24} erg $\cdot$ G^{-1}
Ratio of muon and proton magnetic moments	μ_μ/μ_p	3.183 340 2		
Muon magnetic moment	μ_μ	4.490 474	10^{-26} J $\cdot$ T^{-1}	10^{-23} erg $\cdot$ G^{-1}
Ratio of muon mass to electron mass $\cdot$	m_μ/m_e	206.768 65		
Muon rest mass	m_μ	1.883 566	10^{-28} kg	10^{-25} g
	M_μ	0.113 429 20	u	u
Compton wavelength of the electron, $h/m_e c$	λ_C	2.426 308 9	10^{-12} m	10^{-10} cm
	$\lambda_C/2\pi$	3.861 590 5	10^{-13} m	10^{-11} cm
Compton wavelength of the proton, $h/m_p c$	$\lambda_{C,p}$	1.321 409 9	10^{-15} m	10^{-13} cm
	$\lambda_{C,p}/2\pi$	2.103 089 2	10^{-16} m	10^{-14} cm
Compton wavelength of the neutron, $h/m_n c$	$\lambda_{C,n}$	1.319 590 9	10^{-15} m	10^{-13} cm
	$\lambda_{C,n}/2\pi$	2.100 194 1	10^{-16} m	10^{-14} cm
Standard volume of ideal gas	V_0	22.710 81	10^{5} J $\cdot$ kmol^{-1}	10^{9} erg $\cdot$ mol^{-1}
		22.413 83	m^3 $\cdot$ atm $\cdot$ kmol^{-1}	10^{3} cm^3 $\cdot$ atm mol^{-1}
Gas constant, V_0/T_0	R	8.314 41	10^{3} J $\cdot$ kmol^{-1} $\cdot$ K^{-1}	10^{7} erg $\cdot$ mol^{-1} 10 K^{-1}
$(T_0 = 273.15$ K$)$		8.205 68	10^{-2} m^3 $\cdot$ atm $\cdot$ kmol^{-1} $\cdot$ K^{-1}	10 cm^3 $\cdot$ atm $\cdot$ mol^{-1} $\cdot$ K^{-1}
Boltzmann constant R/N	k	1.380 662	10^{-23} J $\cdot$ K^{-1}	10^{-16} erg K^{-1}
Stefan–Boltzmann constant, $\pi^2 k^4/60\hbar^3 c^2$	σ	5.670 32	10^{-8} W $\cdot$ m^{-2} $\cdot$ K^{-4}	10^{-5} erg $\cdot$ s^{-1} $\cdot$ cm^{-2} $\cdot$ K^{-4}
First radiation constant, $2\pi hc^2$	c_1	3.741 832	10^{-16} W $\cdot$ m^2	10^{-5} erg $\cdot$ cm^2 $\cdot$ s^{-1}
Second radiation constant, hc/k	c_2	1.438 786	10^{-2} m $\cdot$ K	cm $\cdot$ K
Gravitational constant	G	6.672 0	10^{-11} N $\cdot$ m^2 $\cdot$ kg^{-2}	10^{-8} dyn $\cdot$ cm^2 $\cdot$ g^{-2}

SOURCE: Adapted from *CODATA Bulletin No. 11, December 1973*, ICSU Central Office, 19 Westendstrasse, 6 Frankfurt am Main, German Federal Republic (copies of this bulletin are available at no cost from this office).
*For Glossary of Symbols of Units see Appendices B.02 through B.05.

A.03 EARTH DATA

Constant	SI system		FPS system	
	Value	Unit	Value	Unit
Mean radius	6.371 (+03)	km	3.958 (+03)	mi
Equatorial radius	6.378 (+03)	km	3.963 (+03)	mi
Polar radius	6.357 (+03)	km	3.950 (+03)	mi
Volume	1.083 (+21)	m^3	3.825 (+22)	ft^3
Mass	5.975 (+24)	kg	1.317 (+25)	lb
Average mass density	5.522 (+03)	$kg\text{-}m^{-3}$	3.477 (+02)	$lb\text{-}ft^{-3}$
Surface mass density	2.615 (+03)	$kg\text{-}m^{-3}$	1.632 (+02)	$lb\text{-}ft^{-3}$
Period of rotation	8.618 (+04)	s	2.394 (+01)	h
Average linear velocity	2.977 (+04)	$m\text{-}s^{-1}$	9.767 (+04)	$ft\text{-}sec^{-1}$
Average angular velocity	7.279 (−05)	$rad\text{-}s^{-1}$	7.279 (−05)	$rad\text{-}sec^{-1}$
Distance from sun:				
Mean	1.495 (+08)	km	9.289 (+07)	mi
Aphelion	1.521 (+08)	km	9.451 (+07)	mi
Perihelion	1.471 (+08)	km	9.140 (+07)	mi
Sun radiation per 1 m^2 earth surface	1.35 (+00)	$kW\text{-}m^{-2}$	3.171 (+02)	$Btu\text{-}h^{-1}\text{-}ft^{-2}$
Gravitational acceleration	9.807 (+00)	$m\text{-}s^{-2}$	3.217 (+01)	$ft\text{-}sec^{-2}$

A.04 MOON DATA

Constant	SI system		FPS system	
	Value	Unit	Value	Unit
Mean radius	1.740 (+03)	km	1.080 (+03)	mi
Volume	2.205 (+19)	m^3	7.787 (+21)	ft^3
Mass	7.343 (+22)	kg	1.619 (+23)	lb
Average mass density	3.333 (+03)	$kg\text{-}m^{-3}$	2.079 (+02)	$lb\text{-}ft^{-3}$
Period of rotation	2.362 (+06)	s	2.732 (+01)	d
Distance from earth:				
Mean	3.844 (+05)	km	2.389 (+05)	mi
Apogee	4.070 (+05)	km	2.529 (+05)	mi
Perigee	3.570 (+05)	km	2.218 (+05)	mi
Gravitational acceleration	1.569 (+00)	$m\text{-}s^{-2}$	5.147 (+00)	$ft\text{-}sec^{-2}$

A.05 EARTH GRAVITATIONAL ACCELERATION

The variation of the earth-surface gravitational acceleration g_ϕ with latitude is given by the International Gravity Formula adopted in 1930 by the International Geophysical Congress.

$$g_\phi = 978.049\,000(1 + 0.005\,2884\,\sin^2\phi - 0.000\,0059\,\sin^2 2\phi) \qquad \left\{ \begin{array}{l} g \text{ in cm/s}^2 \\ \phi \text{ in deg} \end{array} \right\}$$

where ϕ = latitude of the earth ellipsoid measured from the equator. The numerical evaluation in this formula is given below in m-s^{-2}. For conversion to ft-sec^{-2}, use 1 m-s^{-2} = 3.280 840 ft-sec^{-2}.

$\phi°$	g_ϕ, m-s^{-2}	$\phi°$	g_ϕ, m-s^{-2}	$\phi°$	g_ϕ, m-s^{-2}	$\phi°$	g_ϕ, m-s^{-2}	$\phi°$	g_ϕ, m-s^{-2}
0	9.780 490	18	9.785 409	36	9.798 308	54	9.814 429	72	9.827 254
1	9.780 506	19	9.785 951	37	9.799 170	55	9.815 146	73	9.827 774
2	9.780 553	20	9.786 517	38	9.800 041	56	9.815 990	74	9.828 267
3	9.780 631	21	9.787 107	39	9.800 919	57	9.816 822	75	9.828 734
4	9.780 741	22	9.787 720	40	9.801 805	58	9.817 642	76	9.829 173
5	9.780 831	23	9.788 357	41	9.802 695	59	9.818 448	77	9.829 585
6	9.781 053	24	9.789 015	42	9.803 591	60	9.819 239	78	9.829 968
7	9.781 255	25	9.789 694	43	9.804 490	61	9.820 015	79	9.830 322
8	9.781 487	26	9.790 394	44	9.805 391	62	9.820 774	80	9.830 647
9	9.781 750	27	9.791 113	45	9.806 294	63	9.821 515	81	9.830 942
10	9.782 043	28	9.791 850	46	9.807 166	64	9.822 238	82	9.831 207
11	9.782 364	29	9.792 606	47	9.808 098	65	9.822 941	83	9.831 442
12	9.782 716	30	9.793 378	48	9.808 998	66	9.823 625	84	9.831 645
13	9.783 096	31	9.794 165	49	9.809 894	67	9.824 287	85	9.831 819
14	9.783 504	32	9.794 968	50	9.810 786	68	9.824 927	86	9.831 960
15	9.783 940	33	9.795 785	51	9.811 673	69	9.825 545	87	9.832 071
16	9.784 404	34	9.796 614	52	9.812 554	70	9.826 135	88	9.832 150
17	9.784 893	35	9.797 455	53	9.813 427	71	9.826 709	89	9.832 197
								90	9.832 213

The variation of gravitational acceleration with altitude h above sea level is approximately

$$g = g_\phi - 0.000\,002\,860h \qquad \left\{ \begin{array}{l} h \text{ in m} \\ g \text{ in m-s}^{-2} \end{array} \right\}$$

where $h \leqslant 40{,}000$ m.

A.06 TIME ZONES OF THE WORLD

The map below shows the 24 time zones of the world related to the international date line in the Pacific and the international noon line at Greenwich, England. The number below each time zone is the number of hours by which the zone is preceded by (+) or itself precedes (−) standard Greenwich time (Western European time). The numbers on the map show the zonal time at noon, U.S. Central Standard Time. For example, when the U.S. Central Standard Time Zone is at 12:00 (noon), the Western European Time Zone is at 18:00 (evening) and the Alaska Time Zone is at 8:00 (morning).

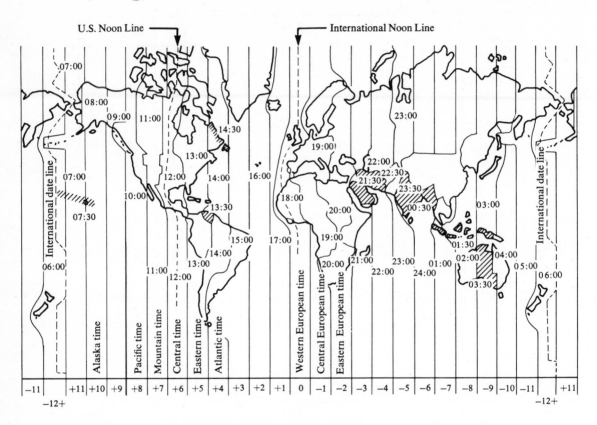

SOURCE: "Values and Measures of the World," Swiss Bank Corporation, Zurich, 1966, p. 28.

A.07 TEMPERATURE ZONES OF THE WORLD

The maps below show the isotherms of the average sea-level temperatures in January and
July, which are considered the months of seasonal extremes. An isotherm is a line drawn
through points of the same sea-level temperature. The difference between the average
temperatures of the warmest and coldest months is called the annual range of temperature,
which can be estimated for a given locality from the same maps.

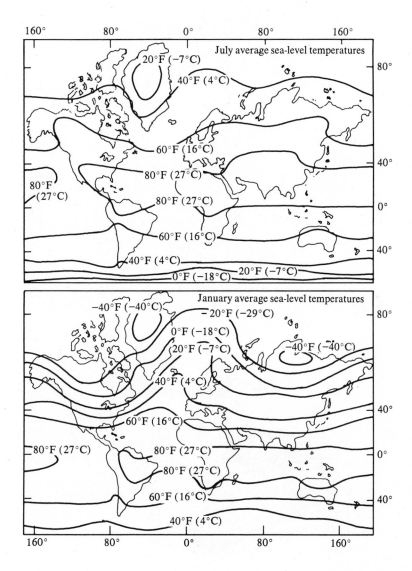

SOURCE: Lounsbury, John F., and Lawrence Ogden: "Earth Science," 3d ed., Harper & Row, New York, 1979, p. 69.

A.08 CONTOUR MAP OF SNOW LOAD IN THE CONTINENTAL UNITED STATES

The contour map below shows the extremes of snow load per square foot of ground area. This map is based on records of the U.S. Weather Bureau summarized in *"Greatest Snow Depth on Ground at Any One Time"* covering the period from 1871 to 1941. Snow loads within the dotted area are to be determined on the basis of local reports and in no case should be less than 45 lbf/ft². All values in the map are given in lbf/ft². For conversion use:

$$1 \text{ lbf/ft}^2 = 4.882 \text{ kgf/m}^2 = 47.880 \text{ Pa}$$

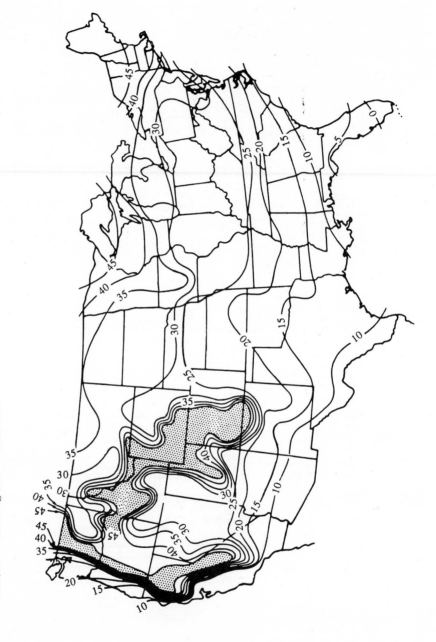

A.09 CONTOUR MAP OF EXTREME FROST PENETRATION IN THE CONTINENTAL UNITED STATES

The variation of temperature in a soil system at a given instant can be represented graphically by a curve called the *temperature profile*. The maximum depth the frost penetrates a soil system is called the *extreme frost penetration*. The extreme frost penetrations in inches based upon state averages are shown on the contour map below. This map is based on records of the U.S. Weather Bureau covering the same period as in Appendix A.08. For conversion use: 1 in. = 2.54 cm.

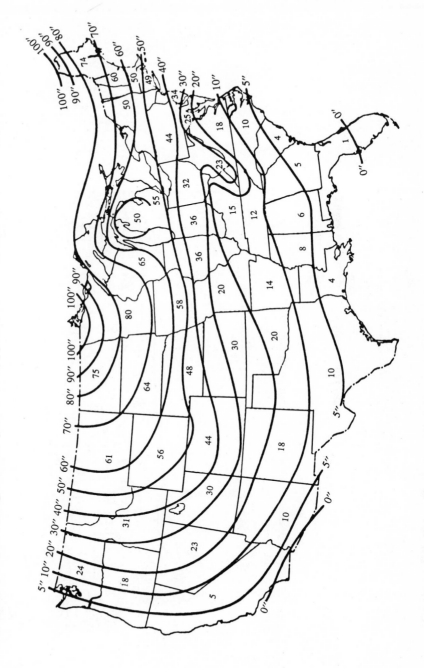

A.10 CONTOUR MAP OF FASTEST WIND VELOCITIES IN THE CONTINENTAL UNITED STATES

The contour map below shows the extremes of wind velocities measured at 30 ft above the ground in miles per hour. This map is based on records of the U.S. Weather Bureau and was presented by the ASCE subcommittee in the final report "Wind Forces on Structures," *Trans. ASCE*, Volume 126, Part II, 1961, p. 1132. Wind velocity increases with elevation by a factor which can be found in the same publication. The static wind pressure on a plane normal to the wind velocity vector is

$$p = \rho v^2/2$$

$$\{\rho \text{ in lb/ft}^3 \qquad v \text{ in ft/sec}\}$$

where ρ = mass density of air = 0.0765 lb/ft³ = 1.225 kg/m³ and v = wind velocity. For conversion use:

$$1 \text{ mile/hour} = 1.467 \text{ ft/sec} = 0.447 \text{ m/s}$$

$$1 \frac{\text{lb}}{\text{ft-sec}^2} = 0.031 \frac{\text{lbf}}{\text{ft}^2} = 0.014 \frac{\text{kgf}}{\text{m}^2} = 0.138 \text{ Pa}$$

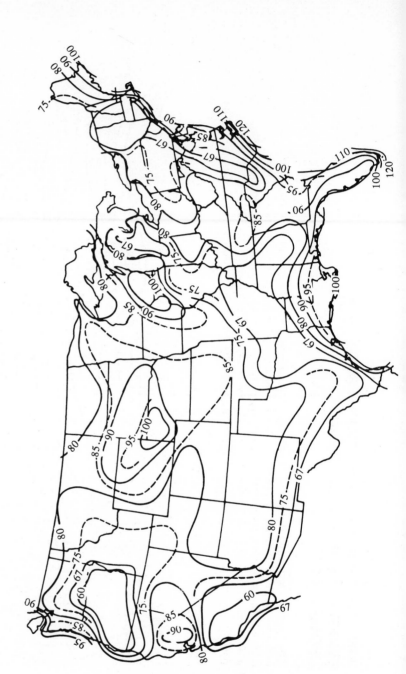

A.11 CONTOUR MAP OF EARTHQUAKE DAMAGE PROBABILITY IN THE CONTINENTAL UNITED STATES

The contour map below shows the classification of the earthquake damage probability in five categories:

Zone 0—Probability of no damage
Zone 1—Probability of minor damage
Zone 2—Probability of moderate damage
Zone 3—Probability of major damage
Zone 4—Areas within zone 3 in proximity to major earth faults

This classification is based on records of the U.S. Geological Survey.

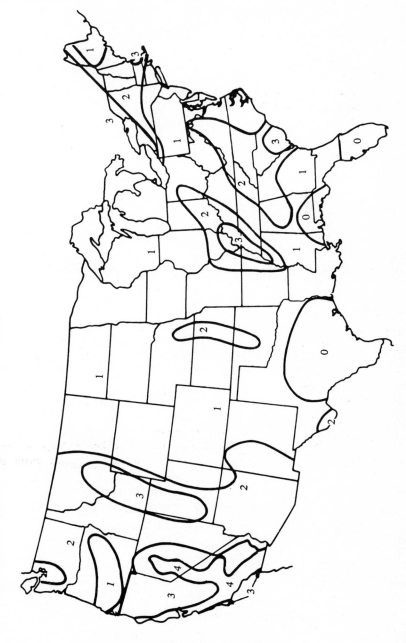

A.12 GEOMETRIC CHARACTERISTICS OF THE PLANETS OF THE SOLAR SYSTEM

The relative sizes of the planets and their relative distances from the sun in selected scales are shown graphically below.

(a) Relative sizes

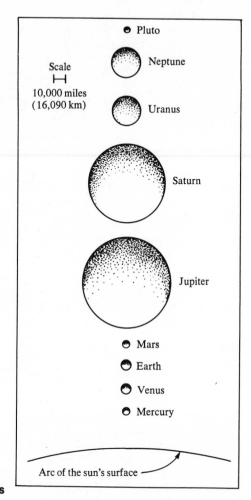

(b) Relative distances

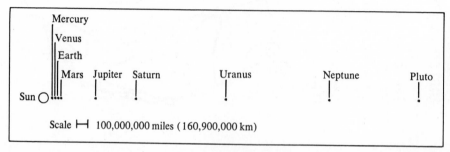

SOURCE: Lounsbury, John F., and Lawrence Ogden: "Earth Science," 3d ed., Harper & Row, New York, 1979, p. 24.

A.13 SUN DATA

Constant	SI units		FPS units	
	Value	Unit	Value	Unit
Mean radius	6.960 (+05)	km	4.325 (+05)	mi
Volume	1.409 (+27)	m^3	4.976 (+28)	ft^3
Mass	1.987 (+30)	kg	4.381 (+30)	lb
Average mass density	1.411 (+03)	$kg\text{-}m^{-3}$	8.802 (+01)	$lb\text{-}ft^{-3}$
Solar power	3.720 (+27)	W	1.27 (+28)	$Btu\text{-}h^{-1}$
Solar temperature	5.714 (+03)	K	9.826 (+03)	°F
Gravitational acceleration	2.739 (+02)	$m\text{-}s^{-2}$	8.986 (+02)	$ft\text{-}sec^{-2}$

A.14 MERCURY DATA

Constant	SI units		FPS units	
	Value	Unit	Value	Unit
Mean radius	2.420 (+03)	km	1.503 (+02)	mi
Volume	5.800 (+19)	m^3	2.048 (+21)	ft^3
Mass	3.167 (+23)	kg	6.982 (+23)	lb
Mean mass density	5.460 (+03)	$kg\text{-}m^{-3}$	3.409 (+02)	$lb\text{-}ft^{-3}$
Period of revolution	8.800 (+01)	d	8.800 (+01)	d
Distance from sun:				
Mean	5.790 (+07)	km	3.598 (+07)	mi
Aphelion	6.980 (+07)	km	4.377 (+07)	mi
Perihelion	4.600 (+07)	km	2.858 (+07)	mi
Gravitational acceleration	3.531 (+00)	$m\text{-}s^{-2}$	1.158 (+01)	$ft\text{-}sec^{-2}$

A.15 VENUS DATA

Constant	SI units		FPS units	
	Value	Unit	Value	Unit
Mean radius	6.260 (+03)	km	3.890 (+03)	mi
Volume	9.818 (+20)	m^3	3.467 (+22)	ft^3
Mass	4.870 (+24)	kg	1.074 (+25)	lb
Mean mass density	4.950 (+03)	$kg\text{-}m^{-3}$	3.096 (+02)	$lb\text{-}ft^{-3}$
Period of revolution	2.247 (+02)	d	2.247 (+02)	d
Distance from sun:				
Mean	1.080 (+08)	km	6.711 (+07)	mi
Aphelion	1.090 (+08)	km	6.773 (+07)	mi
Perihelion	1.070 (+08)	km	6.649 (+07)	mi
Gravitational acceleration	8.532 (+00)	$m\text{-}s^{-2}$	2.799 (+01)	$ft\text{-}sec^{-2}$

A.16 MARS DATA

Constant	SI units		FPS units	
	Value	Unit	Value	Unit
Mean radius	3.390 (+03)	km	2.106 (+03)	mi
Volume	1.550 (+20)	m³	5.473 (+21)	ft³
Mass	6.390 (+23)	kg	1.409 (+24)	lb
Mean mass density	3.950 (+03)	kg-m⁻³	2.466 (+02)	lb-ft⁻³
Period of revolution	6.870 (+02)	d	6.870 (+02)	d
Distance from sun:				
Mean	2.270 (+08)	km	1.411 (+08)	mi
Aphelion	2.490 (+08)	km	1.547 (+08)	mi
Perihelion	2.050 (+08)	km	1.274 (+08)	mi
Gravitational acceleration	3.727 (+00)	m-s⁻²	1.223 (+01)	ft-sec⁻²

A.17 JUPITER DATA

Constant	SI units		FPS units	
	Value	Unit	Value	Unit
Mean radius	6.990 (+04)	km	4.343 (+04)	mi
Volume	1.430 (+24)	m³	5.050 (+25)	ft³
Mass	1.900 (+27)	kg	4.189 (+27)	lb
Mean mass density	1.330 (+03)	kg-m⁻³	8.303 (+01)	lb-ft⁻³
Period of revolution	4.333 (+03)	d	4.333 (+03)	d
Distance from sun				
Mean	7.780 (+08)	km	4.834 (+08)	mi
Aphelion	8.160 (+08)	km	5.070 (+08)	mi
Perihelion	7.400 (+08)	km	4.598 (+08)	mi
Gravitational acceleration	2.589 (+01)	m-s⁻²	8.494 (+01)	ft-sec⁻²

A.18 SATURN DATA

Constant	SI units		FPS units	
	Value	Unit	Value	Unit
Mean radius	5.750 (+04)	km	3.573 (+04)	mi
Volume	8.010 (+23)	m³	2.829 (+25)	ft³
Mass	5.690 (+26)	kg	1.254 (+27)	lb
Mean mass density	7.000 (+02)	kg-m⁻³	4.370 (+01)	lb-ft⁻³
Period of revolution	1.076 (+04)	d	1.076 (+04)	d
Distance from sun:				
Mean	1.430 (+09)	km	8.886 (+08)	mi
Aphelion	1.510 (+09)	km	9.383 (+08)	mi
Perihelion	1.350 (+09)	km	8.389 (+08)	mi
Gravitational acceleration	1.108 (+01)	m-s⁻²	3.635 (+01)	m-sec⁻²

A.19 URANUS DATA

Constant	SI units		FPS units	
	Value	Unit	Value	Unit
Mean radius	2.370 (+04)	km	1.473 (+04)	mi
Volume	5.570 (+22)	m^3	1.967 (+24)	ft^3
Mass	8.690 (+25)	kg	1.916 (+26)	lb
Mean mass density	1.560 (+03)	$kg-m^{-3}$	9.739 (+01)	$lb-ft^{-3}$
Period of revolution	3.069 (+04)	d	3.069 (+04)	d
Distance from sun:				
Mean	2.870 (+09)	km	1.783 (+09)	mi
Aphelion	3.010 (+09)	km	1.870 (+09)	mi
Perihelion	2.730 (+09)	km	1.696 (+09)	mi
Gravitational acceleration	1.049 (+01)	$m-s^{-2}$	3.441 (+01)	$ft-sec^{-2}$

A.20 NEPTUNE DATA

Constant	SI units		FPS units	
	Value	Unit	Value	Unit
Mean radius	2.150 (+04)	km	1.336 (+04)	mi
Volume	4.170 (+22)	m^3	1.473 (+24)	ft^3
Mass	1.030 (+26)	kg	2.271 (+26)	lb
Mean mass density	2.280 (+03)	$kg-m^{-3}$	1.423 (+02)	$lb-ft^{-3}$
Period of revolution	6.018 (+04)	d	6.018 (+04)	d
Distance from sun:				
Mean	4.500 (+09)	km	2.796 (+09)	mi
Aphelion	4.540 (+09)	km	2.881 (+09)	mi
Perihelion	4.460 (+09)	km	2.771 (+09)	mi
Gravitational acceleration	1.383 (+01)	$m-s^{-2}$	4.537 (+01)	$ft-sec^{-2}$

A.21 PLUTO DATA

Constant	SI units		FPS units	
	Value	Unit	Value	Unit
Mean radius	2.900 (+03)	km	1.802 (+03)	mi
Volume	1.193 (+21)	m^3	4.214 (+22)	ft^3
Mass	5.370 (+24)	kg	1.184 (+25)	lb
Mean mass density	4.500 (+03)	$kg-m^{-3}$	2.809 (+02)	$lb-ft^{-3}$
Period of revolution	9.073 (+04)	d	9.073 (+04)	d
Distance from sun:				
Mean	5.910 (+09)	km	3.672 (+09)	mi
Aphelion	7.370 (+09)	km	4.580 (+09)	mi
Perihelion	4.450 (+09)	km	2.765 (+09)	mi
Gravitational acceleration	8.000 (+00)	$m-s^{-2}$	2.625 (+01)	$ft-sec^{-2}$

A.22 CHEMICAL ELEMENTS

Element	Symbol	Atomic number	Av. atomic mass	Element	Symbol	Atomic number	Av. atomic mass
Actinium	Ac	89	[227]	● Mercury	Hg	80	200.59
● Aluminum	Al	13	26.9815	● Molybdenum	Mo	42	95.94
Americium	Am	95	[243]	Neodymium	Nd	60	144.24
● Antimony	Sb	51	121.75	Neon	Ne	10	20.183
Argon	Ar	18	39.948	Neptunium	Np	93	[237]
● Arsenic	As	33	74.9216	● Nickel	Ni	28	58.71
Astatine	At	85	[210]	Niobium	Nb	41	92.906
● Barium	Ba	56	137.34	● Nitrogen	N	7	14.0067
Berkelium	Bk	97	[249]	Nobelium	No	102	[254]
Beryllium	Be	4	9.0122	Osmium	Os	76	190.2
Bismuth	Bi	83	208.980	● Oxygen	O	8	15.9994
Boron	B	5	10.811	Palladium	Pd	46	106.4
Bromine	Br	35	79.909	● Phosphorus	P	15	30.9738
Cadmium	Cd	48	112.40	● Platinum	Pt	78	195.09
● Calcium	Ca	20	40.08	Plutonium	Pu	94	[242]
Californium	Cf	98	[250]	Polonium	Po	84	[210]
● Carbon	C	6	12.011 15	● Potassium	K	19	39.102
Cerium	Ce	58	140.12	Praseodymium	Pr	59	140.907
Cesium	Cs	55	132.905	Promethium	Pm	61	[145]
● Chlorine	Cl	17	35.453	Protactinium	Pa	91	[231]
● Chromium	Cr	24	51.996	Radium	Ra	88	[226]
Cobalt	Co	27	58.9332	Radon	Rn	86	[222]
● Copper	Cu	29	63.543	Rhenium	Re	75	186.2
Curium	Cm	96	[246]	Rhodium	Rh	45	102.905
Dysprosium	Dy	66	162.50	Rubidium	Rb	37	85.47
Einsteinium	Es	99	[254]	Ruthenium	Ru	44	101.07
Erbium	Er	68	167.26	Samarium	Sm	62	150.35
Europium	Eu	63	151.96	Scandium	Sc	21	44.956
Fermium	Fm	100	[255]	Selenium	Se	34	78.96
● Fluorine	F	9	18.9984	● Silicon	Si	14	28.086
Francium	Fr	87	[223]	● Silver	Ag	47	107.870
Gadolinium	Gd	64	157.25	● Sodium	Na	11	22.9898
Gallium	Ga	31	69.72	Strontium	Sr	38	87.62
Germanium	Ge	32	72.59	● Sulfur	S	16	32.064
● Gold	Au	79	196.967	● Tantalum	Ta	73	180.948
Hafnium	Hf	72	178.49	Technetium	Tc	43	[99]
Helium	He	2	4.0026	Tellurium	Te	52	127.60
Holmium	Ho	67	164.930	Terbium	Tb	65	158.924
● Hydrogen	H	1	1.007 97	Thallium	Tl	81	204.37
Indium	In	49	114.82	Thorium	Th	90	232.038
Iodine	I	53	126.9044	Thulium	Tm	69	168.934
● Iridium	Ir	77	192.2	● Tin	Sn	50	118.69
● Iron	Fe	26	55.847	Titanium	Ti	22	47.90
Krypton	Kr	36	83.80	● Tungsten	W	74	183.85
Lanthanum	La	57	138.91	Uranium	U	92	238.03
Lawrencium	Lr	103	[257]	● Vanadium	V	23	50.942
● Lead	Pb	82	207.19	Xenon	Xe	54	131.30
Lithium	Li	3	6.939	Ytterbium	Yb	70	173.04
Lutetium	Lu	71	174.97	Yttrium	Y	39	88.905
● Magnesium	Mg	12	24.312	● Zinc	Zn	30	65.37
● Manganese	Mn	25	54.9380	Zirconium	Zr	40	91.22
Mendelevium	Md	101	[257]				

Notes: Names of the more abundant elements are marked with ● . Bracket values for elements that have no stable isotopes give the mass number of the most stable isotope. Elements 104 and 105 are as yet officially unnamed; their atomic masses are [257] and [262], respectively.

A.23 HARDNESS OF MATERIALS

The hardness H of a particular material can be determined by its resistance to scratching (Mohs test) or to indentation (Brinell test).

(a) Resistance to scratching is determined by measuring the scratch resistance of the particular material against the scratch resistance of 10 selected materials, arranged in order of their hardness, rated from 1 for talc to 10 for diamond. This order is called the *Mohs scale of hardness* given below.

Talc	1	Calcite	3	Apatite	5	Quartz	7	Corundum	9
Gypsum	2	Fluorite	4	Orthoclase	6	Topaz	8	Diamond	10

(b) Resistance to indentation is determined by a hard spherical indenter of diameter D which is pressed into the particular material surface under the load W. The Brinell hardness number, called Bhn, is defined as

$$\text{Bhn} = \frac{W}{A} = \frac{2W}{\pi D(D - \sqrt{D^2 - d^2})} \qquad \begin{cases} W \text{ in kgf} \\ D, d \text{ in mm} \\ \text{Bhn in kgf·mm}^{-2} \end{cases}$$

where A = area of indentation and d = diameter of this area. Typical values of hardness of the most commonly used materials are given below.

Material	Condition	Bhn, kgf·mm^{-2}	Material	Condition	Bhn, kgf·mm^{-2}
Aluminum	Cast	40–80	Molybdenum	Pressed	150–200
	Rolled	25–40	Nickel	Cast	160–210
	Wrought	80–120		Colddrawn	120–200
Bismuth	Cast	9–10		Wrought	150–200
Brass	Cast	90–130	Nitrided steel	Cast + ann.	800–1200
Cadmium	Cast	21–23	Palladium	Cast	40–50
Calcium	Cast	17–18	Plastic	Soft	12–14
Chromium	Cast	90–160		Hard	15–25
Copper	Cast	55–65	Platinum	Cast	90–170
		90–120	Potassium	Cast	0.04
Cutting tool			Silicon	Cast	0.06
steel	Rolled	700–800	Silver	Cast	40–60
Gold	Cast	28–33		Wrought	60–140
	Wrought	70–90	Sodium	Cast	0.07
Iridium	Cast	160	Steel	Cast	100–140
Iron	Cast	160–260		Cast + ann.	200–250
	Wrought	70–90		Rolled	120–160
Lead	Cast	3–5	Tantalum	Ann.	40
	Rolled	9–10	Titanium	Ann.	150
Magnesium	Cast	70–90	Tin	Cast	5–6
	Rolled	50–70	Tungsten	Cast + ann.	300–350
	Wrought	120–150	Zinc	Cast	150
Manganese	Quenched	30–40			

Brinell hardness / Mohs hardness scale:

- 10,000
- 5,000
- 2,000 — Diamond — 10
- 1,000 — Corundum — 9 / Topaz — 8
- 500 — Quartz — 7 / Orthoclase — 6 / Apatite — 5
- 200 — Fluorite — 4 / Calcite — 3
- 100
- 50
- 20 — Gypsum — 2
- 10
- 5 — 1.17 / Talc — 1

Brinell hardness — Mohs hardness

A.24 COEFFICIENTS OF STATIC AND KINETIC FRICTION

Typical values of the coefficients of static friction μ_s, the angles of static friction ϕ_s, the coefficients of kinetic friction μ_k, and the angle of kinetic friction ϕ_k (defined in Sec. 2.08) are given below.

Material	On material	Condition	Static friction		Kinetic friction	
			μ_s	ϕ_s, deg.	μ_k	ϕ_k, deg.
Aluminum	Aluminum	Dry	1.05	46.4	1.40	54.4
		Greasy	0.33	18.3	0.42	22.9
	Mild steel	Dry	0.61	31.5	0.46	24.6
		Greasy	0.20	11.3	0.14	7.9
Brass	Cast iron	Dry	0.42	22.8	0.30	16.7
		Greasy	0.13	7.4	0.08	4.5
	Mild steel	Dry	0.51	27.0	0.44	23.8
		Greasy	0.10	5.8	0.06	3.5
Bronze	Cast iron	Dry	0.40	21.9	0.30	16.8
		Greasy	0.12	6.9	0.06	3.5
Cadmium	Mild steel	Dry	0.57	29.7	0.46	24.7
		Greasy	0.14	7.0	0.09	5.1
Clay	Clay	Dry	1.00	45.0	0.80	38.6
		Wet	0.30	16.7	0.20	11.3
Carbon	Glass	Dry	0.24	13.5	0.18	10.1
	Hard steel	Dry	0.60–0.70	30–34	0.25–0.50	14–26
Cast iron	Cast iron or bronze	Dry	1.10	47.5	0.15	8.5
		Greasy	0.28	15.6	0.08	4.6
	Oak (parallel to grain)	Dry	0.62	31.8	0.49	26.1
		Greasy	0.14	7.9	0.08	4.5
Copper	Cast iron	Dry	1.05	46.4	0.29	16.3
		Greasy	0.26	14.6	0.09	5.1
Glass	Glass	Dry	0.92	42.6	0.40	21.8
		Greasy	0.1–0.3	5–16	0.05–0.1	2–6
	Nickel	Dry	0.78	37.0	0.56	29.3
Hard steel	Babbitt (ASTM 1)	Dry	0.70	34.0	0.33	18.3
		Greasy	0.08–0.23	4–13	0.06–0.16	3–9
	Hard steel	Dry	0.78	37.0	0.42	22.8
		Greasy	0.10–0.50	6–27	0.03–0.08	1.1–4.6
Leather	Cast iron	Dry	0.64	32.6	0.56	29.3
		Greasy	0.42	22.8	0.36	19.8
	Oak	Dry	0.60	31.0	0.50	26.6
		Greasy	0.40	22.0	0.30	17.0

A.24 COEFFICIENTS OF STATIC AND KINETIC FRICTION (Cont.)

Material	On material	Condition	Static friction		Kinetic friction	
			μ_s	ϕ_s, deg	μ_k	ϕ_k, deg.
Mild steel	Cast iron	Dry	0.33	18.3	0.23	13.0
		Greasy	0.18	10.1	0.13	7.3
	Mild steel	Dry	0.52	27.5	0.41	22.3
		Greasy	0.24	13.5	0.12	6.8
	Ice	Dry	0.03	1.6	0.01	0.6
Rubber	Asphalt	Dry	0.60	30.0	0.40	22.0
		Wet	0.30	17.0	0.20	11.3
	Concrete	Dry	0.80	38.6	0.70	34.0
		Wet	0.40	22.0	0.30	17.0
	Ice	Dry	0.02	1.1	0.005	0.3
Stone	Stone	Dry	0.65	33.0	0.60	31.0
Steel wire	Steel pulley	Dry	0.30	16.6	0.15	8.5

A.25 COEFFICIENTS OF ROLLING FRICTION

Rolling friction force, defined as the resistance to the rolling of a wheel (or a roller) of radius R pressed against a horizontal plane, is

$$F_r = \frac{\mu_r}{R} W \qquad \left\{ \begin{array}{l} W,\ F,\ \text{in kgf} \\ \mu_r,\ R\ \text{in cm} \end{array} \right\}$$

where W = compression on the wheel and μ_r = coefficient of rolling friction given below.

Rolling surface	Plane surface	Condition	μ_r, cm	Rolling surface	Plane surface	Condition	μ_r, cm
Oak	Oak	Dry	0.050	Rubber	Concrete	Dry	0.015
		Wet	0.030			Wet	0.005
Rubber	Asphalt	Dry	0.010	Mild steel	Hard steel	Dry	0.001
		Wet	0.003		Mild steel	Dry	0.005

A.26 PROPERTIES OF MINERALS, A–H

The following table includes a list of minerals designated by name, chemical formula, specific mass, crystalline structure, approximate value of hardness on the Mohs scale (Appendix A.23), color, and luster. For a glossary of chemical symbols refer to Appendix A.22.

(a) Abbreviations

Structure		Color and Luster			
amr = amorphous	bit = bituminous	gls = glass	met = metallic
cub = cubic	blk = black	grn = green	nlu = nonlustrous
hex = hexagonal	blu = blue	gry = gray	pal = pale
mon = monoclinic	brn = brown	lbl = light blue	red = red
rho = rhombic	col = colorless	lbr = light brown	smt = semimetallic
tet = tetragonal	dbl = dark blue	lgn = light green	trp = transparent
trl = triclinic	dbr = dark brown	lrd = light red	wht = white
trg = trigonal	dgn = dark green	lus = lustrous	yel = yellow
	drd = dark red		

(b) Table, A–H

Name	Formula	Density	Structure	Hardness	Color	Luster
Anhydrite	$CaSO_4$	2.9–3.0	rho	3.3–3.5	gry, dbl	gls
Antimony	Sb	6.6–6.7	hex	3.2–3.5	gry	wht
Aragonite	$CaCO_3$	2.9–3.0	rho	3.6–3.9	wht, grn	gls
Arsenopyrite	FeAsS	5.9–6.1	mon	5.6–6.0	wht, gry	met
Azurite	$2CuCO_3Cu(OH)_2$	3.8–3.9	mon	3.7–4.1	dbl	gls
Bentonite	$BaTi(SiO_3)_3$	2.5–2.6	rho	5.5–6.0	wht	nlu
Beryl	$Be_3Al_2(SiO_3)_6$	2.6–2.8	hex	7.6–8.0	grn, yel, dbl	gls
Bismutite	Bi_2S_3	6.3–6.6	rho	2.2–2.6	wht, gry	met
Braunite	$(MnSi)_2O_3$	4.7–4.8	tet	6.1–6.4	brn, gry	smt
Calcite	$CaCO_3$	2.6–2.9	trg	2.9–3.1	wht, gry	gls

Mineral	Formula	Sp. gr.	System	Hardness	Color	Luster
Chalcocite	Cu_2S	5.6–5.8	rho	2.6–3.0	gry, brn	met
Chalcopyrite	$CuFeS_2$	4.1–4.3	tet	3.5–4.0	yel, brn	met
Chromite	$FeOCr_2O_3$	4.5–5.0	cub	5.5–6.0	blk	met
Cobaltite	$CoAsS$	6.1–6.2	cub	5.6–5.8	wht, gry	met
Corundum	Al_2O_3	3.9–4.1	trg	9	gry, yel, dbl	lus
Cyanite	Al_2SiO_5	3.6–3.7	trl	4.5–6.5	dbl, grn	gls
Diamond	C	3.4–3.5	cub	10	col, yel	lus
Diaspore	$Al_2O_3(H_2O)$	3.3–3.5	rho	6.7–7.0	brn, yel	gls
Diopside	$CaMg(Si_2O_6)$	5.5–6.2	mon	3.2–3.4	wht, grn, gry	gls
Dolomite	$CaCu_3(MgCO_3)$	2.8–3.0	trg	3.5–4.0	wht, brn	gls
Dyscrasite	Ag_2Sb	9.7–9.8	rho	3.5–4.0	wht, gry	met
Enargite	Cu_2AsS_4	4.4–4.5	rho	3.0–3.1	gry	met
Enstatite	$Mg(Si_2O_6)$	3.1–3.2	rho	5.0–6.0	col, grn, gry	gls
Fluorite	CaF_2	3.1–3.2	cub	4	trp, grn, lbl	gls
Galena	PbS	7.5–7.6	cub	2.5–2.7	gry	met
Gibbsite	$Al(OH)_3$	2.3–2.4	mon	2.6–3.5	wht, gry	met
Glauberite	$Na_2Ca(SO_4)_2$	2.7–2.9	mon	2.5–3.0	trp, wht, yel	gls
Gold	Au	19.3	cub	2.5–3.0	yel	met
Graphite	C	2.0–2.2	hex	1.0–2.0	blk	smt
Gypsum	$CaSO_4 \cdot 2H_2O$	2.3–2.4	mon	1.8–2.0	wht, gry, yel	nlu
Halite	$NaCl$	2.1–2.2	cub	2.3–2.5	trp, wht	gls
Hematite	Fe_2O_2	4.9–5.3	trg	5.5–6.0	gry, red	met
Hercynite	$FeAl_2O_4$	4.3–4.4	cub	7.6–8.0	blk	met
Hessite	Ag_2Te	8.2–8.5	mon	3.6–4.0	gry	met
Huebnerite	$MnWO_4$	7.2–7.3	mon	4.2–4.8	blk, red, brn	met

A.27 PROPERTIES OF MINERALS, I–Z

(a) Notation and Abbreviations

The notation and abbreviations used below are those defined in Appendix A.26.   For a glossary of chemical symbols refer to Appendix A.22.

(b) Table, I–Z

Name	Formula	Density	Structure	Hardness	Color	Luster
Illite	*	2.6–2.7	mon	1.5–2.0	wht	nlu
Kaolinite	$Al_4Si_4O_{11}(OH)_8$	2.6–2.7	trl	2.1–2.5	lgn, lrd, wht	nlu
Limonite	$2Fe_2O_3 \cdot 3H_2O$	3.4–4.0	—	5.2–5.5	brn, yel	nlu
Magnesite	$MgCO_3$	3.0–3.4	trg	3.5–4.5	wht, yel	gls
Magnetite	Fe_3O_4	5.0–5.2	cub	5.5–6.5	blk	met
Malachite	$Cu_2CO_2(OH)_3$	4.0–4.1	mon	3.6–4.0	lgn, dgn	gls
Mica	$KAl_3(OHF)_2(Si_3O_{10})$	2.6–2.9	mon	2.2–3.0	trp, yel	gls
Moissanite	SiC	3.2–3.3	hex	9.4–9.5	grn, blk	met
Montmorillonite	*	2.3–2.6	mon	2.4–2.5	wht, yel, grn	nlu
Molybdenite	MoS_2	4.7–4.8	hex	1.0–1.5	gry	met
Muscovite	$H,KAlSiO_4$	2.7–2.9	mon	2.5–3.0	brn, red, brn	gls
Niccolite	$NiAs$	7.7–7.8	hex	5.1–5.5	red, gry, blk	met
Olivine	$(MgFe)_2SiO_4$	6.5–7.0	rho	3.0–3.4	grn, yel	gls
Opal	$SiO_2 \cdot nH_2O$	1.8–2.2	amr	5.5–6.5	wht, yel, blu, grn	gls
Orthoclase	$KAlSi_3O_8$	2.5–2.6	mon	5.5–6.0	wht, yel, red, grn	gls
Petzite	$(AgAu)_2Te$	8.9–9.1	cub	2.5–3.0	gry, blk	met
Phenakite	Be_2SiO_4	2.9–3.0	trg	7.7–8.0	brn, red	gls
Platinum	Pt	14–19	cub	4.0–4.5	wht, gry	met
Pyrite	FeS_2	4.9–5.1	cub	6.0–6.5	yel	met
Pyrrhotite	FeS	4.5–4.7	hex	3.5–4.5	yel	met

Mineral	Formula	Density	Crystal	Hardness	Color	Luster
Quartz (Silica)	SiO_2	2.6–2.7	trg	7	trp, yel, brn, grn	gls
Raspite	$PbWO_4$	8.5	mon	2.6–3.0	yel, gry, brn	gls
Realgar	AsS	3.4–3.6	mon	1.5–2.0	red	gls
Rutile	TiO_2	4.2–4.3	tet	6.0–6.5	red, yel, blk	met
Scheelite	$CaWO_4$	5.9–6.1	tet	4.5–5.0	wht, yel, lgn	bit
Sellaite	MgF_2	3.0–3.2	tet	5.0–5.2	wht, yel	met
Siderite	$FeCO_3$	3.0–3.8	trg	3.5–4.5	gry, brn	gls
Silver	Ag	10–11	cub	2.5–3.0	wht, gry, blk	met
Sperrylite	$PtAs_2$	10–11	cub	6.0–7.0	wht, yel	met
Spinel	$MgAl_2O_4$	8	cub	3.5–4.0	brn, grn, red	gls
Stibnite	Sb_2S_3	4.5–4.6	rho	2.0–2.5	dbl, gry	met
Stolzite	$PbWO_4$	7.8–8.1	tet	2.7–3.0	gry, grn, brn, red	lus
Sulfur	S	2.0–2.1	rho	1.5–2.5	yel, brn, grn, red	bit
Sylvite	KCl	1.9–2.0	cub	2.0–2.5	trp	gls
Talc	$Mg_3Si_4O_{10}(OH)_2$	2.6–2.8	mon	1.0–1.2	wht, grn, brn	bit
Thorianite	ThO_2	9.3–9.7	cub	6.5	blk, gry	met
Thorite	$ThSiO_4$	4.4–4.5	tet	4.5–5.0	brn, red, blk	gls
Titanite	$CaTiSiO_5$	3.3–3.6	mon	5.0–5.5	yel, grn, brn	lus
Topaz	$Al_2SiO_2(OHF)_2$	3.5–3.6	rho	8.0–8.2	wht, yel, grn, red	gls
Torolite	$SnTa_2O_7$	7.5–8.0	mon	5.5–6.0	brn	lus
Uraninite	UO_2	6.6–10.0	cub	5.0–6.0	blk	bit
Vanadinite	$Pb_5(VO_4)_3Cl$	6.7–7.1	hex	2.5–3.0	yel, brn, red	bit
Wolframite	$(MnFe)WO_4$	6.7–7.4	mon	4.5–5.5	gry, blk	met
Xerotine	YPO_4	4.4–4.6	tet	4.0–5.0	yel, brn, red	lus
Zircon	$ZrSiO_4$	4.7	tet	7.5–8.0	yel, brn, trp	lus

*Illite and montmorillonite are the names of families of similar minerals; hence, no formulas can be given for them.

A.28 STATIC PROPERTIES OF ROCKS AND SOILS*

(a) Notation

p_a = maximum allowable bearing pressure (Pa)

E = modulus of elasticity (Pa)

ν = Poisson's ratio (dimensionless, defined in Sec. 4.03–1f)

(b) Average values of rocks

Designation	p_a, MPa	E, GPa	ν	Designation	p_a, MPa	E, GPa	ν
Amphibolite, solid	13	75	0.15	Greenstone, solid	9	75	0.18
Andesite, solid	8	50	0.20	Limestone, sound	6	55	0.20
Basalt, solid	10	63	0.15	Marblestone, solid	7	75	0.18
Dibase, solid	10	83	0.15	Sandstone, sound	3	27	0.26
Diorite, solid	10	31	0.18	Schist, sound	4	42	0.26
Dolomite, solid	10	52	0.15	Shale, sound	2	20	0.26
Gneiss, solid	9	57	0.17	Siltstone, sound	2	16	0.28
Granite, solid	10	70	0.19	Slate, sound	3	83	0.30

(c) Average values of soils

Designation	p_a, kPa	E, MPa	ν	Designation	p_a, kPa	E, MPa	ν
Gravel, coarse. Loose	720	1,000	0.30	Sand, fine. Loose	140	170	0.20
Dense	960	1,600	0.28	Dense	190	260	0.25
Gravel, medium. Loose	620	800	0.25	Silty	140	190	0.30
Dense	860	1,200	0.20	Clay. Soft	100	30	0.20
Gravel + sand. Loose	480	400	0.25	Medium	200	50	0.15
Dense	720	600	0.28	Hard	400	200	0.10
Sand, coarse. Loose	280	200	0.15	Sandy	300	40	0.25
Dense	380	400	0.20	Silty	200	20	0.30

*Typical average values corresponding to those given in several building codes are given in the tables above. For conversions use:

$$1 \text{ Pa} = 1.020 \times 10^{-1} \text{ kgf-m}^{-2} = 2.089 \times 10^{-2} \text{ lbg-ft}^{-2}$$
$$1 \text{ kgf-m}^{-2} = 9.807 \text{ Pa} = 2.048 \times 10^{-1} \text{ lbf-ft}^{-2}$$
$$1 \text{ lbf-ft}^{-2} = 4.788 \times 10 \text{ Pa} = 4.882 \text{ kgf-m}^{-2}$$

For definitions of symbols, relationships, and additional conversion factors refer to Sec. 5.08-4 and Appendix A.42.

A.29 GRAIN-SIZE CLASSIFICATION OF SOILS

(a) Standard sieve sizes

U.S. standard			British standard			Metric standard		
No.	D (mm)	D (in.)	No.	D (mm)	D (in.)	No.	D (mm)	D (in.)
4	4.76	0.1874	5	3.36	0.1323	5000	5.00	0.1969
6	3.36	0.1323	8	2.06	0.0811	3000	3.00	0.1181
10	2.00	0.0787	12	1.41	0.0555	2000	2.00	0.0787
20	0.84	0.0331	18	0.85	0.0335	1500	1.54	0.0606
40	0.42	0.0165	25	0.60	0.0236	1000	1.00	0.0394
60	0.25	0.0098	36	0.42	0.0165	500	1.00	0.0197
100	0.149	0.0059	60	0.25	0.0098	300	0.50	0.0118
200	0.074	0.0029	100	0.15	0.0059	150	0.15	0.0059
			200	0.076	0.0030	75	0.075	0.0030

(b) Classification table

System	Grain diameter in millimeters
(a)	0.0002 0.0006 0.002 0.006 0.02 0.06 0.2 0.6 2.0 mm — Colloids / Medium / Coarse (Clay); Fine / Medium / Coarse (Silt); Fine / Medium / Coarse (Sand); Gravel
(b)	0.001 0.005 0.074 0.25 2.0 mm — Colloids / Silt / Sand; Fine / Coarse (Sand); Gravel
(c)	0.001 0.005 0.074 0.25 2.0 9 24 76 mm — Colloids / Clay / Silt; Fine / Coarse (Sand); Fine / Medium / Coarse (Gravel); Boulders
(d)	0.002 0.05 0.25 0.5 10 20 76 mm — Clay / Silt; Very fine / Fine / Medium / Coarse / Very coarse (Sand); Fine / Medium (Gravel); Cobbles
(e)	0.005 0.05 0.25 2.0 mm — Clay / Silt; Fine / Coarse (Sand); Gravel
(f)	0.002 0.02 0.2 2.0 mm — Clay / Silt; Fine / Coarse (Sand); Gravel

Sieve sizes: 270 | 140 | 40 | 10 | ½ in. | 3 in. / 200 | 60 | 20 | 4 | ¾ in.

(a) M.I.T. and British Standards Institution; (b) American Society for Testing and Materials; (c) American Association of State Highway Officials; (d) U.S. Department of Agriculture; (e) Federal Aviation Agency; (f) International Society of Soil Science.

SOURCE: Jan J. Tuma and M. Abdel-Hady, "Engineering Soil Mechanics," p. 28, © 1973 by Prentice-Hall, Inc. Reprinted by permission of Prentice-Hall, Inc., Englewood Cliffs, N.J.

A.30 AASHO SOIL CLASSIFICATION SYSTEM

Group	Sub-group	Percent passing U.S. sieve no.			Character of fraction passing no. 40 sieve		Group index no.	Soil description	Subgrade rating
		10	40	200	Liquid limit	Plasticity index			
A–1			50 max	25 max		6 max	0	Well-graded gravel or sand; may include fines.	Excellent to good
	A–1–a	50 max	50 max	15 max		6 max	0	Largely gravel but can include sand and fines.	
	A–1–b		50 max	25 max		6 max	0	Gravelly sand or graded sand; may include fines.	
A–2*				35 max			0–4	Sands and gravels with excessive fines.	
	A–2–4			35 max	40 max	10 max	0	Sands, gravels with low-plasticity silt fines.	
	A–2–5			35 max	41 min	10 max	0	Sands, gravels with elastic silt fines.	
	A–2–6			35 max	40 max	11 min	4 max	Sands, gravels with clay fines.	
	A–2–7			35 max	35 min	11 min	4 max	Sands, gravels with highly plastic clay fines.	
A–3			51 min	10 max		Nonplastic	0	Fine sands.	Fair to poor
A–4				36 min	40 max	10 max	8 max	Low-compressibility silts.	
A–5				36 min	41 min	10 max	12 max	High-compressibility silts, micaceous silts.	
A–6				36 min	40 max	11 min	16 max	Low-to-medium-compressibility clays.	
A–7				36 min	41 min	11 min	20 max	High-compressibility clays.	
	A–7–5			36 min	41 min	11 min†	20 min	High-compressibility silty clays.	
	A–7–6			36 min	41 min	11 min†	20 min	High-compressibility, high-volume-change clays.	
A–8								Peat, highly organic soils.	Unsatisfactory

*Group A–2 includes all soils having 35 percent or less passing a No. 200 sieve that cannot be classed as A–1 or A–3.
†Plasticity index of A–7–5 subgroup is equal to or less than LL–30.   Plasticity index of A–7–6 subgroup is greater than LL–30.

SOURCE: Report of Committee on Classification of Materials for Subgrades and Granular Type Roads, *Proc Highway Res. B.*, vol. 25, Washington, D.C., 1945, pp. 375–388; discussion, pp. 388–392.

A.31 UNIFIED SOIL CLASSIFICATION SYSTEM, DEFINITIONS AND SYMBOLS

(a) Group symbols

Prefix:
- C = clay
- G = gravel
- M = silt
- O = organic clay
- S = sand
- Pt = organic soil

Suffix:
- F = excess of fines
- H = high plasticity (compressibility), $w_L > 50$
- I = intermediate plasticity (compressibility), $w_L = 35\text{--}50$
- L = low plasticity (compressibility), $w_L = 20\text{--}35$
- P = poorly graded
- W = well graded

(b) Special symbols

w_L = liquid limit (Sec. 5.07–1g) w_P = plastic limit (Sec. 5.07–1g)
d = subdivision when $w_L \le 25$ and $w_L - w_P \le 5$
u = subdivision when the conditions stated for d are not satisfied
   For sieve sizes refer to Appendix A.29.

(c) Solid components classification

Cobbles: above 3 in.
Gravels: 3 in. to no. 4 sieve
Coarse sands: no. 4 sieve to no. 10 sieve

Medium sands: no. 10 sieve to no. 40 sieve
Fine sands: no. 40 sieve to no. 20 sieve
Fines: passing no. 200 sieve

(d) Rating

A_1 = excellent
A_2 = good to excellent
A_3 = good
A_4 = fair
A_5 = fair to good
A_6 = fair to poor
A_7 = fair to very poor
A_8 = poor
A_9 = poor to very poor
A_{10} = poor to not suitable
A_{11} = practically impervious
A_{12} = not suitable

B_1 = almost none
B_2 = none to very slight
B_3 = none to slight
B_4 = very slight
B_5 = slight
B_6 = slight to medium
B_7 = slight to high
B_8 = medium
B_9 = medium to high
B_{10} = medium to very high
B_{11} = high
B_{12} = very high

C_1 = none
C_2 = none to very slow
C_3 = none to slow
C_4 = very slow
C_5 = slow
C_6 = slow to quick
C_7 = quick

(e) Compaction equipment

CM = close control of moisture
CT = crawler-type tractor
NO = not practical

SF = sheepsfoot roller
SR = steel-wheeled roller
RT = rubber-tired equipment

SOURCE: A. Casagrande, "Classification and Identification of Soils," *Trans. ASCE*, Vol. 113, 1948, pp. 901–992. For more extensive coverage see "The Unified Soil Classification System," U.S. Army Corps of Engineers Technical Memorandum no. 3-357, Vols. 1 and 2, March 1953.

A.32a   UNIFIED SOIL CLASSIFICATION SYSTEM, DESCRIPTIONS*

Groups			Symbols	Typical names	Field identification procedures		
Coarse-grained soils (More than half of solids do not pass no. 200 sieve.)	Gravels (More than half of solids do not pass no. 4 sieve.)	Gravels with little or no fines	GW	Well-graded gravels, gravel-sand mixtures, little or no fines	Wide range in grain sizes and substantial amounts of all intermediate particle sizes		
			GP	Poorly-graded gravels, gravel-sand mixtures, little or no fines	Predominantly one size or a range of sizes with some intermediate sizes missing		
		Gravels with fines	GM	Silty gravels, gravel-sand-silt mixtures	Nonplastic fines or fines with low plasticity (for identification procedures see ML below)		
			GC	Clayey gravels, gravel-sand-clay mixtures	Plastic fines (for identification procedures see CL below)		
	Sands (More than half of solids pass no. 4 sieve.)	Clean sands with little or no fines	SW	Well-graded sands, gravelly sands, little or no fines	Wide range in grain sizes and substantial amounts of all intermediate particle sizes		
			SP	Poorly graded sands, gravelly sands, little or no fines	Predominantly one size or a range of sizes with some intermediate sizes missing		
		Sands with fines	SM	Silty sands, sand-silt mixtures	Nonplastic fines or fines with low plasticity (for identification procedures see ML below)		
			SC	Clayey sands, sand-clay mixtures	Plastic fines (for identification procedures see CL below)		

Each row in this table continues in Appendix A.32b (opposite page).

Groups			Symbols	Typical names	Identification procedures on fraction smaller than no. 40 sieve size		
					Dry strength (crushing characteristics)	Dilatancy (reaction to shaking)	Toughness (consistency near P_L)
Fine-grained soils (More than half of solids pass no. 200 sieve.)	Silts and clays	$w_L < 50$	ML	Inorganic silts and very fine sands, rock flour, silty or clayey fine sands, or clayey silts	B_3	C_6	B_1
			CL	Inorganic clays of low to medium plasticity, gravelly clays, sandy clays, silty clays, lean clays	B_9	C_2	B_8
			OL	Organic silts and organic silty clays of low plasticity	B_6	C_3	B_6
	Silts and clays	$w_L > 50$	MH	Inorganic silts, micaceous or diatomaceous fine sandy or silty soils, elastic silts	B_6	C_3	B_6
			CH	Inorganic clays of high plasticity, fat clays	B_{11}	C_1	B_{11}
			OH	Organic clays of medium to high plasticity, organic silts	B_9	C_2	B_6
Highly organic soils			Pt	Peat and other highly organic soils	Readily identified by color, odor, spongy feel, and frequently by fibrous texture		

* For definitions of symbols and sources refer to Appendix A.31.

A.32b UNIFIED SOIL CLASSIFICATION SYSTEM, DESIGN VALUES*

Symbols		Value† as			Potential frost action	Compressibility and expansion	Drainage characteristics	Compaction equipment	γ_d† lbf-ft⁻³	CBR§	k¶ lbf-in⁻³
		Subgrade	Subbase	Base							
GW		A_1	A_1	A_3	B_2	B_1	A_1	CT, RT, SR	125–140	40–80	300–500
GP		A_2	A_2	A_4	B_2	B_1	A_1	CT, RT, SR	110–140	30–60	300–500
GM	d	A_2	A_3	A_5	B_6	B_4	A_6	RT, SR, CM	125–145	40–60	300–500
	u	A_3	A_4	A_{10}	B_6	B_5	A_9	RT, SR	115–135	20–30	200–500
GC		A_3	A_4	A_{10}	B_6	B_5	A_9	RT, SR	130–145	20–40	200–500
SW		A_3	A_5	A_8	B_2	B_1	A_1	CT, RT	110–130	20–40	200–400
SP		A_5	A_4	A_{10}	B_2	B_1	A_1	CT, RT	105–135	10–40	150–400
SM	d	A_5	A_5	A_8	B_7	B_4	A_6	RT, SF, CM	120–135	15–40	150–400
	u	A_4	A_6	A_{12}	B_7	B_6	A_9	RT, SF	100–130	10–20	100–300
SC		A_6	A_8	A_{12}	B_7	B_6	A_9	RT, SF	100–135	15–20	100–300
Each row in this table is a continuation of the respective row in Appendix A.32a (opposite page).											
ML		A_6	A_{12}	A_{12}	B_9	B_6	A_6	RT, SF, CM	90–130	<15	100–200
CL		A_6	A_{12}	A_{12}	B_9	B_8	A_{11}	RT, SF	90–150	<15	50–150
OL		A_8	A_{12}	A_{12}	B_9	B_9	A_8	RT, SF	90–105	<5	90–100
MH		A_8	A_{12}	A_{12}	B_{10}	B_{11}	A_6	SF, RT	80–105	<10	50–100
CH		A_6	A_{12}	A_{12}	B_8	B_{11}	A_{11}	SF, RT	90–115	<15	50–150
OH		A_9	A_{12}	A_{12}	B_8	B_{11}	A_{11}	SF, RT	80–110	<5	25–100
Pt		A_{12}	A_{12}	A_{12}	B_5	B_{12}	A_6	NO	—	—	—

*For definitions of symbols and sources refer to Appendix A.31.
†Values when not subjected to frost action.
‡Dry weight density, 1 lbf-ft⁻³ = 1.601 846 (+01) kgf-m⁻³ = 1.570 874 (+02) N-m⁻³.
§California bearing ratio defined in Sec. 5.07–2g.
¶Modulus of subgrade, 1 lbf-in⁻³ = 2.767 991 (+04) kgf-m⁻³ = 2.714 472 (+05) N-m⁻³.

A.33 PERMEABILITY OF SOILS, TABLES

(a) Notation

k = permeability coefficient (m·s^{-1})
D_{10} = grain-size diameter at 10 percent on grain-size distribution curve (mm)
D_{20} = grain-size diameter at 20 percent on grain-size distribution curve (mm)
D = grain-size diameter (mm)
w_L = liquid limit (%)
I_p = plastic limit (%)
γ_d = dry weight density (N·m^{-3})
AASHO = soil system (Appendix A.30)

(b) Typical permeability coefficients*

Flow		Particle-size range				"Effective" size		Permeability coefficient k		
		inches		millimeters		D_{20} (in.)	D_{10} (mm)	ft/year	ft/month	cm/sec
		D_{max}	D_{min}	D_{max}	D_{min}					
Turbulent flow	Derrick stone	120	36	—	—	48	—	100×10^6	100×10^5	100
	One-man stone	12	4	—	—	6	—	30×10^6	30×10^5	30
	Clean, fine to coarse gravel	3	$\frac{3}{4}$	80	10	$\frac{1}{2}$	—	10×10^6	10×10^5	10
	Fine, uniform gravel	$\frac{3}{8}$	$\frac{1}{16}$	8	1.5	$\frac{1}{8}$	—	5×10^6	5×10^5	5
	Very coarse, clean, uniform sand	$\frac{1}{8}$	$\frac{1}{32}$	3	0.8	$\frac{1}{16}$	—	3×10^6	3×10^5	3
Laminar flow	Uniform, coarse sand	$\frac{1}{8}$	$\frac{1}{64}$	2	0.5		0.6	0.4×10^6	0.4×10^5	0.4
	Uniform, medium sand			0.5	0.25		0.3	0.1×10^6	0.1×10^5	0.1
	Clean, well-graded sand and gravel			10	0.05		0.1	0.01×10^6	0.01×10^5	0.01
	Uniform, fine sand			0.25	0.05		0.06	4000	400	40×10^{-4}
	Well-graded, silty sand and gravel			5	0.01		0.02	400	40	4×10^{-4}
	Silty sand			2	0.005		0.01	100	10	10^{-4}
	Uniform silt			0.05	0.005		0.006	50	5	0.5×10^{-4}
	Sandy clay			1.0	0.001		0.002	5	0.5	0.05×10^{-4}
	Silty clay			0.05	0.001		0.0015	1	0.1	0.01×10^{-4}
	Clay (30 to 50% clay sizes)			0.05	0.0005		0.0008	0.1	0.01	0.001×10^{-4}
	Colloidal clay ($-2\mu \leq 50\%$)			0.01	10 Å		40 Å	0.001	10^{-4}	10^{-9}

(c) Permeability coefficients k of soil groups in AASHO system†

AASHO group	w_L %	I_p %	Passing no. 200 sieve %	Consolidating load kips/ft²	Compacted wet		Compacted dry	
					k ft/day	γ_d lbf/ft³	k ft/day	γ_d lbf/ft³
A–1	23	8	26	1	0.000 450	114	0.004 800	117
				2	0.000 360	115	0.002 000	118
				4	0.000 240	117	0.001 700	119
A–2	28	11	40	1	0.000 480	103	0.004 700	99
				2	0.000 280	106	0.001 600	102
				4	0.000 180	109	0.000 900	108
A–4	33	12	99	1	0.000 300	94	0.000 750	90
				2	0.000 260	97	0.000 600	94
				4	0.000 200	100	0.000 240	99
A–5	35	6	35	1	0.021 000	84	0.370 000	78
				2	0.015 000	86	0.250 000	80
				4	0.012 000	90	0.170 000	84
A–6	72	45	86	1	0.000 035	75	0.000 110	89
				2	0.000 019	80	0.000 110	90
				4	0.000 009	86	0.000 010	92
A–7	67	34	71	1	0.000 510	77	0.026 000	78
				2	0.000 280	81	0.008 400	81
				4	0.000 200	88	0.001 200	83
A–8	78	8	38	1	0.003 900	43	0.041 000	47
				2	0.001 000	46	0.018 000	48
				4	0.000 460	49	0.010 000	50

*K. B. Hough, "Basic Soil Engineering," Ronald Press, New York, 1957, p. 69   Reproduced by permission of John Wiley and Sons, New York.
†E. S. Barber and C. L. Sawyer, "Highway Subdrainage," *Proc. U.S. Highway Res. B.*, Vol. 31, 1952.

A.34 CONSOLIDATION OF SOILS, TIME FACTORS

(a) Notation (Sec. 5.07–5)

U = average consolidation ratio (%)
T_v = time factor (1)
u = pore pressure (N·m^{-2})
H = length of drainage path (m)

(b) Particular cases, double drainage

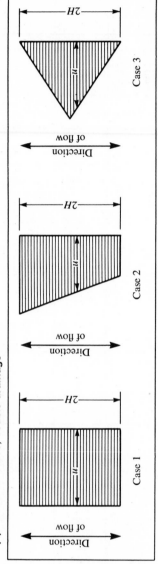

(c) Particular cases, single drainage

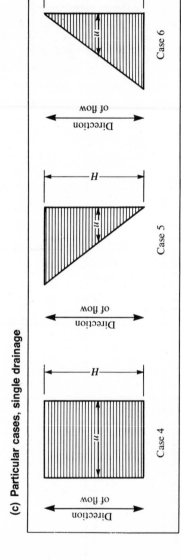

(d) Values of T_v corresponding to desired U

U		Time factor T_v				
	Case 1	Case 2	Case 3	Case 4	Case 5	Case 6
5	1.667 (−03)	1.667 (−03)	2.500 (−02)	1.667 (−03)	2.320 (−02)	6.224 (−04)
10	7.667 (−03)	7.667 (−03)	5.000 (−02)	7.667 (−03)	4.675 (−02)	3.112 (−03)
15	1.667 (−02)	1.667 (−02)	7.500 (−02)	1.667 (−03)	7.402 (−02)	6.163 (−03)
20	3.138 (−02)	3.138 (−02)	1.025 (−01)	3.138 (−02)	1.013 (−01)	9.213 (−03)
25	4.916 (−02)	4.916 (−02)	1.285 (−01)	4.916 (−02)	1.294 (−01)	1.650 (−02)
30	7.071 (−02)	7.071 (−02)	1.562 (−01)	7.071 (−02)	1.574 (−01)	2.378 (−02)
35	9.623 (−02)	9.623 (−02)	1.886 (−01)	9.623 (−02)	1.895 (−01)	3.587 (−02)
40	1.264 (−01)	1.264 (−01)	2.211 (−01)	1.264 (−01)	2.215 (−01)	4.796 (−02)
45	1.587 (−01)	1.587 (−01)	2.566 (−01)	1.587 (−01)	2.577 (−01)	7.055 (−02)
50	1.956 (−01)	1.956 (−01)	2.935 (−01)	1.956 (−01)	2.938 (−01)	9.313 (−02)
55	2.383 (−01)	2.383 (−01)	3.355 (−01)	2.383 (−01)	3.390 (−01)	1.278 (−01)
60	2.856 (−01)	2.856 (−01)	3.795 (−01)	2.856 (−01)	3.842 (−01)	1.625 (−01)
65	3.423 (−01)	3.423 (−01)	4.376 (−01)	3.423 (−01)	4.429 (−01)	2.171 (−01)
70	4.033 (−01)	4.033 (−01)	5.018 (−01)	4.033 (−01)	5.015 (−01)	2.714 (−01)
75	4.766 (−01)	4.766 (−01)	5.756 (−01)	4.766 (−01)	5.833 (−01)	3.563 (−01)
80	5.673 (−01)	5.673 (−01)	6.675 (−01)	5.671 (−01)	6.650 (−01)	4.412 (−01)
85	6.842 (−01)	6.842 (−01)	7.813 (−01)	6.842 (−01)	8.031 (−01)	5.818 (−01)
90	8.481 (−01)	8.481 (−01)	9.455 (−01)	8.481 (−01)	9.412 (−01)	7.223 (−01)
95	1.129 (+00)	1.129 (+00)	1.227 (+00)	1.129 (+00)	1.314 (−01)	1.084 (−01)
100	∞	∞	∞	∞	∞	∞

A.35 DENSITY OF SOLID METALS AT TEMPERATURE T

Name	Density ρ		T, °C	Name	Density ρ		T, °C
	kg-m^{-3}	lb-ft^{-3}			kg-m^{-3}	lb-ft^{-3}	
Aluminum	2,700	168	20	Molybdenum	10,200	637	20
Antimony	6,620	413	20	Nickel	8,800	550	20
Arsenic	5,750	359	14	Osmium	22,500	1,400	20
Barium	3,500	218	20	Palladium	12,100	755	20
Beryllium	1,830	114	20	Platinum	21,400	1,336	20
Bismuth	9,780	610	20	Potassium	870	54.3	20
Cadmium	8,650	540	20	Rhodium	12,440	777	20
Calcium	1,550	96.7	20	Silver	10,500	655	20
Cerium	6,730	420	20	Sodium	971	60.6	20
Cesium	1,873	117	20	Steel	7,721	482	20
Chromium	7,140	446	20	Strontium	2,600	162	20
Cobalt	8,710	544	21	Tantalum	16,800	1,036	20
Copper	8,940	558	20	Thallium	11,850	740	20
Gold	19,300	1,205	20	Thorium	11,500	718	20
Iridium	22,420	1,399	17	Tin	7,300	456	20
Iron	7,860	491	20	Titanium	4,500	281	18
Lead	11,342	708	20	Tungsten	19,300	1,205	20
Lithium	534	33.3	20	Uranium	18,700	1,167	13
Magnesium	1,740	108	20	Vanadium	5,900	368	15
Manganese	7,350	459	20	Zinc	6,920	432	20
Mercury	14,193	886	−38.8	Zirconium	6,440	402	20

A.36 DENSITY OF INDUSTRIAL ALLOYS AT $T = 20°C$

Name	Composition, percent	Density ρ	
		kg-m^{-3}	lb-ft^{-3}
Aluminum alloy	(99.6 Al)	$2,750 \pm 50$	172 ± 3
Aluminum brass	(76 Cu, 22 Zn, 2 Al)	8,330	520
Aluminum bronze	(86 Cu, 4 Fe, 10 Al)	7,500	468
Beryllium copper	(97 Cu, 2 Be)	$8,230 \pm 20$	514 ± 1
Architectural bronze	(57 Cu, 40 Zn, 3 Pb)	8,530	532
Cable sheet lead	(99.8 Pb, 0.028 Ca)	11,350	709
Carbon steel	(99.2 Fe, 0.2 C, 0.45 Mn, 0.15 Si)	7,860	490
Cast iron	(91 Fe, 3.56 C, 2.5 Si)	7,080	442
Cast steel	(97.40 Fe, 0.3 C, 0.7 Mn, 0.6 Si, 0.5 Ni)	7,690	480
Commercial rolled zinc	(99 Zn, 0.3 Pb, 0.3 Cd)	7,140	446
Commercial bronze	(90 Cu, 10 Zn)	8,800	549
Commercial red brass	(85 Cu, 5 Zn, 5 Pb, 5 Sn)	8,600	537
Commercial yellow brass	(71 Cu, 25 Zn, 3 Pb, 1 Sn)	8,400	524
Electrolytic copper	(99.92 Cu, 0.04 O)	8,900	556
Cobalt alloy	(60 Co, 7 W, 10 Ni, 23 Cr)	8,610	538
Grid metal	(91 Pb, 9 Sb)	10,660	665
Bearing lead	(94 Pb, 6 Sb)	10,880	679
Cast nickel	(97 Ni, 1.5 Si, 0.5 Mn, 0.5 C)	8,340	521
Magnesium alloy	(89.9 Mg, 10 Al, 0.1 Mn)	1,810	113
Stainless steel	(70 Fe, 16 Cr, 12 Ni, 2 Mo)	$7,950 \pm 50$	496 ± 3
Tool steel	(77 Fe, 18 W, 4 Cr, 1 V)	8,670	541

SOURCE: "Metal Handbook," 48th ed., American Society for Metals, Metals Park, Ohio, 1961.

A.37 DENSITY OF WOODS AT $T = 20°C$

Common name	Density ρ kg-m⁻³	Density ρ lb-ft⁻³	Common name	Density ρ kg-m⁻³	Density ρ lb-ft⁻³
Applewood	745	46.51	Hemlock:		
Ash:			Eastern	431	26.91
Black	526	32.84	Mountain	480	29.97
Blue	603	37.64	Western	432	26.97
Green	610	38.08	Hickory:		
White	638	39.83	Big leaf	809	50.50
			Mockernut	820	51.19
Aspen:			Pignut	820	51.19
Small leaf	401	25.03	Shagbark	836	52.10
Big leaf	412	25.72	Jacaranda, Rosewood	850	53.06
Balsa*	160 ± 40	9.98 ± 2.5			
Basswood	398	24.85	Larch, western	587	36.65
			Locust:		
Beech:			Black or yellow	765	47.76
Big leaf	655	40.89	Honey	666	41.58
Blue	717	44.76	Magnolia, Cucumber	516	32.21
			Mahogany:		
Birch:			African	668	41.70
Gray	552	34.46	Indian	540	33.71
Paper	600	37.46			
Sweet	714	44.57	Maple:		
Yellow	668	41.70	Black	620	38.71
			Red	546	34.09
Buckeye, yellow	383	23.91	Silver	506	31.59
Butternut	404	25.22	Sugar	676	42.20
Cedar:			Oak:		
Eastern red	492	30.71	Black	669	41.76
Northern white	315	19.66	Bur	671	41.89
Southern white	352	21.97	Chestnut	674	40.08
Tropical American*	535 ± 165	33.40 ± 10.3	Laurel	703	43.89
Western red	344	21.48			

Species			
Cherry:			
Black	534	33.40	
Wild red	425	26.53	
Chestnut	454	28.34	
Corkwood	207	12.92	
Cottonwood:			
Eastern	433	27.03	
Western	452	28.22	
Cypress, southern	482	30.09	
Dogwood	796	49.69	
Douglas Fir:			
Coast	512	31.96	
Mountain	446	27.84	
Ebony:			
Indian	978	61.05	
African	768	47.94	
Elm:			
American	554	34.59	
Chinese	602	37.58	
Rock	658	45.08	
Slippery	568	35.46	
Eucalyptus:			
Karri*	829	51.75	
Hemilampra*	1,058	66.05	
Marginata*	787	49.13	
Fir:			
Balsam	414	25.85	
Silver	415	25.91	
Greenheart*	1,165 ± 115	72.73 ± 7.18	
Gum:			
Black	552	34.46	
Blue	796	49.69	
Red	530	33.09	
Live	977	60.99	
Pin	677	42.26	
Post	738	46.07	
Red	657	41.02	
Scarlet	709	44.26	
Swamp chestnut	756	47.20	
Swamp white	792	49.44	
White	710	44.32	
Pine:			
Jack	461	28.78	
Long needle	638	39.83	
Pitch	542	33.84	
Red	507	31.65	
Short leaf	584	36.46	
White	373	23.29	
Poplar:			
Balsam	331	20.66	
Yellow	427	26.66	
Redwood	436	27.22	
Spruce:			
Black	428	26.72	
Red	413	25.78	
White	431	26.91	
Sycamore	539	33.65	
Tamarack	558	34.83	
Teak*	582	36.33	
Walnut	562	35.08	
Willow	408	25.47	

*Air dry.

SOURCE: "Handbook of Chemistry and Physics," 30th ed., R. C. Weast, Ed., CRC Press, Inc., Cleveland (now located in Boca Raton, Fla.), 1949. Used by permission of CRC Press, Inc.

A.38 DENSITY OF COMMON SOLIDS AND LIQUIDS AT $T = 20°C$

Name	Density ρ kg-m^{-3}	lb-ft^{-3}	Name	Density ρ kg-m^{-3}	lb-ft^{-3}
Agate	2,600 ± 100	162 ± 6	Halite	2,150 ± 50	134 ± 3
Alabaster	2,723 ± 32	170 ± 2	Hematite	5,100 ± 200	318 ± 12
Amber	1,085 ± 25	68 ± 1	Ice	917 ± 0	57.25
Anhydrite	2,900 ± 100	180 ± 6	Ivory	1,870 ± 50	117 ± 3
Anthracite	1,600 ± 200	100 ± 12	Lava, basaltic	2,900 ± 200	181 ± 12
Aragonite	2,950 ± 50	184 ± 3	Lava, trachytic	2,400 ± 200	150 ± 12
Arsenopyrite	6,050 ± 150	378 ± 9	Leather, greased	1,010 ± 20	63 ± 1
Asbestos	2,400 ± 400	145 ± 25	Leather, dry	850 ± 10	53 ± 1
Azurite	3,800 ± 100	230 ± 6	Lime, mortar	1,720 ± 60	107 ± 4
Barite	4,500 ± 200	272 ± 12	Lime, slaked	1,350 ± 50	84 ± 3
Basalt	2,900 ± 100	175 ± 6	Limestone	2,720 ± 40	169 ± 2
Beeswax	960 ± 10	58 ± 1	Limonite	3,650 ± 350	228 ± 22
Beryl	2,750 ± 150	166 ± 9	Loam	1,220 ± 120	76 ± 8
Bone	1,900 ± 100	115 ± 6	Magnesite	3,000 ± 100	182 ± 6
Brick	1,800 ± 400	109 ± 25	Magnetite	5,050 ± 150	315 ± 9
Butter	865 ± 10	53 ± 1	Malachite	3,900 ± 200	243 ± 12
Carbon, diamond	3,500 ± 20	212 ± 1	Marble	2,700 ± 150	169 ± 9
Carbon, graphite	2,240 ± 10	135 ± 1	Mica	2,900 ± 300	181 ± 19
Calcite	2,700 ± 100	163 ± 6	Molybdenite	4,700 ± 100	293 ± 6
Camphor	980 ± 10	59 ± 1	Monazite	5,100 ± 200	183 ± 12
Celluloid	1,400 ± 50	87 ± 3	Mud	1,600 ± 210	100 ± 13
Chalk	2,350 ± 450	147 ± 28	Naphtha	830 ± 30	52 ± 2
Charcoal	420 ± 140	26 ± 9	Olivine	3,320 ± 50	127 ± 3
Clay, hard	2,200 ± 30	137 ± 2	Opal	2,200 ± 0	137 ± 0
Clay, soft	1,350 ± 150	84 ± 9	Paper	950 ± 200	59 ± 12
Coal, bituminous	1,330 ± 80	83 ± 5	Paraffin	890 ± 20	56 ± 1
Coal, lignite	830 ± 30	52 ± 2	Peat	400 ± 80	25 ± 5
Coke	1,280 ± 320	80 ± 20	Plaster of Paris	1,270 ± 60	142 ± 4
Concrete, stone	2,240 ± 50	140 ± 3	Porcelain	2,400 ± 100	150 ± 6
Concrete, cinder	1,760 ± 160	110 ± 10	Pumice stone	640 ± 240	40 ± 15
Cork	240 ± 20	15 ± 1	Pyrite	5,000 ± 100	312 ± 6
Corundum	4,000 ± 100	250 ± 6	Quartz	2,600 ± 50	162 ± 3
Dolomite	2,990 ± 100	181 ± 6	Rock salt	1,050 ± 20	66 ± 1
Ebonite	1,150 ± 20	72 ± 1	Rosin	2,200 ± 20	137 ± 1
Emery	3,950 ± 50	246 ± 3	Rubber, pure	800 ± 160	50 ± 10
Feldspar	2,600 ± 150	162 ± 9	Rubber, hard	1,200 ± 20	75 ± 1
Flint	2,600 ± 50	162 ± 3	Rubber, soft	1,100 ± 20	69 ± 1
Fluorite	3,150 ± 50	197 ± 3	Rutile	4,250 ± 50	265 ± 3
Garnet	3,750 ± 550	234 ± 34	Sand	1,680 ± 240	105 ± 15
Gelatine	125 ± 2	7.8	Sandstone	2,250 ± 100	140 ± 6
Glass, comm'l.	2,600 ± 200	162 ± 12	Slate	2,900 ± 400	181 ± 25
Glass, flint	4,400 ± 1,500	275 ± 93	Starch	1,500 ± 30	94 ± 2
Glue	1,280 ± 40	80 ± 2	Sugar	1,600 ± 20	100 ± 1
Granite	2,700 ± 50	168 ± 3	Talc	2,750 ± 50	172 ± 3
Graphite, comm'l.	2,460 ± 260	154 ± 16	Tar	1,010 ± 10	63 ± 1
Gum arabic	1,350 ± 50	84 ± 3	Tourmaline	3,100 ± 100	194 ± 6
Gypsum	2,320 ± 10	145 ± 1	Wax, mineral	1,050 ± 100	66 ± 6

A.39 DENSITY OF PLASTICS AT $T = 20°C$

Chemical name	Trade name	Density ρ	
		kg-m^{-3}	lb-ft^{-3}
Acrylate and methacrylate	Lucite, Crystalite, Plexiglas	$1{,}180 \pm 30$	73.67 ± 1.87
Casein	Ameroid	$1{,}345 \pm 5$	83.97 ± 3.12
Cellulose acetate (sheet)	Bakelite, Lumarith, Plastacele, Protectoid	$1{,}435 \pm 165$	89.58 ± 10.30
Cellulose acetate (molded)	Fibestos, Hercules, Nixonite, Tenite	$1{,}435 \pm 165$	89.58 ± 10.30
Cellulose acetobutyrate	Tenite II	$1{,}185 \pm 45$	73.98 ± 2.81
Cellulose nitrate	Celluloid, Nitron, Nixonoid, Pyroxylin	$1{,}475 \pm 125$	92.08 ± 7.80
Ethyl cellulose	Ditzler, Ethocel, Ethofoil, Lumarith, Nixon, Hercules	$1{,}150 \pm 100$	71.79 ± 6.24
Phenol-formaldehyde compounds:			
Wood-flour-filled (molded)	Bakelite, Durez, Durite, Micarta, Catalin, Haveg, Indur, Makalot, Resinox, Textolite, Formica	$1{,}385 \pm 135$	86.46 ± 8.43
Mineral-filled (molded)		$1{,}840 \pm 250$	114.87 ± 15.61
Macerated-fabric-filled (molded)		$1{,}415 \pm 55$	88.34 ± 3.43
Paper-base (laminated)		$1{,}350 \pm 50$	84.28 ± 3.12
Fabric base (laminated)		$1{,}350 \pm 50$	84.28 ± 3.12
Cast (unfilled)	Bakelite, Catalin, Gemstone, Marblette, Opalon, Prystal	$1{,}300 \pm 100$	81.16 ± 6.24
Phenolic furfural (filled)	Durite	$1{,}650 \pm 240$	103.01 ± 14.98
Polyvinyl acetals (unfilled)	Alvar, Formvar, Saflex, Butacite, Vinylite X, etc.	$1{,}140 \pm 90$	71.17 ± 5.62
Polyvinyl acetate*	Gelva, Vinylite A, etc.	$1{,}190 + (?)$	$74.29 + (?)$
Copolyvinyl chloride acetate	Vinylite V, etc.	$1{,}355 \pm 15$	84.59 ± 0.94
Polyvinyl chloride (and copolymer) plasticized	Koroseal, Vinylite	$1{,}450 \pm 250$	90.52 ± 15.61
Polystyrene	Bakelite, Loalin, Lustrex, Styron	$1{,}062 \pm 8$	66.29 ± 0.50
Modified isomerized rubber*	Plioform, Pliolite	$1{,}060 + (?)$	$66.17 + (?)$
Chlorinated rubber*	Tornesit, Parlon	$1{,}640 + (?)$	$102.38 + (?)$
Urea formaldehyde	Bakelite, Beetle, Plaskon	$1{,}500 \pm 50$	93.64 ± 3.12
Melamine formaldehyde (filled)	Catalin, Melmac, Plaskon	$1{,}675 \pm 185$	104.57 ± 11.55
Vinylidene chloride	Saran, Velon	$1{,}715 \pm 35$	107.06 ± 2.18

*Upper limit variable.

SOURCE: "Handbook of Chemistry and Physics," 30th ed., R. C. Weast, Ed., CRC Press, Inc., Cleveland (now located in Boca Raton, Fla.), 1949. Used by permission of CRC Press, Inc.

A.40 DENSITY OF LIQUIDS AT TEMPERATURE T

Name	Density ρ kg-m^{-3}	Density ρ lb-ft^{-3}	T, °C	Name	Density ρ kg-m^{-3}	Density ρ lb-ft^{-3}	T, °C
Acetic acid	1,049	65.5	20	Nicotine	1,009	63.0	20
Acetone	791	49.4	20	Nitric acid	1,502	93.8	20
Acrolein	840	52.4	20	Nitrobenzene	1,198	74.8	25
Amyl alcohol	818	51.1	20	Nitrogen dioxide	1,348	84.2	20
Aniline	1,022	63.8	20	Nitroglycerin	1,596	99.6	15
Benzene	879	54.9	20	n-Nonane	714	44.6	20
Bromine	3,119	195.7	20	n-Octane	703	43.9	20
Carbolic acid	1,070	66.8	25	Oleic acid	898	56.1	20
Castor oil	962	60.1	20	Olive oil	918	57.3	20
Chlorine dioxide	1,642	102.5	0	n-Pentane	626	39.1	20
Chloroform	1,489	92.9	20	Petroleum	878	54.8	20
Formic acid	1,226	76.5	15	Phosphorus trichloride	2,877	179.6	20
Gasoline	680	42.5	20	n-Propyl alcohol	804	50.2	20
Glycerin	1,261	78.7	20	Pyridine	982	61.3	20
n-Heptane	684	42.7	20	Sea water	1,025	64	
n-Hexane	660	41.2	20	Selenic acid	2,602	162.4	15
Hydrogen cyanide	688	43.0	20	Silicon tetrachloride	1,481	92.5	20
Hydrogen fluoride	1,002	62.6	0	Stannic chloride	2,229	139.2	20
Hydrogen peroxide	1,442	90.0	20	Sulfuric acid	1,831	114.3	20
Kerosene	800	49.9	20	Turpentine	873	54.5	16
Lubricating oil	900	56.2	20	Water	1,000	62.428	4
Mercury	13,595	848.7	0	Water	958	59.830	100

A.41 DENSITY OF GASES AT $T = 0°C$ AND $p = 1$ atm

Name	Density ρ		Name	Density ρ	
	kg-m^{-3}	lb-ft^{-3}		kg-m^{-3}	lb-ft^{-3}
Acetylene	1.173 9	0.073 28	Hydrogen antimonide	5.302 2	0.330 92
Air (dry)	1.293 0	0.080 72	Hydrogen bromide	3.644 9	0.227 55
Ammonia	0.771 2	0.048 14	Hydrogen chloride	1.639 2	0.102 33
Argon	1.783 6	0.111 35	Hydrogen cyanide	0.901 4	0.056 27
Arsine	3.484 2	0.217 51			
			Hydrogen fluoride	0.922 7	0.057 60
Butane (iso)	2.670 1	0.166 69	Hydrogen iodide	5.789 1	0.361 40
Butane (n)*	2.518 7	0.157 24	Hydrogen selenide	3.670 3	0.229 13
			Hydrogen sulfide	1.538 2	0.096 03
Carbon dioxide	1.976 9	0.123 41	Hydrogen telluride	5.814 7	0.363 00
Carbon monoxide	1.252 0	0.078 16			
Chlorine	3.214 0	0.200 64	Krypton	3.732 4	0.233 00
Chlorine dioxide	3.094 2	0.193 16	Methane	0.716 8	0.044 74
Chlorine monoxide	3.892 0	0.242 97	Methylamine	1.395 8	0.087 14
Cyanogen	2.339 0	0.146 02	Methyl chloride	2.307 6	0.144 06
Ethane	1.356 8	0.084 70	Neon	0.900 2	0.056 20
Ethyl chloride	2.872 5	0.179 32	Nitric oxide	1.340 1	0.083 66
Ethylene	1.260 4	0.078 68	Nitrogen	1.250 5	0.078 07
Fluorine	1.696 6	0.105 91	Oxygen	1.429 0	0.089 21
Helium	0.178 5	0.011 14	Propane	2.009 8	0.125 47
Hydrazine†	1.011 0	0.063 10	Radon	9.958 4	0.621 68
Hydrochloric acid	1.639 1	0.102 33	Silicon hydride	1.441 3	0.089 98
Hydrofluoric acid	0.921 3	0.057 51	Sulfur dioxide	2.926 8	0.182 71
Hydrogen	0.089 88	0.005 611	Xenon	5.851 2	0.365 28

*At $P = 710/760$ atm.   †At $T = 113.3°C$.

A.42 MECHANICAL PROPERTIES OF SOLIDS

Tables A.43 through A.53 display the most commonly used mechanical properties of solids. The symbols, constants, and conversion factors used in these tables are summarized below.

(a) Symbols

f_n = normal stress (Pa)
f_{ni} = normal yield stress (Pa)
f_{nu} = normal ultimate stress (Pa)
E = modulus of elasticity (Pa)
K = bulk modulus (Pa)
ρ = mass density (kg-m^{-3})
Bhn = Brinell hardness (kgf-mm^{-2}, Appendix A.23)
p_a = maximum allowable bearing pressure (Pa)
f_v = shearing stress (Pa)
f_{vi} = shearing yield stress (Pa)
f_{vu} = shearing ultimate stress (Pa)
G = modulus of rigidity (Pa)
λ = Lamé's constant (Pa)
ν = Poisson's ratio

(b) Relationships between elastic constants

	E, ν	G, ν	K, ν	λ, ν
E	E	$2G(1+\nu)$	$3K(1-2\nu)$	$\dfrac{\lambda(1-\nu)(1-2\nu)}{\nu}$
G	$\dfrac{E}{2(1+\nu)}$	G	$\dfrac{2K(1-2\nu)}{2(1+\nu)}$	$\dfrac{\lambda-2\nu}{2\nu}$
K	$\dfrac{E}{3(1-2\nu)}$	$\dfrac{2G(1+\nu)}{3(1-2\nu)}$	K	$\dfrac{\lambda(1+\nu)}{2\nu}$
λ	$\dfrac{\nu E}{(1+\nu)(1-2\nu)}$	$\dfrac{2\nu G}{1-2\nu}$	$\dfrac{3\nu K}{1+\nu}$	λ

(c) Conversion relationships

1 Mg-m^{-3}	$= 1.000 \times 10^3$ kg-m^{-3}	$= 6.243 \times 10$ lb-ft^{-3}	$= 3.613 \times 10^{-2}$ lb-in^{-3}
1 GPa	$= 1.000 \times 10^9$ N-m^{-2}	$= 1.020 \times 10^4$ kgf-cm^{-2}	$= 1.450 \times 10^5$ lbf-in^{-2}
1 MPa	$= 1.000 \times 10^6$ N-m^{-2}	$= 1.020 \times 10$ kgf-cm^{-2}	$= 1.450 \times 10^2$ lbf-in^{-2}
1 kgf-mm^{-2}	$= 9.807 \times 10^6$ N-m^{-2}	$= 1.000 \times 10^2$ kgf-cm^2	$= 1.422 \times 10^3$ lbf-in^{-2}

1 lbf-in^{-2} $= 6.897 \times 10^3$ N-m^{-2} 1 kgf-cm^{-2} $= 9.807 \times 10^4$ N-m^{-2}
$= 7.031 \times 10^{-2}$ kgf-cm^{-2} $= 1.422 \times 10$ lbf-in^{-2}

A.43 MECHANICAL PROPERTIES OF PURE METALS
AT $T = 20°C^*$

Name	Density ρ Mg-m^{-3}	Modulus		Tensile strength f_{nu} MPa	Poisson's ratio ν	Brinell hardness Bhn
		E GPA	G GPa			
Aluminum	2.70	68.6	26.5	58.8	0.34	30
Antimony	6.62	81.4	27.5	10.3	0.33	56
Beryllium	1.83	289.0	140.0	135.0	0.03	170
Bismuth	9.78	31.4	12.7	14.2	0.33	72
Cadmium	8.65	68.6	24.5	82.7	0.30	22
Chromium	7.14	240.0	93.8	480.0	0.28	91
Cobalt	8.90	207.0	——	237.0	——	125
Copper	8.94	107.8	43.1	245.0	0.35	60
Gold	19.30	73.0	35.3	206.0	0.41	30
Iridium	22.40	518.0	208.0	1,172.0	0.26	160
Iron, cast	7.86	78.5	36.4	108.0	0.25	90
Lead	11.34	15.7	5.9	24.8	0.43	4
Magnesium	1.74	43.1	15.7	186.0	0.35	80
Molybdenum	10.20	275.0	104.0	1,860.0	0.32	170
Nickel	8.80	205.0	73.5	490.0	0.30	180
Osmium	22.50	552.0	222.0	——	0.24	240
Palladium	12.10	118.0	44.1	186.0	0.39	45
Platinum	21.40	167.0	58.8	284.0	0.38	95
Plutonium	19.80	98.6	41.8	——	0.18	——
Rhodium	12.44	345.0	——	552.0	——	119
Ruthenium	12.20	410.0	——	620.0	——	220
Silver	10.50	73.5	26.5	284.0	0.37	40
Tantalum	16.80	186.0	68.9	912.0	0.35	40
Tin	7.30	46.1	16.7	196.0	0.33	5
Titanium	4.50	110.0	42.1	241.0	0.31	150
Tungsten	19.30	384.0	130.0	2,940.0	0.27	30
Uranium	18.70	165.0	67.6	1,620.0	0.22	90
Vanadium	5.90	133.0	52.8	372.0	0.26	250
Zinc	6.92	88.3	34.3	157.0	0.30	150
Zirconium	6.44	100.0	——	448.0	——	30

* For definitions of symbols, relationships, and conversion factors refer to Appendix A.42.

A.44 MECHANICAL PROPERTIES OF FERROUS ALLOYS AT $T = 20°C$*

Designation	Density ρ Mg·m^{-3}	Modulus E GPa	Tensile strength		Poisson's ratio ν	Brinell hardness Bhn
			f_{ni} MPa	f_{nu} MPa		
Carbon steels						
(1) AISI 1020—AN	7.86	200	262	350	0.30	110
—HR			206	379	0.30	111
—CD			352	420	0.30	121
(2) AISI 1045—AN	7.86	200	503	551	0.30	170
—HR			310	564	0.30	163
—CD			530	548	0.30	179
(3) AISI 1095—HA	7.86	200	669	745	0.30	248
Low-alloy steels						
(4) AISI 1340—HA	7.86	200	1150	1248	0.29	370
(5) AISI 1140—HA	7.86	200	1124	1240	0.29	370
(6) AISI 8640—HA	7.86	200	1296	1390	0.29	415
Stainless steels						
(7) AISI 304—AN	7.95	196	241	586	0.30	155
(8) AISI 316—AN	7.95	196	241	566	0.30	170
(9) AISI 431—AN	7.95	196	587	860	0.29	260
Cast irons						
(10) ASTM A48—CT	7.08	165	—	172	0.25	180
(11) ASTM A339—CT	7.08	186	365	483	0.25	170

Cast steels

(12) ASTM A27—CT	7.95	196	207	415	0.30	160
(13) ASTM A296—CT	7.95	196	413	655	0.30	170

*For definitions of symbols, relationships, and conversion factors refer to Appendix A.42. Also, AISI = American Iron and Steel Institute designation, SAE = Society of Automotive Engineers designation, ASTM = American Society for Testing and Materials designation. All AISI steels are also SAE steels. The treatment designations are: AN = annealed, HR = hot-rolled, CD = cold-drawn, HA = hardened, CT = cast. The compositions of alloys listed above are given in percents under the corresponding position number below.

	C	Cr	Mn	Mo	Ni	P	S	Si	Fe
(1)	0.20	—	0.45	—	—	—	—	—	99.35
(2)	0.46	—	0.75	—	—	—	—	—	98.79
(3)	0.96	—	0.40	—	—	—	—	—	98.64
(4)	0.41	—	1.75	—	—	—	—	0.28	98.56
(5)	0.41	—	0.93	2.30	—	—	—	0.28	98.38
(6)	0.45	0.50	0.93	0.20	0.55	—	—	0.27	97.10
(7)	0.08	19.00	2.00	—	10.00	—	—	1.00	67.92
(8)	0.08	17.00	2.00	2.30	12.00	—	—	1.00	65.62
(9)	0.20	16.00	1.00	—	1.75	—	—	1.00	80.05
(10)	3.30	—	0.65	—	—	0.25	0.10	2.30	93.40
(11)	Nodular iron, no composition requirements								100.00
(12)	0.30	0.40	0.60	0.20	0.50	—	—	0.80	97.20
(13)	1.30	13.00	1.00	—	1.00	—	—	1.50	82.20

For symbols of chemical elements used in this summary see Appendix A.22.

A.45 MECHANICAL PROPERTIES OF ALUMINUM ALLOYS AT $T = 20°C$*

| Designation | Density ρ Mg·m^{-3} | Strength | | | | Endurance limit MPa | Brinell hardness MPa |
| | | Tension | | Shear | | | |
		f_{ni} MPa	f_{nu} MPa	f_{vi} MPa	f_{vu} MPa		
Wrought alloys							
(1) AA 1060—O	2.70	22.4	68.0	10.3	46.5	20.6	19
—H12		68.9	86.2	34.5	58.6	27.6	23
—H18		—	121.0	—	78.8	44.9	35
(2) AA 1100—O	2.71	29.3	62.7	17.2	58.6	34.5	23
—H12		69.5	103.0	44.9	65.5	41.4	28
—H18		144.0	158.0	76.0	86.2	68.0	44
(3) AA 2014—O	2.60	79.3	176.0	34.5	114.0	89.5	45
—T4		290.0	427.0	137.0	237.0	138.0	105
—T6		413.0	483.0	—	282.0	138.0	135
(4) AA 3003—O	2.73	37.9	104.0	27.6	72.4	49.3	28
—H12		106.0	124.0	48.3	79.3	55.2	35
—H18		179.0	193.0	95.6	110.0	68.9	55
(5) AA 5052—O	2.68	77.6	188.0	41.3	117.0	110.0	47
—H12		176.0	221.0	89.6	134.0	117.0	60
—H18		238.0	279.0	131.0	162.0	138.0	77
(6) AA 6061—O	2.70	44.8	110.0	20.7	75.8	62.1	30
—T4		128.0	224.0	41.4	151.0	95.5	65
—T6		259.0	300.0	138.0	196.0	96.5	95
Cast alloys							
(7) AA 220—SA—T4	2.57	176.0	324.0	—	231.0	55.2	75
—SA—T8		165.0	228.0	—	179.0	58.6	70
—PM—T6		196.0	269.0	—	206.0	89.5	90

*For definitions of symbols, relationships, and conversion factors refer to Appendix A.42. Also, AA = Aluminum Association designation. The treatment designations are: O = annealed temper of wrought alloy, T4 = solution treatment and natural aging, T6 = solution treatment and artificial aging, H12 = cold-worked to desired shape and medium hardness, H18 = cold-worked to desired shape and full hardness, SA = sand casting, PM = permanent mold casting. For all alloys listed above, the elastic constants are approximately: $E \cong 68.9$ GPa, $G \cong 26.2$ GPa, $\nu \cong 0.33$. The endurance limit is defined as the maximum reversal stress which can be applied an indefinite number of times without producing fracture. The compositions of alloys listed above are given in percents under the corresponding position number below.

	Cr	Cu	Fe	Mg	Mn	Si	Si + Fe		Al
(1)	—	0.10	0.35	—	—	0.25	—	—	99.30
(2)	—	0.20	—	—	—	1.00	-	0.10	98.70
(3)	0.10	4.00	1.00	0.50	0.80	0.85	—	0.35	92.40
(4)	—	0.20	0.70	—	1.25	0.60	—	0.10	97.15
(5)	0.20	0.10	—	2.50	0.10	—	0.45	0.10	96.55
(6)	—	0.35	0.70	1.00	0.15	0.60	—	0.35	96.85
(7)	—	0.20	0.20	—	0.10	0.20	—	0.30	99.00
(8)	—	0.10	0.20	0.35	—	7.00	—	0.25	92.10

For symbols of chemical elements used in this summary see Appendix A.22.

A.46 MECHANICAL PROPERTIES OF WROUGHT COPPER ALLOYS AT $T = 20°C$*

Designation	Density ρ Mg·m⁻³	Modulus E GPa	Strength			Brinell hardness Bhn
			Tensile		Shear	
			f_{ni} MPa	f_{nu} MPa	f_{vu} MPa	
(1) Aluminum bronze 8%—AN	7.78	117	172	483	275	80
—HA			448	724	392	210
(2) Architectural bronze—AN	8.47	99	136	421	220	93
(3) Beryllium copper—AN	8.25	124	221	689	353	91
—CR			717	759	—	148
(4) Cartridge brass—AN	8.52	110	205	379	241	155
(5) Cupronickel 10%—AN	8.94	152	128	303	246	111
—CD			393	414	—	241
(6) Cupronickel 30%—AN	8.94	152	172	414	—	70
—CD			414	517	—	150
—CR			490	538	—	157
(7) Manganese bronze—AN	8.36	103	207	448	300	95
—CD			345	552	310	180
(8) Muntz metal—AN	8.39	103	117	372	283	80
(9) Naval brass—AN	8.41	103	207	434	283	90
—CD			276	448	276	150
(10) Nickel silver 18%—AN	8.70	124	172	400	—	70
—CD			483	586	—	170
—CR			641	689	—	220
(11) Phosphor bronze 9%—AN	8.78	114	186	344	268	60
—CR			448	552	—	160
(12) Phosphor bronze 10%—SP	8.75	110	—	841	—	241
(13) Red brass 15%—AN	8.75	103	101	290	221	50
—CD			386	413	290	122
—CR			421	531	303	136
(14) Yellow brass 35%—AN	8.47	103	124	331	248	55
—CD			400	490	300	115
—CR			427	510	317	181

*For definitions of symbols, relationships, and conversion factors refer to Appendix A.42. The treatment designations are: AN = annealed, CD = cold-drawn, CR = cold-rolled, HA = hardened, SP = spring-temper. For all alloys listed above, $\nu = 0.33$–0.34, and their compositions in percents are given under the corresponding position number below.

	Al	Be	Co	Fe	Mn	Ni	P	Pb	Sn	Zn	Cu
(1)	8.00	—	—	—	—	—	—	—	—	—	92.00
(2)	—	—	—	—	—	—	—	3.00	—	40.00	57.00
(3)	—	1.90	0.25	—	—	—	—	—	—	—	97.75
(4)	—	—	—	—	—	—	—	—	—	30.00	70.00
(5)	—	—	—	1.25	0.80	10.00	—	—	—	—	87.95
(6)	—	—	—	0.50	0.60	30.00	—	—	—	—	68.90
(7)	—	—	—	1.00	0.30	—	—	—	1.00	39.00	58.70
(8)	—	—	—	—	—	—	—	—	—	40.00	60.00
(9)	—	—	—	—	—	—	—	—	0.75	39.30	59.95
(10)	—	—	—	—	—	18.00	—	—	17.00	—	65.00
(11)	—	—	—	—	—	—	0.25	—	5.00	—	94.75
(12)	—	—	—	—	—	—	0.25	—	10.00	—	89.75
(13)	—	—	—	—	—	—	—	—	—	15.00	85.00
(14)	—	—	—	—	—	—	—	—	—	35.00	65.00

For symbols of chemical elements used in this summary see Appendix A.22.

A.47 MECHANICAL PROPERTIES OF CAST COPPER ALLOYS AT $T = 20°C$*

Designation	Density ρ Mg^{-m-3}	Modulus E GPa	Strength Tensile f_{ni} MPa	Strength Tensile f_{nu} MPa	Strength Shear f_{vu} MPa	Brinell hardness Bhn
(1) Aluminum bronze—9A	7.78	111	194	524	302	125
—9B	7.60	114	206	593	314	142
—9C	7.50	122	247	516	308	155
—9D	7.67	124	303	572	325	190
(2) Chromium copper 1%	8.70	115	83	214	126	64
(3) Manganese bronze—7A	7.88	103	213	441	305	85
(4) Navy bronze—2A	8.70	92	124	278	190	68
(5) Nickel silver—10A	8.90	112	124	255	182	55
—10B	8.94	108	125	269	184	75
—11A	8.82	132	172	317	208	93
—11B	8.85	148	227	400	316	145
(6) Red brass—4A	8.75	93	103	262	164	55
—4B	8.64	93	90	255	138	50
—5A	8.70	91	107	234	142	50
—5B	8.78	102	105	248	146	55
(7) Silicon bronze—12A	8.30	109	91	359	120	110
(8) Tin bronze—1A	8.72	79	139	314	204	77
—2B	8.77	91	145	255	193	60
—2C	8.77	91	159	276	202	70
—3A	8.96	76	124	259	190	65
—3B	8.91	100	134	242	193	62
—3C	8.86	100	97	213	158	58

(9) Yellow brass—6A	8.50	100	97	262	165	45
—6B	8.44	88	90	228	149	53
—6C	8.37	97	96	277	167	66
—8A	8.32	107	206	517	394	90
—8B	7.89	107	345	555	415	180
—8C	7.75	107	486	790	562	200

*For definitions of symbols, relationships, and conversion factors refer to Appendix A.42. For all alloys listed above, $\nu = 0.33$–0.34, and their compositions are given in percents under the corresponding position number below (balance in Cu):

(1–9A)   9.00 Al, 3.25 Fe
(1–9B)   10.0 Al
(1–9C)   10.5 Al, 4.00 Fe, 2.50 Ni, 0.50 Mn
(1–9D)   10.5 Al, 4.00 Fe, 4.00 Ni, 3.50 Mn
(2)   1.00 Cr
(3–7A)   39.2 Zn, 1.00 Fe, 1.00 Sn, 0.30 Mn
(4–2A)   4.50 Zn, 1.50 Pb, 6.00 Sn
(5–10A)   20.0 Zn, 9.00 Pb, 2.00 Sn, 12.0 Ni
(5–10B)   16.0 Zn, 5.00 Pb, 3.00 Sn, 16.0 Ni
(5–11A)   8.00 Zn, 4.00 Pb, 8.00 Sn, 20.0 Ni
(5–11B)   2.00 Zn, 1.50 Pb, 5.00 Sn, 25.0 Ni
(6–4A)   5.00 Zn, 5.00 Pb, 5.00 Sn
(6–4B)   7.00 Zn, 6.00 Pb, 4.00 Sn
(6–5A)   9.00 Zn, 7.00 Pb, 3.00 Sn
(6–5B)   15.0 Zn, 7.00 Pb, 3.00 Sn

(7–12A)   4.00 Zn, 1.50 Pb, 1.00 Sn, 4.00 Si
(8–1A)   2.50 Zn, 0.30 Pb, 10.0 Sn
(8–2B)   4.00 Zn, 1.00 Pb, 8.00 Sn
(8–2C)   2.50 Zn, 1.00 Pb, 10.0 Sn
(8–3A)   0.80 Zn, 9.00 Pb, 10.0 Sn
(8–3B)   3.00 Zn, 8.00 Pb, 7.00 Sn
(8–3C)   2.00 Zn, 10.0 Pb, 5.00 Sn
(9–6A)   24.0 Zn, 3.50 Pb, 2.00 Sn
(9–6B)   29.0 Zn, 3.50 Pb, 1.50 Sn
(9–6C)   37.0 Zn, 2.00 Pb, 1.50 Sn
(9–8A)   39.0 Zn, 0.50 Pb, 1.00 Sn
(9–8B)   24.0 Zn, 0.80 Ni, 4.00 Fe
(9–8C)   26.0 Zn, 1.00 Ni, 4.60 Fe

For symbols of chemical elements used in this summary see Appendix A.22.

A.48 MECHANICAL PROPERTIES OF NICKEL AND TITANIUM ALLOYS AT $T = 20°C$*

Designation	Density ρ Mg·m⁻³	Modulus E GPa	Strength Tensile f_{ni} MPa	Strength Tensile f_{nu} MPa	Strength Shear f_{vu} MPa	Brinell hardness Bhn
(1) A nickel—AN	8.88	207	137	483	242	105
—HR			179	524	310	110
—CD			470	620	532	175
—CR			620	703	665	200
(2) D nickel—AN	8.77	207	240	524	402	135
—HR			310	620	540	150
—CD			552	690	590	190
(3) Nickel—SC	8.33	148	173	393	286	115
(4) Monel—AN	8.83	179	241	517	384	120
—HR			344	620	550	150
—CD			565	793	645	185
—CR			703	827	780	260
(5) K Monel—AN	8.47	179	345	700	585	160
—AN, AH			690	1,070	860	280
—SP			965	1,034	830	300
—SP, AH			1,100	1,200	1,150	335
(6) Monel—SC	8.64	159	241	517	410	140
(7) Duranickel—AN	8.25	207	310	758	560	160
—AN, AH			862	1,100	—	135
—SP			—	1,200	—	310
—SP, AH			—	1,400	—	360

(8) Hastelloy—B, SC	8.94	179	350	686	510	200
(9) —C, SC	7.81	199	350	515	464	200
(10) —D, SC	8.24	207	—	290	—	300
(11) —F, SC	8.15	200	400	550	—	250
(12) Titanium alloy—Ti-Al-V	4.48	114	880	930	—	—
(13) —Ti-Al-Sn	4.46	114	830	950	—	—
(14) —Ti-Mn	4.70	114	970	1,030	—	—

*For definitions of symbols, relationships, and conversion factors refer to Appendix A.42. The treatment designations are: AN = annealed, AH = age-hardened, CD = cold-drawn, CR = cold-rolled, HR = hot-rolled, SP = spring-tempered, SC = sand-cast. The compositions of alloys listed above are given below in percents under the corresponding position number (balance in Ni or Ti, respectively):

(1) 0.15 Fe, 0.25 Mn, 0.15 Cu, 0.15 Si
(2) 0.15 Fe, 4.50 Mn, 0.05 Cu, 0.05 Si
(3) 0.05 Fe, 0.80 Mn, 0.50 Cu, 1.50 Si
(4) 1.35 Fe, 0.90 Mn, 31.3 Cu, 0.15 Si
(5) 1.00 Fe, 0.70 Mn, 29.0 Cu, 0.15 Si
(6) 1.00 Fe, 1.00 Mn, 29.0 Cu, 1.50 Si
(7) 0.15 Fe, 0.25 Mn, 4.50 Al, 0.55 Si

(8) 5.00 Fe, 28.0 Mo, 0.12 C
(9) 5.00 Fe, 16.0 Mo, 0.15 C, 15.0 Cr
(10) 10.0 Si, 3.0 Cu
(11) 20.0 Fe, 7.0 Mo, 21.0 Cr
(12) 0.40 Fe, 0.3 O, 6.0 Al, 4.00 V
(13) 0.50 Fe, 0.3 O, 5.0 Al, 3.00 Sn
(14) 0.50 Fe, 0.3 O, 0.75 Mn

For symbols of chemical elements used in this summary see Appendix A.22.

A.49 MECHANICAL PROPERTIES OF OTHER NONFERROUS METALS AT $T = 20°C^*$

Designation	Density ρ Mg·m^{-3}	Modulus E GPa	Strength Tensile f_{ni} MPa	Strength Tensile f_{nu} MPa	Strength Shear f_{vu} MPa	Brinell hardness Bhn
(1) Antimonial lead 8%—CR	10.7	13.8	—	28	—	10
(2) 9%—CR	10.7	13.8	—	46	—	14
(3) Arsenical lead—CR	11.3	20.5	—	17	—	5
(4) Calcium lead—EX	11.3	—	—	31	—	—
(5) Lead-base babbitt—SAE13	10.55	29.0	—	70	—	19
(6) —SAE14	9.69	29.0	—	76	—	22
(7) Magnesium alloy—AZ 31B—AN	1.77	44.8	152	255	150	55
—EX			183	255	—	47
—RP			165	258	124	50
—RS			214	300	138	73
(8) Magnesium alloy—AZ 80A—EX	1.80	44.8	250	338	152	64
—FR			235	345	155	72
(9) Magnesium alloy—AZ 91A—DC	1.80	44.8	152	228	140	65
(10) Pewter—AN	7.28	54	—	59	—	10
(11) Soft solder—50-50	8.80	68	33	42	—	14

(12) Tin babbitt alloy 1—CC	7.33	50	—	65	—	17
(13) 2—CC	7.39	53	—	85	—	24
(14) 3—CC	7.45	58	—	70	—	26
(15) Tin die-casting alloy	7.42	—	—	69	—	30
(16) Tin-silver solder	8.45	—	26	32	28	—
(17) Zamak—3—DC	6.64	—	—	283	205	82
(18) —5—DC	6.69	—	—	330	260	90
(19) Zilloy—15—HR	7.17	—	—	200	125	60
—15—CR	7.17	—	—	250	142	80
(20) Zilloy—40—HR	7.17	—	—	165	115	52
—40—CR	7.17	—	—	214	148	60

*For definitions of symbols, relationships, and conversion factors refer to Appendix A.42. The treatment designations are: AN = annealed, CC = chill-cast, CR = cold-rolled, CS = cable sheet, DC = die-cast, EX = extruded, FR = forged, HR = hot-rolled, RP = rolled plate, RS = rolled sheet. The compositions of alloys listed above are given below in percents under the respective position number (balance in Pb or Mg or Sn or Zn, respectively):

(1) 8.00 Sn
(2) 9.00 Sn
(3) 0.10 Sn, 0.15 As, 0.10 Bi
(4) 0.01 Sn, 0.30 Ca, 0.10 Cu
(5) 5.00 Sn, 0.60 As, 0.50 Cu, 10.0 Sb
(6) 10.0 Sn, 0.60 As, 0.50 Cu, 15.0 Sb
(7) 3.00 Al, 0.20 Mn, 1.00 Zn
(8) 8.00 Al, 0.15 Mn, 0.50 Zn
(9) 9.00 Al, 0.12 Mn, 0.60 Zn
(10) 7.00 Sb, 2.00 Cu

(11) 50.0 Pb
(12) 0.35 Pb, 4.50 Sb, 4.50 Cu, 0.10 As
(13) 0.35 Pb, 7.50 Sb, 3.50 Cu, 0.10 Ag
(14) 0.35 Pb, 8.00 Sb, 8.00 Cu, 0.10 Ag
(15) 0.35 Pb, 13.0 Sb, 5.00 Cu
(16) 5.00 Ag
(17) 4.00 Al, 0.30 Cu, 0.10 Fe
(18) 4.00 Al, 1.00 Cu, 0.10 Fe
(19) 1.00 Cu, 0.15 Pb
(20) 1.00 Cu, 0.10 Pb

For symbols of chemical elements used in this summary see Appendix A.22.

A.50 MECHANICAL PROPERTIES OF COMMERCIAL SOFTWOODS*

Designation	Density ρ Mg-m^{-3}	Modulus E GPa	Tensile or flexural f_{nu} MPa	Compression f_{nu} Perpendicular to grain MPa	Compression f_{nu} Parallel to grain MPa	Shear f_{vu} Parallel to grain MPa
Cedar:						
Alaska	0.43	9.8	34.5	4.1	25.2	6.8
Eastern red	0.49	4.8	24.1	2.2	13.3	4.2
Northern white	0.31	6.2	25.5	2.2	13.7	4.8
Port Oxford	0.41	9.7	41.3	3.4	29.6	6.2
Southern white	0.35	6.4	25.1	3.0	17.4	5.1
Western red	0.34	7.6	32.4	3.1	24.1	4.8
Cypress	0.46	8.7	39.3	4.8	26.9	6.2
Douglas fir:						
Coastal	0.50	11.4	45.3	4.9	34.5	7.0
Inland	0.42	9.2	37.4	4.8	27.8	6.9
Mountain	0.45	8.4	34.1	4.4	24.5	6.7
Fir:						
Balsam	0.41	7.0	29.1	2.0	20.7	4.5
Commercial	0.42	7.4	35.2	3.6	22.2	5.8
Hemlock:						
Eastern	0.43	8.0	35.9	2.2	24.0	6.9
Western	0.43	11.4	37.0	3.8	27.6	6.9
Larch	0.50	9.7	44.9	5.8	37.7	8.3
Pine:						
Eastern white	0.38	7.0	34.0	3.9	22.8	6.2
Lodgepole	0.39	7.0	34.3	3.7	22.5	5.7
Northern white	0.35	7.2	34.1	3.2	20.8	5.5
Ponderosa	0.39	8.3	34.1	4.0	21.1	6.8
Southern yellow	0.50	11.4	40.0	5.3	25.0	8.0
Western white	0.37	7.3	33.9	3.0	25.0	4.8
Redwood	0.39	8.4	42.5	5.0	30.0	6.7
Spruce:						
Eastern	0.39	2.6	34.4	3.2	21.8	5.3
Engelmann	0.32	6.5	31.2	2.3	18.7	5.2
Sitka	0.39	8.0	35.2	3.4	25.4	6.4
Tamarack	0.51	9.0	43.1	5.3	27.2	6.8

*For definitions of symbols. relationships, and conversion factors refer to Appendix A.42. For additional information refer to U.S. Department of Agriculture, "Wood Handbook," Handbook No. 72, U.S. Government Printing Office, Washington, D.C., 1955; also, American Institute of Timber Construction, "Timber Construction Manual," Wiley, New York, 1966.

A.51 MECHANICAL PROPERTIES OF COMMERCIAL HARDWOODS*

	Density ρ Mg-m^{-3}	Modulus E GPa	Strength			
			Tensile or flexural f_{nu} MPa	Compression f_{nu}		Shear f_{vu}
				Perpendicular to grain MPa	Parallel to grain MPa	Parallel to grain MPa
Ash:						
Black	0.43	9.6	37.6	4.1	22.1	9.3
White	0.56	10.3	51.5	8.3	31.5	11.6
Aspen	0.36	6.2	31.9	2.5	17.0	5.1
Beech	0.60	11.0	47.1	6.5	22.9	11.0
Birch	0.54	12.0	54.0	5.6	33.7	6.9
Cherry	0.48	10.7	50.6	4.1	33.0	9.7
Chestnut	0.42	7.0	35.0	4.1	21.8	6.2
Cottonwood:						
Eastern	0.38	8.8	33.0	2.8	20.0	5.7
Northern	0.33	8.0	31.4	1.9	15.3	5.1
Elm:						
American	0.48	9.1	39.6	4.7	21.8	8.6
Rock	0.60	9.2	44.6	7.9	26.9	11.2
Gum:						
Black	0.48	7.6	39.3	5.2	20.8	8.6
Red	0.46	9.2	41.0	4.6	24.1	9.2
Hackberry	0.50	7.5	30.6	5.5	20.1	9.4
Hickory	0.65	13.0	59.2	11.8	24.8	12.2
Locust, black	0.68	13.8	74.8	12.7	45.0	14.6
Maple:						
Black	0.54	10.4	43.4	7.0	25.5	10.4
Red	0.52	10.4	43.4	6.0	24.1	10.4
Silver	0.46	7.1	32.0	4.9	22.4	8.7
Sugar	0.60	11.9	50.0	8.0	28.5	12.8
Oak:						
Red	0.60	10.6	44.1	7.0	24.1	10.3
White	0.62	9.6	43.8	7.8	25.1	10.3
Sugarberry	0.48	7.0	32.7	6.3	20.5	7.9
Sycamore	0.48	8.3	34.5	4.4	21.0	8.3
Walnut, black	0.52	10.4	51.7	6.8	32.4	8.5

* For definitions of symbols, relationships, and conversion factors refer to Appendix A.42. For additional information refer to U.S. Department of Agriculture, "Wood Handbook," Handbook No. 72, U.S. Government Printing Office, Washington, D.C., 1955; also, American Institute of Timber Construction, "Timber Construction Manual," Wiley, New York, 1966.

A.52 MECHANICAL PROPERTIES OF COMMERCIAL PLASTICS AT $T = 20°C*$

| | Density ρ Mg-m^{-3} | Modulus E GPa | Strength f_{nu} | | |
| | | | Tensile MPa | Compression MPa | Flexural MPa |
Designation					
Acetal polymer	1.43	2.8	69	110	93
Acrylic:					
Type 1	1.18	2.90 ± 0.5	66 ± 10	104 ± 20	104 ± 20
Type 2	1.19	3.12 ± 0.3	68 ± 12	110 ± 12	120 ± 10
Cellulose acetate:					
Soft	1.20	1.60 ± 1.2	42 ± 10	200 ± 38	68 ± 14
Medium	1.22	2.40 ± 0.4	31 ± 13	134 ± 38	48 ± 28
Hard	1.30	2.40 ± 0.2	50 ± 9	210 ± 10	57 ± 33
Chlorinated polymer	1.40	1.10 ± 0.1	43 ± 1	—	35 ± 1
Epoxy, cast:					
General	1.80 ± 0.6	2.40 ± 0.2	48 ± 34	206 ± 70	97 ± 41
Heat-resistant	2.20 ± 1.0	1.20 ± 0.3	66 ± 31	274 ± 52	97 ± 41
Aluminum filler	1.60 ± 0.2	4.10 ± 1.0	62 ± 41	165 ± 61	110 ± 55
Silica filler	1.90 ± 0.3	3.80 ± 0.8	70 ± 34	200 ± 76	79 ± 24
Epoxy, reinforced:					
Carbon fiber	1.40 ± 0.1	2.80 ± 0.2	90 ± 20	210 ± 42	180 ± 20
Glass fiber	2.10 ± 0.1	21.20 ± 1.0	137 ± 69	224 ± 52	241 ± 73
Fluorocarbon:					
CTFE	2.00 ± 0.2	1.70 ± 0.4	36 ± 5	40 ± 1	43 ± 20
TFE	2.20 ± 0.1	0.40 ± 0.1	13 ± 1	11 ± 1	11 ± 1
FEB	2.16 ± 0.1	0.40 ± 0.1	20 ± 1	11 ± 1	——
Ionomer:					
Polymer	0.94	3.40 ± 0.1	28 ± 6	——	——
Elastomer	1.23	0.70 ± 0.1	45 ± 10	137 ± 10	7 ± 1
Nylon:					
Type 66	1.14	2.00 ± 0.8	72 ± 10	66 ± 17	90 ± 20
Type 11	1.11	1.50 ± 0.4	58 ± 10	——	——
Type 6	1.14	1.90 ± 0.9	66 ± 17	69 ± 20	76 ± 21
Melamine-formaldehyde:					
Asbestos filler	1.66	13.20 ± 0.3	42 ± 1	193 ± 10	70 ± 4
Cellulose filler	1.54	7.20 ± 0.1	62 ± 28	242 ± 60	89 ± 24
Glass filler	1.92	15.90 ± 0.1	53 ± 17	190 ± 50	130 ± 28
Phenol-furfural:					
Unfilled	1.30 ± 0.1	6.20 ± 1.1	55 ± 10	145 ± 35	97 ± 13
Asbestos filler	1.70 ± 0.2	13.80 ± 6.5	40 ± 2	190 ± 42	76 ± 21
Glass filler	1.80 ± 0.2	24.10 ± 3.0	93 ± 41	300 ± 160	250 ± 180
Mineral filler	1.80 ± 0.1	26.30 ± 7.0	48 ± 3	193 ± 20	70 ± 14

A.52 MECHANICAL PROPERTIES OF COMMERCIAL PLASTICS AT $T = 20°C$ (Cont.)

Designation	Density ρ Mg-m^{-3}	Modulus E GPa	Strength f_{nu}		
			Tensile MPa	Compression MPa	Flexural MPa
Phenolic:					
Unfilled	1.20	7.50 ± 2.8	52 ± 9	110 ± 14	96 ± 21
Asbestos filler	1.80 ± 0.1	24.40 ± 9.8	345 ± 70	345 ± 30	485 ± 100
Glass filler	1.80 ± 0.1	17.20 ± 6.2	345 ± 70	258 ± 13	555 ± 100
Polycarbonate polyester:					
Allyl	1.40 ± 0.1	1.70 ± 0.3	40 ± 9	159 ± 21	70 ± 28
Styrene	1.30 ± 0.2	2.70 ± 1.7	48 ± 21	169 ± 66	90 ± 41
Asbestos filler	1.50 ± 0.1	——	34 ± 14	150 ± 70	62 ± 15
Glass filler	1.80 ± 0.2	13.80 ± 2.0	276 ± 103	241 ± 68	396 ± 121
Mineral filler	2.00 ± 0.3	13.80 ± 7.0	36 ± 16	193 ± 7	90 ± 30
Polyethylene:					
Low density I	0.90	0.20	13 ± 3	——	——
Medium density II	0.93	0.30 ± 0.1	14 ± 1	——	41 ± 2
High density III	0.95	$0.70 \pm 0,3$	24 ± 4	——	7 ± 1
Ethyl acrylate	0.94	——	10 ± 2	——	28 ± 4
Vinyl acetate	0.94	——	14 ± 4	——	24 ± 2
Polypropylene polymer	0.90	2.00 ± 0.1	33 ± 5	48 ± 7	48 ± 7
Polystyrene:					
General	1.05	3.10 ± 0.3	45 ± 10	93 ± 15	79 ± 24
Heat-resistant	1.07	3.04 ± 0.7	76 ± 6	100 ± 18	96 ± 22
Polyvinyl chloride:					
Flexible	1.40	——	16 ± 8	10 ± 2	——
Rigid	1.40	2.60 ± 0.2	50 ± 14	72 ± 17	90 ± 20
Silicone:					
Glass fiber	0.15	——	34 ± 3	86 ± 15	83 ± 14
Mineral fiber	0.10	——	21 ± 1	113 ± 20	113 ± 20

*For definitions of symbols, relationships, and conversion factors refer to Appendix A.42. The modulus of elasticity E given above is for tension and flexure. For additional information refer to C. Harper, "Handbook of Plastics and Elastomers," McGraw-Hill, New York, 1975; and C. T. Lynch, "Handbook of Materials Science," vol. 3, CRC Press, Cleveland (now located in Boca Raton, Fla.) 1975.

A.53 MECHANICAL PROPERTIES OF BRICKS, CONCRETE, AND ROCKS*

Designation	Density ρ Mg-m^{-3}	Modulus E GPa	Strength f_u			
			Tension MPa	Compression MPa	Shear MPa	Flexure MPa
Amphibolite	3.10	75.0	41.0	344.0	56.0	41.0
Andesite	2.67	50.0	15.4	130.0	27.0	9.0
Basalt	2.94	63.0	19.2	166.0	32.0	17.2
Brick:						
Soft	1.60	10.0	1.9	13.0	2.5	2.7
Hard	1.90	13.0	5.1	34.0	6.0	6.2
Vitrified	2.25	24.0	10.0	68.0	11.0	12.4
Concrete:						
7.5 gal/sack	2.40	21.0	3.8	25.0	6.9	3.6
7.0 gal/sack	2.40	21.4	4.0	27.0	7.1	3.9
6.5 gal/sack	2.40	22.7	4.3	30.0	8.3	4.1
6.0 gal/sack	2.40	24.1	4.9	35.0	10.2	4.8
Dibase	2.80	83.0	27.0	241.0	40.0	27.0
Diorite	2.65	31.0	21.0	179.0	13.0	21.0
Dolomite	2.90	52.0	7.0	83.0	8.0	7.0
Feldspar	2.55	12.4	3.2	36.0	8.5	3.2
Gneiss	2.60	57.0	21.0	173.0	17.0	21.0
Granite	2.80	70.0	12.8	172.0	22.4	12.8
Greenstone	3.00	75.0	20.0	170.0	22.0	20.0
Limestone	2.64	55.0	8.0	88.0	15.0	8.7
Marblestone	2.72	75.0	8.3	82.0	8.9	8.3
Quartzite	2.64	58.0	20.0	156.0	20.0	20.0
Sandstone	2.30	27.0	14.0	41.0	11.0	4.0
Schist	2.70	42.0	7.0	68.0	8.7	7.0
Shale	2.80	20.0	7.0	68.0	8.3	7.0
Siltstone	2.20	16.0	3.4	36.0	9.8	3.4
Slate	2.85	83.0	10.3	103.0	11.7	10.3

*For definitions of symbols, relationships, and conversion factors refer to Appendix A.42. The strength of concrete is given for the water-cement ratio given above at the end of 28 days. Poisson's ratio for concrete is 0.10–0.15 and for bricks 0.25, and for stones it is given in Appendix A.24.

A.54 GENERAL STRESS-STRAIN EQUATIONS

(a) Notation

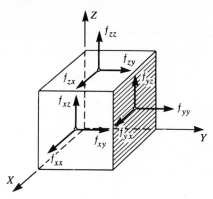

f_{xx}, f_{yy}, f_{zz} = normal stresses (Pa)

$\left.\begin{array}{l}f_{xy}, f_{yz}, f_{zx} \\ f_{yx}, f_{zy}, f_{xz}\end{array}\right\}$ = shearing stresses, (Pa)

$\varepsilon_{xx}, \varepsilon_{yy}, \varepsilon_{zz}$ = normal strains

$\left.\begin{array}{l}\varepsilon_{xy}, \varepsilon_{yz}, \varepsilon_{zx} \\ \varepsilon_{yx}, \varepsilon_{zy}, \varepsilon_{xz}\end{array}\right\}$ = shearing strains

$f_{xy} = f_{yx}, f_{yz} = f_{zy}, f_{zx} = f_{xz}$

$\varepsilon_{xy} = \varepsilon_{yx}, \varepsilon_{yz} = \varepsilon_{zy}, \varepsilon_{zx} = \varepsilon_{xz}$

E, ν, λ = material constants defined in Appendix A.42

(b) Stress equations in three dimensions

$$
\begin{bmatrix} f_{xx} \\ f_{yy} \\ f_{zz} \\ \hline f_{xy} \\ f_{yz} \\ f_{zx} \end{bmatrix} = \lambda
\left[\begin{array}{ccc|ccc}
(1-\nu)/\nu & 1 & 1 & 0 & 0 & 0 \\
1 & (1-\nu)/\nu & 1 & 0 & 0 & 0 \\
1 & 1 & (1-\nu)/\nu & 0 & 0 & 0 \\
\hline
0 & 0 & 0 & (1-2\nu)/2\nu & 0 & 0 \\
0 & 0 & 0 & 0 & (1-2\nu)/2\nu & 0 \\
0 & 0 & 0 & 0 & 0 & (1-2\nu)/2\nu
\end{array}\right]
\begin{bmatrix} \varepsilon_{xx} \\ \varepsilon_{yy} \\ \varepsilon_{zz} \\ \hline \varepsilon_{xy} \\ \varepsilon_{yz} \\ \varepsilon_{zx} \end{bmatrix}
$$

(c) Strain equations in three dimensions

$$
\begin{bmatrix} \varepsilon_{xx} \\ \varepsilon_{yy} \\ \varepsilon_{zz} \\ \hline \varepsilon_{xy} \\ \varepsilon_{yz} \\ \varepsilon_{zy} \end{bmatrix} = \frac{\nu}{E}
\left[\begin{array}{ccc|ccc}
1/\nu & -1 & -1 & 0 & 0 & 0 \\
-1 & 1/\nu & -1 & 0 & 0 & 0 \\
-1 & -1 & 1/\nu & 0 & 0 & 0 \\
\hline
0 & 0 & 0 & 2(1+\nu)/\nu & 0 & 0 \\
0 & 0 & 0 & 0 & 2(1+\nu)/\nu & 0 \\
0 & 0 & 0 & 0 & 0 & 2(1+\nu)/\nu
\end{array}\right]
\begin{bmatrix} f_{xx} \\ f_{yy} \\ f_{zz} \\ \hline f_{xy} \\ f_{yz} \\ f_{zx} \end{bmatrix}
$$

(d) Stress equations in two dimensions

$$
\begin{bmatrix} f_{xx} \\ f_{yy} \\ f_{xy} \end{bmatrix} = \frac{E\nu}{1-\nu^2}
\begin{bmatrix}
1/\nu & 1 & 0 \\
1 & 1/\nu & 0 \\
0 & 0 & (1-\nu)/2\nu
\end{bmatrix}
\begin{bmatrix} \varepsilon_{xx} \\ \varepsilon_{yy} \\ \varepsilon_{xy} \end{bmatrix}
$$

$f_{zz} = 0, f_{zx} = 0, f_{yz} = 0$

(e) Strain equations in two dimensions

$$
\begin{bmatrix} \varepsilon_{xx} \\ \varepsilon_{yy} \\ \varepsilon_{xy} \end{bmatrix} = \frac{\nu}{E}
\begin{bmatrix}
1/\nu & -1 & 0 \\
-1 & 1/\nu & 0 \\
0 & 0 & 2(1+\nu)/\nu
\end{bmatrix}
\begin{bmatrix} f_{xx} \\ f_{yy} \\ f_{xy} \end{bmatrix}
$$

$\varepsilon_{zz} = -\dfrac{\nu}{E}(f_{xx} + f_{yy}) = -\dfrac{\nu}{1-\nu}(\varepsilon_{xx} + \varepsilon_{yy})$

$\varepsilon_{yz} = 0 \qquad \varepsilon_{zx} = 0$

A.55 VISCOSITY OF LIQUIDS AND GASES AT ATMOSPHERIC PRESSURE*

The numerical values of the coefficient of kinematic viscosity ν (Sec. 5.01–5g) and of the mass density ρ of the most common liquids and gases at $p = 1$ atm and at a specified temperature T are given below.   The coefficients of absolute viscosity μ (Sec. 5.01–5c) of the same substances are given by the relation $\mu = \rho \times \nu$.

Substance	T, °C	ρ, kg-m^{-3}	ν, m^2-s^{-1}	Substance	T, °C	ρ, kg-m^{-3}	ν, m^2-s^{-1}
Air	−20	1.392 (+00)	1.152 (−05)	Ethyl alcohol	20	7.893 (+02)	1.532 (−06)
	−10	1.340 (+00)	1.245 (−05)	Gasoline (regular)	4	7.380 (+02)	7.524 (−07)
	0	1.293 (+00)	1.319 (−05)		20	7.325 (+02)	6.500 (−07)
	10	1.247 (+00)	1.412 (−05)		40	7.080 (+02)	5.481 (−07)
	20	1.201 (+00)	1.486 (−05)	Glycerin	20	1.263 (+03)	6.605 (−04)
	30	1.160 (+00)	1.607 (−05)	Machine oil	20	9.075 (+02)	1.366 (−04)
	40	1.129 (+00)	1.691 (−05)	Medium fuel oil	4	8.550 (+02)	6.085 (−06)
Ammonia	20	7.177 (−01)	1.533 (−05)		20	8.550 (+02)	3.948 (−06)
Benzene	20	8.790 (+02)	7.451 (−07)		40	8.421 (+02)	2.388 (−06)
Carbon dioxide	20	1.836 (+00)	8.454 (−07)	Medium lubricating oil	4	9.050 (+02)	4.431 (−04)
Carbon tetrachloride	4	1.620 (+03)	7.525 (−07)		20	8.930 (+2)	1.353 (−04)
	10	1.607 (+03)	6.978 (−07)		40	8.785 (+02)	4.136 (−05)
	20	1.584 (+03)	6.131 (−07)	Methane	20	6.666 (−01)	1.793 (−05)
	30	1.561 (+03)	5.871 (−07)	Nitrogen	20	1.163 (+00)	1.589 (−05)
	40	1.533 (+03)	4.812 (−07)	Oxygen	20	1.330 (+00)	1.589 (−05)
Castor oil	20	9.650 (+02)	1.031 (−03)	Turpentine	20	8.624 (+02)	1.728 (−06)

*For conversion to ft^2-sec^{-1}, use 1 m^2-s^{-1} = 10.764 ft^2-sec^{-1}; to lb-ft^{-3}, use 1 kg-m^{-3} = 0.06243 lb-ft^{-3}.

A.56 BULK MODULUS OF LIQUIDS

The numerical values of the bulk modulus K (Sec. 5.01–6j) of selected liquids at specified pressure p and temperature T are given below. In general, K increases with rising pressure and decreases with rising temperature. However, water reaches a maximum K at 50°C.

Substance	T, °C	p, MPa	K, GPa	Substance	T, °C	p, MPa	K, GPa
Benzene	20	1	1.1	Water	0	0.1	1.99
Carbon tetrachloride	20	0.1–20	1.1		20	0.1	2.17
		21–40	1.3		50	0.1	2.28
		41–60	1.5		80	0.1	2.19
Ethyl alcohol	20	0.1–20	1.3	Water	0	0.5	2.03
		21–50	1.5		20	0.5	2.24
		50–90	2.1		50	0.5	2.31
Glycerin	20	0.1–20	4.1		80	0.5	2.27
Methyl alcohol	20	0.1–20	1.3	Water	0	1.0	2.06
		21–50	1.5		20	1.0	2.25
		51–90	2.2		50	1.0	2.36
Propyl alcohol	20	0.1–20	1.5		80	1.0	2.28
		21–50	1.8	Seawater	10	0.1	2.36
		51–90	2.4		20	0.1	2.32

For conversions to other units use: 1 lbf-in^{-2} = 1 psi = 6.8948 × 10^{-3} Pa = 6.8046 × 10^{-2} atm; 1 atm = 1.0133 × 10^5 Pa = 1.4696 × 10 psi; 1 Pa = 9.8692 × 10^{-6} atm = 1.4504 × 10^{-4} psi; and 1 MPa = 10^6 Pa, 1 GPa = 10^9 Pa.

A.57 VISCOSITY OF WATER AT ATMOSPHERIC PRESSURE*

T °C	ρ kg-m^{-3}	ν 10^{-6} m^2-s^{-1}	T °C	ρ kg-m^{-3}	ν 10^{-6} m^2-s^{-1}	T °C	ρ kg-m^{-3}	ν 10^{-6} m^2-s^{-1}	T °C	ρ kg-m^{-3}	ν 10^{-6} m^2-s^{-1}
0	999.9	1.794	25	997.1	0.898	50	988.1	0.555	75	974.9	0.391
1	999.9	1.732	26	996.8	0.878	51	987.6	0.547	76	974.3	0.386
2	1,000.0	1.674	27	996.5	0.858	52	987.2	0.539	77	973.7	0.381
3	1,000.0	1.619	28	996.3	0.839	53	986.7	0.531	78	973.1	0.376
4	1,000.0	1.568	29	996.0	0.821	54	986.2	0.522	79	972.4	0.372
5	1,000.0	1.519	30	995.7	0.803	55	985.7	0.514	80	971.8	0.367
6	1,000.0	1.473	31	995.4	0.787	56	985.2	0.506	81	971.2	0.362
7	999.9	1.429	32	995.1	0.771	57	984.8	0.499	82	970.6	0.358
8	999.9	1.387	33	994.7	0.755	58	984.2	0.492	83	969.9	0.355
9	999.8	1.348	34	994.4	0.740	59	983.8	0.485	84	969.3	0.351
10	999.7	1.310	35	994.1	0.735	60	983.2	0.478	85	968.6	0.347
11	999.6	1.274	36	993.7	0.710	61	982.7	0.471	86	968.0	0.343
12	999.5	1.239	37	993.4	0.697	62	982.2	0.464	87	967.3	0.339
13	999.4	1.206	38	993.0	0.684	63	981.7	0.458	88	966.7	0.335
14	999.3	1.175	39	992.6	0.671	64	981.1	0.452	89	966.0	0.331
15	999.1	1.145	40	992.2	0.659	65	980.6	0.446	90	965.3	0.328
16	999.0	1.116	41	991.9	0.647	66	980.0	0.440	91	964.7	0.324
17	998.8	1.088	42	991.5	0.635	67	979.5	0.433	92	964.0	0.322
18	998.6	1.062	43	991.1	0.624	68	978.9	0.428	93	963.3	0.318
19	998.4	1.036	44	990.7	0.613	69	978.4	0.422	94	962.6	0.315
20	998.2	1.011	45	990.2	0.603	70	977.8	0.416	95	961.9	0.311
21	998.0	0.986	46	989.8	0.593	71	977.2	0.411	96	961.2	0.308
22	997.8	0.963	47	989.4	0.583	72	976.7	0.405	97	960.5	0.305
23	997.6	0.941	48	989.0	0.574	73	976.1	0.400	98	959.8	0.302
24	997.3	0.918	49	988.5	0.565	74	975.5	0.396	99	959.1	0.299
									100	958.4	0.296

*For conversion to ft^2-sec^{-1}, use 1 m^2-s^{-1} = 10.764 ft^2-sec^{-1}; to lb-ft^{-3}, use 1 kg-m^{-3} = 0.06243 lb-ft^{-3}.

A.58 SURFACE TENSION OF LIQUID METALS AT TEMPERATURE T*

Metal	Gas	T °C	σ gf-cm^{-1}	Metal	Gas	T °C	σ gf-cm^{-1}
Aluminum	Air	700	0.8413	Lead	H_2	400–700	0.4440
Antimony	H_2	650–750	0.3671	Platinum	Air	2,000	1.8400
Bismuth	H_2	300–600	0.3733	Silver	Air	1,000	0.8060
Cadmium	H_2	300	0.6271	Sodium	Vacuum	100	0.2040
Copper	H_2	1,000–1,200	1.1220	Tin	H_2	300–800	0.5250
Gold	H_2	1,000	0.8750†	Zinc	Air	550	0.7200

*For conversion to dyne-cm^{-1}, use 1 gf-cm^{-1} = 9.80665 × 10^2 dyne-cm^{-1}. For definition and applications, refer to Sec. 5.02–6.
†Variation of ± 15%.

A.59 CHART OF ABSOLUTE VISCOSITY μ IN FPS UNITS

The absolute viscosity μ, defined in Sec. 5.01–5c for a fluid of specific weight G at temperature T in degrees Fahrenheit, is

$$\mu = \text{(factor from chart)} \times 10^{-6} \text{ lbf-sec-ft}^{-2}$$

For conversion to other units, use Appendix B.37.

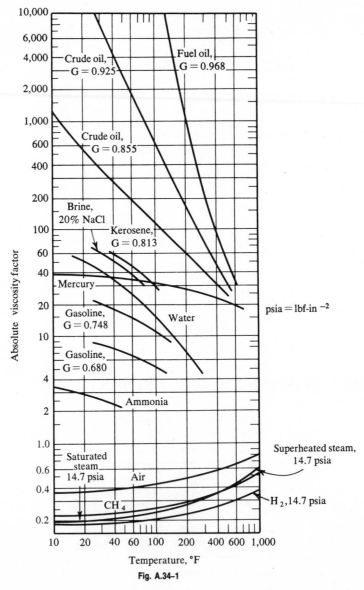

Fig. A.34–1

SOURCE: Daugherty, R. L., Some Properties of Water and Other Fluids, *Trans. ASME*, vol. 57, July 1935.

A.60 CHART OF KINEMATIC VISCOSITY ν IN FPS UNITS

The kinematic viscosity ν, defined in Sec. 5.01–5g for a fluid of specific weight G at temperature T in degrees Fahrenheit, is

$$\nu = (\text{factor from chart}) \times 10^{-6} \text{ ft}^2\text{-sec}^{-1}$$

For conversion to other units, use Appendix B.38.

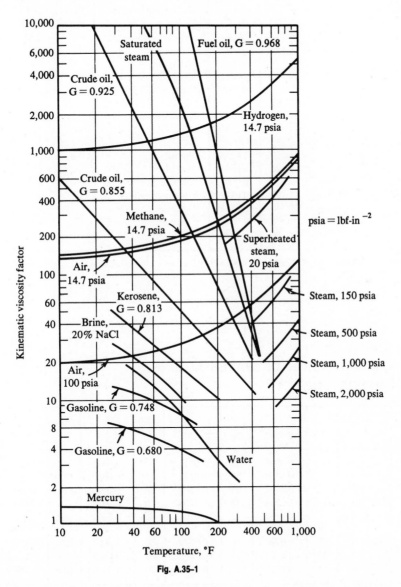

Fig. A.35–1

SOURCE: Daugherty, R. L., Some Properties of Water and Other Fluids, *Trans. ASME*, vol. 57, July 1935.

A.61 SURFACE TENSION OF LIQUIDS AT TEMPERATURE T*

Liquid	In contact with	T °C	σ gf-cm^{-1}	Liquid	In contact with	T °C	σ gf-cm^{-1}
Acetic acid	Vapor	20	0.0283	Glycerol	Air	20	0.0644
Acetone	Vapor†	20	0.0239	n-Hexane	Air	20	0.0186
Ammonia	Vapor	10	0.0244	Hydrogen peroxide	Vapor	20	0.0773
Aniline	Air	20	0.0416	Mercury	Vacuum	0	0.4895
Benzene	Air	11	0.0298	Mercury	Air	20	0.4806
Benzene	Air	20	0.0283	Mercury	Air	40	0.4772
Bromine	Vapor†	20	0.0423	Methyl alcohol	Air	0	0.0250
n-Butyric acid	Air	20	0.0272	Methyl alcohol	Vapor	20	0.0234
Carbon disulfide	Vapor†	20	0.0329	Nitric acid (70%)	Air	20	0.0605
Carbon tetrachloride	Vapor	20	0.0274	Nitric acid (98%)	Air	20	0.0377
Castor oil	Air	20	0.0335	Nitrobenzene	Vapor	20	0.0389
Chlorine	Air	20	0.0189	Olive oil	Air	20	0.0316
Chlorobenzene	Vapor	20	0.0342	n-Octane	Vapor	20	0.0219
Chloroform	Air	20	0.0276	Phenol	Vapor	10	0.0283
Ethyl alcohol	Air	20	0.0232	Phenol	Air	20	0.0579
Ethyl alcohol	Air	40	0.0207	Sulfuric acid (85%)‡	Vapor	20	0.0415
Ethyl ether	Vapor	20	0.0172	n-Propylamine	Air	20	0.0228
Ethyl ether	Vapor	40	0.0145	Toluene	Vapor	20	0.0280
Glycerol	Vapor	20	0.0645	Turpentine	Air	20	0.0272
Glycerol	Vapor	100	0.0549	Xylene (meta)	Vapor	20	0.0296

*For conversion to dyne-cm^{-1}, use 1 gf-cm^{-1} = 9.80665 × 10^2 dyne-cm^{-1}. For definition and applications, refer to Sec. 5.02-6.
†Or in contact with air.
‡For (98.5%), use 0.0562.

A.62 SURFACE TENSION OF WATER AGAINST AIR AT TEMPERATURE T*

T °C	σ gf-cm^{-1}	T °C	σ gf-cm^{-1}	T °C	σ gf-cm^{-1}	T °C	σ gf-cm^{-1}	T °C	σ gf-cm^{-1}
0	0.0769	18	0.0745	36	0.0716	54	0.0685	72	0.0652
2	0.0766	20	0.0740	38	0.0712	56	0.0682	74	0.0649
4	0.0766	22	0.0737	40	0.0709	58	0.0678	76	0.0645
6	0.0764	24	0.0734	42	0.0706	60	0.0675	78	0.0641
8	0.0761	26	0.0732	44	0.0702	62	0.0671	80	0.0638
10	0.0757	28	0.0729	46	0.0699	64	0.0667	85	0.0629
12	0.0754	30	0.0726	48	0.0695	66	0.0663	90	0.0620
14	0.0751	32	0.0723	50	0.0692	68	0.0659	95	0.0610
16	0.0748	34	0.0719	52	0.0689	70	0.0656	100	0.0600

*For conversion to dyne-cm^{-1}, use 1 gf-cm^{-1} = 9.80665 × 10^2 dyne-cm^{-1}. For definition and applications, refer to Sec. 5.02-6.

A.63 FRICTION FACTORS f FOR CLEAN CAST-IRON PIPES WITH FLOW OF WATER AT $T = 15°C$ IN SI UNITS*

f = tabulated value $\times 10^{-4}$

Diameter, m	Velocity of water, m-s^{-1}										
	0.5	1	1.5	2	2.5	3	3.5	4	5	10	15
0.10	277	250	236	224	215	209	207	204	202	187	173
0.15	260	236	223	213	205	200	199	197	192	180	167
0.20	243	222	210	203	196	192	191	190	188	174	161
0.25	235	215	202	196	191	187	196	195	183	170	158
0.30	228	207	194	190	186	182	181	180	179	166	155
0.35	224	203	191	186	182	179	178	177	176	164	153
0.40	220	199	188	182	179	176	175	174	173	162	152
0.45	216	196	185	180	176	173	172	171	170	160	150
0.50	212	193	182	178	173	171	170	169	168	158	149
0.55	208	190	180	175	172	169	168	167	166	156	147
0.60	205	187	178	173	170	167	166	165	164	154	146
0.65	203	185	176	171	168	165	164	163	162	153	145
0.70	201	183	174	169	166	164	163	162	161	152	144
0.75	199	181	172	168	165	163	162	161	160	151	143
0.80	197	180	171	167	164	162	161	160	159	150	143

*For definition and applications, refer to Sec. 5.04–1.

A.64 FRICTION FACTORS f FOR OLD CAST-IRON PIPES WITH FLOW OF WATER AT $T = 15°C$ IN SI UNITS*

f = tabulated value $\times 10^{-4}$

Diameter, m	Velocity of water, m-s^{-1}										
	0.5	1	1.5	2	2.5	3	3.5	4	5	10	15
0.10	422	410	404	400	396	392	390	389	387	369	351
0.15	415	404	398	394	390	386	384	383	381	362	344
0.20	409	398	392	388	384	380	378	377	375	356	337
0.25	406	395	389	384	380	376	374	373	371	353	334
0.30	403	393	387	381	377	373	371	369	368	350	332
0.35	400	390	384	378	375	371	369	367	366	348	331
0.40	398	387	381	376	373	369	367	365	364	346	330
0.45	396	385	379	374	371	367	365	363	362	345	329
0.50	394	383	377	372	369	365	363	362	360	344	328
0.55	392	381	375	370	367	363	361	360	359	343	327
0.60	390	379	373	368	365	361	360	359	358	342	326
0.65	388	377	371	366	363	360	359	358	357	341	326
0.70	386	375	369	365	362	359	358	357	356	340	325
0.75	385	374	368	364	361	358	357	356	355	339	325
0.80	384	373	367	363	360	357	356	355	354	338	324

*For definition and applications, refer to Sec. 5.04–1.

A.65 MOODY DIAGRAM FOR FRICTION FACTOR
f AND Re $\sqrt{f}$ FOR PIPE FLOW

The diagram given below yields the values of the friction factor f for any type of pipe flow (Sec. 5.04). The value of Re is computed by the formula given below the chart. The value of absolute roughness is estimated from the table in the chart. Once Re and f are known, Re $\sqrt{f}$ can be read from the top scale by means of the inclined lines connecting the point of intersection of the vertical of Re and the curve of ϵ/d. For notation see Table A.66.

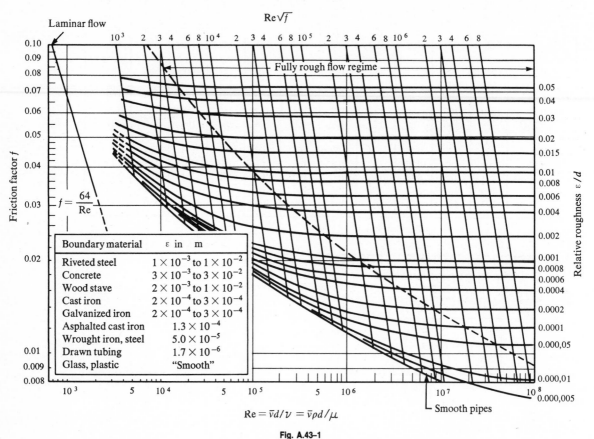

Fig. A.43–1

SOURCE: Moody, L. F., Friction Factors for Pipe Flow, *Trans. ASME*, vol. 66, Nov. 1944.

A.66 MOODY DIAGRAM FOR FRICTION FACTOR
f AND $\mathrm{Re}\sqrt[5]{f}$ FOR PIPE FLOW

The diagram given below is identical to that in Table A.65, but the top scale can be used for finding the values of $\mathrm{Re}\sqrt[5]{f}$.

Notation: d = diameter of pipe, f = friction factor, $\bar{v}$ = average velocity of flow, μ = absolute viscosity of liquid, ρ = mass density of liquid, ν = kinematic viscosity of fluid, and Re = Reynolds number.

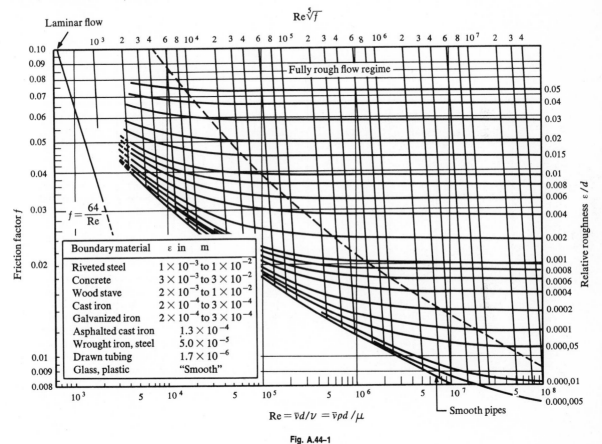

Fig. A.44–1

SOURCE: Moody, L. F., Friction Factors for Pipe Flow, *Trans. ASME*, vol. 66, Nov. 1944.

A.67 THERMAL CONSTANTS OF METALLIC SOLIDS AT TEMPERATURE T

Name	Average specific heat capacity		Average thermal coefficient of linear expansion		Thermal conductivity	
	T °C	$\bar{c}*$ kcal-kg^{-1}-°C^{-1}	T °C	$\bar{\alpha}$† 10^{-6}°C^{-1}	T °C	k‡ kcal-m^{-1}-s^{-1}-°C^{-1}
Aluminum	0–700	0.214	20–300	23.8	0–100	0.052 ± 0.0040
Antimony	20–300	0.049	20–100	10.9	0–50	0.005 ± 0.0003
Arsenic	0–100	0.082	20–100	3.8	. . . . .	
Barium	0–100	0.069	20–100	13.5	. . . . .	
Beryllium	0–100	0.442	0–100	12.2	0–100	0.040 ± 0.0070
Bismuth	0–100	0.030	20–300	13.2	0–100	0.002 ± 0.0004
Cadmium	0–100	0.055	0–300	31.0	0–100	0.022 ± 0.0012
Calcium	0–100	0.157	0–300	22.7	. . . . .	
Cerium	20–100	0.051	0–200	6.4	. . . . .	
Cesium	0–30	0.052	20	98.9	. . . . .	
Chromium	20–100	0.111	20–200	6.1	0–100	0.017 ± 0.0022
Cobalt	20–100	0.102	20–200	13.8	0–100	0.016 ± 0.0010
Copper	20–100	0.094	20–100	16.8	0–100	0.088 ± 0.0161
Gold	18–100	0.031	20–200	14.3	0–100	0.073 ± 0.0042
Iridium	18–100	0.032	20–200	6.7	0–100	0.014 ± 0.0015
Iron	20–100	0.107	20–200	11.6	0–100	0.014 ± 0.0023
Lead	0–100	0.031	20–200	29.0	0–100	0.008 ± 0.0002
Lithium	0–100	0.082	0–180	55.0	0–100	0.017 ± 0.0010
Magnesium	0–100	0.249	20–300	26.4	0–100	0.036 ± 0.0020
Manganese	20–100	0.107	20–200	23.3	. . . . .	
Mercury	0–100	0.033	(−40)–(−70)	41.0	0–100	0.002 ± 0.0005
Molybdenum	0–100	0.065	0–300	5.2	0–100	0.035 ± 0.0006
Nickel	0–100	0.111	20–200	13.1	0–200	0.019 ± 0.0015
Osmium	0–100	0.031	20–200	6.2	0–100	0.014 ± 0.0016
Palladium	0–100	0.059	20–200	11.8	0–100	0.016 ± 0.0012
Platinum	20–100	0.032	0–300	9.2	0–200	0.016 ± 0.0010
Potassium	0–100	0.180	20–40	84.0	0–100	0.023 ± 0.0025
Rhodium	10–100	0.059	0–200	8.4	0–100	0.021 ± 0.0014
Silver	0–100	0.056	0–200	19.3	0–100	0.101 ± 0.0023
Sodium	0–100	0.300	0–50	70.0	0–100	0.032 ± 0.0016
Steel	0–100	0.115	0–100	11.6	0–100	0.011 ± 0.0012
Strontium	15–100	0.065	. . . . .	. . . .	. . . . .	
Tantalum	15–100	0.034	0–200	6.6	0–100	0.017 ± 0.0009
Thorium	0–100	0.028	20–100	11.2	. . . . .	
Tin	0–100	0.060	20–200	20.9	0–100	0.016 ± 0.0007
Titanium	0–100	0.143	20–100	18.2	. . . . .	
Tungsten	0–100	0.035	0–300	4.6	0–100	0.037 ± 0.0025
Uranium	0–100	0.028	20–100	14.9	. . . . .	
Vanadium	0–100	0.117	20–100	7.9	. . . . .	
Zinc	0–100	0.093	0–100	26.3	0–100	0.030 ± 0.0035

*For conversion to Btu-lb^{-1}-°F^{-1}, note 1 kcal-kg^{-1}-°C^{-1} = 1 Btu-lb^{-1}-°F^{-1}. For definition and application, refer to Sec. 6.01–5.

†For conversion to °F^{-1}, use 1°C^{-1} = $\frac{5}{9}$°F^{-1}. For definition and application, refer to Sec. 6.03–1.

‡For conversion to Btu-in.-ft^{-2}-h^{-1}-°F^{-1}, use 1 kcal-m^{-1}-s^{-1}-°C^{-1} = 29,000 Btu-in.-ft^{-2}-h^{-1}-°F^{-1}. For definition and application, refer to Sec. 6.04–2.

A.68 THERMAL CONSTANTS OF NONMETALLIC SOLIDS AT TEMPERATURE T*

Name	Average specific heat capacity		Average thermal coefficient of linear expansion		Thermal conductivity	
	T °C	$\bar{c}$ kcal-kg^{-1}-°C^{-1}	T °C	$\bar{\alpha}$ 10^{-6}°C^{-1}	T °C	k 10^{-4} kcal-m^{-1}-s^{-1}-°C^{-1}
Agate	0–100	0.193	· · · · · ·	· · · · ·	0–100	26.000 ± 0.500
Asbestos	0–200	0.195	20–60	12.4	0–200	0.600 ± 0.050
Bakelite	0–40	0.350	20–60	28.0	0–40	1.200 ± 0.100
Brick	0–200	0.225	0–200	9.3	0–200	1.700 ± 0.020
Carbon, diamond	0–100	0.120	20–100	1.2	0–100	20.000 ± 0.100
Carbon, graphite	0–200	0.250	20–100	7.8	0–100	12.000 ± 1.500
Celluloid	· · · · · ·	· · · · ·	20–80	108.0	20–80	0.090 ± 0.010
Chalk	0–200	0.220	20–200	7.6	0–200	2.000 ± 0.100
Clay, dry	20–100	0.220	0–200	10.2	20–200	2.000 ± 0.200
Coal	0–100	0.205	· · · · · ·	· · · · ·	0–100	0.300 ± 0.050
Coke	20–400	0.265	· · · · · ·	· · · · ·	0–400	0.500 ± 0.050
Concrete	20–300	0.210	20–300	12.0	20–300	3.000 ± 0.500
Cork	20–100	0.480	· · · · ·	· · · · ·	20–100	0.100 ± 0.050
Glass	0–100	0.180	0–100	9.0	0–100	2.000 ± 0.600
Granite	0–100	0.195	0–100	8.4	0–100	4.500 ± 0.250
Gypsum	0–100	0.210	· · · · · ·	· · · · ·	0–100	3.200 ± 0.100
Ice	(−20)–(0)	0.485	(−20)–(0)	51.0	(−20)–(0)	3.900 ± 0.005
Leather	0–100	0.360	· · · · · ·	· · · · ·	0–100	0.300 ± 0.150
Limestone	0–100	0.216	0–100	9.0	· · · · · ·	· · · · · · · · · · ·
Marble	0–100	0.210	0–100	12.2	0–60	8.400 ± 0.400
Mica	20–100	0.207	20–100	7.55	20–100	0.850 ± 0.010
Paper	20–60	0.310	· · · · · ·	· · · · ·	20–60	0.300 ± 0.020
Paraffin	0–20	0.710	10–40	115.00 ± 15	10–40	0.600 ± 0.020
Plaster of Paris	0–100	0.250	· · · · · ·	· · · · ·	20–150	0.400 ± 0.020
Porcelain	10–900	0.260	10–400	3.55	10–900	3.200 ± 1.000
Quartz	10–100	0.189	10–100	0.50	10–100	95.000 ± 60.000
Rubber, hard	10–100	0.425	20–60	80.00 ± 10	20–60	0.400 ± 0.030
Sand, dry	0–100	0.195	· · · · · ·	· · · · ·	20–160	0.800 ± 0.040
Sandstone	0–100	0.215	10–400	4.25	0–100	5.000 ± 0.500
Slate	0–100	0.184	20–60	8.00 ± 1	20–100	4.500 ± 0.300
Silicon	0–100	0.176	20–60	7.50 ± 1	20–100	2.500 ± 0.500
Lumber, hard	10–60	0.700	10–60	4.00 ± 0.5	10–60	0.300 ± 0.100
Lumber, soft	10–60	0.300	10–60	3.00 ± 0.5	10–60	0.250 ± 0.100

A.69 THERMAL CONDUCTIVITY k OF INSULATORS AT TEMPERATURE T*

Name	T °C	k 10^{-4} kcal-m^{-1}-s^{-1}-°C^{-1}	Name	T °C	k 10^{-4} kcal-m^{-1}-s^{-1}-°C^{-1}
Asbestos-paper	0–200	0.600 ± 0.005	Fiber insulation	0–200	0.090 ± 0.010
Corkboard	0–160	0.100 ± 0.020	Glass wool	0–200	0.090 ± 0.010
Feathers	0–160	0.150 ± 0.020	Sawdust	0–100	0.140 ± 0.010

*For definitions, applications, and conversions, see footnotes, Appendix A.67.

A.70 THERMAL CONSTANTS OF LIQUIDS AT $p = 1$ atm AND TEMPERATURE T*

Name	Average specific heat capacity		Average thermal coefficient of volume expansion		Thermal conductivity	
	T °C	$\bar{c}$ kcal-kg^{-1}-°C^{-1}	T °C	$\bar{\beta}$ 10^{-4}°C^{-1}	T °C	k 10^{-4} kcal-m^{-1}-s^{-1}-°C^{-1}
Acetic acid	20–90	0.542	5–55	10.7 ± 0.5	5–55	0.475 ± 0.015
Acetone	3–23	0.515	5–60	14.8 ± 0.7	5–60	0.425 ± 0.025
Alcohol, amyl	20–120	0.685	5–40	9.2 ± 0.2	5–40	0.305 ± 0.025
Alcohol, ethyl	0–12	0.574	5–40	11.1 ± 0.4	5–40	0.400 ± 0.025
Aniline	10–90	0.513	5–60	8.5 ± 0.1	5–60	0.400 ± 0.030
Benzene	5–60	0.418	5–40	12.4 ± 0.6	5–40	0.300 ± 0.015
Chloroform	10–20	0.228	5–60	12.7 ± 0.2	5–25	0.270 ± 0.025
Gasoline	0–100	0.500	5–60	4.9 ± 0.2	5–60	0.280 ± 0.030
Glycerol	15–50	0.575	5–40	5.1 ± 0.1	5–40	0.560 ± 0.030
Kerosene	0–100	0.500	5–40	9.5 ± 0.2	5–40	0.310 ± 0.040
Nitrobenzene	0–30	0.340	5–40	9.1 ± 0.2	5–40	0.350 ± 0.040
Olive oil	20–100	0.470	5–60	9.8 ± 0.6	—	
n-Pentane	0–50	0.510	5–40	16.1 ± 0.5	5–40	0.285 ± 0.025
n-Propyl alcohol	0–50	0.508	5–40	9.4 ± 0.4	—	
Sea water	0–80	0.930	0–100	2.4 ± 0.3	—	
Turpentine	0–50	0.410	5–50	9.7 ± 0.4	5–50	0.255 ± 0.015
Water	0–100	1.000	0–100	2.0 ± 0.2	5–100	1.430 ± 0.170

*For definitions, applications, and conversions, see footnotes, Appendix A.67.

A.71 THERMAL CONSTANTS OF GASES AT TEMPERATURE $T = 293$ K*

Name	Formula	Molecular mass kg-kmol^{-1}	Gas constant R/Mkg J-kg^{-1}-K^{-1}	Specific heat capacity		α c_p/c_v dimensionless	C_p $c_p Mkg$	C_v $c_v Mkg$
				At constant pressure c_p kcal-kg^{-1}-K^{-1}	At constant volume c_v		kcal-kmol^{-1}-K^{-1}	
Acetylene	C_2H_2	26.036	319.365	0.3833	0.3066	1.25	9.980	7.993
Air	$O_2 + N_2$	28.952	287.200	0.2385	0.1704	1.40	6.905	4.918
Ammonia	NH_3	17.031	488.227	0.5362	0.4222	1.27	9.132	7.145
Argon	Ar	39.944	208.166	0.1232	0.0738	1.67	4.921	2.934
Carbon dioxide	CO_2	44.010	188.934	0.2012	0.1560	1.29	8.855	6.868
Carbon monoxide	CO	28.010	296.858	0.2479	0.1770	1.40	6.944	4.957
Ethane	C_2H_6	30.060	276.613	0.4025	0.3354	1.20	12.099	10.112
Freon (F-12)	CCl_2F_2	120.925	68.762	0.1372	0.1204	1.14	16.591	14.604
Helium	He	4.002	2,077.711	1.2500	0.7530	1.66	5.003	3.015
Hydrogen	H_2	2.016	4,124.504	3.3920	2.4063	1.41	6.838	4.851
Methane	CH_4	16.038	518.456	0.5929	0.4706	1.26	9.509	7.522
Nitrogen	N_2	28.016	296.795	0.2480	0.1771	1.40	6.948	4.961
Oxygen	O_2	32.000	259.844	0.2186	0.1561	1.40	6.995	5.008
Propane	C_2H_8	44.080	188.634	0.3975	0.3549	1.12	17.522	15.535
Water vapor	H_2O	18.016	416.534	0.4550	0.3445	1.32	8.197	6.210

*For definitions, applications, and conversions, refer to Sec. 6.02.

A.72 SPECIFIC HEAT CAPACITY c_p OF GASES AT PRESSURE p AND TEMPERATURE T*

c_p = tabulated value in kcal-kg^{-1}-K^{-1}

p °K	Air ($O_2 + N_2$) 1 atm	4 atm	10 atm	100 atm	p °K	Hydrogen (H_2) 0 atm	1 atm	10 atm	100 atm
100	0.2459	. . . . . .	. . . . . .		20	2.4644	. . . . . .	. . . . . .	
200	0.2385	0.2415	0.2478	0.4136	40	2.4644	2.5276	3.4136	
300	0.2385	0.2395	0.2418	0.2752	60	2.4831	2.5075	2.7401	3.9005
400	0.2425	0.2431	0.2442	0.2600	80	2.5678	2.5678	2.6844	3.5127
500	0.2463	0.2466	0.2473	0.2566	100	2.6755	2.6826	2.7501	3.2479
600	0.2514	0.2516	0.2521	0.2583	150	3.0115	3.0129	3.0101	3.2485
800	0.2627	0.2629	0.2632	0.2665	200	3.2327	3.2356	3.2485	3.3648
1000	0.2739	0.2732	0.2733	0.2754	250	3.3519	3.3562	3.3705	3.4450
1500	0.2938	0.2945	0.2945	0.2953	300	3.4193	3.4208	3.4265	3.4753
2000	0.3199	0.3175	0.3166	0.3161	400	3.4596	3.4610	3.4639	3.4883
2500	0.4037	0.3688	0.3560	0.3412	500	3.4682	3.4682	3.4711	3.4854
3000	0.6836	0.5220	0.4615	0.3845	600	3.4768	3.4768	3.4782	3.4854

p °K	Nitrogen (N_2) 1 atm	4 atm	10 atm	100 atm	p °K	Oxygen (O_2) 1 atm	4 atm	10 atm	100 atm
100	0.2563	. . . . . .	. . . . . .	. . . . . .	200	0.2185	0.2216	0.2282	0.4719
200	0.2492	0.2523	0.2587	0.4000	300	0.2199	0.2209	0.2233	0.2587
300	0.2488	0.2500	0.2523	0.2852	400	0.2251	0.2256	0.2268	0.2435
400	0.2498	0.2503	0.2514	0.2675	500	0.2323	0.2327	0.2334	0.2435
500	0.2525	0.2529	0.2235	0.2632	600	0.2398	0.2400	0.2405	0.2472
600	0.2569	0.2572	0.2576	0.2640	700	0.2465	0.2465	0.2469	0.2516
800	0.2682	0.2683	0.2686	0.2712	800	0.2521	0.2522	0.2523	0.2559
1000	0.2789	0.2790	0.2792	0.2812	1000	0.2605	0.2606	0.2607	0.2628
1500	0.2973	0.2973	0.2973	0.2984	1500	0.2731	0.2731	0.2731	0.2740
2000	0.3069	0.3049	0.3069	0.3073	2000	0.2817	0.2822	0.2822	0.2826
2500	0.3124	0.3124	0.3134	0.3126	2500	0.2907	0.2907	0.2907	0.2911
3000	0.3159	0.3159	0.3160	0.3161	3000	0.2985	0.2985	0.2979	0.2986

*For definitions, applications, and conversions, refer to Sec. 6.02.

SOURCE: J. Hilsenrath et al., Tables of Thermal Properties of Gases, *Natl. Bur. Std. Circ.* 564, 1956.

A.73 THERMAL CONDUCTIVITY OF GASES k
AT $p = 1$ atm AND TEMPERATURE T*

k = tabulated value $\times 10^{-6}$ kcal-m^{-1}-s^{-1}-K^{-1}

Name	Formula	T, K								
		100	200	273	300	400	500	600	800	1000
Air	$O_2 + N_2$	2.20	4.35	5.73	6.24	7.90	9.39	10.89	13.60	16.06
Argon	Ar	1.54	2.99	3.90	4.23	5.33	6.22	7.10	8.63	10.00
Carbon monoxide	CO	2.09	4.16	5.52	6.02	7.72	9.16	10.61	13.12	15.39
Carbon dioxide	CO_2	.....	2.28	3.51	3.96	5.88	7.48	9.08	12.91	16.01
Freon (F-12)	CCL_2F_2	.....	.....	.....	2.32	3.61	.....	.....	.....	
Helium	He	17.45	27.49	33.61	35.85	43.02	51.04	59.03	73.39	86.77
Hydrogen	H_2	16.16	30.59	40.02	43.51	54.53	62.03	69.55	86.04	102.29
Krypton	Kr	.....	1.55	2.08	2.27	2.92	3.47	4.02	4.95	
Nitrogen	N_2	2.33	4.47	5.76	6.24	7.74	9.12	10.49	13.20	15.78
Neon	Ne	5.19	8.84	10.92	11.69	14.12	16.23	18.33	21.93	25.10
Oxygen	O_2	2.16	4.37	5.84	6.38	8.15	9.74	11.33	14.20	17.16
Water vapor	H_2O	.....	.....	.....	4.36	6.31	8.70	11.09	16.25	
Xenon	Xe	.....	0.91	1.21	1.32	1.71	2.08	2.44	3.06	3.61

A.74 CHANGE IN PHASE OF SOLIDS AT PRESSURE $p = 1$ atm*

Name	T_f °C	L_f kcal/kg	T_v °C	L_v kcal/kg	Name	T_f °C	L_f kcal/kg	T_v °C	L_v kcal/kg
Aluminum	659	93.00	2050	2,000	Molybdenum	2621	68.90	4699	1,333
Antimony	630	38.30	1420	373	Nickel	1452	71.70	2600	1,487
Arsenic	815	.....	625	74	Osmium	2500	.....	4400	
Barium	850	345.00	1562	625	Palladium	1550	34.2	3500	610
Beryllium	1284	285.00	2767		Platinum	1765	26.10	4300	610
Bismuth	271	12.50	1470	220	Potassium	63	14.30	765	512
Cadmium	321	13.00	767	225	Rhodium	1960	42.3	3700	620
Calcium	810	62.00	1400	912	Silver	960	25.50	1980	575
Cerium	645	15.10	1418		Sodium	97	27.50	886	1,070
Cesium	28.5	3.80	670	130	Steel	1350	63.20	2500	1,390
Chromium	1420	43.50	2200		Strontium	781	33.80	1383	1,045
Cobalt	1480	62.10	2900		Sugar	160	.....	....	
Copper	1083	50.60	2595	1,760	Sulfur	112	13.2	445	
Glass	700	.....	....		Tantalum	2850	.....	4100	
Gold	1063	16.10	2966	446	Thallium	303	7.2	1550	228
Iridium	2400	26.50	4350	342	Thorium	1845	.....	3200	
Iron	1539	65.00	2740	1,620	Tin	231	14.4	2270	653
Lead	327	6.30	1744	222	Titanium	1800	.....	3200	1,340
Lithium	186	158.30	1352		Tungsten	3400	44.00	5927	1,180
Magnesium	650	78.20	1120	1,400	Uranium	1700	.....	4140	
Manganese	1270	63.20	1900	1,000	Vanadium	1710	.....	3000	
Mercury	−39	2.78	357	71	Zinc	419	24.10	907	362

*For definitions, applications, and conversions, refer to Sec. 6.03.

A.75 CHANGE IN PHASE OF LIQUIDS AT PRESSURE $p = 1\ \text{atm}^*$

Name	T_f °C	L_f $\dfrac{\text{kcal}}{\text{kg}}$	T_v °C	L_v $\dfrac{\text{kcal}}{\text{kg}}$	Name	T_f °C	L_f $\dfrac{\text{kcal}}{\text{kg}}$	T_v °C	L_v $\dfrac{\text{kcal}}{\text{kg}}$
Acetic acid	17	44.8	118	96.8	Nitroglycerin	3	—	260exp†	
Acetone	−95	19.6	56	124.5	Nitrotoluene	−4	—	222	
Acrolein	−88	....	52		n-Nonane	−51	—	151	
Amyl alcohol	−78	—	138	120.0	n-Octane	−57	43.2	125	71.2
Aniline	−6	21.0	184	110.0	Oleic acid	16	—	286	
Benzene	6	30.3	80	94.2	Olive oil	20	—	300	
Bromine	−27	16.2	61	45.0	n-Pentane	−130	27.8	36	85.3
Carbolic acid	41	30.2	182	61.2	Petroleum	−70	—	250	
Chloroform	−68	22.8	−34	65.3	Phosphorus				
Ethyl alcohol	−115	25.2	79	204.1	trichloride	−112	—		
Formic acid	8	60.2	101	121.2	n-Propyl alcohol	−127		97	
Gasoline	—	—	72–90		Pyridine	−42		115	
n-Heptane	−90	36.6	98	75.3	Silicon tetrachloride	−70		60	
n-Hexane	−95	37.8	86	79.5	Stannic chloride	−33	—	114	
Hydrogen fluoride	−93	—	20		Styrene	−31	—	145d†	
Hydrogen peroxide	−2	—	152		Sulfuric acid	10	24.1	340d†	122.1
Methyl alcohol	−98	22.5	65	263.2	Toluene	−95	—	111	87.2
Nitric acid	−47	9.5	86d†	114.9	Turpentine	−10	—	160	168.7
Nitrobenzene	7	22.5	211	78.9	Water	0	79.7	100	539.2
Nitrogen peroxide	−9	—	22d†		o-Xylene	−26	—	143	83.1

A.76 CHANGE IN PHASE OF GASES AT PRESSURE $p = 1\ \text{atm}^*$

Name	T_f °C	L_f $\dfrac{\text{kcal}}{\text{kg}}$	T_v °C	L_v $\dfrac{\text{kcal}}{\text{kg}}$	Name	T_f °C	L_f $\dfrac{\text{kcal}}{\text{kg}}$	T_v °C	L_v $\dfrac{\text{kcal}}{\text{kg}}$
Acetylene	−81	28.6	−75	148.8	Hydrogen chloride	−111	13.1	−84	105.8
Ammonia	−76	108.0	−34	327.1	Hydrogen cyanide	−14	74.2	26	223.0
Argon	−189	7.2	−185	38.8	Hydrogen fluoride	−92	54.7	19	372.8
Arsine	−114	....	−55		Hydrogen iodide	−51	5.7	−35	33.8
Butane (iso)	−144	—	−10	87.2	Hydrogen selenide	−64	—	−42	
Butane (n)	−138	—	−0.5	91.5	Hydrogen sulfide	−84	16.7	−61	130.9
Carbon dioxide	−57	44.2	−80	137.4	Hydrogen telluride	−48	—	−2	
Carbon monoxide	−207	7.6	−190	51.3	Krypton	−163	—	−153	28.2
Carbon tetrachloride	−23	41.6	76	46.4	Methane	−183	14.4	−161	146.6
Chlorine	−101	22.7	−35	75.2	Methyl chloride	−98	—	−24	102.2
Chlorine dioxide	−59	—	100exp		Methyl ether	−138	—	−24	
Chlorine monoxide	−20	—	4exp		Methyl fluoride	−112	—	−78	
Cyanogen	−27	—	−21		Neon	−249	2.8	−246	20.6
Divinyl	−109	—	−4		Nitric oxide	−164	18.3	−152	110.2
Ethane	−178	—	−87	258.6	Nitrogen	−210	6.1	−196	47.7
Ethyl chloride	−139	—	12	92.5	Nitrous oxide	−102	35.5	−89	89.7
Ethylene	−169	28.6	−104	115.5	Oxygen	−219	3.3	−183	50.9
Fluorine	−223	—	−188	40.7	Propane	−189	19.1	−44	101.3
Helium	−272	—	−269	6.0	Radon	−71	—	−62	
Hydrochloric acid	−112	13.1	−84	105.8	Sulfur dioxide	−75	—	−10	95.1
Hydrogen	−259	13.9	−253	107.1	Vinyl chloride	−160	—	−14	95.1
Hydrogen bromide	−86	7.4	−68	50.4	Xenon	−126	3.7	−110	24.6

*T_f = melting point, L_f = latent heat of fusion, T_v = boiling point, L_v = latent heat of vaporization; for definition of T_f, L_f, T_v, L_v, refer to Sec. 6.01–6; for conversion to Btu/lb, use 1 kcal/kg = 1.8 Btu/lb.   †d = decomposes at T_v; exp = explodes at T_v.

A.77 AVERAGE HEAT OF COMBUSTION*

Name	$\dfrac{10^3 \text{ kcal}}{\text{kg}}$	$\dfrac{10^3 \text{ Btu}}{\text{lb}}$	$\dfrac{10^3 \text{ kcal}}{\text{m}^3}$	$\dfrac{10^3 \text{ Btu}}{\text{ft}^3}$	Name	$\dfrac{10^3 \text{ kcal}}{\text{kg}}$	$\dfrac{10^3 \text{ Btu}}{\text{lb}}$	$\dfrac{10^3 \text{ kcal}}{\text{m}^3}$	$\dfrac{10^3 \text{ Btu}}{\text{ft}^3}$
Acetylene	11.90	21.50	13.30	1.50	Gasoline (0.71)	11.50	20.70	—	
Anthracite coal	8.40	15.20	—		Gasoline (0.77)	11.90	21.50	—	
Benzene (0.88)	10.30	18.50	—		Heating oil	13.90	25.00	—	
Benzene (0.78)	10.10	18.20	33.80	3.80	Hydrogen	33.90	61.00	2.85	0.32
Benzene (0.68)	9.90	17.90	24.90	2.80	Kerosene (0.78)	11.10	20.00	—	
Bituminous coal	7.50	13.50	—		Kerosene (0.80)	11.70	21.00	—	
Carbohydrate	4.00	7.20	—		Lignite	5.00	9.00	—	
Carbon to CO	2.50	4.50	—		Methyl alcohol	5.30	9.50	7.10	0.80
Carbon to CO_2	8.10	14.50	—		Peat	2.80	5.00	—	
Carbon disulfide	3.20	5.80	—		Petroleum (0.70)	11.70	21.00	—	
Carbon monoxide	2.40	4.40	2.70	0.30	Petroleum (0.80)	11.10	20.00	—	
Charcoal	6.70	12.00	—		Petroleum (0.90)	10.60	19.00	—	
Coke	6.40	11.50	—		Petroleum (1.00)	10.00	18.00	—	
Diesel oil	10.60	19.00	—		Propane	11.70	21.00	20.50	2.30
Ethyl alcohol	7.30	13.20	14.20	1.60	Protein	4.00	7.20	—	
Fat	9.40	17.00	—		Pure alcohol	6.90	12.40	—	
Gas, coal	—	—	5.80	0.65	Semibituminous coal	7.20	13.00	—	
Gas, coke	—	—	4.90	0.55	Straw	3.10	5.50	—	
Gas, natural	—	—	10.70	1.20	Sulfur	2.20	4.00	—	
Gas, oil	—	—	7.10	0.80	Wood, air-dried	2.80	5.00	—	

*The amount of heat released by the substance by its complete burndown is the heat of combustion. For example, the complete combustion of 1,000 kg of bituminous coal releases $7,500 \times 1,000 = 7.5 \times 10^6$ kcal of heat.

A.78 DIELECTRIC CONSTANT c OF LIQUIDS AT $p = 1$ atm AND TEMPERATURE T*

Name	T °C	Dielectric constant c dimensionless	Name	T °C	Dielectric constant c dimensionless
Acetone	20	21.25	n-Heptane	20	1.92
Amyl acetate	20	4.83	n-Hexane	20	1.88
Aniline	20	7.15	Kerosene	20	2.16
Benzene	20	2.28	Methyl alcohol	20	34.68
Bromine	20	3.38	Nitrobenzene	20	36.15
Carbon disulfide	20	2.56	n-Octane	20	1.95
Carbon tetrachloride	20	2.64	Olive oil	20	2.25
Chlorine dioxide	20	6.29	Paraffin oil	20	2.50
Chloroform	20	4.80	n-Pentane	20	1.84
Ethyl alcohol	20	24.25	Toluene	15	2.56
Formic acid	20	58.12	Transformer oil	20	2.28
Glycerol	15	39.24	Water	20	79.63

*For definition and applications, refer to Sec. 7.02–5.

A.79 DIELECTRIC CONSTANT c OF SOLIDS AT TEMPERATURE T*

Name	T °C	Dielectric constant c dimensionless	Name	T °C	Dielectric constant c dimensionless
Amber	20	2.8 ± 0.1	Ivory	20	7.0 ± 0.2
Apatite	20	8.6 ± 0.1	Marble	20	8.3 ± 0.1
Asphalt	20	2.7 ± 0.1	Mica	20	6.2 ± 0.7
Balsam, Canada	20	2.9 ± 0.3	Nylon	20	5.0 ± 0.1
Barium titanite	20	1,300	Oiled paper	20	4.1 ± 0.1
Beeswax	20	2.7 ± 0.1	Paper	20	2.2 ± 0.2
Casting mold	20	2.8 ± 0.4	Paraffin	20	2.3 ± 0.2
Celluloid	20	4.0 ± 0.1	Phenol-formaldehyde		
Diamond	20	15.0 ± 0.2	resin	20	4.0 ± 0.6
Ebonite	20	2.6 ± 0.2	Phenolic resin	20	8.0 ± 0.4
Epoxy	25	3.4 ± 0.2	Plexiglas	20	3.1 ± 0.1
Forsterite	20	6.4 ± 0.1	Plywood	20	1.7 ± 0.1
Fused silica	20	4.5 ± 0.5	Polyethylene	20	2.6 ± 0.2
High alum. ceramic	20	8.1 ± 0.1	Polypropylene	20	2.5 ± 0.2
Glass, crown	20	6.2 ± 1.0	Polystyrene	20	2.8 ± 0.1
Glass, flint	20	8.5 ± 1.5	Quartz, crystalline	20	4.4 ± 0.3
Gutta-percha	20	3.3 ± 0.7	Quartz, fused	20	3.7 ± 0.3
Granite	20	8.0 ± 1.0	Shellac	24	3.6 ± 0.1
Ice	-10	12.5 ± 0.1	Slate	20	6.8 ± 0.7
Ice	-20	3.5 ± 0.1	Steatite	20	5.7 ± 0.1
Insulation, elec.			Sulfur	20	3.9 ± 0.4
cable	20	4.1 ± 0.1	Rubber, vulcanized	20	3.0 ± 0.2
Insulation, tel. cable	20	1.7 ± 0.1	Teflon	20	2.0 ± 0.1
Insulator porcelain	20	5.8 ± 0.2	Wood	20	3.0 ± 0.5

A.80 DIELECTRIC CONSTANT c OF GASES AT $p = 1$ atm AND TEMPERATURE T*

Name	T °C	Dielectric constant c dimensionless	Name	T °C	Dielectric constant c dimensionless
Air	0	1.000 590	Hydrogen	0	1.000 264
Air	20	1.000 537	Hydrogen	20	1.000 254
Carbon dioxide	0	1.000 945	Nitrogen	0	1.000 605
Carbon monoxide	0	1.000 691	Nitrogen	20	1.000 548
Helium	0	1.000 068	Oxygen	0	1.000 524
Helium	20	1.000 065	Oxygen	20	1.000 495

*For definition and applications, refer to Sec. 7.02–5.

A.81 RESISTIVITY ρ OF PURE METALS
AT TEMPERATURE T^*

Name	ρ, 10^{-8} Ω-m		Name	ρ, 10^{-8} Ω-m	
	$T = 0°C$	$T = 20°C$		$T = 0°C$	$T = 20°C$
Aluminum	2.54	2.78	Nickel	7.00	7.82
Antimony	37.60	40.96	Osmium	8.47	9.24
Arsenic	26.10	28.83	Palladium	9.85	10.62
Bismuth	105.10	115.00	Platinum	9.62	10.57
Cadmium	7.21	7.42	Plutonium	144.00	142.95
Calcium	3.92	4.25	Potassium	6.25	6.92
Chromium	12.24	13.03	Rhodium	4.39	4.77
Cobalt	8.72	9.46	Silver	1.47	1.60
Copper	1.55	1.72	Sodium	4.29	4.72
Gold	2.23	2.44	Strontium	19.90	22.80
Iridium	5.15	5.68	Tantalum	12.15	12.95
Iron	8.68	9.75	Thorium	13.80	15.20
Lead	19.26	20.80	Tin	10.20	11.30
Lithium	8.50	9.30	Titanium	43.42	43.85
Magnesium	3.96	4.45	Tungsten	5.05	5.55
Mercury	94.10	95.80	Uranium	24.20	25.70
Molybdenum	5.08	5.38	Zinc	5.48	5.87

A.82 RESISTIVITY ρ OF INDUSTRIAL
ALLOYS AT $T = 20°C^*$

Name	Composition	ρ 10^{-8} Ω-m
Aluminum (1)	AIEE, hand-drawn	2.83
Brass (2)	60% Cu + 40% Zn	6.80
Brass (3)	66% Cu + 34% Al	7.18
Bronze	96% Cu + 4% Al	18.25
Bronze (5)	90% Cu + 10% Al	16.37
Constantan (6)	55% Cu + 45% Ni	49.63
German silver (7)	65% Cu + 18% Ni + 17% Zn	28.97
Manganin (8)	84% Cu + 12% Mn + 4% Ni	47.88
Nichrome (9)	60% Ni + 15% Cr + 25% Fe	112.00
Iron (10)	Cast	60.00
Iron (11)	Wire	12.00
Steel (12)	Rail	18.00

*For definitions and applications, refer to Sec. 7.03–4.

A.83 RESISTIVITY ρ OF NONMETALLIC SOLIDS
AND LIQUIDS AT $T = 20°C$*

Name	ρ, Ω-m	Name	ρ, Ω-m
Alcohol, ethyl	3.2 (+04)	Phenol-formaldehyde	
Alcohol, methyl	1.5 (+04)	resin	1.0 (+10)
Amber	1.0 (+15)	Plexiglas	1.0 (+12)
Asbestos, paper	1.5 (+09)	Polystyrene	1.0 (+15)
Asphalt insulation	1.2 (+09)	Porcelain	1.0 (+15)
Carbon	4.0 (−05)	Quartz	1.0 (+12)
Diamond	1.0 (+10)	Rubber	1.0 (+16)
Ebonite	1.0 (+16)	Selenium	5.0 (+02)
Glass	5.0 (+12)	Shellac	1.0 (+14)
Gutta-percha	3.0 (+08)	Silica	6.0 (+14)
Ice	7.2 (+06)	Slate	5.0 (+07)
Ivory	2.5 (+06)	Sulfur	1.0 (+15)
Marble	1.0 (+08)	Turpentine	2.0 (+10)
Mica	1.0 (+13)	Water	5.0 (+03)
Paper, sheet	1.0 (+08)	Wood, paraffined	5.0 (+10)
Paraffin	5.0 (+15)		

A.84 THERMAL COEFFICIENT α_0
OF RESISTANCE OF SOLIDS*

Name†	α_0, °C^{-1}	Name†	α_0, °C^{-1}
Aluminum (1)	+ 4.03 (−03)	German silver (7)	+ 5.00 (−04)
Brass (2)	+ 1.00 (−03)	Manganin (8)	+ 1.00 (−06)
Brass (3)	+ 2.10 (−03)	Mercury	+ 9.00 (−04)
Bronze (4)	+ 1.00 (−03)	Nichrome (9)	+ 4.00 (−04)
Bronze (5)	+ 1.10 (−03)	Iron (10)	+ 6.50 (−03)
Carbon	− 3.00 (−04)	Iron (11)	+ 6.50 (−03)
Constantan (6)	+ 2.80 (−05)	Steel (12)	+ 5.00 (−03)

*For definitions and applications, refer to Sec. 7.03–4.
†For composition, refer to Table A.82.

A.67 U.S. STANDARD ATMOSPHERE*

h, km	T, °K	p, N-m^{-2}	ρ, kg-m^{-3}	ν, m^2-s^{-1}	k, kcal-m^{-1}-s^{-1}-K^{-1}	v_L, m-s^{-1}
−5	320.676	1.7776 (+05)	1.9311 (+00)	1.0058 (−05)	6.6545 (−06)	358.986
−4	314.166	1.1560 (+05)	1.7697 (+00)	1.0806 (−05)	6.5356 (−06)	355.324
−3	307.659	1.4297 (+05)	1.6189 (+00)	1.1625 (−05)	6.4161 (−06)	351.625
−2	301.154	1.2778 (+05)	1.4782 (+00)	1.2525 (−05)	6.2958 (−06)	347.888
−1	294.651	1.1393 (+05)	1.3470 (+00)	1.3516 (−05)	6.1748 (−06)	344.111
0	288.150	1.0133 (+05)	1.2250 (+00)	1.4607 (−05)	6.0530 (−06)	340.294
0.5	284.900	9.5461 (+04)	1.1673 (+00)	1.5195 (−05)	5.9919 (−06)	338.370
1.0	281.651	8.8976 (+04)	1.1117 (+00)	1.5813 (−05)	5.9305 (−06)	336.435
1.5	278.402	8.4596 (+04)	1.0581 (+00)	1.6463 (−05)	5.8690 (−06)	334.489
2.0	275.154	7.9501 (+04)	1.0066 (+00)	1.7147 (−05)	5.8073 (−06)	332.532
2.5	271.906	7.4692 (+04)	9.5695 (−01)	1.7868 (−05)	5.7454 (−06)	330.563
3	268.659	7.0121 (+04)	9.0925 (−01)	1.8628 (−05)	5.6833 (−06)	328.583
4	262.166	6.1660 (+04)	8.1935 (−01)	2.0275 (−05)	5.5586 (−06)	324.589
5	255.676	5.4048 (+04)	7.3643 (−01)	2.2110 (−05)	5.4331 (−06)	320.545
6	249.187	4.7218 (+04)	6.6011 (−01)	2.4160 (−05)	5.3068 (−06)	316.452
7	242.700	4.1105 (+04)	5.9002 (−01)	2.6461 (−05)	5.1798 (−06)	312.306
8	236.215	3.5652 (+04)	5.2579 (−01)	2.9044 (−05)	5.0520 (−06)	308.105
9	229.733	3.0801 (+04)	4.6706 (−01)	3.1957 (−05)	4.9235 (−06)	303.848
10	223.252	2.6500 (+04)	4.1351 (−01)	3.5251 (−05)	4.7942 (−06)	299.532
11	221.931	2.3622 (+04)	3.6976 (−01)	4.2800 (−05)	4.7677 (−06)	298.638
12	220.611	2.0745 (+04)	3.2601 (−01)	5.0349 (−05)	4.7412 (−06)	297.745
13	219.309	1.7867 (+04)	2.8223 (−01)	5.7898 (−05)	4.7147 (−06)	296.851
14	217.977	1.4990 (+04)	2.3850 (−01)	6.5447 (−05)	4.6882 (−06)	295.958
15	216.650	1.2112 (+04)	1.9475 (−01)	7.2995 (−05)	4.6617 (−06)	295.069
16	215.336	1.0796 (+04)	1.7358 (−01)	9.0374 (−05)	4.6352 (−06)	294.466
17	214.016	9.4790 (+03)	1.5241 (−01)	1.0775 (−04)	4.6094 (−06)	294.148
18	214.016	8.1625 (+03)	1.3124 (−01)	1.2513 (−04)	4.6257 (−06)	294.452
19	215.336	6.8460 (+03)	1.1008 (−01)	1.4251 (−04)	4.6436 (−06)	294.726
20	216.650	5.5293 (+03)	8.8910 (−02)	1.5989 (−04)	4.6617 (−06)	295.069
25	221.552	2.5492 (+03)	4.0084 (−02)	3.6135 (−04)	4.7602 (−06)	298.389
30	226.509	1.9703 (+03)	1.8410 (−02)	8.0134 (−04)	4.8593 (−06)	301.709
35	238.430	1.1287 (+03)	1.1203 (−02)	2.4038 (−03)	5.0944 (−06)	309.449
40	250.350	2.8714 (+02)	3.9957 (−03)	4.0063 (−03)	5.3295 (−06)	317.189
45	260.500	1.8346 (+02)	2.5113 (−03)	1.0299 (−02)	5.5255 (−06)	323.494
50	270.650	7.9779 (+01)	1.0269 (−03)	1.6591 (−02)	5.7214 (−06)	329.799
55	263.211	5.1120 (+01)	6.6640 (−04)	3.4916 (−02)	5.5782 (−06)	325.203
60	255.772	2.2461 (+01)	3.0592 (−04)	5.3241 (−02)	5.4349 (−06)	320.606
65	237.736	7.3991 (+01)	1.9673 (−04)	1.0878 (−01)	5.0789 (−06)	308.873
70	219.700	5.5205 (+00)	8.7535 (−05)	1.6431 (−01)	4.7230 (−06)	297.139
80	180.650	1.0366 (+00)	1.9990 (−05)	6.0850 (−01)	3.9250 (−06)	269.440

*h = altitude, T = temperature, p = pressure, ρ = mass density, ν = kinematic viscosity, k = thermal conductivity, and v_L = speed of sound.
SOURCE: "U.S. Standard Atmosphere 1962," published by the U.S. Committee on the Extension of the Standard Atmosphere, Washington, D.C., 1962.

A.86 SPEED OF SOUND v_L IN SOLIDS
AT TEMPERATURE T*

Name	T °C	v_L m-s^{-1}	Name	T °C	v_L m-s^{-1}
Aluminum, rolled	20	5,050	Lead, rolled	20	1,230
Beryllium	20	2,840	Lucite	20	1,850
Brass, yellow	20	3,420	Magnesium, annealed	20	4,770
Brick	20	3,640	Marble	20	3,800
Bronze	20	3,530	Nickel	20	4,900
Cadmium	20	2,320	Nylon	20	1,800
Clay, dry	10	1,600	Platinum	20	2,800
Cobalt	20	4,500	Polyethylene	20	940
Constantan	20	4,280	Polystyrene	20	1,830
Copper, rolled	20	3,600	Rubber, soft	20	400
Copper, rolled	100	3,250	Rubber, hard	20	1600
Copper, rolled	200	2,950	Silver, annealed	20	2,680
Cork	20	500	Slate	20	4,550
Glass, flint	20	4,400	Steel, low-strength	20	5,000
Glass, crown	20	5,250	Steel, high-strength	20	6,000
Gold, hard	20	2,100	Tin, rolled	20	2,690
Granite	20	6,000	Titanium	20	5,050
Gypsum	20	2,300	Tungsten, drawn	20	4,950
Ice	0	3,250	Tungsten carbide	20	6,180
Iron, cast	20	5,200	Wood, soft	20	3,200
Iron, cast	100	5,000	Wood, hard	20	4,800
Iron, cast	200	4,800	Zinc, rolled	20	3,780

A.87 SPEED OF SOUND v_L IN LIQUIDS
AT PRESSURE $p = 1$ atm AND TEMPERATURE T*

Name	T, °C	v_L, m-s^{-1}	Name	T, °C	v_L, m-s^{-1}
Acetone	20	1,203	Petroleum	20	1,222
n-Butyl alcohol	0	1,327	n-Propyl alcohol	0	1,295
n-Butyl alcohol	20–50	1,208	n-Propyl alcohol	20–50	1,222
Chloroform	20	975	Seawater	0	1,450
Ethyl alcohol	0	1,232	Seawater	20	1,520
Ethyl alcohol	20–50	1,115	Turpentine	20	1,325
Mercury	20	1,425	Water	0	1,401
Methyl alcohol	0	1,192	Water	20	1,478
Methyl alcohol	20–50	1,084	Water	50–100	1,543

*For definition and applications, refer to Sec. 9.03-2.

A.88 SPEED OF SOUND v_L IN GASES AT PRESSURE $p = 1$ atm AND TEMPERATURE $T = 0°C$*

Name	Formula	v_L, m-s^{-1}	Name	Formula	v_L, m-s^{-1}
Ammonia	NH_2	415	Illuminating gas	....	449
Argon	A	319	Methane	CH_4	430
Carbon dioxide	CO_2	260	Neon	Ne	434
Carbon monoxide	CO	338	Nitrogen	N_2	334
Chlorine	Cl_2	206	Nitric oxide	NO	334
Ethylene	C_2H_4	317	Oxygen	O_2	317
Helium	He	968	Steam, saturated	H_2O	402

*For definition and applications, refer to Sec. 9.03–2.

A.89 NORMAL-INCIDENCE REFLECTION OF POLISHED SURFACES AT $T = 20°C$

The reflection coefficients μ_r (Sec. 10.02–2d) of light rays of wavelength λ at normal incidence ($\theta_i = 0$) from polished plane surfaces are given below. μ_r is dimensionless.

Surface	λ, 10^{-9} m						
	400	450	500	550	600	650	700
Aluminum	0.915	0.925	0.920	0.915	0.910	0.905	0.890
Antimony	0.620	0.650	0.635	0.660	0.665	0.655	0.660
Beryllium	0.475	0.465	0.460	0.465	0.470	0.475	0.480
Bismuth	0.575	0.565	0.575	0.600	0.635	0.680	0.700
Cadmium	0.730	0.735	0.760	0.770	0.765	0.780	0.750
Chromium	0.420	0.490	0.520	0.540	0.560	0.565	0.570
Cobalt	0.530	0.560	0.590	0.620	0.625	0.630	0.635
Copper	0.300	0.380	0.435	0.480	0.715	0.800	0.830
Gold	0.360	0.385	0.415	0.740	0.845	0.890	0.925
Iridium	0.570	0.540	0.515	0.540	0.570	0.585	0.810
Iron	0.530	0.560	0.575	0.590	0.594	0.595	0.600
Lead	0.550	0.590	0.620	0.650	0.660	0.670	0.675
Magnesium	0.680	0.730	0.780	0.840	0.875	0.900	0.935
Molybdenum	0.440	0.450	0.455	0.465	0.475	0.485	0.495
Nickel	0.530	0.565	0.595	0.620	0.645	0.660	0.675
Palladium	0.460	0.475	0.520	0.490	0.550	0.610	0.680
Platinum	0.540	0.580	0.560	0.570	0.590	0.640	0.690
Rhodium	0.820	0.810	0.800	0.795	0.790	0.810	0.830
Silver	0.875	0.915	0.950	0.950	0.955	0.955	0.960
Steel	0.530	0.540	0.550	0.555	0.560	0.567	0.570
Tantalum	0.290	0.335	0.380	0.410	0.450	0.500	0.550
Tungsten	0.410	0.425	0.435	0.450	0.480	0.485	0.490

A.90 NORMAL-INCIDENCE REFLECTION
OF MIRROR SURFACES AT $T = 20°C$

The reflection coefficients μ_r (Sec. 10.02–2d) of light rays of wavelength λ at normal incidence ($\theta_i = 0$) from freshly evaporated mirror coatings are given below. μ_r is dimensionless.

Coating	λ, 10^{-9} m						
	400	450	500	550	600	650	700
Aluminum	0.925	0.925	0.915	0.915	0.910	0.905	0.895
Copper	0.390	0.500	0.580	0.620	0.715	0.850	0.910
Gold	0.360	0.380	0.475	0.780	0.880	0.920	0.950
Mirror alloy	0.835	0.830	0.830	0.825	0.830	0.830	0.835
Nickel	0.570	0.595	0.610	0.625	0.650	0.665	0.685
Platinum	0.590	0.620	0.650	0.670	0.700	0.715	0.730
Rhodium	0.775	0.760	0.765	0.780	0.795	0.810	0.820
Silver	0.910	0.940	0.950	0.970	0.975	0.980	0.985

A.91 NORMAL-INCIDENCE REFLECTION
OF NATURAL SURFACES AT $T = 20°C$

The reflection coefficients μ_r (Sec. 10.02–2d) of incandescent light rays at normal incidence ($\theta_i = 0$) from natural plane surfaces are given below.

Material	μ_r	Material	μ_r
Black paper, mat	0.05	Red paper, glossy	0.08
Black paper, glossy	0.06	Snow, fresh	0.92
Blue paper, mat	0.50	Stellite, specular	0.62
Blue paper, glossy	0.60	Talcum, specular	0.88
Green paper, mat	0.45	White paper, mat	0.82
Green paper, glossy	0.50	White paper, glossy	0.90
Porcelain, mat	0.73	Yellow paper, mat	0.40
Porcelain, glossy	0.78	Yellow paper, glossy	0.50
Red paper, mat	0.07	Wheat flour, specular	0.90

A.92 INDEX OF REFRACTION μ_m OF GLASSES AT $T = 20°C$

The refractive indices μ_m (Sec. 10.02–4) of commercial and optical glasses for light rays of wavelength $\lambda_R = 656 \times 10^{-9}$ m, $\lambda_Y = 589 \times 10^{-9}$ m, and $\lambda_B = 486 \times 10^{-9}$ m are given below. μ_m is dimensionless.

Glass	λ_R	λ_Y	λ_B	Glass	λ_R	λ_Y	λ_B
Barium crown	1.567	1.569	1.577	Heavy crown	1.596	1.610	1.619
Barium flint	1.577	1.580	1.587	Heavy flint	1.663	1.672	1.687
Borosilicate crown	1.514	1.517	1.523	Light crown	1.516	1.517	1.524
Borosilicate flint	1.521	1.528	1.534	Light flint	1.571	1.574	1.585
Commercial plate	1.516	1.523	1.529	Vycor optical	1.457	1.458	1.457

A.93 AVERAGE INDEX OF REFRACTION $\bar{\mu}_m$ OF INDUSTRIAL PLASTICS AT $T = 20°C$

Plastic	$\bar{\mu}_m$, dimensionless	Plastic	$\bar{\mu}_m$, dimensionless
Acetal polymer	1.48	Polyester, allyl	1.55
Acrylic, type I	1.50	Polyester,	
Acrylic, type II	1.57	styrene	1.59
Epoxy, general	1.61	Polyethylene	1.52
Epoxy,		Polyvinyl	1.48
heat-resistant	1.56	Silicone,	
Nylon, 6/6, 11, 6	1.58	mineral	1.48
Polycarbonate	1.59	Silicone, glass	1.43
Phenolic	1.61	Vinyl polymer	1.49

A.94 INDEX OF REFRACTION μ_m OF WATER AT TEMPERATURE T

λ, 10^{-9} m	T, °C	μ_m	λ, 10^{-9} m	T, °C	μ_m	λ, 10^{-9} m	T, °C	μ_m
300	20	1.357	480	0	1.338	580	40	1.335
430	0	1.342	480	20	1.337	650	0	1.332
430	20	1.340	480	40	1.335	650	20	1.331
430	40	1.338	580	0	1.334	760	20	1.329
430	80	1.330	580	20	1.333	1020	20	1.325

A.95 AVERAGE INDEX OF REFRACTION $\bar{\mu}_m$ OF LIQUIDS AT $T = 20$°C

Liquid	$\bar{\mu}_m$	Liquid	$\bar{\mu}_m$	Liquid	$\bar{\mu}_m$
Acetaldehyde	1.332	Chloroform	1.447	Naphthalene (90°C)	1.583
Acetone	1.359	Ethyl acetate	1.373	Nitrobenzene	1.553
Alcohol, allyl	1.413	Ethyl bromide	1.424	n-Octane	1.401
Alcohol, ethyl	1.370	Ethyl iodide	1.518	n-Pentane	1.358
Alcohol, methyl	1.331	Ethyl nitrate	1.386	n-Propyl alcohol	1.386
Aniline	1.586	Formic acid	1.371	Seawater	1.442
Benzene	1.501	Glycerol	1.473	Styrene	1.548
Carbon disulfide	1.669	n-Heptane	1.384	Toluene	1.497
Carbon tetrachloride	1.465	n-Hexane	1.395	Turpentine	1.474

A.96 AVERAGE INDEX OF REFRACTION $\bar{\mu}_m$ OF GASES AT $p = 1$ atm AND $T = 20$°C

Gas	$\bar{\mu}_m$	Gas	$\bar{\mu}_m$	Gas	$\bar{\mu}_m$
Air	1.000 292	Carbon monoxide	1.000 337	Methane	1.000 442
Acetylene	1.001 850	Chlorine	1.000 770	Nitrogen	1.000 298
Ammonia	1.000 376	Helium	1.000 036	Oxygen	1.000 271
Argon	1.000 280	Hydrogen	1.000 132	Steam	1.000 259
Carbon dioxide	1.000 450	Hydrogen sulfide	1.000 642	Sulfur dioxide	1.000 686

A.97 PROPERTIES OF PLANE SECTIONS

Section	Centroid	Area	Moments of inertia
(a) Rectangle	$x_{0c} = b/2$	$A = bh$	$I_{cx} = Ah^2/12$
	$y_{0c} = h/2$		$I_{cy} = Ab^2/12$
(b) Triangle	$x_{0c} = (b + b)/3$	$A = bh/2$	$I_{cx} = Ah^2/18$
	$y_{0c} = h/3$	$b = b_1 + b_2$	$I_{0x} = Ah^2/6$
(c) Trapezoid	$x_{0c} = a/2$	$A = ch/2$	$I_{cx} = \dfrac{A(c^2 + 2ab)h}{18c}$
	$y_{0c} = \dfrac{h(b + c)}{3c}$	$c = a + b$	$I_{cy} = \dfrac{A(c^2 - 2ab)}{24}$
(d) Regular polygon	$x_{0c} = 0$	$A = \dfrac{na^2}{4 \tan \alpha}$	$I_{cx} = \dfrac{A(12r^2 + a^2)}{48}$
n = number of sides	$y_{0c} = 0$	$a = 2\sqrt{R^2 - r^2}$	$I_{cy} = I_{cx}$
(e) Circle	$x_{0c} = 0$	$A = \pi R^2$	$I_{cx} = Ar^2/4$
	$y_{0c} = 0$		$I_{cy} = I_{cx}$
(f) Ellipse	$x_{0c} = 0$	$A = ab\pi$	$I_{cx} = Ab^2/4$
	$y_{0c} = 0$		$I_{cy} = Aa^2/4$

Section	Centroid	Area	Moments of inertia
(a) Quarter of circular area	$x_{0c} = \dfrac{4R}{3\pi}$ $y_{0c} = \dfrac{4R}{3\pi}$	$A = \pi R^2/4$	$I_{0x} = A^2/\pi$ $I_{0y} = I_{0x}$
(b) Quarter of elliptic area	$x_{0c} = \dfrac{4a}{3\pi}$ $y_{0c} = \dfrac{4b}{4\pi}$	$A = ab\pi/4$	$I_{0x} = A^2 b/a\,\pi$ $I_{0y} = A^2 a/b\,\pi$
(c) Half of parabolic area	$x_{0c} = 3a/5$ $y_{0c} = 3h/8$	$A = 2ah/3$	$I_{0x} = Ah^2/5$ $I_{0y} = 3Aa^2/7$
(d) Circular segment	$x_{0c} = \dfrac{2R \sin \alpha}{3\alpha}$ $y_{0c} = 0$	$A = \alpha R^2$ α in rad	$I_{0x} = \dfrac{AR^2}{8}\left(2 - \dfrac{\sin 2\alpha}{\alpha}\right)$ $I_{0y} = \dfrac{AR^2}{8}\left(2 + \dfrac{\sin 2\alpha}{\alpha}\right)$
(e) Circular sector	$x_{0c} = \dfrac{2R \sin^3 \alpha}{3(\alpha - \sin \alpha \cos \alpha)}$ $y_{0c} = 0$	$A = R^2(\alpha - \sin \alpha \cos \alpha)$ α in rad	$I_{0x} = \dfrac{AR^2}{4}\left(1 - \dfrac{2B}{3}\right)$ $I_{0y} = \dfrac{AR^2}{4}(1 + 2B)$ $B = \dfrac{R^2 \sin^3 \alpha \cos \alpha}{A}$
(f) Sector of annulus	$x_{0c} = \dfrac{2(R^3 - r^3)\sin \alpha}{3\alpha(R^2 - r^2)}$ $y_{0c} = 0$	$A = \pi(R^2 - r^2)$ α in rad	$I_{0x} = \dfrac{A(R^2 + r^2)}{8}\left(2 - \dfrac{\sin 2\alpha}{\alpha}\right)$ $I_{0y} = \dfrac{A(R^2 + r^2)}{8}\left(2 + \dfrac{\sin 2\alpha}{\alpha}\right)$

A.99 PROPERTIES OF HOMOGENEOUS SOLIDS

Solid	Centroid	Mass*	Mass moments of inertia
(a) Rectangular prism	$x_{0c} = 0$ $y_{0c} = 0$ $z_{0c} = h/2$	$m = abh\gamma$	$I_{cx} = m(b^2 + h^2)/12$ $I_{cy} = m(a^2 + h^2)/12$ $I_{cz} = m(a^2 + b^2)/12$
(b) Rectangular pyramid	$x_{0c} = 0$ $y_{0c} = 0$ $z_{0c} = h/4$	$m = abh\gamma/3$	$I_{cx} = m(4b^2 + 3h^2)/80$ $I_{cy} = m(4a^2 + 3h^2)/80$ $I_{cz} = (a^2 + b^2)/20$
(c) Rectangular tetrahedron	$x_{0c} = a/4$ $y_{0c} = b/4$ $z_{0c} = c/4$	$m = abc\gamma/6$	$I_{cx} = 3m(b^2 + c^2)80$ $I_{cy} = 3m(a^2 + c^2)80$ $I_{cz} = 3m(a^2 + b^2)80$
(d) Straight rod	$x_{0c} = 0$ $y_{0c} = 0$ $z_{0c} = 0$	$m = $ given value	$I_{cx} = 0$ $I_{cy} = m\ell^2/12$ $I_{cz} = m\ell^2/12$
(e) Quarter circular rod	$x_{0c} = 2R/\pi$ $y_{0c} = 2R/\pi$ $z_{0c} = 0$	$m = $ given value	$I_{cx} = mR^2/10$ $I_{cy} = mR^2/10$ $I_{cz} = mR^2/5$

*γ = mass density.

A.100 PROPERTIES OF HOMOGENEOUS SOLIDS

Solid	Centroid	Mass*	Mass moments of inertia
(a) Hemisphere	$x_{0c} = 0$ $y_{0c} = 0$ $z_{0c} = 3R/8$	$m = 2\pi R^3 \gamma/3$	$I_{cx} = 83mR^2/320$ $I_{cy} = 83mR^2/320$ $I_{cz} = 2mR^2/5$
(b) Right circular cylinder	$x_{0c} = 0$ $y_{0c} = 0$ $z_{0c} = h/2$	$m = \pi R^2 h \gamma$	$I_{cx} = (3R^2 + h^2)m/12$ $I_{cy} = (3R^2 + h^2)m/12$ $I_{cz} = mR^2/2$
(c) Right circular cone	$x_{0c} = 0$ $y_{0c} = 0$ $z_{0c} = h/3$	$m = \pi R^2 h \gamma/3$	$I_{cx} = 3(4R^2 + h^2)m/80$ $I_{cy} = 3(4R^2 + h^2)m/80$ $I_{cz} = 3mR^2/10$
(d) Rotational paraboloid	$x_{0c} = 0$ $y_{0c} = 0$ $z_{0c} = h/3$	$m = \pi R^2 h \gamma/2$	$I_{cx} = (3R^2 + h^2)m/18$ $I_{cy} = (3R^2 + h^2)m/18$ $I_{cz} = 2R^2 m/6$
(e) Semiellipsoid	$x_{0c} = 0$ $y_{0c} = 0$ $z_{0c} = 3h/8$	$m = 2\pi abh\gamma/3$	$I_{cx} = (64b^2 + 19h^2)m/320$ $I_{cy} = (64a^2 + 19h^2)m/320$ $I_{cz} = (a^2 + b^2)m/5$

*γ = mass density.

A.101 PROPERTIES OF LARGE ROLLED WIDE-FLANGE STEEL SECTIONS IN FPS UNITS*

The wide-flange sections introduced below are designated by the letter W followed by the nominal depth in inches and the weight in pound-force per foot. For the definition of I, Z, and k refer to Secs. 2.07–6 and 4.05–1.

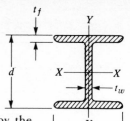

Designation	Area A, in^2	Depth d, in	Flange Width b_f, in	Flange Thickness t_f, in	Web Thickness t_w, in	Axis X-X I_{xx} in^4	Axis X-X Z_{xx} in^3	Axis X-X k_{xx} in	Axis Y-Y I_{yy} in^4	Axis Y-Y Z_{yy} in^3	Axis Y-Y k_{yy} in
W36 × 300	88.3	36.74	16.655	1.680	0.945	20300	1110	15.2	1300	156	3.83
135	39.7	35.55	11.950	0.790	0.600	7800	439	14.0	225	37.7	2.38
W33 × 201	59.1	33.68	15.745	1.150	0.715	11500	684	14.0	749	95.2	3.56
118	34.7	32.86	11.480	0.740	0.550	5900	359	13.0	187	32.6	2.32
W30 × 173	50.8	30.44	14.985	1.065	0.655	8200	539	12.7	598	79.8	3.43
99	29.1	29.65	10.450	0.670	0.520	3990	269	11.7	128	24.5	2.10
W27 × 146	42.9	27.38	13.965	0.975	0.605	5630	411	11.4	443	63.5	3.21
84	24.8	26.71	9.960	0.640	0.460	2850	213	10.7	106	21.2	2.07
W24 × 104	30.6	24.06	12.750	0.750	0.500	3100	258	10.1	259	40.7	2.91
68	20.1	23.73	8.965	0.585	0.415	1830	154	9.55	70.4	15.7	1.87
W21 × 101	29.8	21.36	12.290	0.800	0.500	2420	227	9.02	248	40.3	2.89
62	18.3	20.99	8.240	0.615	0.400	1330	127	8.54	57.5	13.9	1.77
44	13.0	20.66	6.500	0.450	0.350	843	81.6	8.06	20.7	6.36	1.26
W18 × 106	31.1	18.73	11.200	0.940	0.590	1910	204	7.84	220	39.4	2.66
76	22.3	18.21	11.035	0.680	0.425	1330	146	7.73	152	27.6	2.61
50	14.7	17.99	7.495	0.570	0.355	800	88.9	7.38	40.1	10.7	1.65
35	10.3	17.70	6.000	0.425	0.300	510	57.6	7.04	15.3	5.12	1.22
W16 × 77	22.6	16.52	10.295	0.760	0.455	1110	134	7.00	138	26.9	2.47
57	16.8	16.43	7.120	0.715	0.430	758	92.2	6.72	43.1	12.1	1.60
40	11.8	16.01	6.995	0.505	0.305	518	64.7	6.63	28.9	8.25	1.57
31	9.12	15.88	5.525	0.440	0.275	375	47.2	6.41	12.4	4.49	1.17
26	7.68	15.69	5.500	0.345	0.250	301	38.4	6.26	9.59	3.49	1.12
W14 × 370	109	17.92	16.475	2.660	1.655	5440	607	7.07	1990	241	4.27
145	42.7	14.78	15.550	1.090	0.680	1710	232	6.33	677	87.3	3.98
82	24.1	14.31	10.130	0.855	0.510	882	123	6.05	148	29.3	2.48
68	20.0	14.04	10.035	0.720	0.415	723	103	6.01	121	24.2	2.46
53	15.6	13.92	8.060	0.660	0.370	541	77.8	5.89	57.7	14.3	1.92
43	12.6	13.66	7.995	0.530	0.305	428	62.7	5.82	45.2	11.3	1.89
38	11.2	14.10	6.770	0.515	0.310	385	54.6	5.88	26.7	7.88	1.55
30	8.85	13.84	6.730	0.385	0.270	291	42.0	5.73	19.6	5.82	1.49
26	7.69	13.91	5.025	0.420	0.255	245	35.3	5.65	8.91	3.54	1.08
22	6.49	13.74	5.000	0.335	0.230	199	29.0	5.54	7.00	2.80	1.04

SOURCE: F. P. Beer and E. R. Johnston, Jr., "Mechanics of Materials," McGraw-Hill, New York, 1981, p. 586, based on "Manual of Steel Construction," 8th ed., American Institute of Steel Construction, Chicago, 1980.

A.102 PROPERTIES OF LARGE ROLLED WIDE-FLANGE STEEL SECTIONS IN SI UNITS*

The wide-flange sections introduced below are designated by the letter W followed by the nominal depth in millimeters and the weight in newtons per meter. For the definition of I, Z, and k refer to Secs. 2.07–6 and 4.05–1.

Designation	Area A, mm^2	Depth d, mm	Flange Width b_f, mm	Flange Thickness t_f, mm	Web Thickness t_w, mm	Axis X-X I_{xx} 10^6 mm^4	Axis X-X Z_{xx} 10^3 mm^3	Axis X-X k_{xx} mm	Axis Y-Y I_{yy} 10^6 mm^4	Axis Y-Y Z_{yy} 10^3 mm^3	Axis Y-Y k_{yy} mm
W920 × 4378	57000	933	423	42.7	24.0	8450	18110	386	541	2560	97.3
1970	25600	903	304	20.1	15.2	3250	7200	356	93.7	616	60.5
W840 × 2933	38100	855	400	29.2	18.2	4790	11200	356	312	1560	90.4
1722	22400	835	292	18.8	14.0	2460	5890	330	77.8	533	58.9
W760 × 2525	32800	773	381	27.1	16.6	3410	8820	323	249	1307	87.1
1445	18800	753	265	17.0	13.2	1660	4410	297	53.3	402	53.3
W690 × 2131	27700	695	355	24.8	15.4	2340	6730	290	184.4	1039	81.5
1226	16000	678	253	16.3	11.7	1186	3500	272	44.1	349	52.6
W610 × 1518	19700	611	324	19.0	12.7	1290	4220	256	107.8	665	73.9
992	13000	603	228	14.9	10.5	762	2530	243	29.3	257	47.5
W530 × 1474	19200	543	312	20.3	12.7	1007	3710	229	103.2	662	73.4
904	11800	533	209	15.6	10.2	554	2080	217	23.9	229	45.0
642	8390	525	165	11.4	8.9	351	1337	205	8.62	104.5	32.0
W460 × 1547	20100	476	284	23.9	15.0	795	3340	199.1	91.6	645	67.6
1109	14400	463	280	17.3	10.8	554	2390	196.3	63.3	452	66.3
730	9480	457	190	14.5	9.0	333	1457	187.5	16.69	175.7	41.9
511	6650	450	152	10.8	7.6	212	942	178.8	6.37	83.8	31.0
W410 × 1124	14600	420	261	19.3	11.6	462	2200	177.8	57.4	440	62.7
832	10800	417	181	18.2	10.9	316	1516	170.7	17.94	198.2	40.6
584	7610	407	178	12.8	7.7	216	1061	168.4	12.03	135.2	39.9
452	5880	403	140	11.2	7.0	156.1	775	162.8	5.16	73.7	29.7
379	4950	399	140	8.8	6.4	125.3	628	159.0	3.99	57.0	28.4
W360 × 5400	70300	455	418	67.6	42.0	2260	9930	179.6	828	3960	108.5
2116	27500	375	394	27.7	17.3	712	3800	160.8	282	1431	101.1
1197	15500	363	257	21.7	13.0	367	2020	153.7	61.6	479	63.0
992	12900	357	255	18.3	10.5	301	1686	152.7	50.4	395	62.5
773	10100	354	205	16.8	9.4	225	1271	149.6	24.0	234	48.8
628	8130	347	203	13.5	7.7	178.1	1027	147.8	18.81	185.3	48.0
555	7230	358	172	13.1	7.9	160.2	895	149.4	11.11	129.2	39.4
438	5710	352	171	9.8	6.9	121.1	688	145.5	8.16	95.4	37.8
379	4960	353	128	10.7	6.5	102.0	578	143.5	3.71	58.0	27.4
321	4190	349	127	8.5	5.8	82.8	474	140.7	2.91	45.8	26.4

*Ibid., p. 434.

A.103 PROPERTIES OF SMALL ROLLED WIDE-FLANGE STEEL SECTIONS IN FPS UNITS*

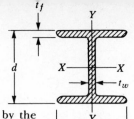

The wide-flange sections introduced below are designated by the letter W followed by the nominal depth in inches and the weight in pound-force per foot. For the definition of I, Z, and k refer to Secs. 2.07–6 and 4.05–1.

Designation	Area A, in^2	Depth d, in	Flange Width b_f, in	Flange Thickness t_f, in	Web Thickness t_w, in	Axis X-X I_{xx} in^4	Axis X-X Z_{xx} in^3	Axis X-X k_{xx} in	Axis Y-Y I_{yy} in^4	Axis Y-Y Z_{yy} in^3	Axis Y-Y k_{yy} in
W12 × 96	28.2	12.71	12.160	0.900	0.550	833	131	5.44	270	44.4	3.09
72	21.1	12.25	12.040	0.670	0.430	597	97.4	5.31	195	32.4	3.04
50	14.7	12.19	8.080	0.640	0.370	394	64.7	5.18	56.3	13.9	1.96
40	11.8	11.94	8.005	0.515	0.295	310	51.9	5.13	44.1	11.0	1.93
35	10.3	12.50	6.560	0.520	0.300	285	45.6	5.25	24.5	7.47	1.54
30	8.79	12.34	6.520	0.440	0.260	238	38.6	5.21	20.3	6.24	1.52
26	7.65	12.22	6.490	0.380	0.230	204	33.4	5.17	17.3	5.34	1.51
22	6.48	12.3¹	4.030	0.425	0.260	156	25.4	4.91	4.66	2.31	0.848
16	4.71	11.99	3.990	0.265	0.220	103	17.1	4.67	2.82	1.41	0.773
W10 × 112	32.9	11.36	10.415	1.250	0.755	716	126	4.66	236	45.3	2.68
68	20.0	10.40	10.130	0.770	0.470	394	75.7	4.44	134	26.4	2.59
54	15.8	10.09	10.030	0.615	0.370	303	60.0	4.37	103	20.6	2.56
45	13.3	10.10	8.020	0.620	0.350	248	49.1	4.33	53.4	13.3	2.01
39	11.5	9.92	7.985	0.530	0.315	209	42.1	4.27	45.0	11.3	1.98
33	9.71	9.73	7.960	0.435	0.290	170	35.0	4.19	36.6	9.20	1.94
30	8.84	10.47	5.810	0.510	0.300	170	32.4	4.38	16.7	5.75	1.37
22	6.49	10.17	5.750	0.360	0.240	118	23.2	4.27	11.4	3.97	1.33
19	5.62	10.24	4.020	0.395	0.250	96.3	18.8	4.14	4.29	2.14	0.874
15	4.41	9.99	4.000	0.270	0.230	68.9	13.8	3.95	2.89	1.45	0.810
W8 × 58	17.1	8.75	8.220	0.810	0.510	228	52.0	3.65	75.1	18.3	2.10
48	14.1	8.50	8.110	0.685	0.400	184	43.3	3.61	60.9	15.0	2.08
40	11.7	8.25	8.070	0.560	0.360	146	35.5	3.53	49.1	12.2	2.04
35	10.3	8.12	8.020	0.495	0.310	127	31.2	3.51	42.6	10.6	2.03
31	9.13	8.00	7.995	0.435	0.285	110	27.5	3.47	37.1	9.27	2.02
28	8.25	8.06	6.535	0.465	0.285	98.0	24.3	3.45	21.7	6.63	1.62
24	7.08	7.93	6.495	0.400	0.245	82.8	20.9	3.42	18.3	5.63	1.61
21	6.16	8.28	5.270	0.400	0.250	75.3	18.2	3.49	9.77	3.71	1.26
18	5.26	8.14	5.250	0.330	0.230	61.9	15.2	3.43	7.97	3.04	1.23
15	4.44	8.11	4.015	0.315	0.245	48.0	11.8	3.29	3.41	1.70	0.876
13	3.84	7.99	4.000	0.255	0.230	39.6	9.91	3.21	2.73	1.37	0.843
W6 × 25	7.34	6.38	6.080	0.455	0.320	53.4	16.7	2.70	17.1	5.61	1.52
20	5.87	6.20	6.020	0.365	0.260	41.4	13.4	2.66	13.3	4.41	1.50
16	4.74	6.28	4.030	0.405	0.260	32.1	10.2	2.60	4.43	2.20	0.967
12	3.55	6.03	4.000	0.280	0.230	22.1	7.31	2.49	2.99	1.50	0.918
9	2.68	5.90	3.940	0.215	0.170	16.4	5.56	2.47	2.20	1.11	0.905
W5 × 19	5.54	5.15	5.030	0.430	0.270	26.2	10.2	2.17	9.13	3.63	1.28
16	4.68	5.01	5.000	0.360	0.240	21.3	8.51	2.13	7.51	3.00	1.27
W4 × 13	3.83	4.16	4.060	0.345	0.280	11.3	5.46	1.72	3.86	1.90	1.00

*Ibid., p. 439.

A.104 PROPERTIES OF SMALL ROLLED WIDE-FLANGE STEEL SECTIONS IN SI UNITS*

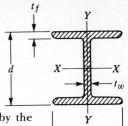

The wide-flange sections introduced below are designated by the letter W followed by the nominal depth in millimeters and the weight in newtons per meter. For the definition of I, Z, and k refer to Secs 2.07–6 and 4.05–1.

Designation	Area A, mm²	Depth d, mm	Flange Width b_f, mm	Flange Thickness t_f, mm	Web Thickness t_w, mm	Axis X–X I_{xx} 10⁶ mm⁴	Axis X–X Z_{xx} 10³ mm³	Axis X–X k_{xx} mm	Axis Y–Y I_{yy} 10⁶ mm	Axis Y–Y Z_{yy} 10³ mm³	Axis Y–Y k_{yy} mm
W310 × 1401	18200	323	309	22.9	14.0	347	2150	138.2	112.4	728	78.5
1051	13600	311	306	17.0	10.9	248	1595	134.9	81.2	531	77.2
730	9480	310	205	16.3	9.4	164.0	1058	131.6	23.4	228	49.8
584	7610	303	203	13.1	7.5	129.0	851	130.3	18.36	180.9	49.0
511	6650	317	167	13.2	7.6	118.6	748	133.4	10.20	122.2	39.1
438	5670	313	166	11.2	6.6	99.1	633	132.3	8.45	101.8	38.6
379	4940	310	165	9.7	5.8	84.9	548	131.3	7.20	87.3	38.4
321	4180	313	102	10.8	6.6	64.9	415	124.7	1.940	38.0	21.5
234	3040	305	101	6.7	5.6	42.9	281	118.6	1.174	23.2	19.63
W250 × 1635	21200	289	265	31.8	19.2	298.0	2060	118.4	98.2	741	68.1
992	12900	264	257	19.6	11.9	164.0	1242	112.8	55.8	434	65.8
788	10200	256	255	15.6	9.4	126.1	985	111.0	42.8	336	65.0
657	8580	257	204	15.7	8.9	103.2	803	110.0	22.2	218	51.1
569	7420	252	203	13.5	8.0	87.0	690	108.5	18.73	184.5	50.3
482	6260	247	202	11.0	7.4	70.8	573	106.4	15.23	150.8	49.3
438	5700	266	148	13.0	7.6	70.8	532	111.3	6.95	93.9	34.8
321	4190	258	146	9.1	6.1	49.1	381	108.5	4.75	65.1	33.8
277	3630	260	102	10.0	6.4	40.1	308	105.2	1.796	35.2	22.2
219	2850	254	102	6.9	5.8	28.7	226	100.3	1.203	23.6	20.6
W200 × 846	11000	222	209	20.6	13.0	94.9	855	92.7	31.3	300	53.3
701	9100	216	206	17.4	10.2	76.6	709	91.7	25.3	246	52.8
584	7550	210	205	14.2	9.1	60.8	579	89.7	20.4	199.0	51.8
511	6650	206	204	12.6	7.9	52.9	514	89.2	17.73	173.8	51.6
452	5890	203	203	11.0	7.2	45.8	451	88.1	15.44	152.1	51.3
407	5320	205	166	11.8	7.2	40.8	398	87.6	9.03	108.8	41.1
350	4570	201	165	10.2	6.2	34.5	343	86.9	7.62	92.4	40.9
306	3970	210	134	10.2	6.4	31.3	298	88.6	4.07	60.7	32.0
263	3390	207	133	8.4	5.8	25.8	249	87.1	3.32	49.9	31.2
219	2860	206	102	8.0	6.2	20.0	194.2	83.6	1.419	27.8	22.3
190	2480	203	102	6.5	5.8	16.48	162.4	81.5	1.136	22.3	21.4
W150 × 365	4740	162	154	11.6	8.1	22.2	274	68.6	7.12	92.5	38.6
292	3790	157	153	9.3	6.6	17.23	219	67.6	5.54	72.4	38.1
234	3060	160	102	10.3	6.6	13.36	167.0	66.0	1.844	36.2	24.6
175	2290	153	102	7.1	5.8	9.20	120.3	63.2	1.245	24.4	23.3
131	1730	150	100	5.5	4.3	6.83	91.1	62.7	0.916	18.32	23.0
W130 × 277	3590	131	128	10.9	6.9	10.91	166.6	55.1	3.80	59.4	32.5
236	3040	127	127	9.1	6.1	8.87	139.7	54.1	3.13	49.3	32.3
W100 × 190	2470	106	103	8.8	7.1	4.70	88.7	43.7	1.607	31.2	25.4

*Ibid., p. 434.

A.105 PLASTIC PROPERTIES OF PLANE SECTIONS (Sec. 4.05–3)

Section	Plastic section modulus Z_P	Shape factor α_P
(a) Rectangle	$\dfrac{bh^2}{4}$	1.5
(b) Triangle	$\dfrac{2-\sqrt{2}}{3}\,bh^2$	2.34
(c) Circle	$\dfrac{4r^3}{3}$	1.7
(d) Hollow circle	$\dfrac{4R^3}{3}\left[1-\left(1-\dfrac{R-r}{R}\right)^3\right]$	$\dfrac{16}{3\pi}\dfrac{1-\left(\dfrac{r}{R}\right)^3}{1-\left(\dfrac{r}{R}\right)^4}$
(e) Hollow rectangle	$t<<h \qquad w<<b$ $\cong \dfrac{bh^2}{4}$	$\cong 1.12$
(f) I-section	$bt\,(h-t)+\dfrac{w}{4}\,(h-2t)^2$	$\cong 1.12$

Appendix B
CONVERSION TABLES

B.01 SHORT TABLE OF EQUIVALENTS*

$\pi \cong 22/7$	$e \cong 19/7$	$g \cong 510/52$ meters/second^2
1 foot $\cong 64/210$ meter	1 pound $\cong 29/64$ kilogram	
1 bar = 10 newtons/centimeter^2 $\cong 52/51$ technical atmosphere $\cong 73/74$ physical atmosphere	1 Btu $\cong 778$ feet $\times$ pound-force $\cong 252$ calories $\cong 1,055$ joules	
1 kilogram-force/meter$^2 \cong 1/703$ pound-force/inch^2		

*The error of the equivalents equated by $\cong$ is less than 0.2%.

B.02 GLOSSARY OF SYMBOLS OF UNITS

Symbols beginning with lowercase letters

Symbol	Name	Unit	Section Definition	Section Conversion
a	atto = 10^{-18}	Prefix	B.07	B.07
abA	abampere	Electric current	B.40	B.40
abC	abcoulomb	Electric charge	B.40	B.40
abF	abfarad	Electric capacitance	B.42	B.42
abH	abhenry	Inductance	B.42	B.42
abV	abvolt	Voltage, potential	B.41	B.41
abΩ	abohm	Electric resistance	B.41	B.41
amu (or u)	atomic mass unit	Mass	6.02–1	B.16
at	technical atmosphere	Pressure	5.01–6	B.25
atm	physical atmosphere	Pressure	5.01–6	B.25
b	bar	Pressure	5.01–6	B.25
bbl	barrel	Volume	B.14	B.14
c	centi = 10^{-2}	Prefix	B.07	B.07
cal	calorie	Energy, work	6.01–4	B.29
cd	candela	Luminous intensity	10.01–3	B.10
cg	centigram = 10^{-2} g	Mass	B.16	B.16
cL	centiliter = 10^{-2} L	Volume	B.13	B.13
cm	centimeter = 10^{-2} m	Length	2.01–2	B.11
cm^2	square centimeter	Area	B.12	B.12
cm^3	cubic centimeter	Volume	B.13	B.13
cm-s^{-1}	centimeter per second	Speed	B.23	B.23
cm-s^{-2}	centimeter per second^2	Acceleration	B.24	B.24
cs^{-1}	cycle per second = 1 Hz	Frequency	8.04–1	—
cP	centipoise	Absolute viscosity	5.01–5	B.37
cS	centistoke	Kinematic viscosity	5.01–5	B.38
d	day	Time	3.01–2	B.21
d	deci = 10^{-1}	Prefix	B.07	B.07
dB	decibel = 10^{-1}B	Sound intensity	9.04–2	—
deg	degree	Plane angle	2.01–3	B.22
dg	decigram = 10^{-1} g	Mass	B.16	B.16
dL	deciliter = 10^{-1} L	Volume	B.14	B.14
dm	decimeter = 10^{-1} m	Length	2.01–2	B.11
dyn	dyne	Force	3.02–4	B.15
dyn-cm	dyne $\times$ centimeter	Moment	B.17	B.17
dyn-cm^{-2}	dyne per centimeter^2	Pressure	B.25	B.25
erg	erg = dyne $\times$ centimeter	Energy, work	3.05–1	B.29
erg-s^{-1}	erg per second	Power	B.30	B.30
erg-cm^{-1}	erg per centimeter	Force	B.15	B.15
eV	electronvolt	Energy, work	7.02–3	B.29
f	femto = 10^{-15}	Prefix	B.07	B.07
ft	foot	Length	2.01–2	B.11
ft^2	square foot	Area	B.12	B.12
ft^3	cubic foot	Volume	B.13	B.13
ft-lbf	foot $\times$ pound-force	Energy, work	3.05–1	B.29
ft-sec^{-1}	foot per second	Speed	3.01–3	B.23
ft-sec^{-2}	foot per second^2	Acceleration	3.01–3	B.24
g	gram	Mass	3.02–3	B.16
g-cm^{-3}	gram per centimeter^3	Mass density	5.01–2	B.26

B.03 GLOSSARY OF SYMBOLS OF UNITS

Symbols beginning with lowercase letters

Symbol	Name	Unit	Section Definition	Section Conversion
gal	gallon	Volume	B.14	B.14
gf	gram-force	Force	2.02–1	B.15
gf-cm^{-3}	gram-force per centimeter^3	Weight density	5.01–3	B.26
grad	grad = 0.9 degree	Plane angle	2.01–3	B.22
h	hecto = 10^2	Prefix	B.07	B.07
h	hour = 60 min	Time	3.01–2	B.21
hg	hectogram = 10^2 g	Mass	3.02–3	B.16
hl	hectoliter = 10^2 l	Volume	B.14	B.14
hm	hectometer = 10^2 m	Length	2.01–2	B.11
hp	horsepower	Power	3.05–6	B.30
hp-h	horsepower × hour	Work, energy	3.05–6	B.29
in. (or in)	inch	Length	2.01–2	B.11
in.^{2}	square inch	Area	2.01–2	B.12
in.^{3}	cubic inch	Volume	B.13	B.13
in.-s^{-1}	inch per second	Speed	B.23	B.23
in.-s^{-2}	inch per second^2	Acceleration	B.24	B.24
k	kilo = 10^3	Prefix	B.07	B.07
kcal	kilocalorie = 10^3 cal	Energy, work	6.01–4	B.29
kg	kilogram = 10^3 g	Mass	3.02–3	B.16
kg-m	kilogram × meter	Moment of mass	B.17	B.17
kg-m^2	kilogram × meter^2	Mass moment of inertia*	3.04–3	B.20
kg-m-s^{-1}	kilogram × meter per second	Power	3.05–6	B.30
kg-m^{-3}	kilogram per meter^3	Mass density	5.01–2	B.26
kgf	kilogram-force	Force	2.02–1	B.15
kgf-m	kilogram-force × meter	Moment of force, work	2.03–1	B.17
kgf-m-s^{-1}	kilogram-force × meter per second	Power	3.05–6	B.30
kgf-m^{-1}	kilogram-force per meter	Intensity of force	4.04–1	——
kgf-m^{-2}	kilogram-force per meter^2	Pressure	5.01–6	B.25
kgf-m^{-3}	kilogram-force per meter^3	Weight density	5.01–3	B.26
kipf	kilopound-force = 10^3 lbf	Force	2.02–1	B.15
kipf-ft	kilopound-force × foot	Moment, work	2.03–1	B.17
kipf-ft^2	kilopound-force per foot^2	Pressure	5.01–6	B.25
kL	kiloliter =	Volume	B.14	B.14
km	kilometer = 10^3 m	Length	2.01–2	B.11
km^2	square kilometer	Area	B.12	B.12
km-h^{-1}	kilometer per hour	Speed	B.23	B.23
kmol	kilomole = 10^3 mol	Mass	6.02–1	——
kN	kilonewton = 10^3 N	Force	B.15	B.15
kW	kilowatt = 10^3 W	Power	3.05–6	B.30
kW-h	kilowatt × hour	Work, energy	3.05–6	B.29
lb	pound	Mass	3.02–3	B.16
lb-ft	pound × foot	Moment of mass	B.17	B.17
lb-ft^2	pound × foot^2	Mass moment of inertia*	3.04–3	B.20
lb-ft^{-3}	pound per foot^3	Mass density	5.01–2	B.26
lbf	pound-force	Force	2.02–1	B.15
lbf-ft	pound-force × foot	Moment of force, work	2.03–1	B.17
lbf-ft-s^{-1}	pound-force × foot per second	Power	3.05–6	B.30

*Also mass product of inertia.

B.04 GLOSSARY OF SYMBOLS OF UNITS

Symbols beginning with lowercase letters

Symbol	Name	Unit	Section Definition	Section Conversion
lbf-ft^{-1}	pound-force per foot	Intensity of force	4.04–1	——
lbf-ft^{-2}	pound-force per foot^2	Pressure	5.01–6	B.25
lbf-ft^{-3}	pound-force per foot^3	Weight density	5.01–3	B.26
lm	lumen	Luminous flux	10.01–3	B.10
lx	lux	Illumination	10.01–4	B.10
m	milli $= 10^{-3}$	Prefix	B.07	B.07
m	meter $= 10^2$ cm	Length	2.01–2	B.11
m^2	square meter	Area	2.01–2	B.12
m^3	cubic meter	Volume	2.01–2	B.13
m-kgf	meter $\times$ kilogram-force	Energy, work	3.05–1	B.29
m-N	meter $\times$ newton	Energy, work	3.05–1	B.29
mg	milligram $= 10^{-3}$ g	Mass	3.02–3	B.16
mi	mile	Length	2.01–2	B.11
mi^2	square mile	Area	2.01–2	B.12
mi-h^{-1}	mile per hour	Speed	B.23	B.23
min	minute $= 60$ s	Time	3.01–2	B.21
mL	mililiter $= 10^{-3}$ L	Volume	B.14	B.14
mm	millimeter $= 10^{-3}$ m	Length	2.01–2	B.11
mol	mole	Mass (amount of substance)	6.02–1	——
n	nano $= 10^{-9}$	Prefix	B.07	B.07
p	pico $= 10^{-12}$	Prefix	B.07	B.07
pd	poundal	Force	3.02–4	B.15
pd-ft	poundal $\times$ foot	Moment of force, work	2.03–1	B.17
psi	pound-force per inch^2	Pressure	5.01–6	B.25
rad	radian	Plane angle	2.01–3	B.22
rad-s^{-1}	radian per second	Angular speed	3.01–5	——
rad-s^{-2}	radian per second^2	Angular acceleration	3.01–5	——
rev	revolution	Rotation	2.01–3	B.22
rev-s^{-1}	revolution per second	Rotational speed	3.03–2	——
s	second	Time	3.01–2	B.21
sl	slug $= 32.17$ lb	Mass	3.02–3	B.16
sl-in^{-3}	slug per inch^3	Mass density	B.26	B.26
sl-ft^{-3}	slug per foot^3	Mass density	B.26	B.26
sr	steradian	Solid angle	10.01–2	——
sA	statampere	Electric current	B.40	B.40
sC	statcoulomb	Electric charge	B.40	B.40
sF	statfarad	Electric capacitance	B.42	B.42
sH	stathenry	Inductance	B.42	B.42
sV	statvolt	Voltage, potential	B.41	B.41
sΩ	statohm	Electric resistance	B.41	B.41
t	ton $= 10^3$ kg	Mass	3.02–3	B.16
tf	ton-force $= 10^3$ kgf	Force	2.02–1	B.15
tf-m^{-2}	ton-force per meter^2	Pressure	5.01–6	B.25
torr	torr	Pressure	5.01–6	B.25
yd	yard	Length	2.01–2	B.11
yd^2	square yard	Area	B.12	B.12
yd^3	cubic yard	Volume	B.13	B.13

B.05 GLOSSARY OF SYMBOLS OF UNITS
Symbols beginning with capital letters or with Greek letters

Symbol	Name	Unit	Section	
			Definition	Conversion
A	ampere	Electric current	7.03–1	B.40
A-s	ampere × second = 1 C	Electric charge	7.03–1	B.40
A-m^{-1}	ampere per meter	Magnetic field strength	8.01–3	B.40
Å	angstrom = 10^{-10} m	Length	2.01–2	B.11
B	bel	Sound intensity	9.04–2	——
Btu	British thermal unit	Energy, work	6.01–4	B.29
Btu-h^{-1}	British thermal unit per hour	Power	6.01–4	B.30
C	coulomb	Electric charge	7.01–2	B.40
C-s^{-1}	coulomb per second = 1 A	Electric current	7.01–2	B.40
°C	degree Celsius	Temperature	6.01–2	B.27
D (or da)	deka = 10	Prefix	B.07	B.07
E	exa = 10^{18}	Prefix	B.07	B.07
F	farad	Electric capacitance	7.02–4	B.42
F-m^{-1}	farad per meter	Dielectric strength	7.02–7	B.42
°F	degree Fahrenheit	Temperature	6.01–2	B.27
G	gauss	Magnetic flux density	8.01–4	B.43
G	giga = 10^9	Prefix	B.07	B.07
H	henry	Inductance	8.03–2	B.42
H-m^{-1}	henry per meter	Permeability	8.03–2	B.42
Hz	hertz = s^{-1}	Frequency	8.04–1	——
J	joule	Energy, work	3.05–1	B.29
J-m^{-1}	joule per meter = 1 N	Force	3.05–6	B.15
J-s^{-1}	joule per second = 1 W	Power	3.05–6	B.30
K	kelvin	Absolute temperature	6.01–2	B.27
L	liter	Volume	B.14	B.14
M	mega = 10^6	Prefix	B.07	B.07
N	newton = kg × m × s^{-2}	Force	3.02–4	B.15
N-m*	newton × meter	Moment of force	2.03–1	B.17
N-m^{-2}	newton per meter^2 = Pa	Pressure	5.01–6	B.25
P	peta = 10^{15}	Prefix	B.07	B.07
P	poise	Absolute viscosity	5.01–5	B.37
Pa	pascal = N-m^{-2}	Pressure	5.01–6	B.25
°R	degree Rankine	Temperature	6.01–2	B.27
°Re	degree Reaumur	Temperature	6.01–2	B.27
S	siemens = $1/\Omega$	Electric conductance	7.03–2	
St	stoke(s)	Kinematic viscosity	5.01–5	B.38
T	tera = 10^{12}	Prefix	B.07	B.07
T	tesla	Magnetic flux density	8.01–3	B.43
T-m^2	tesla × meter^2 = Wb	Magnetic flux	8.01–4	B.43
V	volt	Voltage potential	7.02–3	B.41
V-m^{-1}	volt per meter	Electric field strength	7.02–3	B.41
W	watt = J × s^{-1}	Power	3.05–6	B.30
Wb	weber	Magnetic flux	8.01–4	B.43
Wb-m^{-2}	weber per meter^2	Magnetic flux density	8.01–4	B.43
W-h	watt × hour = W × h	Energy, work	3.05–6	B.29
μ	micro = 10^{-6}	Prefix	B.07	B.07
μm	micrometer = 10^{-6} m	Length	10.01–1	B.11
Ω	ohm	Electric resistance	7.03–2	B.41

*m-N = meter × newton = joule = unit of energy (work).

Conversion Tables   437

B.06 SYSTEMS OF BASIC UNITS

Designation	Dimensions	System		
		English (FPS)	Metric (MKS)	International (SI)
Length	(L)	foot (ft)	meter (m)	meter (m)
Mass	(M)	pound (lb)	kilogram (kg)	kilogram (kg)
Time	(t)	second (sec)	second (s)	second (s)
Electric current	(A)	ampere (A)	ampere (A)	ampere (A)
Temperature	(T)	degree Fahrenheit (°F)	degree Celsius (°C)	kelvin (K)
Luminous intensity	(l)	candela (cd)	candela (cd)	candela (cd)

B.07 DECIMAL MULTIPLES AND FRACTIONS OF SI UNITS

Factor	Prefix	Symbol	Factor	Prefix	Symbol
10^1	deka	D (da)	10^{-1}	deci	d
10^2	hecto	h	10^{-2}	centi	c
10^3	kilo	k	10^{-3}	milli	m
10^6	mega	M	10^{-6}	micro	μ
10^9	giga	G	10^{-9}	nano	n
10^{12}	tera	T	10^{-12}	pico	p
10^{15}	peta	P	10^{-15}	femto	f
10^{18}	exa	E	10^{-18}	atto	a

B.08 SYSTEMS OF DERIVED UNITS, GEOMETRY, MASS

Designation	Dimensions	FPS	MKS	SI
Area	$(L)^2$	ft^2	m^2	m^2
Static moment of area	$(L)^3$	ft^3	m^3	m^3
Moment of inertia of area	$(L)^4$	ft^4	m^4	m^4
Product of inertia of area	$(L)^4$	ft^4	m^4	m^4
Polar moment of inertia of area	$(L)^4$	ft^4	m^4	m^4
Volume	$(L)^3$	ft^3	m^3	m^3
Static moment of volume	$(L)^4$	ft^4	m^4	m^4
Moment of inertia of volume	$(L)^5$	ft^5	m^5	m^5
Product of inertia of volume	$(L)^5$	ft^5	m^5	m^5
Polar moment of inertia of volume	$(L)^5$	ft^5	m^5	m^5
Mass = M = W/g*	(M)	lb	kg	kg
Static moment of mass	$(M)(L)$	lb-ft	kg-m	kg-m
Moment of inertia of mass	$(M)(L)^2$	lb-ft^2	kg-m^2	kg-m^2
Product of inertia of mass	$(M)(L)^2$	lb-ft^2	kg-m^2	kg-m^2
Polar moment of inertia of mass	$(M)(L)^2$	lb-ft^2	kg-m^2	kg-m^2
Curvature of a curve	$(L)^{-1}$	1/ft	1/m	1/m
Torsion of a curve	$(L)^{-1}$	1/ft	1/m	1/m
Plane angle†	(R)	rad	rad	rad
Solid angle†	(S)	sr	sr	sr

*In the English system (FPS) and in the metric system (MKS), the mass M is a derived unit, defined as the weight W divided by the acceleration due to gravity g. lb = pound mass, kg = kilogram mass.

†The unit of plane angle called *radian* (rad) and the unit of solid angle called *steradian* (sr) are supplementary units.

B.09 SYSTEMS OF DERIVED UNITS, MECHANICS

Designation	Dimensions	FPS	MKS	SI
Linear velocity	$(L)(t)^{-1}$	ft/sec	m/s	m/s
Angular velocity	$(R)(t)^{-1}$	rad/sec	rad/s	rad/s
Linear acceleration	$(L)(t)^{-2}$	ft/sec²	m/s²	m/s²
Angular acceleration	$(R)(t)^{-2}$	rad/sec²	rad/s²	rad/s²
Linear momentum*	$(M)(L)(t)^{-1}$	lb-ft/sec	kg-m/s	N-s
Angular momentum	$(M)(L)^2(t)^{-1}$	lb-ft²/sec	kg-m²/s	N-m-s
Force†	$(M)(L)(t)^{-2}$	lbf	kgf	N
Moment of force	$(M)(L)^2(t)^{-2}$	lbf-ft	kgf-m	N-m
Linear impulse	$(M)(L)(t)^{-1}$	lbf-sec	kgf-s	N-s
Angular impulse	$(M)(L)^2(t)^{-1}$	lbf-m-sec	kgf-m-s	N-m-s
Stress (pressure)	$(M)(L)^{-1}(t)^{-2}$	lbf/ft²	kgf/m²	N/m²
Work (energy)‡	$(M)(L)^2(t)^{-2}$	ft-lbf	m-kgf	J
Linear power§	$(M)(L)^2(t)^{-3}$	lbf-ft/sec	kgf-m/s	W
Angular power	$(M)(L)^2(R)(t)^{-3}$	lbf-ft-rad/sec	kgf-m-rad/s	W-rad
Viscosity, absolute	$(M)(L)^{-1}(t)^{-1}$	lb/(ft-sec)	kg/(m-s)	N-s/m²
Viscosity, kinematic	$(L)^2(t)^{-1}$	ft²/sec	m²/s	m²/s

*N = newton = kg-m/s² = 10⁵ dyne.

†lbf = pound-force = lb(32.174 ft/sec²); 1 kgf = kilogram-force = kg(9.80655 m/s²).

‡J = joule = N-m = kg-m²/s² = 10⁷ erg.   §Watt = N-m/s = kg-m²/s³ = J/s.

B.10 SYSTEMS OF DERIVED UNITS, HEAT, ELECTRICITY, MAGNETISM

Designation	Dimensions	FPS	MKS	SI
Quantity of heat[a]	$(M)(L)^2(t)^{-2}$	ft-lbf	m-kgf	J
Heat capacity	$(M)(L)^2(t)^{-2}(T)^{-1}$	(ft-lbf)/°F	(m-kgf)/°C	J/K
Specific heat capacity	$(L)^2(t)^{-2}(T)^{-1}$	(ft-lbf)/(lb-°F)	(m-kgf)/(kg-°C)	J/(kg-K)
Heat flow[b]	$(M)(L)^2(t)^{-3}$	ft-lbf/sec	m-kgf/s	W
Heat conductivity	$(M)(L)(t)^{-3}(T)^{-1}$	lbf/(sec-°F)	kgf/(s-°C)	W/(m-K)
Heat transfer coefficient	$(M)(t)^{-3}(T)^{-1}$	lbf/(ft-sec-°F)	kgf/(m-s-°C)	W/(m²-K)
Electric charge[c]	$(A)(t)$	C	C	C
Voltage potential[d]	$(M)(A)^{-1}(L)^2(t)^{-3}$	V	V	V
Electric resistance[e]	$(M)(A)^{-2}(L)^2(t)^{-3}$	Ω	Ω	Ω
Electric capacitance[f]	$(M)^{-1}(A)^2(L)^{-2}(t)^4$	F	F	F
Electric inductance[g]	$(M)(A)^{-2}(L)^2(t)^{-2}$	H	H	H
Electric field strength	$(M)(A)^{-1}(L)(t)^{-3}$		V/m	V/m
Permeability	$(M)(A)^{-2}(L)(t)^{-2}$		H/m	H/m
Magnetic field strength	$(L)^{-1}(A)$		A/m	A/m
Magnetic flux[h]	$(M)(A)^{-1}(L)^2(t)^{-2}$	Wb	Wb	Wb
Magnetic flux density[i]	$(M)(A)^{-1}(t)^{-2}$	T	T	T
Luminous flux[j]	$(l)(S)$	lm	lm	lm
Luminance	$(l)(L)^{-2}$	cd/ft²	cd/m²	cd/m²
Illumination[k]	$(l)(S)(L)^{-2}$	lm/ft²	lx	lx

[a] J = joule = N-m = kg-m²/s² = 10⁷ ergs.

[b] W = watt = J/s = N-m/s = kg-m²/s³.

[c] C = coulomb = A-s = ampere × second.

[d] V = volt = W/A = (N-m)/(A-s) = (kg-m²)/(A-s³).

[e] Ω = ohm = V/A = (N-m)/(A²-s) = (kg-m²)/(A²-s³).

[f] F = farad = A-s/V = (A²-s²)/(N-m) = (A²-s⁴)/(kg-m²).

[g] H = henry = V-s/A = (N-m)/A² = (kg-m²)/(A²-s²).

[h] Wb = weber = V-s = (N-m)/A = (kg-m²)/(A-s²).

[i] T = tesla = Wb/m² = N/A-m = kg/(A-s²).

[j] lm = lumen = cd-sr.

[k] lx = lux = lm/m².

B.11 LENGTH (Sec. 2.01–2)

	km	mi	m	ft	in
1 kilometer	1	6.213 711 (−01)	1.000 000 (+03)	3.280 839 (+03)	3.937 007 (+04)
1 mile (U.S. st)	1.609 344 (+00)	1	1.609 344 (+03)	5.280 000 (+03)	6.336 000 (+04)
1 meter	1.000 000 (−03)	6.213 711 (−04)	1	3.280 839 (+00)	3.937 007 (+01)
1 foot	3.048 000 (−04)	1.893 939 (−04)	3.048 000 (−01)	1	1.200 000 (+01)
1 inch	2.540 000 (−05)	1.578 283 (−05)	2.540 000 (−02)	8.333 333 (−02)	1

1 angstrom (Å)	= 1.000 000 (−10) meter	
1 astronomical unit	= 1.495 980 (+08) kilometers	= 9.295 587 (+07) mile (U.S. st)
1 centimeter (cm)	= 3.937 007 (−01) inch	
1 decimeter (dm)	= 3.937 007 (+00) inches	
1 light-year	= 9.460 550 (+12) kilometers	= 5.878 512 (+12) mile (U.S. st)
1 micron (μ)*	= 1.000 000 (−06) meter	
1 millimicron (mμ)*	= 1.000 000 (−09) meter	
1 mile (U.S. naut)	= 1.852 000 (+00) kilometers	= 1.150 780 (+00) miles (U.S. st)
1 millimeter (mm)	= 1.000 000 (−03) meter	
1 parsec	= 3.083 740 (+13) kilometers	= 1.916 147 (+13) miles (U.S. st)
1 rod	= 5.029 200 (+00) meters	= 1.650 000 (+01) feet
1 yard (yd)	= 9.144 000 (−01) meter	= 3.000 000 (+00) feet

*Current usage substitutes the micrometer (μm) for the micron (μ); and the nanometer (nm) for the millimicron mμ; use of the older units is to be avoided.

B.12 AREA (Sec. 2.07–2)

	mi^2	acre	m^2	ft^2	in^2
1 mile^2	1	6.400 000 (+02)	2.589 988 (+06)	2.787 840 (+07)	4.014 490 (+09)
1 acre	1.562 500 (−03)	1	4.046 856 (+03)	4.356 000 (+04)	6.272 640 (+06)
1 meter^2	3.861 021 (−07)	2.471 054 (−04)	1	1.076 391 (+01)	1.550 003 (+03)
1 foot^2	3.587 006 (−08)	2.295 684 (−05)	9.290 304 (−02)	1	1.440 000 (+02)
1 inch^2	2.490 976 (−10)	1.594 225 (−07)	6.451 600 (−04)	6.944 444 (−03)	1

1 are	= 1.000 000 (+02) meters^2	= 1.076 391 (+03) feet^2
1 barn	= 1.000 000 (−28) meter^2	
1 centimeter^2 (cm^2)	= 1.000 000 (−04) meter^2	= 1.550 003 (−01) inch^2
1 circular mil	= 5.067 075 (−10) meter^2	= 7.853 981 (−07) inch^2
1 decimeter^2 (dm^2)	= 1.000 000 (−02) meter^2	= 1.550 003 (+01) inches^2
1 hectare	= 1.000 000 (+04) meters^2	= 1.076 391 (+05) feet^2
1 kilometer^2 (km^2)	= 3.861 021 (−01) mile^2	= 2.471 054 (+02) acres
1 micrometer^2 (μm^2)	= 1.000 000 (−12) meter^2	
1 millimeter^2 (mm^2)	= 1.000 000 (−06) meter^2	
1 section	= 2.589 988 (+06) meters^2	= 1.000 000 (+00) miles^2
1 township	= 9.323 957 (+07) meters^2	= 3.600 000 (+01) sections
1 yard^2 (yd^2)	= 8.361 274 (−01) meter^2	= 9.000 000 (+00) feet^2

	m^3	yd^3	ft^3	in^3	cm^3
1 meter³	1	1.307 950 (+00)	3.531 466 (+01)	6.102 373 (+04)	1.000 000 (+06)
1 yard³	7.645 549 (−01)	1	2.700 000 (+01)	4.665 600 (+04)	7.645 549 (+05)
1 foot³	2.831 685 (−02)	3.703 703 (−02)	1	1.728 000 (+03)	2.831 685 (+04)
1 inch³	1.638 706 (−05)	2.143 347 (−05)	5.787 037 (−04)	1	1.638 706 (+01)
1 centimeter³	1.000 000 (−06)	1.307 950 (−06)	3.531 466 (−05)	6.102 373 (−02)	1

1 board foot	= 2.359 737 (−03) meter³	= 1.440 000 (+02) inches³
1 bushel (U.S.)	= 3.523 907 (−02) meter³	= 8.000 000 (+00) gallons (U.S. dry)
1 cord	= 3.624 556 (+00) meters³	= 1.280 000 (+02) feet³
1 cup	= 2.365 882 (−04) meter³	
1 gallon (U.S. dry)	= 4.404 883 (−03) meter³	= 1.555 570 (−01) foot³
1 hogshead	= 2.384 809 (−01) meter³	
1 liter (L)	= 1.000 000 (−03) meter³	= 2.270 207 (−01) gallon (U.S. dry)
1 peck (U.S. dry)	= 8.809 768 (−03) meter³	= 2.000 000 (+00) gallons (U.S. dry)
1 pint (U.S. dry)	= 5.506 105 (−04) meter³	= 1.250 000 (−01) gallon (U.S. dry)
1 quart (U.S. dry)	= 1.101 221 (−03) meter³	= 2.500 000 (−01) gallon (U.S. dry)
1 stere	= 1.000 000 (+00) meter³	
1 ton (reg)	= 2.831 685 (+00) meters³	= 1.000 000 (+02) feet³

*Meter³, decimeter³, centimeter³, liter, yard³, foot³, inch³, and cup are also used as liquid measures.

	bbl	gal	m^3	L	ft^3
1 barrel (U.S. oil)	1	4.200 000 (+01)	1.589 873 (−01)	1.589 873 (+02)	5.614 583 (+00)
1 gallon (U.S. fluid)	2.380 952 (−02)	1	3.785 412 (−03)	3.785 412 (+00)	1.336 805 (−01)
1 meter³	6.289 810 (+00)	2.641 720 (+02)	1	1.000 000 (+03)	3.531 466 (+01)
1 liter	6.289 810 (−03)	2.641 720 (−01)	1.000 000 (−03)	1	3.531 466 (−02)
1 foot³	1.781 076 (−01)	7.480 522 (+00)	2.831 685 (−02)	2.831 685 (+01)	1

For pure water at 4°C (39.2°F), 1 decimeter³ = 1 liter = 1 kilogram.
For any other fluid, 1 decimeter³ = 1 liter.

1 acre-foot	= 1.233 482 (+03) meters³	= 4.356 000 (+04) feet³
1 dram (U.S. fluid)	= 3.696 691 (−06) meter³	
1 fluid ounce (U.S.)	= 2.957 353 (−05) meter³	
1 gallon (British)	= 4.546 087 (−03) meter³	= 1.200 949 (+00) gallons (U.S. fluid)
1 gill (U.S. fluid)	= 1.182 941 (−04) meter³	= 3.125 000 (−02) gallon (U.S. fluid)
1 ounce (U.S. fluid)	= 2.957 353 (−05) meter³	
1 pint (U.S. fluid)	= 4.731 765 (−04) meter³	= 1.250 000 (−01) gallon (U.S. fluid)
1 quart (U.S. fluid)	= 9.463 530 (−04) meter³	= 2.500 000 (−01) gallon (U.S. fluid)
1 tablespoon	= 1.478 676 (−05) meter³	
1 teaspoon	= 4.928 922 (−06) meter³	

*Meter³, decimeter³, centimeter³, yard³, foot³, inch³, and cup are also used as liquid measures.

B.15 FORCE (Secs. 2.02–1, 3.02–4)*

	N	kgf	lbf	pd	dyn
1 newton	1	1.019 716 (−01)	2.248 089 (−01)	7.233 011 (+00)	1.000 000 (+05)
1 kilogram-force	9.806 650 (+00)	1	2.204 622 (+00)	7.093 160 (+02)	9.806 650 (+05)
1 pound-force	4.448 222 (+00)	4.535 924 (−01)	1	3.217 404 (+01)	4.448 222 (+05)
1 poundal	1.382 550 (−01)	1.409 808 (−02)	3.108 094 (−02)	1	1.382 550 (+04)
1 dyne	1.000 000 (−05)	1.019 716 (−06)	2.248 089 (−06)	7.233 011 (−05)	1

1 dram-force (av.)	= 1.737 586 (−02) newton	= 3.906 248 (−03) pound-force (av.)
1 dram-force (tr.)	= 3.812 762 (−02) newton	= 2.194 286 (+00) dram-force (av.)
1 grain-force	= 6.354 602 (−04) newton	= 1.428 571 (−04) pound-force (av.)
1 gram-force	= 9.806 650 (−03) newton	
1 hundredweight-force (long)	= 4.982 000 (+02) newtons	
1 hundredweight-force (short)	= 4.448 221 (+02) newtons	
1 ounce-force (av.)	= 2.780 138 (−01) newton	= 6.250 000 (−02) pound-force (av.)
1 ounce-force (tr.)	= 3.050 209 (−01) newton	= 1.097 143 (+00) ounce-force (av.)
1 pound-force (tr.)	= 3.660 251 (+00) newtons	= 8.228 569 (−01) pound-force (av.)
1 pennyweight-force	= 1.525 105 (−02) newton	
1 ton-force (MKS)	= 9.806 650 (+03) newtons	= 1.102 311 (+00) ton-force (FPS)
1 ton-force (FPS)	= 8.896 443 (+03) newtons	= 9.071 850 (−01) ton-force (MKS)

*av. = avoirdupois; tr. = troy or apothecaries'.

B.16 MASS (Sec. 3.02–3)*

	kg	lb	sl	g	t
1 kilogram	1	2.204 622 (+00)	6.852 177 (−02)	1.000 000 (+03)	1.102 311 (−03)
1 pound (av.)	4.535 924 (−01)	1	3.108 095 (−02)	4.535 924 (+02)	5.000 000 (−04)
1 slug	1.459 390 (+01)	3.217 404 (+01)	1	1.459 390 (+04)	1.608 702 (−02)
1 gram	1.000 000 (−03)	2.204 622 (−03)	6.852 177 (−05)	1	1.102 311 (−06)
1 ton (FPS)	9.071 847 (+02)	2.000 000 (+03)	6.216 190 (+01)	9.071 847 (+05)	1

1 atomic mass unit	= 1.660 566 (−27) kilogram	= 3.660 920 (−27) pound (av.)
1 carat	= 2.000 000 (−04) kilogram	= 4.409 244 (−04) pound (av.)
1 dram (av.)	= 1.771 845 (−03) kilogram	= 3.906 248 (−03) pound (av.)
1 dram (tr.)	= 3.887 935 (−03) kilogram	= 2.194 286 (+00) drams (av.)
1 grain	= 6.479 891 (−05) kilogram	= 1.428 571 (−04) pound (av.)
1 hundredweight (long)	= 5.080 235 (+01) kilograms	= 1.120 000 (+02) pounds (av.)
1 hundredweight (short)	= 4.535 924 (+01) kilograms	= 1.000 000 (+02) pounds (av.)
1 milligram	= 1.000 000 (−06) kilogram	= 2.204 622 (−06) pound (av.)
1 ounce (av.)	= 2.834 952 (−02) kilogram	= 6.250 000 (−02) pound (av.)
1 ounce (tr.)	= 3.110 348 (−02) kilogram	= 1.097 143 (+00) ounces (av.)
1 pennyweight	= 1.555 174 (−03) kilogram	= 3.182 228 (−03) pound (av.)
1 pound (tr.)	= 3.732 417 (−01) kilogram	= 8.228 568 (−01) pound (av.)
1 ton (MKS)	= 1.000 000 (+03) kilograms	= 1.102 311 (+00) tons (FPS)

*av. = avoirdupois; tr. = troy or apothecaries'.

B.17 MOMENT OF FORCE (Sec. 2.03–1)

	N-m	kgf-m	lbf-ft	lbf-in	dyn-cm
1 newton × meter	1	1.019 716 (−01)	7.375 622 (−01)	8.850 746 (+00)	1.000 000 (+07)
1 kilogramf × meter	9.806 650 (+00)	1	7.233 016 (+00)	8.679 619 (+01)	9.806 650 (+07)
1 poundf × foot	1.355 818 (+00)	1.382 550 (−01)	1	1.200 000 (+01)	1.355 818 (+07)
1 poundf × inch	1.129 848 (−01)	1.152 124 (−02)	8.333 333 (−02)	1	1.129 848 (+06)
1 dyne × centimeter	1.000 000 (−07)	1.019 716 (−08)	7.375 622 (−08)	8.850 746 (−07)	1

1 tonf (MKS) × meter = 9.806 650 (+03) newton × meter
1 tonf (FPS) × foot = 2.765 100 (+02) kilogramf × meter
1 poundal × foot = 3.108 094 (−02) poundf × foot
1 newton × meter = 1.019 716 (−04) tonf (MKS) × meter
1 newton × meter = 3.687 811 (−04) tonf (FPS) × foot

B.18 MOMENT OF INERTIA OF AREA (Secs. 2.07–6, 5.02–3)

	m⁴	dm⁴	cm⁴	ft⁴	in⁴
1 meter⁴	1	1.000 000 (+04)	1.000 000 (+08)	1.158 618 (+02)	2.402 509 (+06)
1 decimeter⁴	1.000 000 (−04)	1	1.000 000 (+04)	1.158 618 (−02)	2.402 509 (+02)
1 centimeter⁴	1.000 000 (−08)	1.000 000 (−04)	1	1.158 618 (−06)	2.402 509 (−02)
1 foot⁴	8.630 972 (−03)	8.630 912 (+01)	8.630 972 (+05)	1	2.073 600 (+04)
1 inch⁴	4.162 315 (−07)	4.162 315 (−03)	4.162 315 (+01)	4.822 531 (−05)	1

B.19 MOMENT OF INERTIA OF VOLUME (Sec. 3.04–3)

	m⁵	dm⁵	cm⁵	ft⁵	in⁵
1 meter⁵	1	1.000 000 (+05)	1.000 000 (+10)	3.801 239 (+02)	9.458 698 (+07)
1 decimeter⁵	1.000 000 (−05)	1	1.000 000 (+05)	3.801 239 (−03)	9.458 698 (+02)
1 centimeter⁵	1.000 000 (−10)	1.000 000 (−05)	1	3.801 239 (−08)	9.458 698 (−03)
1 foot⁵	2.630 721 (−03)	2.630 721 (+02)	2.630 721 (+07)	1	2.488 320 (+05)
1 inch⁵	1.057 230 (−08)	1.057 230 (−03)	1.057 230 (+02)	4.018 775 (−06)	1

B.20 MOMENT OF INERTIA OF MASS (Sec. 3.04–3)

	m²-kg	cm²-kg	ft²-lb	in²-lb	cm²-g
1 meter² × kilogram	1	1.000 000 (+04)	2.373 034 (+01)	3.417 171 (+03)	1.000 000 (+07)
1 centimeter² × kilogram	1.000 000 (−04)	1	2.373 034 (−03)	3.417 171 (−01)	1.000 000 (+03)
1 foot² × pound	4.214 013 (−02)		1	1.440 000 (+02)	4.214 013 (+05)
1 inch² × pound	2.926 397 (−04)	2.926 397 (+00)	6.944 444 (−03)	1	2.926 397 (+03)
1 centimeter² × gram	1.000 000 (−07)	1.000 000 (−03)	2.373 034 (−06)	3.417 171 (−04)	1

B.21 TIME (Sec. 3.01–2)

	y	d	h	min	s
1 year (mean solar year)	1	3.652 422 (+02)	8.765 813 (+03)	5.259 488 (+05)	3.155 693 (+07)
1 day (mean solar day)	2.737 909 (−03)	1	2.400 000 (+01)	1.440 000 (+03)	8.640 000 (+04)
1 hour	1.140 795 (−04)	4.166 667 (−02)	1	6.000 000 (+01)	3.600 000 (+03)
1 minute	1.901 326 (−06)	6.944 444 (−04)	1.666 667 (−02)	1	6.000 000 (+01)
1 second	3.168 876 (−08)	1.157 407 (−05)	2.777 778 (−04)	1.666 667 (−02)	1

1 mean solar day	= 8.640 000 (+04) seconds
1 sidereal day	= 8.616 409 (+04) seconds = period of rotation of the earth
1 mean calendar year	= 3.153 600 (+07) seconds
1 sidereal year	= 3.155 815 (+07) seconds = 3.662 564 (+02) sidereal days
1 mean solar year	= 3.652 422 (+02) solar days = period of revolution of the earth

B.22 PLANE ANGLE (Sec. 2.01–3)

	deg	min	sec	rad	rev
1 degree	1	6.000 000 (+1)	3.600 000 (+03)	1.745 329 (−02)	2.777 778 (−03)
1 minute	1.666 667 (−02)	1	6.000 000 (+01)	2.908 882 (−04)	4.629 629 (−05)
1 second	2.777 778 (−04)	1.666 667 (−02)	1	4.848 137 (−06)	7.716 048 (−07)
1 radian	5.729 578 (+01)	3.437 747 (+03)	2.062 648 (+05)	1	1.591 549 (−01)
1 revolution	3.600 000 (+02)	2.160 000 (+04)	1.296 000 (+06)	6.283 185 (+00)	1

B.23 SPEED (Sec. 3.01–3)

	km-h⁻¹	mi-h⁻¹	m-s⁻¹	ft-sec⁻¹	knot
1 kilometer per hour	1	6.213 711 (−01)	2.777 778 (−01)	9.113 444 (−01)	5.400 000 (−01)
1 mile per hour (U.S. st)	1.609 344 (+00)	1	4.470 400 (−01)	1.466 667 (+00)	8.690 389 (−01)
1 meter per second	3.600 000 (+00)	2.236 935 (+00)	1	3.280 839 (+00)	1.943 844 (+00)
1 foot per second	1.097 280 (+00)	6.818 179 (−01)	3.048 000 (−01)	1	5.925 264 (−01)
1 knot	1.852 000 (+00)	1.150 696 (+00)	5.144 444 (−01)	1.687 688 (+00)	1

1 knot = 1 U.S. nautical mile per hour. 1 r/min = 1.047 198 (−01) rad/s

The speed table uses units km-h^{-1}, mi-h^{-1}, m-s^{-1}, ft-sec^{-1}.

B.24 ACCELERATION (Sec. 3.01–3)

	m-s⁻²	m-min⁻²	ft-sec⁻²	ft-min⁻²	in-sec⁻²
1 meter per second²	1	3.600 000 (+03)	3.280 840 (+00)	1.181 102 (+04)	3.937 008 (+01)
1 meter per minute²	2.777 778 (−04)	1	9.113 444 (−04)	3.280 839 (+00)	1.093 613 (−02)
1 foot per second²	3.048 000 (−01)	1.097 280 (+03)	1	3.600 000 (+03)	1.200 000 (+01)
1 foot per minute²	8.466 667 (−05)	3.048 000 (−01)	2.777 778 (−04)	1	3.333 333 (−03)
1 inch per second²	2.540 000 (−02)	9.144 000 (+01)	8.333 333 (−02)	3.000 000 (+02)	1

1 galileo (gal) = 1 centimeter per second².

B.25 PRESSURE (Sec. 2.07–1, 5.01–6)

	atm	kgf-cm^{-2}	lbf-in^{-2}	N-m^{-2}	b
1 physical atmosphere	1	1.033 227 (+00)	1.469 594 (+01)	1.013 250 (+05)	1.013 250 (+00)
1 kilogramf per centimeter^2	9.678 415 (−01)	1	1.422 334 (+01)	9.806 650 (+04)	9.806 650 (+01)
1 poundf per inch^2	6.804 601 (−02)	7.030 097 (−02)	1	6.894 757 (+03)	6.894 757 (−02)
1 newton per meter^2 (pascal)	9.869 233 (−06)	1.019 716 (−05)	1.450 377 (−04)	1	1.000 000 (−05)
1 bar	9.869 233 (−01)	1.019 716 (+00)	1.450 377 (+01)	1.000 000 (+05)	1

1 dyne per centimeter^2 = 1.019 716 (−06) kilogramf per centimeter^2
1 technical atmosphere = 1.000 000 (+00) kilogramf per centimeter^2
1 centimeter of mercury (0°C) = 1.359 506 (+02) kilogramf per meter^2
1 centimeter of water (4°C) = 1.000 000 (+01) kilogramf per meter^2
1 inch of mercury (0°C) = 3.453 155 (+02) kilogramf per meter^2
1 inch of water (4°C) = 2.539 929 (+01) kilogramf per meter^2
1 torr = 1 mm mercury (°C) = 1.359 506 (−03) kilogramf per centimeter^2

B.26 DENSITY (Secs. 3.04–1, 5.01–2)

	kg-m^{-3}	lb-ft^{-3}	g-cm^{-3}	lb-in^{-3}	sl-ft^{-3}
1 kilogram per meter^3	1	6.242 800 (−02)	1.000 000 (−03)	3.612 728 (−05)	1.940 319 (−03)
1 pound per foot^3	1.601 846 (+01)	1	1.601 846 (−02)	5.787 034 (−04)	3.108 092 (−02)
1 gram per cm^3	1.000 000 (+03)	6.242 797 (+01)	1	3.612 728 (−02)	1.940 319 (+00)
1 pound per inch^3	2.767 991 (+04)	1.728 000 (+03)	2.767 991 (+01)	1	5.370 788 (+01)
1 slug per foot^3	5.153 790 (+02)	3.217 407 (+01)	5.153 791 (−01)	1.861 925 (−02)	1

B.27 TEMPERATURE EQUATIONS (Sec. 6.01–3)

Degrees Celsius = C	Degrees Fahrenheit = F	Kelvins = K	Degrees Rankine = R	Degrees Reaumur = Re
$C = 5/9(F - 32)$	$F = 9/5C + 32$	$K = C + 273.15$	$R = 9/5C + 491.67$	$Re = 4/5C$
$= K - 273.15$	$= 9/5(K - 255.37)$	$= 5/9(F + 459.67)$	$= F + 459.67$	$= 4/9(F - 32)$
$= 5/9(K - 491.67)$	$= R - 459.67$	$= 5/9R$	$= 9/5K$	$= 4/5(K - 273.15)$
$= 5/4Re$	$= 9/4Re + 32$	$= 5/4Re + 273.15$	$= 9/4Re + 491.67$	$= 4/9(R - 491.67)$

Absolute zero temperature = −273.15°C = −459.67°F = 0.00 K = 0.00°R = −218.52°Re
Freezing point of water = 0.00°C = +32.00°F = +273.15 K = +491.67°R = 0.00°Re
Boiling point of water = +100.00°C = +212.00°F = +373.15 K = +671.67°R = +80.00°Re

B.28 COEFFICIENT OF THERMAL EXPANSION (Sec. 6.03–1)

	1/1°C	1/1°F	1/1 K	1/1°R	1/1°Re
1/1° Celsius	1	5/9	1	5/9	5/4
1/1° Fahrenheit	9/5	1	9/5	1	9/4
1/1 Kelvin	1	5/9	1	5/9	5/4
1/1° Rankine	9/5	1	9/5	1	9/4
1/1° Reaumur	4/5	4/9	4/5	4/9	1

B.29 ENERGY, WORK (Secs. 3.05–1, 6.01–4)

	J	kcal	Btu	m-kgf	ft-lbf
1 joule	1	2.388 915 (−04)	9.478 672 (−04)	1.019 716 (−01)	7.375 616 (−01)
1 kilocalorie (technical)	4.186 000 (+03)	1	3.967 773 (+00)	4.268 532 (+02)	3.087 433 (+03)
1 British thermal unit	1.055 000 (+03)	2.520 305 (−01)	1	1.075 800 (+02)	7.781 275 (+02)
1 meter × kilogramf	9.806 650 (+00)	2.342 725 (−03)	9.295 402 (−03)	1	7.233 008 (+00)
1 foot × poundf	1.355 818 (+00)	3.238 934 (−04)	1.285 135 (−03)	1.382 549 (−01)	1

1 electronvolt = 1.602 189 (−19) joule = 1.633 778 (−20) meter × kilogramf
1 kilocalorie (thermochemical) = 9.995 522 (−01) kilocalorie (technical)
1 kilocalorie (thermochemical) = 3.965 877 (+00) British thermal unit
1 British thermal unit = 2.521 279 (−01) kilocalorie (thermochemical)
1 kilowatt × hour = 8.604 207 (+02) kilocalorie (thermochemical)
1 kilowatt × hour = 3.600 000 (+06) joule
1 kilowatt × hour = 2.655 212 (+06) foot × poundf

B.30 POWER (Secs. 3.05–6, 5.06–3)

	W	kcal-s⁻¹	Btu-h⁻¹	hp	ft-lbf-sec⁻¹
1 watt	1	2.388 915 (−04)	3.412 322 (+00)	1.359 621 (−03)	7.375 621 (−01)
1 kilocalorie (tech) per second	4.186 000 (+03)	1	1.428 398 (+04)	5.691 374 (+00)	3.087 435 (+03)
1 Btu per hour	2.930 000 (−01)	7.000 000 (−05)	1	3.983 690 (−04)	2.161 057 (−01)
1 horsepower (MKS)	7.354 988 (+02)	1.757 044 (−01)	2.510 257 (+03)	1	5.424 760 (+02)
1 foot-poundf per second	1.355 818 (+00)	3.238 934 (−04)	4.627 407 (+00)	1.843 340 (−03)	1

B.31 HORSEPOWER (Secs. 3.05–6, 5.06–3)

	hp (MKS)	hp (FPS)	W	m-kgf-s⁻¹	ft-lbf-sec⁻¹
1 horsepower (MKS)	1	9.863 200 (−01)	7.354 988 (+02)	7.500 000 (+01)	5.424 760 (+02)
1 horsepower (FPS)	1.013 870 (+00)	1	7.456 999 (+02)	7.604 025 (+01)	5.500 000 (+02)
1 horsepower (boiler)	1.333 721 (+01)	1.315 476 (+01)	9.809 500 (+03)	1.000 291 (+02)	7.235 111 (+02)
1 horsepower (electric)	1.014 278 (+00)	1.000 403 (+00)	7.460 000 (+02)	7.607 085 (+01)	5.502 210 (+02)
1 horsepower (water)	1.014 336 (+00)	1.000 460 (+00)	7.460 430 (+02)	7.607 520 (+01)	5.502 527 (+02)

B.32 SPECIFIC HEAT CAPACITY (Sec. 6.01–5b)

	J-kg⁻¹-°C⁻¹	kcal-kg⁻¹-°C⁻¹	Btu-lb⁻¹-°F⁻¹	m-kgf-kg⁻¹-°C⁻¹	ft-lbf-lb⁻¹-°F⁻¹
1 $\frac{joule}{kilogram \times °Celsius}$	1	2.390 057 (−04)	2.390 057 (−04)	1.019 716 (−01)	1.858 624 (−01)
1 $\frac{kilocalorie (thchem)}{kilogram \times °Celsius}$	4.184 000 (+03)	1	1	4.266 492 (+02)	7.776 484 (+02)
1 $\frac{British thermal unit}{pound \times °Fahrenheit}$	4.184 000 (+03)	1	1	4.266 492 (+02)	7.776 484 (+02)
1 $\frac{meter \times kilogramf}{kilogram \times °Celsius}$	9.806 650 (+00)	2.343 845 (−03)	2.343 845 (−03)	1	1.822 687 (+00)
1 $\frac{foot \times poundf}{pound \times °Fahrenheit}$	5.380 324 (+00)	1.285 928 (−03)	1.285 928 (−03)	5.486 406 (−01)	1

B.33 HEAT CONDUCTION (Sec. 6.04–2e)

	$\text{W-m}^{-1}\text{-}°\text{C}^{-1}$	$\text{kcal-m}^{-1}\text{-s}^{-1}\text{-}°\text{C}^{-1}$	$\text{Btu-in-ft}^{-2}\text{-sec}^{-1}\text{-}°\text{F}^{-1}$	$\text{Btu-in-ft}^{-2}\text{-sec}^{-1}\text{-}°\text{R}^{-1}$
$1 \dfrac{\text{watt}}{\text{meter} \times °\text{Celsius}}$	1	2.390 057 (−04)	1.927 253 (−03)	1.927 253 (−03)
$1 \dfrac{\text{kilocalorie (thchem)}}{\text{meter} \times \text{second} \times °\text{Celsius}}$	4.184 000 (+03)	1	8.063 627 (+00)	8.063 627 (+00)
$1 \dfrac{\text{British thermal unit} \times \text{inch}}{\text{foot}^2 \times \text{second} \times °\text{Fahrenheit}}$	5.188 732 (+02)	1.240 136 (−01)	1	1
$1 \dfrac{\text{British thermal unit} \times \text{inch}}{\text{foot}^2 \times \text{second} \times °\text{Rankine}}$	5.188 732 (+02)	1.240 136 (−01)	1	1

$1 \dfrac{\text{British thermal unit} \times \text{inch}}{\text{inch}^2 \times \text{second} \times °\text{Rankine}} = 7.471\,774\,(+04)\ \dfrac{\text{watt}}{\text{meter} \times °\text{Celsius}}$

$1 \dfrac{\text{British thermal unit} \times \text{foot}}{\text{foot}^2 \times \text{hour} \times °\text{Rankine}} = 1.729\,577\,(+00)\ \dfrac{\text{watt}}{\text{meter} \times °\text{Celsius}}$

$1 \dfrac{\text{kilocalorie (thchem)} \times \text{centimeter}}{\text{centimeter}^2 \times \text{second} \times °\text{Celsius}} = 6.719\,689\,(+01)\ \dfrac{\text{British thermal unit} \times \text{foot}}{\text{foot}^2 \times \text{second} \times °\text{Fahrenheit}}$

$1 \dfrac{\text{kilocalorie (thchem)} \times \text{meter}}{\text{meter}^2 \times \text{hour} \times °\text{Celsius}} = 1.866\,549\,(-04)\ \dfrac{\text{British thermal unit} \times \text{foot}}{\text{foot}^2 \times \text{second} \times °\text{Fahrenheit}}$

$1 \dfrac{\text{watt} \times \text{centimeter}}{\text{centimeter}^2 \times °\text{Celsius}} = 1.606\,044\,(-02)\ \dfrac{\text{British thermal unit} \times \text{foot}}{\text{foot}^2 \times \text{second} \times °\text{Fahrenheit}}$

B.34 HEAT CONVECTION (Sec. 6.04–4c)

	$\text{W-m}^{-2}\text{-}°\text{C}^{-1}$	$\text{kcal-m}^{-2}\text{-s}^{-1}\text{-}°\text{C}^{-1}$	$\text{Btu-ft}^{-2}\text{-sec}^{-1}\text{-}°\text{F}^{-1}$	$\text{m-kgf-m}^{-2}\text{-sec}^{-1}\text{-}°\text{C}^{-1}$
$1 \dfrac{\text{watt}}{\text{meter}^2 \times °\text{Celsius}}$	1	2.390 057 (−04)	4.895 141 (−05)	1.019 716 (−01)
$1 \dfrac{\text{kilocalorie (thchem)}}{\text{meter}^2 \times \text{second} \times °\text{Celsius}}$	4.184 000 (+03)	1	2.048 127 (−01)	4.266 492 (+02)
$1 \dfrac{\text{British thermal unit}}{\text{foot}^2 \times \text{second} \times °\text{Fahrenheit}}$	2.042 842 (+04)	4.882 509 (+00)	1	2.083 119 (+03)
$1 \dfrac{\text{meter} \times \text{kilogramf}}{\text{meter}^2 \times \text{second} \times °\text{Celsius}}$	9.806 650 (+00)	2.343 845 (−03)	4.800 493 (−04)	1

B.35 HEAT OF FUSION, VAPORIZATION AND COMBUSTION (Sec. 6.01–6c,e)

	J-kg^{-1}	kcal-kg^{-1}	Btu-lb^{-1}	m-kgf-kg^{-1}
$1 \dfrac{\text{joule}}{\text{kilogram}}$	1	2.390 057 (−04)	4.302 103 (−04)	1.019 716 (−01)
$1 \dfrac{\text{kilocalorie (thchem)}}{\text{kilogram}}$	4.184 000 (+03)	1	1.800 000 (+00)	4.266 492 (+02)
$1 \dfrac{\text{British thermal unit}}{\text{pound}}$	2.324 444 (+03)	5.555 556 (−01)	1	2.370 273 (+02)
$1 \dfrac{\text{meter} \times \text{kilogramf}}{\text{kilogram}}$	9.806 650 (+00)	2.343 846 (−03)	4.218 923 (−03)	1

B.36 GAS CONSTANT UNITS (Sec. 6.02–3d)

	atm-m^3-kmol^{-1}-K^{-1}	J-kmol^{-1}-K^{-1}	kcal-kmol^{-1}-K^{-1}	m-kgf-kmol^{-1}-K^{-1}
1 $\dfrac{\text{atmosphere (phys)} \times \text{meter}^3}{\text{kilomole} \times \text{kelvin}}$	1	1.013 250 (+05)	2.421 725 (+01)	1.033 227 (+04)
1 $\dfrac{\text{joule}}{\text{kilomole} \times \text{kelvin}}$	9.869 233 (−06)	1	2.390 057 (−04)	1.019 716 (−01)
1 $\dfrac{\text{kilocalorie (thchem)}}{\text{kilomole} \times \text{kelvin}}$	4.129 288 (−02)	4.184 000 (+03)	1	4.266 492 (+02)
1 $\dfrac{\text{meter} \times \text{kilogramf}}{\text{kilomole} \times \text{kelvin}}$	9.678 414 (−05)	9.806 650 (+00)	2.343 846 (−03)	1

B.37 ABSOLUTE VISCOSITY (Sec. 5.01–5c)

	kg-m^{-1}-s^{-1}	lb-ft^{-1}-sec^{-1}	poise	kgf-s-m^{-2}	lbf-sec-ft^{-2}
1 $\dfrac{\text{kilogram}}{\text{meter} \times \text{second}}$	1	6.719 689 (−01)	1.000 000 (+01)	1.019 716 (−01)	2.088 543 (−02)
1 $\dfrac{\text{pound}}{\text{foot} \times \text{second}}$	1.488 164 (+00)	1	1.488 164 (+01)	1.517 504 (−01)	3.108 093 (−02)
1 poise	1.000 000 (−01)	6.719 689 (−02)	1	1.019 716 (−02)	2.088 543 (−03)
1 $\dfrac{\text{kilogramf} \times \text{second}}{\text{meter}^2}$	9.806 650 (+00)	6.589 768 (+00)	9.806 650 (+01)	1	2.048 162 (−01)
1 $\dfrac{\text{poundf} \times \text{second}}{\text{foot}^2}$	4.788 026 (+01)	3.217 405 (+01)	4.788 026 (+02)	4.882 427 (+00)	1

1 centipoise $= 1.000\,000\,(-02)$ poise $= 1.000\,000\,(-03)\,\dfrac{\text{newton} \times \text{second}}{\text{meter}^2}$

1 centipoise $= 2.088\,543\,(-05)\,\dfrac{\text{slug}}{\text{foot} \times \text{second}}$ $= 1.450\,376\,(-07)$ reyn

1 $\dfrac{\text{kilogram}}{\text{meter} \times \text{second}}$ $= 2.088\,543\,(-02)\,\dfrac{\text{slug}}{\text{foot} \times \text{second}}$ $= 1.450\,376\,(-04)$ reyn

1 $\dfrac{\text{kilogramf} \times \text{second}}{\text{meter}^2}$ $= 2.048\,162\,(-01)\,\dfrac{\text{slug}}{\text{foot} \times \text{second}}$ $= 1.422\,333\,(-03)$ reyn

1 $\dfrac{\text{pound}}{\text{foot} \times \text{second}}$ $= 3.108\,095\,(-02)\,\dfrac{\text{slug}}{\text{foot} \times \text{second}}$ $= 2.158\,399\,(-04)$ reyn

1 $\dfrac{\text{poundf} \times \text{second}}{\text{foot}^2}$ $= 1.000\,000\,(+00)\,\dfrac{\text{slug}}{\text{foot} \times \text{second}}$ $= 6.944\,444\,(-03)$ reyn

1 reyn $= 1.440\,000\,(+02)\,\dfrac{\text{slug}}{\text{foot} \times \text{second}}$ $= 1.440\,000\,(+02)\,\dfrac{\text{poundf} \times \text{second}}{\text{foot}^2}$

1 reyn $= 6.894\,756\,(+03)\,\dfrac{\text{kilogram}}{\text{meter} \times \text{second}}$ $= 7.030\,695\,(+02)\,\dfrac{\text{kilogramf} \times \text{second}}{\text{meter}^2}$

1 $\dfrac{\text{slug}}{\text{foot} \times \text{second}}$ $= 4.788\,026\,(+02)$ poise $= 6.944\,444\,(-03)$ reyn

1 $\dfrac{\text{slug}}{\text{foot} \times \text{second}}$ $= 4.788\,026\,(+01)\,\dfrac{\text{kilogram}}{\text{meter} \times \text{second}}$ $= 4.882\,427\,(+00)\,\dfrac{\text{kilogramf} \times \text{second}}{\text{meter}^2}$

1 $\dfrac{\text{slug}}{\text{foot} \times \text{second}}$ $= 3.217\,406\,(+01)\,\dfrac{\text{pound}}{\text{foot} \times \text{second}}$ $= 1.000\,000\,(+00)\,\dfrac{\text{poundf} \times \text{second}}{\text{foot}^2}$

B.38 KINEMATIC VISCOSITY (Sec. 5.01–5g)

	$m^2 \cdot h^{-1}$	$m^2 \cdot s^{-1}$	stoke	$ft^2 \cdot h^{-1}$	$ft^2 \cdot sec^{-1}$
$1 \dfrac{meter^2}{hour}$	1	2.777 778 (−04)	2.777 778 (+00)	1.076 391 (+01)	2.989 975 (−03)
$1 \dfrac{meter^2}{second}$	3.600 000 (+03)	1	1.000 000 (+04)	3.875 008 (+04)	1.076 391 (+01)
1 stoke	3.600 000 (−01)	1.000 000 (−04)	1	3.875 008 (+00)	1.076 391 (−03)
$1 \dfrac{foot^2}{hour}$	9.290 303 (−02)	2.580 639 (−05)	2.580 639 (−01)	1	2.777 778 (−04)
$1 \dfrac{foot^2}{second}$	3.344 509 (+02)	9.290 303 (−02)	9.290 303 (+02)	3.600 000 (+03)	1

1 stoke $= 1.000\,000\,(+00) \dfrac{centimeter^2}{second} = 1.000\,000\,(+02)$ centistoke

B.39 DAMPING VISCOSITY (Sec. 9.01–5b)

	$kg \cdot s^{-1}$	$lb \cdot sec^{-1}$	$slug \cdot sec^{-1}$	$kgf \cdot s \cdot m^{-1}$	$lbf \cdot sec \cdot ft^{-1}$
$1 \dfrac{kilogram}{second}$	1	2.204 622 (+00)	6.852 177 (−02)	1.019 716 (−01)	6.852 172 (−02)
$1 \dfrac{pound}{second}$	4.535 925 (−01)	1	3.108 096 (−02)	4.625 355 (−02)	3.108 096 (−02)
$1 \dfrac{slug}{second}$	1.459 390 (+01)	3.217 404 (+01)	1	1.488 163 (+00)	1.000 000 (+00)
$1 \dfrac{kilogram \times second}{meter}$	9.806 650 (+00)	2.161 997 (+01)	6.719 694 (−01)	1	6.719 694 (−01)
$1 \dfrac{poundf \times second}{foot}$	1.459 389 (+01)	3.217 404 (+01)	1.000 000 (+00)	1.488 163 (+00)	1

B.40 ELECTRIC CHARGE AND CURRENT (Secs. 7.01–2, 7.03–1, 8.02–2)

abA = abampere   A = ampere   sA = statampere
abC = abcoulomb   C = coulomb   sC = statcoulomb

	abC	C	sC		abA	A	sA
abC	1	10	2.997 925 (+10)	abA	1	10	2.997 925 (+10)
C	10^{-1}	1	2.997 925 (+09)	A	10^{-1}	1	2.997 925 (+09)
sC	3.335 640 (−11)	3.335 640 (−10)	1	sA	3.335 640 (−11)	3.335 640 (−10)	1

1 electron charge $= 1e = 1.602\,189\,(−19)$ C

1 coulomb $=$ ampere $\times$ second $= 2.997\,925\,(+09)$ centimeter $\times \sqrt{dyne}$

B.41 ELECTRIC POTENTIAL AND RESISTANCE (Secs. 7.02–3, 7.03–2)

abV = abvolt   V = volt   sV = statvolt
abΩ = abohm   Ω = ohm   sΩ = statohm

	abV	V	sV		abΩ	Ω	sΩ
abV	1	10^{-8}	$3.335\,640\,(-11)$	abΩ	1	10^{-9}	$1.112\,650\,(-21)$
V	10^{8}	1	$3.335\,640\,(-03)$	Ω	10^{9}	1	$1.112\,650\,(-12)$
sV	$2.997\,925\,(+10)$	$2.997\,925\,(+02)$	1	sΩ	$8.987\,554\,(+20)$	$8.987\,554\,(+11)$	1

$$1\ \text{volt} = \frac{\text{joule}}{\text{coulomb}} = \frac{\text{watt}}{\text{ampere}} = \text{ohm} \times \text{ampere}$$

$$1\ \text{ohm} = \frac{\text{volt}}{\text{ampere}} = \frac{\text{joule} \times \text{second}}{\text{coulomb}^2} = \frac{\text{watt}}{\text{ampere}^2}$$

B.42 CAPACITANCE AND INDUCTANCE (Secs. 7.02–4, 8.03–2)

abF = abfarad   F = farad   sF = statfarad
abH = abhenry   H = henry   sH = stathenry

	abF	F	sF		abH	H	sH
abF	1	10^{9}	$8.987\,554\,(+20)$	abH	1	10^{-9}	$1.112\,650\,(-21)$
F	10^{-9}	1	$8.987\,554\,(+11)$	H	10^{9}	1	$1.112\,650\,(-12)$
sF	$1.112\,650\,(-21)$	$1.112\,650\,(-12)$	1	sH	$8.987\,554\,(+20)$	$8.987\,554\,(+11)$	1

$$1\ \text{farad} = \frac{\text{ampere} \times \text{second}}{\text{volt}} = \frac{\text{coulomb}^2}{\text{joule}} = \frac{\text{ampere}^2 \times \text{second}}{\text{watt}} = \frac{\text{second}}{\text{ohm}}$$

$$1\ \text{henry} = \frac{\text{volt} \times \text{second}}{\text{ampere}} = \frac{\text{joule}}{\text{ampere}^2} = \frac{\text{second}^2}{\text{farad}} = \text{ohm} \times \text{second}$$

B.43 MAGNETIC FLUX AND INTENSITY (Secs. 8.01–3, 8.01–4)

G = gauss   Mx = maxwell   T = tesla   Wb = weber

	Mx	Wb	T-m^2		G	T	Wb-m^{-2}
Mx	1	10^{-8}	10^{-8}	G	1	10^{-4}	10^{-4}
Wb	10^{8}	1	1	T	10^{4}	1	1
T-m^2	10^{8}	1	1	Wb-m^{-2}	10^{4}	1	1

$$1\ \text{weber} = \text{tesla} \times \text{meter}^2 = \frac{\text{joule}}{\text{ampere}} = \frac{\text{kilogram} \times \text{meter}^2}{\text{coulomb} \times \text{second}}$$

$$1\ \text{tesla} = \frac{\text{weber}}{\text{meter}^2} = \frac{\text{newton}}{\text{ampere} \times \text{meter}} = \frac{\text{kilogram} \times \text{meter}^2}{\text{coulomb} \times \text{second}}$$

REFERENCES AND BIBLIOGRAPHY

Theory and Methods of Solution

All governing relations and methods of solution in this handbook are given in a simple mathematical form without proof. For the derivations of these relations, more advanced forms of solution, and additional information the following books are suggested:

Abbot, M. N., and M. C. Ness: "Thermodynamics," McGraw-Hill, New York, 1972 (Differential Relations, Advanced Topics).

Edmister, J. A.: "Electric Circuits," McGraw-Hill, New York, 1965 (Differential Relations, Vector Methods, Polyphase Systems, Laplace Transforms).

Giles, R. V.: "Fluid Mechanics and Hydraulics," 2d ed., McGraw-Hill, New York, 1962 (Differential Relations, Dimensional Analysis, Pipe Systems, Fluid Machinery).

Hecht, E.: "Optics," McGraw-Hill, New York, 1975 (Differential Relations, Polarization, Diffraction).

Hughes, F. W., and J. A. Brighton: "Fluid Dynamics," McGraw-Hill, New York, 1967 (Differential Relations, Vector Methods, Compressible Fluids).

Pilkey, W. D., and D. H. Pilkey: "Mechanics of Solids," Quantum, New York, 1974 (Differential Relations, Transport Matrices and Buckling of Bars).

Seto, W. W.: "Acoustics," McGraw-Hill, New York, 1972 (Differential Relations, Advanced Topics).

———, "Mechanical Vibration," McGraw-Hill, New York, 1964 (Several Degrees of Freedom, Nonlinear Vibration, Electric Analogies).

Tuma, J. J.: "Dynamics," Quantum, New York, 1974 (Differential Relations, Vector and Matrix Methods, Space Motion).

———, "Statics," Quantum, New York, 1974 (Differential Relations, Vector and Matrix Methods, Space Transformations).

——— and M. Abdel-Hady: "Engineering Soil Mechanics," Prentice-Hall, Englewood Cliffs, N.J., 1973 (Properties and Classification, Rheology, Stress and Strength Analysis).

Tables of Physical Constants

The most frequently used physical constants based on standard conditions (temperature, pressure, etc.) are given in Appendix A. More extensive tables of physical constants may be found in the following publications:

'Bolz, R. E., and G. L. Tuve (eds.): "Handbook of Tables for Applied Engineering Science," 2d ed., CRC Press, Cleveland (now located in Boca Raton, Fla.), 1973.

Brady, G. S., and H. Clauser: "Materials Handbook," 11th ed., McGraw-Hill, New York, 1980.

Dean, J. (ed.): "Lange's Handbook of Chemistry," 12th ed., McGraw-Hill, New York, 1979.

Forsythe, W. E. (ed.): "Smithsonian Physical Tables," 9th ed., The Smithsonian Institution, Washington, D.C., 1956.

Gray, D. E. (ed.): "American Institute of Physics Handbook," 3d ed., McGraw-Hill, New York, 1972.

Harper, C. (ed.,): "Handbook of Plastics and Elastomers," McGraw-Hill, New York, 1975.

Mantell, C. (ed.): "Engineering Materials Handbook," McGraw-Hill, New York, 1958.

Parker, E. R.: "Materials Data Book," McGraw-Hill, New York, 1967.

Shand, E. B. (ed.): "Glass Engineering Handbook," 2d ed., McGraw-Hill, New York, 1958.

Weast, R. C. (ed.): "Handbook of Chemistry and Physics," 61st ed., CRC Press, Boca Raton, Florida, 1980.

"Metal Handbook," American Society of Metals, Metals Park, Ohio, 1961.

"Wood Handbook," Forest Products Laboratory, U.S. Department of Agriculture, Washington, D.C., 1974.

INDEX

References are made to page numbers for the text material and to table numbers for the appendix material. In the designation of systems of units the following abbreviations are used:

SI = international system
MKS = metric system
FPS = English system

AP = absolute practical system
EM = electromagnetic system
ES = electrostatic system